R00028 94023

D1506831

DUE DATE	RETURN DATE	DUE DATE	RETURN DATE

Indicators of Genotoxic Exposure

Row 1: I. B. Weinstein; A. Wyrobek; D. Herron; M. Rajewsky.
Row 2: C. J. Calleman; B. E. Butterworth; L.B. Russell; R. D. Combes.
Row 3: R. J. Preston, H. J. Evans; N. Petrakis; S. Parodi; E. Eisenstadt.
Row 4: R. J. Albertini; W. Thilly, J. A. Heddle; G. Zetterberg; B. A. Bridges.

13

INDICATORS OF
GENOTOXIC EXPOSURE

Edited by

BRYN A. BRIDGES
MRC Cell Mutation Unit   University of Sussex

BYRON E. BUTTERWORTH
Chemical Industry Institute of Toxicology

I. BERNARD WEINSTEIN
Columbia University College of Physicians and Surgeons

COLD SPRING HARBOR LABORATORY
1982

BANBURY REPORT SERIES

Banbury Report 13
Indicators of Genotoxic Exposure

© 1982 by Cold Spring Harbor Laboratory
All rights reserved

Printed in the United States of America

Cover and book design by Emily Harste

Library of Congress Cataloging in Publication Data

Main entry under title:

Indicators of genotoxic exposure.
   (Banbury report, ISSN 0198-0068 ; 13)
   Bibliography: p.
   Includes index.
   1. Medical genetics. 2. Chemical
mutagenesis. 3. Mutagenicity testing.
I. Bridges, B. A. II. Butterworth, Byron E.
III. Weinstein, I. Bernard. IV. Series.
[DNLM: 1. Mutagenicity tests—Congresses.
2. Body fluids—Analysis—Congresses. 3. DNA—
Congresses. 4. Carcinogens—Congresses.
BA 19 v. 13 / QH 460 I39 1982]
RB155.I523 1982 616'.042 82-12972
ISBN 0-87969-212-X

L-485634

PARTICIPANTS

Richard J. Albertini, Department of Medicine, College of Medicine, University of Vermont

James W. Allen, Health Effects Research Laboratory, U.S. Environmental Protection Agency

Bryn A. Bridges, Medical Research Council, Cell Mutation Unit, University of Sussex, England

Paul Brubaker, Exxon Biomedical Sciences

Byron E. Butterworth, Chemical Industry Institute of Toxicology

C. J. Calleman, Department of Radiobiology, University of Stockholm, Sweden

Anthony V. Carrano, Biomedical Sciences Division, Lawrence Livermore National Laboratory

Leigh M. Henderson, School of Biological Sciences, University of Sussex, England

Robert D. Combes, Department of Biological Sciences, Portsmouth Polytechnic, Hants, England

Eric Eisenstadt, Department of Microbiology, Harvard University School of Public Health

H. John Evans, Medical Research Council, Clinical and Population Cytogenetics Unit, Western General Hospital, Edinburgh, Scotland

Peter B. Farmer, Medical Research Council Toxicology Unit, MRC Laboratories, Surrey, England

Chie Furihata, Department of Molecular Oncology, Institute of Medical Science, University of Tokyo, Japan

Hector D. Garcia, Research Center, Philip Morris U.S.A.

William A. Haseltine, Sidney Farber Cancer Institute

John A. Heddle, Ludwig Institute for Cancer Research, Toronto, Canada

Deborah C. Herron, Toxicology and Product Safety, Atlantic Richfield Company

Ernest B. Hook, Birth Defects Institute, Division of Laboratories and Research, New York State Department of Health

Abraham W. Hsie, Biology Division, Oak Ridge National Laboratory

Andrew Kligerman, Chemical Industry Institute of Toxicology

Werner K. Lutz, Institut für Toxicologie der Eidgenossischen Technischen Hochschule und der Universitat Zurich, Schwerzenbach bei Zurich, Switzerland

Wendell H. McKenzie, Department of Genetics, North Carolina State University

Heinrich V. Malling, National Institute of Environmental Health Sciences

Jon C. Mirsalis, Life Sciences Division, SRI International

Harvey Mohrenweiser, Department of Human Genetics, University of Michigan

Nan Newell, Office of Technology Assessment, Congress of the United States

Silvio Parodi, Istituto Scientifico per lo Studio e la Cura dei Tumori, Genoa, Italy

Michael A. Pereira, Health Effects Research Laboratory, U.S. Environmental Protection Agency

Nicholas L. Petrakis, Department of Epidemiology and International Health, University of California at San Francisco Medical School

R. Julian Preston, Biology Division, Oak Ridge National Laboratory

Manfred J. Rajewsky, Institut für Zellbiologie (Tumorforschung), Universitat Essen, Federal Republic of Germany

Liane B. Russell, Biology Division, Oak Ridge National Laboratory

Gary Sega, Biology Division, Oak Ridge National Laboratory

Michael J. Skinner, Toxicology Division, Mobil Oil Corporation

Thomas R. Skopek, Department of Molecular Biophysics and Biochemistry, Yale University

Marja Sorsa, Institute of Occupational Health, Helsinki, Finland

Gary H. S. Strauss, Medical Research Council, Cell Mutation Unit, University of Sussex, England

James A. Swenberg, Chemical Industry Institute of Toxicology

Gail Theall, Institute of Cancer Research, Columbia University College of Physicians & Surgeons

William G. Thilly, Department of Nutrition and Food Science, Massachusetts Institute of Technology

Guylyn R. Warren, Department of Chemistry, Montana State University

Michael D. Waters, Genetic Toxicology Division, Health Effects Research Laboratory, U.S. Environmental Protection Agency

I. Bernard Weinstein, Institute of Cancer Research, Columbia University College of Physicians & Surgeons

Gerald N. Wogan, Nutrition and Food Science, Massachusetts Institute of Technology

Andrew J. Wyrobek, Biomedical Sciences Division, Lawrence Livermore National Laboratory

Gösta Zetterberg, Department of Genetics, The Gustaf Werner Institute, University of Uppsala, Sweden

PREFACE

It has at times been suggested that perhaps '*Homo sapiens*' is not the most appropriate name for our species, and that something similar to '*Homo fabricans*' might be more to the point. For it is humans alone among the species of the earth who truly 'make' the world in which they live. Technology, the conscious transformation of nature toward the accomplishment of ends which we ourselves deem desirable, is central to the evolutionary development of our species. There is even compelling evidence for concluding that while excelling as 'makers,' human sapient faculties for guiding this technological virtuosity is a later and still relatively underdeveloped species characteristic.

Multivariate analyses encompassing prospective social, political, economic, environmental, and biological impacts of technological innovations are well beyond our grasp. Even within just one circumscribed area of potential impact, analysis remains inordinately complex and frequently inconclusive. Yet from both individual and societal perspectives, such analyses are becoming ever more imperative. From the individual's point of view, one of the most basic concerns is what such innovations may portend for each of us as biologically functional organisms.

Transformation of natural substances into new forms and combinations often are accompanied by inadvertent concomitant changes that were neither suspected nor intended. In this respect, many modern industrial products and processes have proven to be of a biologically highly reactive nature. Some of these may interact with and cause changes in our basic genetic material. These changes may be passed on to future generations or may give rise to new populations of cells in our own bodies, leading to invasive tumors and their metastases. Many tests have been devised to try to identify which chemicals may be capable of bringing about such changes and how their effects may become exacerbated with more extensive exposure. But predictive extrapolations from inbred animal or microbial assays to human exposure conditions and responses can be highly problematic. From 18-21 April, 1982 a conference was held at Banbury Center of Cold Spring Harbor Laboratory to assess the scientific capabilities of what might be described as the next stage in this effort—the direct monitoring of the individual for the effects of such exposure. This book is a compilation of both the formal presentations of that conference along with the discussions which they elicited. It is the hope that such quantitative indices of exposure effects in the very individuals so placed at risk will be a further step toward the goal of accurate assessment of health risks which may be implicit in any of the environments which we may choose to construct.

It is with the greatest of pleasure that I wish to acknowledge the invaluable role played by Bryn A. Bridges, Byron E. Butterworth, I. Bernard Weinstein, and my predecessor at Banbury Center, Victor K. McElheny, in organizing and implementing the conference on *Indicators of Genotoxic Exposure* as well as in the production of this book. The unflagging yet ever-cheerful assistance of Lynda Moran and Bea Toliver of the Banbury Staff has been indispensable to all phases of the project, and the additional editing contributions of Joanne DeOliveira have been instrumental in enabling on-schedule production of this book within 7 months of the original conference. I would particularly like to acknowledge the financial assistance of the American Petroleum Institute in support of both the meeting and this subsequent publication. Additional funding was derived from the following Banbury core supporters: Bristol-Myers Fund, The Chevron Fund, Conoco Inc., The Dow Chemical Company, E. I. du Pont de Nemours & Company, Exxon Education Foundation, Getty Oil Company, International Business Machines Corporation, Eli Lilly and Company, New York Life Insurance Company, Phillips Petroleum Foundation, Inc., and Texaco Philanthropic Foundation Inc.

Many thanks, as always, are due Dr. J. D. Watson, Director of the Cold Spring Harbor Laboratory, for his continual guidance and encouragement throughout all phases of this effort.

Michael Shodell
Director
October, 1982

CONTENTS

SESSION 7: GERM CELL EFFECTS

Indicators of Genotoxic Exposure

SESSION I:
CLINICAL PERSPECTIVES

Molecular Cancer Epidemiology

I. BERNARD WEINSTEIN AND FREDERICA P. PERERA
Division of Environmental Sciences and
Cancer Center/Institute of Cancer Research
Columbia University College of Physicians and Surgeons
New York, New York 10032

It is a pleasure to welcome all of you to the Banbury Center. This conference has been in the planning stage for at least 2 years. Victor McElheny, the Director of the Banbury Center, the Board of Advisors to the Banbury Center, and the organizers of this conference, Drs. Bryn Bridges, Byron Butterworth, and myself, are certain that this will prove to be a stimulating and productive meeting. The field of environmental mutagenesis and carcinogenesis is currently at an interesting transition point in terms of concepts and methods and we are hopeful that this conference will help to set the course for future progress in these areas.

Within the past few decades there has been a remarkable change in the major causes of death in the United States and Western Europe. Deaths due to tuberculosis, pneumonia, and other infectious diseases have decreased dramatically, although parasitic diseases remain major health problems in many of the economically underdeveloped countries. In the Western world, cancer is now the second major cause of death. There is considerable evidence that most human cancers are due to environmental factors, including our diet and life-style (Hiatt et al. 1977). Thus, in theory it should be possible to make a major impact on the prevention of this disease by identifying specific environmental factors involved in its causation.

I believe that our current approaches to cancer causation are somewhat analogous to the state of the field of the infectious diseases about 100 years ago. At that time the clinical patterns of certain major infectious diseases were well known and the science of microbiology was burgeoning. What was required, and what eventually happened, was a merging of these two disciplines. In the area of infectious diseases it is now an accepted fact that one doesn't go out into the field as an epidemiologist or into the clinic as a physician without "bringing along" one's laboratory colleagues, collecting samples from clinical materials, and then assaying these samples back in the laboratory. Koch's postulates, formulated almost exactly 100 years ago, formalized the criteria by which one could utilize clinical findings and laboratory methods to establish the role of microbes in the causation of specific diseases (Koch 1882). In a sense, at this conference we are searching for a set of Koch's postulates that will be applicable to studies on human cancer causation, especially with respect to environmental agents that act via genotoxicity. The ground rules are, however, quite different because we are not dealing with self-replicating microbes as causative agents. In addition, in studies on the causation of cancer, the interplay of multiple factors (both endogenous and exogenous), the long latent period between exposure and the occurrence of the disease, and the evolution of the disease through a multistage process, considerably complicate the elucidation of etiologic factors. This suggests that we will need a multitude of sophisticated laboratory approaches and that we must be prepared to simultaneously assess a number of parameters.

Our failure to merge epidemiologic, clinical, and laboratory methods in studies on human cancer causation has led to what might be called a crisis in risk extrapolation (Weinstein 1981a). Long-term animal bioassays are being done with increasing frequency, and the results are widely publicized. The field of in vitro tests (i.e., mutagenesis assays,etc.) has grown rapidly and is producing an abundance of data. On the other hand, we lack precise methods for extrapolating the results obtained from animal bioassays and in vitro tests to humans, particularly if we are attempting to make quantitative risk estimates. Conventional epidemiologic methods also have several important limitations. These include the fact that they are extremely time consuming, they are not very sensitive, and they are usually retrospective. In addition, although they may provide evidence for an association they often do not provide direct proof of a cause and effect relationship. Figure 1 emphasizes the need to develop new approaches which can bridge the disciplines of epidemiology, animal bioassays,

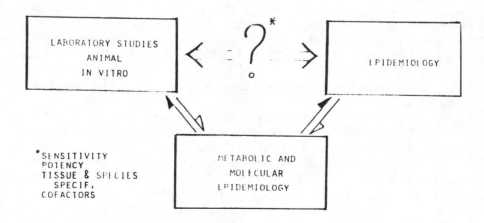

Figure 1
The problem of risk extrapolation

and in vitro tests. A number of factors that must be considered in attempting to extrapolate results obtained in rodent bioassays or in vitro tests to humans are:

1. appropriate dose and method of administration;
2. species, organ, sex and age specificity;
3. extrapolation from high to low dose, shape of dose-response curve, possible existence of a threshold;
4. sensitivity of the assay, limitations of confidence in a negative result;
5. synergy and cofactors.

In the enthusiasm to develop assays that are rapid, efficient, and highly sensitive, they may have been simplified to the point where it is difficult to do quantitative risk extrapolations with the data obtained. This leads us to the major theme of this meeting, i.e., the development of new indicators of genotoxicity that can be applied to the intact individual, thus avoiding some of the pitfalls of extrapolating from simpler systems.

In view of the above, we believe that it is necessary to develop an entirely new type of methodology which combines epidemiologic methods with laboratory techniques that measure specific biochemical and molecular parameters in human tissues and biologic fluids. We refer to this approach as "molecular cancer epidemiology" and have discussed this subject in greater detail elsewhere (Perera and Weinstein 1982). I will give only a brief survey of the approach. This survey will serve as an introduction to a number of the subjects which will be discussed in detail by various speakers at this conference.

MOLECULAR EPIDEMIOLOGY

Definition of Terms and Types of Laboratory Methods

It may be useful to define certain specific terms to be used within the context of our discussions. We use the term "molecular cancer epidemiology" to describe an approach in which advanced laboratory methods are used in combination with analytical epidemiology to identify at the biochemical or molecular level specific exogenous agents and (or) host factors that play a role in human cancer causation. The laboratory methods may include: 1) techniques to assess specific host factors that may influence susceptibility to carcinogens; 2) assays that detect carcinogens in human tissues, cells, or fluids; 3) assays at the cellular level of biologically effective doses of carcinogens; and 4) methods to measure early biologic and biochemical responses to carcinogenic agents (Perera and Weinstein 1982).

We use "exposure" to mean the concentration of a particular chemical substance to which an individual is subjected, based upon estimates of the amount of that agent in the individual's immediate environment (i.e., air, water, food, cigarette smoke, etc.). "Dose" refers to the amount of a substance that actually enters the body as a consequence of ingestion, inhalation, skin absorption, etc. Dose can be estimated indirectly through knowledge of exposure but this may be quite imprecise. More precise information on dose of a chemical carcinogen can be obtained by direct measurement of the concentration of the parent compound or its metabolites in body tissues, fluids, and excretions. Rapid advances in analytical techniques have made it possible to detect chemical agents with extremely high sensitivity and specificity. The development of antibodies to specific carcinogens and their metabolites should also greatly facilitate the detection and quantitation of these materials in environmental samples, in tissues, and in body fluids. It is surprising that this approach has not been exploited more extensively. The *Salmonella typhimurium* mutagenesis assay to detect mutagens in body fluids or excretions (urine and feces) can be a very useful tool in epidemiologic studies (Bruce et al. 1977; Yamasaki and Ames 1977; Falck et al. 1979).

It is important to distinguish between "dose" and "biologically effective dose," i.e., the amount of the activated agent that has actually reacted with critical cellular targets, such as DNA, protein, or RNA, that are presumed to be directly involved in the carcinogenic process (Fig. 2). The difference between "dose" and "biologically effective dose" is somewhat comparable to that between the roentgen and the rem in radiation biology. Simple measurements of the dose of a carcinogen based on tissue concentrations of the parent compound may be misleading because many of the known chemical carcinogens are not active as such but require metabolic activation to "ultimate carcinogens" (Miller and Miller 1981). In addition, metabolic processes leading to detoxification and excretion can avert or mitigate the effects of carcinogen exposure.

Definition:

> The amount of the activated substance that is capable of interacting with critical cellular targets presumed to be directly involved in the carcinogenic process.

Examples of Potential Markers:

- Adduct formation with proteins

- Adduct formation with nucleic acids

- Urinary excretion of excised DNA adducts

- DNA damage and repair

Figure 2
Biologically effective dose of a chemical carcinogen. (For additional details see text and Perera and Weinstein 1982).

Unlike radiation whose penetration into the cell is governed by predictable physical laws, biochemical rules by which one might predict the metabolic activation or detoxification of a chemical carcinogen cannot yet be formulated. In Figure 2 we list several biochemical markers that might be used to estimate the biologically effective dose of certain chemical carcinogens, thus serving as tissue dosimeters. However, although they have the distinct advantage of accounting for metabolic activation and detoxification processes, these markers are subject to other uncertainties which will be discussed later.

"Response" denotes early biologic or biochemical changes in the target tissue that result from the action of the carcinogen and are thought to be either a step in the carcinogenic process or to correlate closely with that process. Several examples are given in Figure 3. A drawback shared by markers for biologically effective dose and markers for response is that at the present time there are little or no data on "background" levels of these markers in various human populations. It is necessary, therefore, to develop this information by selecting a suitable normal control group and then determining whether the differences between levels of these markers observed in the exposed population and in the control group are statistically significant.

Studies in both experimental animals and in humans clearly indicate that a variety of host factors can alter individual responses to environmental carcinogens (Slaga et al. 1978; Miller 1980; Weinstein 1982). These include various acquired diseases (infectious and/or chronic inflammatory); physiologic or

Definition:

> Biological or biochemical changes in a target tissue or cells that result from the action of a carcinogen and are thought to be either a step in the carcinogenic process or correlate closely with that process

Examples of Potential Markers:

- Chromosomal Abnormalities

- Markers for Point Mutations

- Markers for Altered Gene Expression

- Reproductive Toxicity

- Monoclonal Antibodies to Carcinogen-altered cells

Figure 3
Early biological or biochemical responses to chemical carcinogens. (For additional details, see text and Perera and Weinstein 1981).

Table 1
Host Factors in Carcinogenesis[a]

1. Age
2. Sex
3. Hormones, growth factors and receptors
4. Immunologic factors
5. Nutritional status
6. Acquired diseases
7. Prior exposure to environmental carcinogens (initiators, etc.)
8. Carcinogen metabolism
9. DNA repair
10. Chromosomal defects
11. Cellular proliferation
12. State of differentiation and gene expression
13. Oncogenes[b]

[a]*Note*: 1) Host factors can be either inherited or acquired, and 2) that host factors can operate at the systemic level or at the target cell level.
[b]In this context oncogenes refers to genes that may be normally present in the target cell of the host and play a role in establishment and (or) maintenance of the transformed state.

nutritional factors (e.g., hormone imbalance, vitamin deficiency); and relatively rare genetic or inherited factors. This subject has been reviewed in detail elsewhere (Mulvihill 1980). Several genetic and acquired host factors are listed in Table 1. A recently developed method permits analysis of polymorphism in humans directly at the DNA level, by using restriction enzyme analysis and DNA cloning techniques (Wyman and White 1980; Lewin 1981). Although not yet used to identify genetic factors in human cancer, this is an exciting advance and may facilitate linkage and other types of studies on the role of specific genes in the causation of human cancer. Perhaps the most important host factors are the genes that reside in the target cell that will ultimately be responsible for the development and maintenance of the transformed state (i.e., the "oncogenes") (Weinstein 1981b; Gattoni et al. 1982). I will return to this subject later.

AHH (aryl hydrocarbon hydroxylase) inducibility has been proposed as an indicator of human genetic susceptibility to lung cancer (Kellerman et al. 1973). However, subsequent studies (Paigen et al. 1978) have indicated that the assay is subject to considerable variability. Moreover, recent identification of the major metabolites of benzo[a]pyrene (B[a]P) involved in nucleic acid binding suggests that the conventional AHH-assay may not identify aspects of B[a]P metabolism that are critical to the carcinogenic process (Jeffrey et al. 1980).

Assays for specific hormones in blood and urine have played an important role in assessing risk factors in human breast cancer, although the precise roles of steroid hormones and prolactin in the causation of human breast cancer are not clear at the present time (Weinstein 1980b). The development of radio-immunoassays for these hormones and their cellular receptors will no doubt greatly facilitate further epidemiological studies on this and other endocrine-related cancers.

Most of the methods mentioned above are readily applicable to both laboratory animals and humans. Thus they can be extremely useful in comparative studies between experimental animals and humans, thereby providing a data base for extrapolation between species. We should stress, however, that at the present time we do not know whether and to what degree individual risk of cancer is associated with the presence and level of any of the above described markers. Thus, extensive studies will be required to not only standardize but validate the usefulness of these markers in experimental animals and in humans.

It should be noted that complex interactions occurring between viruses and chemical carcinogens may be responsible for certain human cancers (Fisher and Weinstein 1979). Indeed this appears to be the case for some forms of liver and nasopharyngeal cancer (for review, see Weinstein 1980b). In certain epidemiological studies it may be critical, therefore, to assess the role of both chemical and viral agents. Methods to detect viral agents include antibody titers, viral antigens, tissue culture techniques, and nucleic acid hybridization. In addition, recent studies suggest that a series of cellular genes (designated proto-oncogenes) that are homologous to the oncogenes present in the acute transforming retroviruses may play a role in human cancer (for review, see Rigby 1982). Should this prove to be the case, then specific DNA probes to these oncogenes could be used to analyze the DNA and RNA of human tissues, thus providing an extremely powerful tool for molecular cancer epidemiology.

Immunoassays to Detect Carcinogen-DNA Adducts

As mentioned above, a molecular marker with great potential as an indicator of the effective biological dose of an environmental carcinogen is the covalently bound adduct formed between that substance, or its metabolite, and DNA. Covalent binding assays have demonstrated that a number of carcinogens are metabolically activated to electrophilic species capable of reacting with electron-rich atoms in cellular DNA (Miller and Miller 1981). Brookes and Lawley (1964) demonstrated that for a series of polycyclic aromatic hydrocarbons, the ability of a chemical substance to bind covalently to DNA of mouse skin correlated with its carcinogenic potency. In more recent in vitro studies of five polycyclic hydrocarbons, values for the extent of metabolite-nucleoside adduct formation also reflected the carcinogenic potency of the compounds studied (Pelkonen et al. 1980). It seems reasonable, therefore, to assume that DNA binding assays can be used as a particularly relevant measure of the biologically effective dose of initiating carcinogens, although other factors (DNA repair, cell proliferation, tumor promotion, etc.) also influence the carcinogenic process.

Extensive studies from several laboratories have implicated a specific diol epoxide derivative of B[a]P, 7β,8α-dihydroxy-9α,10α-epoxy 7,8,9,10-tetrahydrobenzo[a]pyrene (BPDE I), as the major electrophilic, mutagenic, and carcinogenic metabolite of B[a]P involved in covalent binding to DNA (for review see Jeffrey et al. 1980). The complete chemical structure and conformation of the predominant adduct formed in vitro between BPDE I and nucleic acids of mammalian (including human) cells and tissues has been elucidated. In this adduct, the C-10 position of BPDE I is linked to the 2-amino group of guanine residues in DNA and RNA. This structure, designated BPDE-I-dG, has been detected as the major DNA adduct formed when a variety of human, bovine, and rodent cells were exposed to B[a]P in culture (Weinstein et al. 1976; Autrup et al. 1980; Jeffrey et al. 1980).

Highly sensitive methods are available for detecting carcinogen-DNA adducts utilizing radioactively labeled precursors, fluorescence, and high pressure liquid chromatography (for review see Jeffrey et al. 1980; Lutz 1979). For obvious reasons, the administration of radioactively labeled carcinogens is not applicable to studies in humans. However, immunologic methods can be safely used to detect and quantitate carcinogen-DNA adducts in human tissues obtained from individuals who were exposed to the compound because of its presence in the workplace, diet, general environment, etc.

Extremely specific and sensitive antibodies have been developed to several diverse carcinogen-DNA adducts (for review see Rajewsky, this volume). In collaboration with investigators at the National Cancer Institute, our research group at Columbia University has developed immunoassays capable of detecting extremely low levels of the above described BPDE-I-dG adduct in the DNA of cells and tissues exposed to B[a]P in vivo and in vitro. One method is a competitive radioimmunoassay (RIA) using rabbit antiserum specific for BPDE-I-dG. This method can detect picomole levels of this adduct (Poirier et al. 1977; 1980). Another method utilizes the same antiserum in an enzyme-linked immunosorbent assay (ELISA), which has been modified by Hsu et al. (1981) to provide an ultrasensitive enzymatic radioimmunoassay (USERIA). The latter assays are capable of detecting as little as 3 femtomole of this adduct in 1 μg of DNA (Hsu et al. 1981). This is equivalent to approximately 1 carcinogen residue per 10^7 nucleotides. There are about 2×10^9 nucleotides of DNA in a human cell; and in experimental systems, carcinogens often modify DNA to the extent of one residue per 10^5 nucleotides. Thus these assays appear to be sufficiently sensitive to detect carcinogen-DNA adducts at levels that would be biologically significant.

The techniques described are also quite versatile and can, in principle, be applied to the detection and quantitation of any carcinogen-DNA adduct. Normally the assays are performed using 10-100 μg of human DNA—a quantity that can be obtained from 0.2-1 gm of tissue or the buffy coat (white cells) in 25-50 ml of human blood. A limitation of this approach in human studies is that the relevant tissue may not be readily available. Lung or liver tissue, for example, can only be obtained at autopsy or as a surgical biopsy. On the other hand, certain biological samples such as white blood cells, skin biopsies, and placental tissue are more easily obtained and could be employed in such studies. Furthermore, if the carcinogen-nucleoside adduct is excised from cellular DNA during DNA repair and not further metabolized, then these adducts might be detected and quantitated in urine samples, using the appropriate antibodies.

Important variables to be considered in a study of B[a]P-DNA binding are the possible effects of other environmental agents and endogenous factors (e.g. age, hormonal status, etc.) on B[a]P activation and detoxification, as well as interindividual variation in metabolism. Immunoassays for B[a]P-DNA adducts could, in principle, determine the degree of variation in DNA adduct formation that might occur between individuals undergoing similar environmental exposures. While interindividual variation complicates the picture, an understanding of this variation is necessary in order to extrapolate from limited animal or human studies to a heterogeneous human population. The effects of DNA repair must also be taken into account since the level of carcinogen adducts in DNA is a function not only of their extent of formation but also of the efficiency of specific DNA excision mechanisms. Very little is known about the rates of DNA repair systems in intact human tissues.

In collaboration with Drs. Miriam Poirier, and Stuart Yuspa we are currently carrying out a pilot study to determine whether we can detect and quantitate BPDE-DNA adducts in tissue and leukocyte DNA samples obtained from individuals heavily exposed to B[a]P (i.e., cigarette smokers, coke oven workers, etc.). Our initial results look promising. Hopefully, this type of approach can be extended to other compounds and also correlated with the actual cancer risks in individual subjects.

LIMITATIONS OF ASSAYS FOR GENOTOXICITY

Although we are enthusiastic about the use of some of the above described indicators in studies on human cancer causation, let us briefly discuss certain major limitations. First, most of the methods discussed earlier have been examined only to a limited extent in humans. Thus, further study is needed to validate their usefulness in human studies. Second, most human cancers result from a complex interaction between multiple factors, only some of which may be genotoxic. Thus the exposure of an individual to tumor promoters, hormones, dietary macronutrients (fat, fiber, protein), vitamins (A, C, E, etc.), and other cofactors may markedly enhance, and in some cases inhibit, the carcinogenic process (Hiatt et al. 1977). Almost all of the assays that will be discussed at this meeting do not assess these parameters. Furthermore, certain compounds that can act as complete carcinogens, for example B[a]P, may exert both tumor initiating and tumor promoting activity (Ivanovic and Weinstein 1982). Assays for genotoxicity may not detect the latter effects and thus will not reflect the full range of activity or potency of the compound in question. Hopefully,

→ → → Heritable, Constitutive Program of Aberrant Gene Expression

Mechanisms

1. Point Mutation.
2. Gene Rearrangements.
   a. Structural genes—Oncogenes or other host genes
   b. Regulatory elements—Long Terminal Repeat (LTR) sequences, enhancer sequences, etc.
3. Altered DNA Methylation.
4. Altered Chromatin Structure and Other Epigenetic Mechanisms.

Figure 4
Multistage carcinogenesis

current advances in our understanding of the biochemical and cellular effects of tumor promoters and various cofactors will eventually provide markers that can also be used in studies in the intact animal and in human studies. These markers might include in vitro assays for the presence of tumor promoters or growth factors in tissues or biologic fluids, assays for growth factor receptors, and assays for alterations in membrane structure and function (Weinstein 1980a).

Another major limitation relates to the fact that although the covalent binding of several environmental carcinogens to cellular DNA appears to be a critical event in the carcinogenic process, the subsequent biochemical events that lead to the development of neoplasia are poorly understood. Possible molecular mechanisms are listed in Figure 4. The initiating event in carcinogenesis is often thought of as a simple random point mutation resulting from errors in DNA replication at the sites of carcinogen damage. Several features of the carcinogenic process, however, including the high efficiency of carcinogen-induced cell transformation when compared to specific locus mutations, the lengthy latency period needed for the expression of the transformed state, and the multistep nature of the carcinogenic process are not consistent with this mechanism (Weinstein 1981b). Alternative mechanisms include the possibilities that carcinogen-induced DNA damage might induce more complex genomic changes, for example, gene rearrangements or gene amplification, altered DNA methylation or alterations in chromatin structure (Figure 4; Weinstein 1981b). Progress in our understanding of these events may lead to the development of molecular probes for detecting genotoxic events specifically related to carcinogenesis.

SUMMARY

At present there is a paucity of data and methods that can be used for precise quantitative assessments of human risks from specific environmental carcinogens. Further, only crude measures of the relative potency of chemical carcinogens are available by which to rank relative risks to humans. Conventional cancer epidemiology and animal bioassays have several major limitations. Similarly, although in vitro studies and recently developed short term assays for mutagens and carcinogens provide important information on the mechanism of action and the potential hazard of chemical agents, it is not clear that they can be used to assess the magnitude of human risks. By combining epidemiological methods with laboratory procedures such as the assay for the actual extent of in vivo covalent binding of certain activated carcinogens to cellular DNA it may be possible to predict human risks more precisely than hitherto possible. A particularly promising approach is the use of highly sensitive immunologic techniques to detect the levels of carcinogen-DNA adducts in tissues, white blood cells, urine, and other body fluids of individuals exposed to a specific agent. This technique warrants intensive study since it could provide a dosimeter at the cellular level of the biologically effective dose of a carcinogen under the actual conditions of human exposure. This and other laboratory methods when combined with epidemiologic studies will lead to more effective methods of assessing quantitatively the risks to humans of various environmental chemicals.

REFERENCES

Autrup, H., F.C. Wefald, A.M. Jeffrey, H. Tate, R.D. Schwartz, B.F. Trump, and C.C. Harris. 1980. Metabolism of benzo(a)pyrene by cultured tracheobronchial tissues from mice, rats, hamsters, bovines, and humans. *Int. J. Cancer* 25:293.

Brookes, P. and P.D. Lawley. 1964. Evidence for binding of polynuclear aromatic hydrocarbons to the nucleic acids of mouse skin: Relation between carcinogenic power of hydrocarbons and their binding to DNA. *Nature* 202:781.

Bruce, W.R., A.J. Varghese, R. Furrer, and P.C. Land. 1977. A mutagen in the feces of normal humans. *Cold Spring Harbor Conf. Cell Proliferation* 4:1641.

Falck, K., P. Grohn, M. Sorsa, H. Vainio, E. Heinonen, and L.R. Holsti. 1979. Mutagenicity in urine of nurses handling cytostatic drugs. *Lancet* i:1250.

Fisher, P.B. and I.B. Weinstein. 1979. Chemical-viral interactions and multistep aspects of cell transformation. *IARC Sci. Publ.* 27:113.

Gattoni, S., P. Kirschmeier, I.B. Weinstein, J. Escobedo, and D. Dino. 1982. Cellular Moloney murine sarcoma ("c-mos") sequences are hypermethylated and transcriptionally silent in normal and transformed rodent cells. *Mol. Cell. Biol.* 2:42.

Hiatt, H.H., J.D. Watson, and J.A. Winsten (eds). 1977. *Cold Spring Harbor Conf. Cell Prolif.* 4.

Hsu, I.C., M.C. Poirier, S.H. Yuspa, D. Grunberger, I.B. Weinstein, R.H. Yolken, and C.C. Harris. 1981. Measurement of benzo(a)pyrene-DNA adducts by enzyme immunoassays and radioimmunoassay. *Cancer Res.* 41:1090.

Ivanovic, V. and I.B. Weinstein. 1982. Benzo(a)pyrene and other inducers of cytochrome P_1-450 inhibit binding of epidermal growth factor to cell surface receptors. *Carcinogenesis* 3:505.

Jeffrey, A.M., T. Kinoshita, R.M. Santella, D. Grunberger, L. Katz, and I.B. Weinstein. 1980. The chemistry of polycyclic aromatic hydrocarbon-DNA adducts. In *Carcinogenesis: Fundamental Mechanisms and Environmental Effects* (eds. B. Pullman et al.), p. 565. R. Reidel Publishing, Amsterdam.

Kellerman, G., C.R. Shaw, and M. Luyten-Kellerman. 1973. Aryl hydrocarbon hydroxylase activity and bronchogenic carcinoma. *N. Engl. J. Med.* **289**: 934.

Koch, R. 1882. Die Aetiologie der Tuberculose. *Berl. Klin. Wochenschr.* **19**:221.

Lewin, R. 1981. Jumping genes help trace inherited diseases. *Science* **211**:690.

Lutz, W.K. 1979. In vivo covalent binding of organic chemicals to DNA as a quantitative indicator in the process of chemical carcinogenesis. *Mutat. Res.* **65**:289.

Miller, D.G. 1980. On the nature of susceptibility to cancer. *Cancer* **46**:1307.

Miller, E.C. and J.A. Miller. 1981. Mechanisms of chemical carcinogenesis. *Cancer* **47**:1055.

Mulvihill, J.J. 1980. Clinical observations of ecogenetics in human cancer. *Ann. Intern. Med.* **92**:809.

Paigen, B., H.L. Gurtoo, J. Minowada, E. Ward, L. Houten, K. Paigen, A. Reilly, and R. Vincent. 1978. Genetics of aryl hydrocarbon hydroxylase in human population and its relationship to lung cancer. In *Polycyclic hydrocarbons and cancer* (eds. H. Gelboin et al.), vol. 2, p. 391. Academic Press, New York.

Pelkonen, O., K. Vahakangas, and D.W. Nebert. 1980. Binding of polycyclic aromatic hydrocarbons to DNA: Comparison with mutagenesis and tumorigenesis. *J. Toxicol. Environ. Health* **6**:1009.

Perera, F.P. and I.B. Weinstein. 1982. Molecualr epidemiology and carcinogen-DNA adduct detection: New approaches to studies of human cancer causation. *J. Chronic Dis.* **35**:581.

Poirier, M.C., S.H. Yuspa, I.B. Weinstein, and S. Blobstein. 1977. Detection of carcinogen-DNA adducts by radioimmunoassay. *Nature* **270**:186.

Poirier, M.C., R. Santella, I.B. Weinstein, D. Grunberger, and S.H. Yuspa. 1980. Quantitation of benzo(a)pyrene-deoxyguanosine adducts by radioimmunoassay. *Cancer Res.* **40**:412.

Rigby, P.W.J. 1982. The oncogenic circle closes. *Nature* **297**:451.

Slaga, T.J., A. Sivak, and R.K. Boutwell (eds). 1978. *Carcinogenesis—A comprehensive survey.* Vol. 2, Raven Press, New York.

Weinstein, I.B. 1980a. Cell culture systems for studying multifactor interactions in carcinogenesis. In *Mechanisms of Toxicity and Hazard Evaluation* (eds. B. Holmstedt et al.), p. 49. Elsevier/North Holland Biomedical Press, Amsterdam.

————. 1980b. Studies on the mechanism of action of tumor promoters and their relevance to mammary carcinogenesis. In *Cell Biology of Breast Cancer* (eds. C.M. McGrath et al.), p. 425. Academic Press, New York.

————. 1981a. The scientific basis for carcinogen detection and primary cancer prevention. *Cancer* **47**:1133.

_____. 1981b. Current Concepts and Controversies in Chemical Carcinogenesis. *J. Supramol. Struct.* **17**:99.

_____. 1982. Carcinogenesis as a multistage process-experimental evidence. In *Symposium on Host Factors in Carcinogenesis*, International Agency for Research on Cancer, Cape Sunion, Greece. (In press).

Weinstein, I.B., A.M. Jeffrey, K.W. Jenette, and S.H. Blobstein. 1976. Benzo(a)-pyrene diol epoxides as intermediates in nucleic acid binding in vitro and in vivo. *Science* **193**:592.

Wyman, A. and R. White. 1980. A highly polymorphic locus in human DNA. *Proc. Natl. Acad. Sci. U.S.A.* **77**:6754.

Yamasaki, E. and B.N. Ames. 1977. Concentration of mutagens from urine by adsorption with the nonpolar resin XAD-2: Cigarette smokers have mutagenic urine. *Proc. Natl. Acad. Sci. U.S.A.* **74**:3555.

COMMENTS

EISENSTADT: I thought I just might comment that bacteria might not be so far off of the mark either in comparison to mammalian cells. For example, following exposure to a modest dose of UV in *Salmonella*, for example, you can detect unstable duplications of many regions of the chromosome in something like 10% of the survivors. I think that might relate to the amplification phenomenon you were describing for mammalian cells.

WEINSTEIN: Yes. Dr. Evans and I recently participated in an interesting IARC workshop in which we scored about 200 carcinogens for their performance in a variety of genotoxicity assays. I was struck by the fact that there is also a subcategory of agents which are negative in the Ames *Salmonella* test, yet in years and *Drosophila* they may be positive for mutagenic chromosome effects. For example, diethylstilbestrol is negative in the Ames test and yet it produces chromosome effects in mammalian cells. So there may be more complex responses to carcinogens in higher organisms.

EVANS: I don't think that is surprising. I think the point you are making about the tests we have—certainly the bacterial tests are designed as tests to pick up mutations in bacteria. They might not have all that much relevance to mutation induction in man. They are specific for picking up changes in bacteria. I think when you talk about potency and transfer of information from, let's say, an Ames test to an in vivo human effect, it is a very long distance apart. It is not surprising, I think, to find differences of the sort you are referring to.

BRIDGES: I don't think you will find many people who would disagree with that.

ZETTERBERG: Wouldn't one expect initiation to be a rare phenomenon, in frequency, like mutation is? If you are looking for something, you would look for something rare, like transformation in tissue culture cells, if you don't put on another restriction, saying that the initiation is a frequent phenomenon.

WEINSTEIN: Actually, we really don't know the frequency of the initiation of the carcinogenic process in vivo. In a number of tissue culture studies comparing the frequency of cell transformation to that of specific mutations, for example, thioguanine or ouabaine resistance, the frequency of transformation has usually been much higher than that of random mutation. Depending upon the way the studies are done, it can be ten- to

1,000-fold higher. It is a striking fact that cell transformation frequencies (morphologic transformation) come out higher than the frequency of random point mutation.

The other point is that the frequency of rearrangements of insertion elements in bacteria is in the range of 10^{-4}. So there could be responses to carcinogens which occur with about the same frequency as random point mutations but occur by different mechanisms.

So, one, I don't know what the frequency of initiation is in vivo, but cell transformation can occur in vitro with a frequency that is considerably higher than that of point mutation; and, two, I think the frequency argument is not very strong evidence that one is dealing with random point mutation.

The Value and Limitations of Clinical Observations in Assessing Chemically-Induced Genetic Damage in Humans

ERNEST B. HOOK
Bureau of Maternal and Child Health
New York State Department of Health
Albany, New York 12237
and
Department of Pediatrics
Albany Medical College
Albany, New York 12208

The proportion of human morbidity and mortality attributable to mutation and thus by inference at least potentially subject to influence by chemical mutagens is large. Table 1 lists estimated proportions of various categories of morbidity associated with chromosomal "mutations," most of which occur in the first generation after their induction. (This is a summary and updated correction of estimates discussed in greater detail elsewhere [Hook 1982a]). This list only includes defined cytogenetic disorders associated with obvious numerical and gross structural abnormalities detectable with contemporary techniques. Table 2 lists estimates of the livebirth prevalence of specific types of cytogenetic abnormalities detected in population surveys and the proportion attributable to mutation in the most recent generation.

With regard to specific locus mutations, there were in 1978, 1364 "definite" or proven entries in McKusick's catalog of Mendelian traits in humans (McKusick 1978). These apparently distinct loci[1] are classified as follows: 736 autosomal dominant, 521 autosomal recessive, and 107 X-linked. About 50% of the dominant loci and most of the loci listed in the other two categories (about 90%) have at least one allele associated with significant pathology.

---

[1] It is not generally realized that McKusick's catalog and enumeration is actually of *loci* not traits or phenotypes, despite its subtitle. (*See* p. xiii of McKusick 1978.) Thus the actual number of recognized *alleles* associated with pathology are greater, probably far greater than the number of loci which I estimated as associated with pathology. Moreover, this does lead to a certain inconsistency in classification of loci at which there are alleles with autosomal recessive *and* dominant effects. Thus, code 14190, a single entry in the autosomal dominant section includes all known alleles at the β-globin locus, over 200. Sickle-cell disease, a recessive disorder produced by the presence of two alleles for Hgb S at this locus is indexed *only* to this entry in the dominants, as are some other hemoglobinopathies.

Table 1

Estimated Contribution of Germinal Cytogenetic Abnormalities to Various Categories of Human Morbidity and Mortality[a]

Fetal deaths	
up to 28 weeks	17% to 50% depending on gestational stage; about 30% total
28 weeks and later	5-6%
Infant deaths	5-6%
Later childhood deaths	? 7%
Congenital defects	4%-8%
Congenital heart defects	10%
Mental retardation	
severe, IQ < 50:	15%-30% (depending on maternal age structure of population and survival patterns)
moderate, IQ 50-69	10%, higher in males
Developmental disabilities in those without retardation	? 1%-5%
Deviance (presence in security setting)	0.8%-3%
Male infertility (all)	2%
Oligospermia	3-6%
Azoospermia	15%
Multiple miscarriages	< 1% to 13% depending on selection criteria
Hypogonadism	
females	25%
(all with hypergonadotropic	
hypogonadism)	(> 50%)
males	? 10%
Malignancy[b]	< 1%

[a]Estimates are updated from Hook 1982 in which further references and discussion will be found. I include also with the germinal abnormalities the somatic abnormalities that occurred sufficiently early in embryogenesis to involve a significant proportion of the organism, such as XY/XO gonadal dysgenesis. However, somatic chromosome breaks and rearrangements in specific tissues or somatic mutations giving rise to clonal outgrowth associated with malignancies are excluded.

[b]A significant proportion of malignancies however, are associated with somatic chromosome abnormalities.

Table 2
Livebirth Prevalence of Chromosome Abnormalities Associated with Significant Probability of Pathology at Birth or Defect or Developmental Disability Manifesting Later in Life

	Rate (per 10,000)[a]	Proportion associated with germin "mutation" in first generation	
Numerical			
47,+21	10	?	~90%[b]
47,+18	1	??	~90%[b]
47,+13	1	??	~90%[b]
47,XXY	5	>99%	
47,XYY	5	>99%	
47,XXX	5	>99%	
45,X	0.5	>99%	
mos 46,XX/45,X	~5[c]	<1%	
All others	0.2	??	~90%
	32.7		~80%
Structural			
Interchange Down's syndrome	0.5		~75%
Interchange Patau's syndrome	0.3		~60%
Cri du chat, 5p-	0.2	?	~80%
46,X,i(Xq)	0.2		>99%
All others including supernumerary except "fragile" sites	3	?	75%
Subtotal	4.2		~85%
Xq28—"fragile" and other "fragile" chromosomes[d]	??5	??	<10%
	? 9.2	??	35%

[a]The estimates for livebirth prevalence are derived from Hook and Hamerton (1977) and assume no selective abortion because of prenatal diagnosis. Use of prenatal diagnosis is increasing rapidly but future trends cannot be predicted now. The estimates for the trisomies moreover, assume maternal age distributions similar to that observed in developed countries in the late 1970s.

[b]The estimates of germinal "mutation" for 47,+21, including the few mosaics, are derived from the assumption that about 80% of 47,+21 result from meiotic first division nondisjunction, and that the remainder are equally divided between cases resulting from meiotic second division nondisjunction and mitotic nondisjunction. The actual rate attributable to germinal cell events is thus about 80% to 100% if the estimates for the proportion at first division is correct. (It is also assumed that the proportion attributable to inheritance of an extra chromosome [secondary nondisjunction] is relatively trivial, no more than 5%.) The estimates for 47,+18 and 47,+13 assume that the proportions mutant are the same as for 47,+21, for which there is no evidence either way at present.

[c]The observed rate of mos 45,X/46,XX in newborn studies is about 2.6/10,000 (Hook and Hamerton 1977) but the exact proportion depends upon how thoroughly a search is made for mosaicism. It is not clear moreover, how many of those with this genotype have significant probability of pathology. This is therefore, the most uncertain entry in the list. Almost all of these cases result from somatic anaphase loss of an X chromosome, not germinal mutation.

[d]The livebirth prevalence of the so called "fragile" chromosomes is not known, nor is it entirely clear what proportion of these chromosomes are associated with significant pathology. The best evidence of pathology is for the association of the "fragile" X with mental retardation.

Table 3 lists the estimated collative (potential) livebirth prevalence[2] of conditions in each of these categories and the estimated prevalence resulting from mutation in the most recent generation.

INVESTIGATION OF HYPOTHESIZED CAUSES OF HUMAN MUTATION

In investigation of the putative causes of human mutations there are several straightforward strategies that may be adopted. (See, for further discussion, texts such as Lilienfeld 1976.) The first is a cohort study, either retrospective or prospective, in which one searches for the consequences of germinal mutations in individuals who have been exposed to a known or suspected hazard, and compares them with the unexposed group in the same cohort.

Case-control studies are an alternative and often more efficient method of searching for causal associations, although they are methodologically more suspect than cohort or cross-sectional studies because among other reasons it is often easier for unsuspected biases to enter in design or analysis, for instance in picking appropriate controls. With regard to investigations of mutations, one starts with a group of individuals with mutations—preferably a homogeneous group of mutations—and an appropriate control group and searches for a preconceptual history of known or putative mutagens.

But these systematic methodological approaches are the domain of the epidemiologist who will usually turn to the cytogeneticist or biochemist for confirmation of a mutation. Where does the clinician enter in evaluation of mutation?

THE ROLE OF THE CLINICIAN

R.W. Miller has been a strong proponent of the "alert clinician" approach to detection of effects of environmental hazards upon chronic disease (Miller 1978, 1981, 1982). He has urged that the most outstanding need in stimulating discovery of adverse environmental agents is the training and encouragement of clinicians to seek out environmental factors in the histories of patients. Indeed the claim has been made that "virtually all known human teratogens and carcinogens were first recognized by alert practitioners" and that, in contrast, epidemiologic studies in this area have been useful mainly for testing hypotheses not for generating them (Miller 1982).

---

[2] Individuals may be born with adverse genes but not manifest their effects phenotypically until later in life. A term is needed for the proportion of livebirths who have or will eventually manifest genetic disorders. The term "gene frequency" is inappropriate because i) affected individuals with recessive disorders have two doses of the allele, and ii) the frequency of "adverse" genes associated with polygenic or multifactional conditions cannot now be estimated. It would appear "potential livebirth prevalence rate" would be useful.

Table 3

Estimated (potential) Livebirth Prevalence of Specific Locus Mutations Associated with Significant Pathology[a]

	Potential livebirth prevalence (per 1000)[b]	Frequency (per 1000) resulting from mutation in first generation
Autosomal dominant	9	0.5[c] (0.2 to 2.0)
Autosomal recessive	2 to 3	very small, < .001
X-linked	0.5	? < .01[d]

[a]The potential livebirth prevalence is the proportion of livebirths that will eventually manifest pathological findings associated with the allele(s).

[b]Based on estimates given in Carter (1982) on European populations. Thus these apply only to such groups.

[c]The figure of 0.5 per 1000 is derived by assuming that about 2/3 of the total of those with autosomal dominant disorders are attributable to conditions such as monogenic hypocholesterolemia, neurofibromatosis, and Huntington's chorea, almost all of which are inherited and that about 50% of the remainder are attributable to mutation in the most recent generation. If alternatively one assumes that there are *at least* 10^3 loci at which autosomal dominant alleles associated with significant pathology can occur—the total of all definite autosomal dominants in McKusick's catalog in 1978 was 736—and that the average mutation rate resulting in livebirths who will be affected at some time in their life is between 10^{-7} and 10^{-6} per gamete, then the livebirth prevalence rate associated with mutation in the first generation is, at a minimum, between 0.2 and 2 per 1000.

[d]Certainly less than this figure.

There are indeed impressive examples of such discoveries and it is worthwhile to review some of them briefly. The major episode that essentially reshaped clinical thinking about environmental hazards to the fetus was the discovery of rubella embryopathy by Gregg, an Australian ophthalmologist (Gregg 1941). Radiation had already been known to induce fetal damage but in doses large enough to be toxic to the mother also, and more subtle damage associated with lower doses had not to my knowledge been discovered in humans by that time. Gregg reported the association of cataracts (as well as deafness and congenital heart defects) with maternal rubella exposure 6 to 9 months prior to birth. It is still a mystery to many why the association had not been picked up many years earlier. Of course, rubella embryopathy has a much larger spectrum of the effect than the induction of cataracts (as was noted by Gregg) and not all with the embryopathy develop these lesions, but it was the fortunate (for science) concentration of individuals with cataracts in Dr. Gregg's office that led to his discovery.[3]

[3]Of interest, there is the apocryphal tale, which I have not been able to confirm, that it was not Dr. Gregg himself, but an anonymous mother who made the original discovery. Allegedly, she had this insight in Dr. Gregg's waiting room while waiting with her child, after finding that other mothers there had had children with cataracts and other problems similar to those of her own and had experienced rubella during pregnancy. Although I can't confirm this, I find this tale at least credible. So perhaps this is another example of "anonymous was a woman."

Other examples of discoveries of teratogens by practicing physicians are the associations of thalidomide with limb reduction defects, tetracycline with teeth staining, warfarin with bone defects, alcohol with the "fetal alcohol syndrome," and androgens with fetal masculinization, among others. (See Miller 1978, 1981 for references.) In many instances (e.g., rubella and alcohol) the spectrum of effect is quite extensive, but the original associations that were made were with features that are suggestive of the syndromes but are not necessarily the most characteristic or severe. A cluster of a group of children with a specific facies (thin lips, long philtrum, etc.) suggested to Jones et al. (1973) that maternal alcohol ingestion is a human teratogen, but mental retardation and psychological disabilities were found to be the most severe consequences. Cataract is frequent in rubella embryopathy but deafness may be more frequent, and mental retardation and congenital heart disease are usually the most significant clinical lesions.

But what led to the recognition of these human teratogens (most of whose effects incidentally had not been anticipated in animal studies) was the consistency of their effects, the induction so to speak of sentinel lesions with relatively high frequency. Many recognized teratogens—and carcinogens as well—have target organs and tissues so that specificity of response is not unexpected.

But the "target" for most germinal, specific locus mutagens is primarily DNA of the germ cell, damage to which may result in quite nonspecific lesions in the resulting organism.[4] Thus the paradigm of "alert clinicians" reporting multiple instances of the *same* phenotype attributable to environmental mutagens is likely to be much less useful than for discovery of environmental teratogens and carcinogens. It is not surprising that Miller's review of discovery of human teratogens, carcinogens, and mutagens cites many instances of discoveries of teratogens and carcinogens by alert clinicians, but not of mutagens.

THE ROLE OF THE MEDICAL GENETICIST

Despite the limitations discussed above there is at least one group particularly well-situated to suspect the possibility of effects of mutagens. Medical geneticists encounter patients from a large number of different disciplines. Logistically, they are in the best position to recognize an increase in instances of (dominant) mutations. (Mutations at recessive loci would not be particularly useful for this

---

[4] It is an interesting question as to the pertinence to studies of human mutagenesis of i) observations of mutagens with specific biochemical effects in lower organisms; and ii) of "hot spots" within the genome and even within loci in microorganisms. Eventually I suspect we may in some sense be able to suspect specific types of mutagens from observations of specific mutagenic phenotypes in humans—but probably only by broadening our concept of "phenotype" to include a much more detailed biochemical description of the mutation than is generally available today. My main point here is that without denying the possibility of mutagen specificity in at least some senses in humans, compared to teratogens and carcinogens mutagens are likely to have much broader effects upon "crude" human phenotypes.

goal.) At present however, many medical geneticists regard instances of mutation as at best, examples of a curiosity from the viewpoint of etiology. While they are likely to elicit detailed histories of possible environmental hazards during the gestation of children with birth defects, they are much less likely to inquire systematically concerning the preconceptual exposure of both parents to possible mutagens. (See below.) We need to sensitize medical geneticists to the importance of systematically eliciting such histories. It could be argued that many cases of new mutations are not recognized until later in life, by which time it is not possible to obtain useful data from the parents. But there are numerous autosomal dominant conditions that may present in the newborn period including such disorders as Achondroplasia, Apert's syndrome, and aniridia. (Some have suggested these three would be useful sentinel phenotypes for systematic surveillance of mutation rates in newborn mutations [Sutton 1969] but there are still serious logistic problems with using these disorders for surveillance.) Homes et al. (1981) for instance, in examination of 24,418 infants found 12 with malformations attributable to autosomal dominant conditions of which 6 were presumptive new mutations. (Unfortunately the only environmental histories elicited from the parents were those concerning the pregnancy, not the preconceptual history!) Thus medical geneticists, if they can be made aware of the need for systematic inquiry in evaluation of those with specific locus mutant conditions, may well be the first to suspect the effects of new mutagens in humans.

THE ROLE OF THE CYTOGENETICIST

In the section above I considered specific locus mutations exclusively. For most such mutations associated with pathology in humans there is at present no straightforward test for determining if the observed event is the result of a mutation in the most recent generation or has been inherited from an unaffected carrier parent. But investigation of parents of those with chromosome abnormalities can usually determine if the structural rearrangement is the result of a new mutation, and do so with some but not complete confidence for numerical abnormalities. Thus the clinical cytogeneticist is at present probably best situated to detect effects of agents which produce germinal chromosomal rearrangements or numerical abnormalities. Unfortunately, few clinical cytogeneticists inquire routinely of their patients concerning parental preconceptual exposure to possible mutagens.

Some case-control studies have been done on Down's syndrome (trisomy 21) both by cytogeneticists and epidemiologists, particularly with regard to possible effects of radiation and of oral contraceptives. (For references and discussion see Hook 1981, 1982b.) A rough generalization is that the published studies of cytogeneticists have tended to be positive, the studies by epidemiologists tend (albeit, not uniformly) to be negative. I believe this only reflects the fact that cytogeneticists do such studies less well than epidemiologists. Probably cytogeneticists should confine themselves to generating hypotheses, not trying to confirm them unless they have expert epidemiological and statistical assistance.

At the present time there is no *strong* evidence for any environmental agent causing trisomy 21 in humans. The most suggestive evidence for an environmental effect upon any numerical chromosome abnormalities is that of spermacides upon tetraploidy (92 chromosomes) a condition almost uniformly fatal in embryos and fetuses (Warburton et al. 1980; Strobino et al. 1980). It probably arises by the induction by either the active compound (or the vehicle) of endoreduplication, i.e. redoubling of chromosomes before cell division, and has little relevance to the production of monosomy, trisomy, or structural chromosome rearrangements which are responsible for major human toll. This association and its' putative mechanism is moreover, still to be confirmed.

With regard to *germinal* chromosome structural rearrangements, there is at present only one study to my knowledge indicating effects of environmental hazards. These are the results of a controversial protocol involving the irradiation of gonads of volunteer prisoners whose testes were then biopsied. Clear evidence was found for dose response of chromosome rearrangements, although the doses used were much larger than we are usually concerned about today (Brewen et al. 1975). There are of course numerous studies of putative environmental effects upon somatic chromosome rearrangements in humans but in my opinion these provide little ground for inferences about germinal effects.

It is also possible using various strategies to systematically collect data from cytogenetic laboratories, data which may be used as a starting point for investigation. For example, our unit in the Health Department is responsible for the New York State Chromosome Registry (Hook et al. 1981) and also is affiliated with the U.S. Interregional Chromosome Register System (Prescott et al. 1978). Both of these get summary data from affiliated laboratories on all women who received prenatal cytogenetic diagnosis. The reasons for study, maternal age, and residence are reported. In the approximately 27,000 women reported to date, most, about 80%, have of course been studied because of advanced maternal age. But there are also 65 women studied because of exposure to chemicals or drugs of suspicion, and 71 because of radiation exposure. In the 136 total there was one known structural chromosome mutant, a deletion, in the radiation series and one possible structural mutant, an inversion, in the drug series. (The father had taken multiple illicit drugs including LSD and vanished before it could be determined if he was a carrier.) The rate of known or possible mutants in these two groups combined is 1.5%, 10-fold higher than the rate of nonmosaics in all others studied 0.14% ($p < .02$, Fisher's exact test) (Hook et al. 1982). The case associated with radiation exposure is of interest in that it involved a father who had a child after radiotherapy for Hodgkin's disease. One advantage of the amniocentesis data is that they are gathered prospectively. Some information about exposure is available before the outcome is known, so selective memory is less likely to bias analyses. Use of this data source will eventually expand greatly our knowledge of human mutation.

But perhaps the recent discovery of methods for examining *directly* chromosomes of human sperm provide the best new prospects for cytogeneticists in discovery of human mutagens. (Rudak et al. 1978; Martin et al. 1981, 1982). This is, I believe, the most attractive new frontier for human environmental mutagenesis. Currently, the tedious methodology prevents its widespread application to studies of many populations.

INFERTILITY AND STERILITY: THE ROLE OF THE OBSTETRICIAN AND UROLOGIST

Observations of infertility by physicians come closest to meeting Miller's paradigm of the alert clinician in the field of human mutation. The effect of chlordecone (kepone) upon sterility for example, was first detected by an alert clinician. But the discovery of the effects of dibromochloropropane (DBCP) upon fertility are less flattering to the medical profession. This was an effect discovered by the exposed workers themselves after they failed to impress their skeptical physicians and independently sought evaluation of their seminal fluids by a clinical laboratory. (See discussion and references in Miller 1981.) While these effects, important as they are, have resulted from direct or indirect damage to germ cells or their precursors, it is not clear that they have been mediated by *mutational* mechanisms and not through some other toxic pathway.

Nevertheless, infertility can clearly be produced by at least some known mutagens, e.g. ionizing irradiation, and whatever the cause, it would be valuable to encourage urologists and obstetricians, the subspecialists usually investigating infertility, to inquire systematically concerning possible environmental hazards.

THE "DOMINANT LETHAL" PARADIGM IS PROBABLY NOT USEFUL FOR INVESTIGATION OF HUMAN MUTATION

In experimental animals, under strictly controlled conditions, an increase in embryonic lethality after paternal exposure to a substance is often assumed to result from mutagenic effects resulting in "dominant lethals." In humans however, such an inference is often difficult or impossible, particularly for chemicals. The main difficulty is exclusion of the possible effect of an embryotoxin acting during gestation upon the fetus. Even if only the father has been exposed, if his exposure continued after the time of conception then indirect transmission of the substance to the mother, e.g. through work clothes, may result in fetal toxicity. Only if 1) exposure to either parent ceased before conception, and 2) it could be safely inferred that substances were not stored in, say maternal body fat depots and slowly released into the blood stream over a long period of time, would inferences about *mutagenic* effects be warranted. Such circumstances are very rare, and thus it seems likely that the rates of embryonic and (or) fetal deaths per se are not likely to be useful plausible markers for mutagenic events in humans as they are in experimental animals. Cytogenetic or biochemical study would be necessary to identify a mutation in fetuses.

It should also be emphasized that when there have been outbreaks of marked increase in embryonic and fetal deaths, almost always these have resulted from embryotoxic factors of significance during gestation, such as Rubella virus epidemics.

CONCLUSION

Thus the most important role of primary clinicians, and their allies—cytogeneticists, biochemists, and medical geneticists among others—is to identify those with mutations. If these professionals are alert moreover, they may well suggest etiologic hypotheses for further investigation in collaboration with epidemiologists and biostatisticians.

But to play the devil's advocate for the moment, I wonder if we do not have already a glut of etiologic hypotheses. In the field of birth defects for example, for any substance x one may find an article in some clinical journal reporting the association of x with the occurrence of some birth defect (occasionally even several occurrences) and a claim that "perhaps the association is not coincidental." Admittedly, many such reports of a *rare* substance with the same *rare* defect may lend to an important discovery, as for the putative recent association of maternal dilantin exposure and neuroblastoma, but usually the defects and (or) the substances involved are sufficiently common to make interpretation of even multiple case reports involving the same defect and substance very difficult.

What are needed are not just hypotheses but *good* hypotheses, soundly based and plausible. Before entering into an epidemiologic investigation of a putative mutagen in human populations one wants to be sure that there are good presumptive grounds for human effects of the substance and a significant social impact. Systematic investigation of the effects of putative environmental mutagens in humans can be time-consuming, frustrating, and expensive, as the experience in trying to document the mutagenic effects of ionizing irradiation—a known mutagen—at Hiroshima and Nagasaki demonstrate. (See, e.g., Schull et al. 1981a,b).

Moreover, not only are human geneticists and epidemiologists who are interested in mutation confronted with numerous case reports of uncertain significance from clinicians, but they also must evaluate which of the many reports from lower organisms should be followed up in human populations. Additionally, in this source there is a huge number of suspicious substances, only some of which can be readily investigated. What would be very useful for the human geneticist is a set of criteria generated by those working in mutation in lower organisms as to what type of data provide the strongest grounds for suspicion of effect on germinal mutation in humans. There are voluminous reports of mutagens in microorganisms. Should the epidemiologists await confirmation of such effects in mice or other mammals before studying such substances in people? What if any evidence on mutagenicity derived only from microorganisms should warrant a major effort in epidemiological studies of germinal mutations in human populations?

It is hoped that the deliberations of this conference can provide useful answers to queries such as these.

REFERENCES

Brewen, J.G., R.J. Preston, and N. Gengozian. 1975. Analyses of x-ray induced chromosomal translocations in human and marmoset spermatogonial cells. *Nature* 253:468.

Carter, C.O. 1982. Contribution of gene mutations to genetic disease in humans. In *Chemical mutagenesis, human population monitoring, and genetic risk assessment.* (ed. K.C. Bora, G.R. Douglas, and E.R. Nestmann). vol. 3, p. 1. Elsevier Biomedical Press, Amsterdam.

Gregg, N.M. 1941. Congenital cataract following German measles in the mother. *Trans Ophthalmol Soc. Australia* 3:35.

Holmes, L.B., S.E. Vincent, C. Cook, and K.R. Cote. 1981. Surveillance of newborn infants for malformations due to spontaneous germinal mutations. In *Population and biological aspects of human mutation.* (eds. E.B. Hook and I.H. Porter), p. 351. Academic Press, New York.

Hook, E.B. 1981. Down syndrome: Frequency in human populations and factors pertinent to variation in rates. In *Trisomy 21 (Down syndrome): Research perspectives.* (ed. P.S. Gerald and F. de la Cruz), p. 3. University Park Press, Baltimore.

_____. 1982. Contribution of chromosome abnormalities to human morbidity and mortality and some comments upon surveillance of chromosome mutation rates. In *Chemical mutagenesis, human population monitoring, and genetic risk assessment.* (ed. K.C. Bora, G.R. Douglas, and E.R. Nestmann). (Progress in Mutation Research). vol. 3, p. 9. Elsevier-North Holland, Amsterdam.

_____. 1982b. The epidemiology of Down syndrome. In *Down syndrome: Advances in biomedicine and the behavioral sciences.* (ed. S.M. Pueschel). Ware Press, Cambridge (In press).

_____. 1982a. The epidemiology of human chromosome abnormalities. In *Perinatal epidemiology.* (ed. M.B. Bracken). Oxford University Press, New York.

Hook, E.B. and J.L. Hamerton. 1977. The frequency of chromosome abnormalities detected in consecutive newborn studies—Differences between studies—Results by sex and severity of phenotypic involvement. In *Population cytogenetics: studies in humans.* (ed. E.B. Hook and I.H. Porter), p. 63. Academic Press, New York.

Hook, E.B., P.K. Cross, and D. Schreinemachers. 1981. The evolution of the New York State Chromosome Registry. In *Population and biological aspects of human mutation.* (ed. E.B. Hook and I.H. Porter), p. 389. Academic Press, New York.

Hook, E.B., D.M. Schreinemachers, A.M. Willey, and P.K. Cross. 1982. Rates of mutant structural chromosome rearrangements in human fetuses: Data from prenatal cytogenetic studies and associations with maternal age and parental mutagen exposure. *Am. J. Hum. Genet.* (in press).

Jones, K.L., D.W. Smith, C.N. Ulleland, and A.P. Streissguth. 1973. Pattern of malformation in offspring of chronic alcoholic mothers. *Lancet* 1:1267.

Lilienfeld, A. 1976. *Foundations of epidemiology*, p. 283. Oxford University Press, New York.

Martin, R.H., W. Balkan, K. Burns, and C.C. Lin. 1981. Direct chromosomal analysis of human spermatozoa: Results from 15 normal men. *Abstracts of the 6th International Congress of Human Genetics*, p. 46. Jerusalem.

Martin, R.H., C.C. Lin, W. Balkan, and K. Burns. 1982. Direct chromosomal analysis of human spermatozoa: Preliminary results from 18 normal men. *Am. J. Hum. Genet.* 34:459.

McKusick, V.A. 1978. *Mendelian inheritance in man: Catalogs of autosomal dominant, autosomal recessive, and X-linked phenotypes*, p. 975. Johns Hopkins University Press, Baltimore.

Miller, R.W. 1978. The discovery of human teratogens, carcinogens, and mutagens: Lessons for the future. In *Chemical mutagens: Principles and methods for their detection* (ed. A. Hollaender and F.J. deSerres), vol. 5, p. 101. Plenum Press, New York.

———. 1981. Pollutants and children: lessons from case histories. In *Guidelines for studies of human populations exposed to mutagenic and reproductive hazards.* (ed. A.D. Bloom) March of Dimes-Birth Defects Foundation, New York.

———. 1982. The clinician's role in monitoring for diseases from chemicals in the environment. In *Environmental mutagens and carcinogens.* (ed. T. Sugimura, S. Kondo, and H. Takebbe), p. 655. University of Tokyo Press and Alan R. Liss, Inc., Tokyo and New York.

Prescott, G.H., M.K. Rivas, L. Shanbeck, D.W. Macfarlane, H.E. Wyandt, W.R. Breg, H.A. Lubs, R.E. Magenis, R.L. Summitt, C.G. Palmer, F. Hecht, W. Kimberling, and D. Clow. 1978. *Birth Defects Original Article Series 14* (6C):269. The Interregional Cytogenetic Register System (ICRS).

Rudak, E., P.A. Jacobs, and R. Yanagimachi. 1978. Direct analysis of the chromosome constitution of human spermatozoa. *Nature* 274:911.

Schull, W.J., M. Otake, and J.V. Neel. 1981a. Hiroshima and Nagasaki: A reassessment of the mutagenic effect of exposure to ionizing radiation. In *Population and biological aspects of human mutation* (eds. E.B. Hook and I.H. Porter), p. 277. Academic Press, New York.

———. 1981b. Genetic effects of the atomic bomb: A reappraisal. *Science* 213:1220.

Strobino, B., J. Kline, Z. Stein, M. Susser, and D. Warburton. 1980. Exposure to contraceptive creams, jellies, and douches and their effects on the zygote. *Am. J. Epidemiol.* 112:434.

Sutton, H.E. 1971. Report for the committee for the study of monitoring of human mutagenesis. *Teratology* 4:103.

Warburton, D., Z. Stein, J. Kline, and B. Strobino. 1980. Environmental influences on rates of chromosome anomalies in spontaneous abortions. *Am. J. Hum. Genet.* 32:27A.

COMMENTS

BRIDGES: I am aware of your very great skepticism about the use of fetal loss. Yet it seems to me that if only one could use it, the advantages are so great. The numbers are reasonable. The levels of spontaneous fetal loss mean that you haven't got to use tens of thousands of cases, and also you don't really have a latent period. If you suspect genotoxic exposure in a population of males, for example, you don't have to wait 20 years before you can get some evidence that there is some sort of genetic effect.

The criticism that you can never have confidence that you have excluded female exposure, I would have thought that some of the techniques that we are now going to talk about at this meeting could well be used to study that particular point.

HOOK: Let me clarify a few issues. If you are not concerned about confounding teratologic-embryotoxic effects with mutagenic effect, then I agree. If you are interested in germinal mutagenic effect, then there is a major problem.

The other question, of course, is whether you look only at the overall rate of embryonic and fetal death, or you want to do something more interesting, such as actually score chromosomal aberrations in fetal wastage. This will give greater precision as to mutations, and you can make inferences, presumably, about the relevance of preconceptual agents. I might note that whenever there have been significant and marked alterations in the overall rate of embryonic and fetal death in populations, these have been attributed to embryotoxic factors such as Rubella epidemics, famine, etc. and not to defined mutagens.

THILLY: Could you go through for our benefit some of the analyses that you have published on in utero selection against errors of chromosome number and chromosome structure? It seems to me that we can grow cells in culture which have various forms of monosomies all the way down to almost haploidy for human cells, yet there is this enormous filter keeping genetically damaged liveborns from becoming live-born. It seems to me that without a general discussion of that filter, we don't have a feeling for exactly what the actual rate of gene mutations or chromosomal changes is in humans.

HOOK: It is not just my work. There are many others who have done such work. In fact, I am not sure that my own work is actually the most recent or the best available on this topic. I am particularly excited about the recent reported success in looking at chromosome aberrations in human sperm first by Rudak et al. (1978), and then refined by Martin (Martin et al. 1981, 1982). There are reports by Martin of a very high rate of aneuploidy in human sperm about 8%-10% and much higher than that I had estimated in detectable conceptuses at around 5 weeks (from the LMP).

I will just draw three points starting with the gamete, and then 5 weeks, which is usually the first time of recognizable pregnancy in human populations, and live birth usually at 40 weeks. Five weeks from the LMP (first day of last menstrual period) is really 3 weeks from the time of conception. The frequency of abnormalities, all chromosome abnormalities, in live births is around 0.6%. In recognizable conceptuses at age 3 weeks I estimated a rate of about 5%. In the sperm, 2 weeks before conception, Martin et al. (1981, 1982) observed a rate of abnormality of about 8%-10% and we may report at least as great a contribution from ova. If there is no prezygote selection then at least 15%-20% of zygotes would be expected to be cytogenetically abnormal. Yet 3 weeks later the proportion drops. It is estimated to about 5%, and to 1/10 of this figure by the end of gestation. These reflect great intra-term lethality of these chromosomal mutations.

SESSION II:
DETECTION OF MUTAGENS IN BODY FLUIDS

Detecting Mutagens in Cigarette Smokers' Urine

ERIC EISENSTADT, NORMAN V. KADO, AND RESHA M. PUTZRATH
Harvard School of Public Health
Boston, Massachusetts 02115

RATIONALE AND ADVANTAGES

Our exposure to mutagens can be estimated by determining the mutagenicity of either environmental samples (e.g., food, air, or water) or human samples (e.g., urine or blood). Human samples assayed for mutagens can provide estimates of mutagen exposure from the many environments people encounter. Analyzing human samples requires no assumptions about the source of the activity; in fact, endogenous activities can be detected as well. Although chemical detection of certain mutagens in human samples can be accomplished with high sensitivity and specificity, methods must be tailored to particular compounds or classes of compounds. Mutagenicity assays, on the other hand, respond to a wide variety of chemicals (McCann et al. 1975). Thus, a single assay can respond to many different kinds of unidentified mutagens. This potential to detect unanticipated activities is an especially attractive and useful feature of mutagenicity tests. Mutagenicity tests can also be used to guide the purification and characterization of new mutagens (Sugimura et al. 1977; Varghese et al. 1977; Putzrath et al. 1981).

LIMITATIONS AND APPLICABILITY

The feasibility of detecting mutagens in human body fluids—urine, in particular—has been well established for samples collected from cigarette smokers (Varghese et al. 1977; Yamasaki and Ames 1977; Dolara et al. 1981; Putzrath et al. 1981), certain occupational groups (Falck et al. 1979, 1980; Dolara et al. 1981), and patients receiving chemotherapy (Siebert and Simon 1973; (Connor et al. 1977). Smoking provides perhaps the heaviest regular exposure (via the respiratory tract) to mutagens; the health consequences of smoking are well-documented (DHEW 1979). Smokers constitute a noisy background from which other mutagenic activities, if found, have to be distinguished. Smokers also provide a population of individuals whose samples might be exploited to develop monitoring procedures. Nonsmoking or nonoccupationally exposed subjects generally have nonmutagenic urine extracts, suggesting either that present methods detect only certain kinds or degrees of exposure or that our routine exposure to mutagens is not monitorable in urine. Obviously, it is important that we learn to distinguish between these explanations.

The fraction of an exposure which is recovered in urine will be influenced by a large number of factors including absorption, biotransformation, tissue distribution, and excretion of chemicals (which, themselves, are subject to environmental and genetic influences). Sample collection, work-up, and assay conditions also influence detectability. The influence of such factors can be expected to differ in magnitude from one chemical to another. For example, biotransformation of aflatoxin B_1 results in the excretion of metabolites with diminished mutagenicity (Campbell et al. 1970), whereas biotransformation of 2-acetylaminofluorene results in the excretion of metabolites which are more mutagenic, following deconjugation, than the parent compound (Weisburger et al. 1964; Durston and Ames 1974). Because of these factors, the ability to monitor exposures will vary significantly among different agents.

DETECTION AND CHARACTERIZATION OF MUTAGENS IN CIGARETTE SMOKERS' URINE

Yamasaki and Ames (1977) showed that XAD-2 resin could be used to concentrate the mutagenic activity in smokers' urine and to separate the mutagens from histidine present in the urine. Our protocol for characterizing the mutagens in smokers' urine employed standard chemical separation procedures subsequent to XAD-2 extraction (Fig. 1). Details of this protocol have been published by Putzrath et al. (1981).

Since XAD-2 resin is widely used to prepare fluids for mutagenicity testing, it is worth mentioning some of the limitations of this procedure. The resin selectively adsorbs nonpolar molecules from aqueous solutions. Since human tissue or body fluid samples may contain conjugated mutagens which are highly polar, these mutagens may be systematically underestimated or lost from XAD-2 extracts. Even if conjugated compounds are extracted by XAD-2 (Bradlow 1968), their mutagenicity could go undetected without either chemical or enzymatic hydrolysis. Our study examined the mutagens extracted by XAD-2 from untreated smokers' urine. We recognize, therefore, that we may be extracting a subset of the mutagenic activity in smokers' urine. Furthermore, some of the material from urine which was retained by the resin is not released by standard organic eluting solvents. A colored material remained on the column and could not be removed except by harsh cleaning treatments which would not be appropriate for sample preparation. In spite of these limitations, XAD-2 remains a useful tool for rapidly concentrating nonpolar mutagens from aqueous solutions.

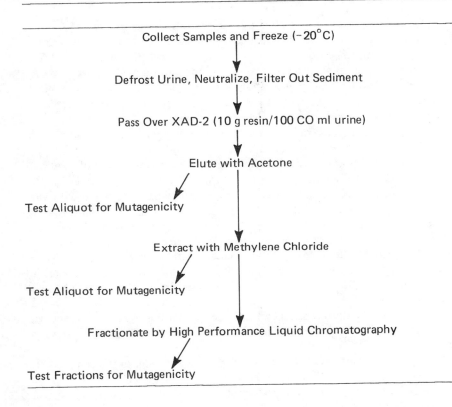

Collect Samples and Freeze ($-20°C$)

Defrost Urine, Neutralize, Filter Out Sediment

Pass Over XAD-2 (10 g resin/100 CO ml urine)

Elute with Acetone

Test Aliquot for Mutagenicity

Extract with Methylene Chloride

Test Aliquot for Mutagenicity

Fractionate by High Performance Liquid Chromatography

Test Fractions for Mutagenicity

Figure 1
Preparation of urine extracts for mutagenicity testing

The acetone eluate from an XAD-2 column loaded with smokers' urine gives a positive dose-response curve with several of the Ames tester strains. Strain TA1538 was most frequently used to monitor mutagenic activity in our studies. We found that application of a small number of additional separation techniques subsequent to XAD-2 extraction can further purify and characterize the mutagens in smokers' urine. By resuspending dried XAD-2 eluate in water and extracting this solution with methylene chloride, the mutagens in smokers' urine could be isolated from nonmutagenic material. We recovered approximately 95% of the mutagenic activity in the methylene chloride fraction while 95% of the dry weight remained in the aqueous phase. This purification procedure has an additional benefit since nonmutagenic materials that are toxic to the *Salmonella* system appear to be removed. Dolara et al. (1981) have recently observed a similar reduction in toxicity of urine extracts if the XAD-2 column is eluted with methylene chloride instead of acetone.

The mutagens purified by methylene chloride extraction of an acetone eluate from XAD-2 were sufficiently concentrated to be analyzed by reverse phase high performance liquid chromatography (HPLC). Fractions from HPLC were assayed for mutagenic activity (Fig. 2). The elution gradient was 35% to 100% methanol:water, changing at a rate such that 100% methanol was reached in fraction number 21. As could be expected from the preparative procedure, the mutagens were relatively nonpolar. A large number of the fractions were mutagenic, suggesting a complex mixture of mutagens reminiscent of the multiple mutagens observed with cigarette smoke condensates (Kier et al. 1974).

If this broad distribution of mutagenic activity is characteristic of smokers' urine extracts, HPLC analysis might be useful in distinguishing smokers' urine from other mutagenic urine. Although a general pattern of activity in the fractions emerged, we also observed variation between samples. Samples collected on different days from one smoker showed as much variation as samples collected from many different smokers. The differences observed in urine from one smoker from Boston, Massachusetts (Fig. 2A through 2E) appeared to be no greater than differences between his urine and urine pooled from smokers in Akron, Ohio (Fig. 2F) or urine from a smoker from the Republic of San Marino (Fig. 2G). However, when a single sample of urine concentrate was analyzed twice by HPLC, the mutagenic activity of the two elutions was nearly identical. Therefore, the HPLC fractionation procedure is not responsible for the variation observed.

Identification of these mutagens and their relationship to the components of cigarette smoke would provide valuable information. On a mass basis they are not present in great quantities. For example, the UV-adsorption of this part of the eluate is very low. We detect them because of their biological activity. It would, however, be an enormous undertaking for analytical chemists to purify and identify the mutagens in cigarette smoker's urine.

We have also been interested in improving the sensitivity of detecting activity in urine. One method that seems to work well involves combining bacteria and urine extract in a small volume and incubating them in the presence of activating enzymes for up to 2 hours. This procedure results in a > 20-fold increase in sensitivity (Kado et al., unpubl. results).

PROSPECTS

Monitoring human samples for mutagenicity has potential as a measure of at least some degrees or kinds of human exposure to mutagens. However, much more exploration is required before adoption of this approach as a general screening procedure. Adaptation of current testing procedures which simplified assays of human samples would make it easier to screen large numbers of samples. This would establish base-line data and reliably document the range of activity that is normally found in people. With further development and in combination with other screening procedures, mutagenicity testing of human

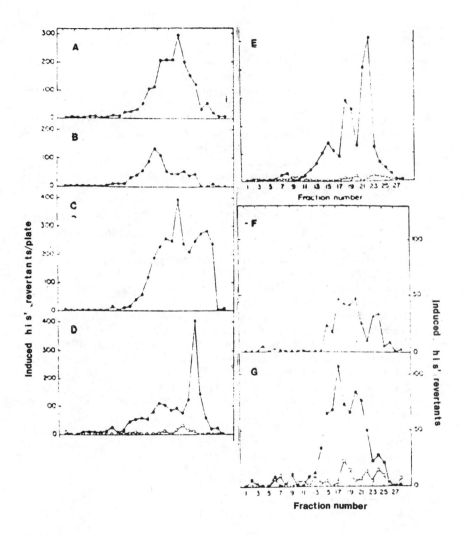

Figure 2

HPLC fractionation of smokers' and nonsmokers' urine. Methylene chloride extracts of XAD-2 concentrates were dried, resuspended in methanol, and fractionated by reverse phase HPLC using a Whatman Partisil Magnum-9 column and a Perkin-Elmer Series 2 programmed for an elution gradient from 35% to 100% methanol:water changing at 3%/min with a flow rate of 5 ml/min. Fractions, collected at 1-min intervals, were tested for mutagenicity. Volumes from each fraction, equivalent to 250 ml of fractionated urine, were tested. *B*: a 5-liter preparation; all other preparations are 1-2 liters. A-E: urine of a smoker (●) and pooled urines from nonsmokers (○) from Boston, Massachusetts. F: pooled urines from smokers from Akron, Ohio (▲); G: urine from a smoker (■) and nonsmoking spouse (□) from the Republic of San Marino. (Reprinted, with permission, from Putzrath et al. 1981.)

samples could contribute to a program for monitoring human exposure to mutagens.

REFERENCES

Bradlow, H.R. 1968. Extraction of steroid conjugates with a neutral resin. *Steroids* 11:265.

Campbell, T.C., J.P. Caedo, Jr., J. Bulatao-Jayme, L. Salamat, and R.W. Engel. 1970. Aflatoxin M_1 in human urine. *Nature* 227:403.

Connor, T.H., M. Stoeckel, J. Evrard, and M.S. Legator. 1977. The contribution of metronidazole and two metabolites to the mutagenic activity detected in urine of treated humans and mice. *Cancer Res.* 37:629.

Dolara, P., S. Mazzoli, D. Rosi, E. Buiatti, S. Baccetti, A. Turchi, and V. Vannucci. 1981. Exposure to carcinogenic chemicals and smoking increases urinary excretion of mutagens in humans. *J. Toxicol. Environ. Health* 8: 95.

Durston, W.E. and B.N. Ames. 1974. A simple method for the detection of mutagens in urine: Studies with the carcinogen 2-acetylaminofluorene. *Proc. Natl. Acad. Sci. USA* 71:737.

Falck, K., P. Gröhn, and L.R. Holsti. 1979. Mutagenicity in urine of nurses handling cytostatic drugs. *Lancet* i:1250.

Falck, K., M. Sorsa, H. Vainio, and I. Kilpikari. 1980. Mutagenicity in urine of workers in rubber industry. *Mutat. Res.* 79:45.

Kier, L.D., E. Yamasaki, and B.N. Ames. 1974. Detection of mutagenic activity in cigarette smoke condensates. *Proc. Natl. Acad. Sci. U.S.A.* 71:4159.

McCann, J., E. Choi, E. Yamasaki, and B.N. Ames. 1975. Detection of carcinogens as mutagens in the Salmonella/microsome test: Assay of 300 chemicals. *Proc. Natl. Acad. Sci. USA* 72:5135.

Putzrath, R.M., D. Langely, and E. Eisenstadt. 1981. Analysis of mutagenic activity in cigarette smokers' urine by high performance liquid chromatography. *Mutat. Res.* 85:97.

Siebert, D. and U. Simon. 1973. Cyclophosphamide: Pilot study of genetically active metabolites in the urine of a treated human patient. Induction of mitotic gene conversions in yeast. *Mutat. Res.* 19:65.

U.S. Department of Health, Education, and Welfare. 1979. Smoking and health, a report of the Surgeon General. DHEW Publ. No. (PHS)79-50066.

Sugimura, T., T. Kawachi, M. Nagao, T. Yahagi, Y. Seino, T. Okamoto, K. Shudo, T. Kosuge, K. Tsuji, K. Wakabayashi, Y. Iitaka, and A. Itai. 1977. *Proc. Jpn. Acad.* 53:58.

Varghese, A.J., P.C. Land, R. Furrer, and W.R. Bruce. 1977. Evidence for the formation of mutagenic N-nitroso compounds in the human body. *Proc. Am. Assoc. Cancer Res.* 18:80.

Weisburger, J.H., P.H. Grantham, E. Vanhorn, N.H. Steigbigel, D.P. Rall, and E.K. Weisburger. 1964. Activation and detoxification of N-2-fluorenylacetamide in man. *Cancer Res.* 24:475.

Yamasaki, E. and B.N. Ames. 1977. Concentration of mutagens from urine by adsorption with the non-polar resin XAD-2: Cigarette smokers have mutagenic urine. *Proc. Natl. Acad. Sci. U.S.A.* 74:3555.

Urine Mutagen Screening as a Population Monitoring Technique: Children in an Isolated, High Lung Cancer Mortality Area

GUYLYN R. WARREN AND SAMUEL J. ROGERS
Department of Biochemistry
Montana State University
Bozeman, Montana 59717

According to death certificate studies done by the National Cancer Institute (NCI) for the period 1950-1969 and the Montana Department of Health and Environmental Sciences (1969-1973), the Montana lung cancer death rate was 16% lower than the national rate. However, certain age groups living in Deer Lodge and Silver Bow counties had rates twice the national average. These counties have both been sites of mining and smelting copper ores since the early 1800s. Nearly the entire population of Silver Bow county lives in or near the town of Butte, the county seat. Anaconda is one of the major populated areas in Deer Lodge County. Both males and females were at increased risk in Butte-Silver Bow while only male rates were elevated in Deer Lodge. Histologic studies of fatal lung cancers in both areas led Newman et al. (1976) to suspect street dust as a possible cause in Butte-Silver Bow because of long-term presence in the area of known carcinogens such as cadmium, chromium and beryllium. The Montana State Legislature (1977 and 1979) funded a special air monitoring and health effects project, Montana Air Pollution Study (MAPS) which updated the death certificate studies and allowed basic research on urine and air mutagen content to be done.

We were searching for a technique allowing dynamic assessment of cancer risk in human populations. Advantages of such a method would be:

1. immediate confirmation of presence of causative agents;
2. possibility to correlate such data with death certificate statistics after the latent period of cancer has passed;
3. actual correlations with environmental exposure.

Geographic isolation of the study area and restriction of the cancer to a specific primary site (lung) made this study uniquely suited for identifying environmental carcinogens.

Experimental Procedure

The method we chose initially was a recently published urine analysis for mutagenic substances using the Ames Salmonella assay (Yamasaki and Ames 1977). We chose urine analysis for the reasons delineated by Eisenstadt (this volume) and especially because it was a noninvasive technique not requiring extensive legal protection and one which could be done relatively easily. One of the greatest problems with the death certificate studies was the confounding effects of cigarette smoking as a factor in lung cancer. Nearly all those who died of cancer were also smokers. We assumed that if we chose to monitor children below normal smoking age we could eliminate this variable and attempt to correlate actual mutagen levels with other readily measured parameters such as mutagens in the air. Air filter samples were monitored over a period of a year in each of the test cities and a profile of mutagen levels was established. Another portion of the MAPS study dealt with lung function in school children and they seemed to be easily accessed due to cooperation from school officials and the community. We chose to sample students in the 3rd and 4th grades since significant numbers of older children smoked and younger ones did not fully comprehend instructions necessary for the procedure. Our samples were not 24-hour samples but were collected for the 6 hours of actual school residence. Urines were analyzed for sugar, ketone, protein, pH, bilirubin, creatinine content, and specific gravity. Volumes were recorded before and after extraction by XAD-2 resin column chromatography. Concentrates from the columns were then tested by use of two bioassays. One was the Ames test for mutagens, the other was a DNA repair assay detecting agents which interfere with DNA replication or repair (Warren 1981). The latter method was included expressly to allow screening for inorganics. Tests were done in groups of ten and backgrounds plus positive controls were done for each group. Only background for the particular group were subtracted from that group. High volume or dichotomous air filters were extracted by descending dimethylsulfoxide (DMSO) chromatography.

Problems of sample size, S9 enzyme optimization, toxicity of mixed samples, optimal column function, and sample storage are common to all environmental monitoring.

Results and Discussion

Monthly averages of mutagens in the air presented as number of revertants per cubic meter of air sampled in Butte and Anaconda during a year are presented in Figure 1. Butte has much higher air filter mutagen levels than the other two sites while all sites contain more mutagen in winter than in summer. The mutagens are predominantly direct-acting in the summer while those present in the winter require S9-enzyme activation. We therefore chose to sample students in the early spring, due initially to time constraints and finally to school district procedures.

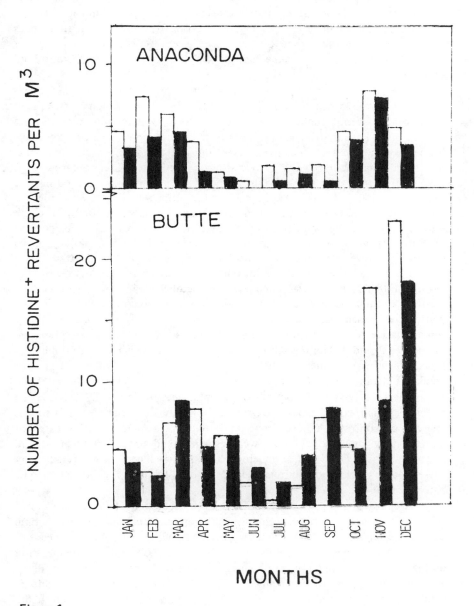

Figure 1
High volume air filter His[+] revertants of TA98 induced by extracts sampled monthly in Butte and Anaconda. Filter sampled had the monthly mean particulate load. (■) without S9; (□) with S9.

One school in Butte and one in Anaconda were sampled with concomitant air filter sampling during May of 1978. We had been sampling college students and found that we could not detect urine mutagens in persons who smoked < 10-low tar and nicotine brand cigarettes/day. All nonsmokers and low-tar smokers as well as those who rolled their own cigarettes gave a very flat and reproducible negative background. We expected to see the same results from school children.

The Ames test results are presented in Figures 2 and 3. Figure 2 lists results on TA1535 and Figure 3 shows TA1538. S9 activation was required for effect in TA1538 (data not presented). Toxicity was found upon activation on TA1535. To our surprise a number of positive samples (> 2x background) were assayed. All of the positives were from Butte and none from Anaconda. The Anaconda smelter was not operating at the time. Filter mutagen levels run at the same time showed Butte much higher than Anaconda during the sampling period (see Table 1). Both urine and filter mutagens were frameshift type. A repair assay detected several more active samples. A total of 0 of 29 samples Ames tested were positive in Anaconda while 8 of 22 were positive in Butte. A group of urine samples were also analyzed for content of 10 genetically active inorganics. It is not surprising that some students in Butte were high in copper or cobalt while some in Anaconda were slightly elevated in arsenic. Further testing was conducted in October adding another school in Butte to expand the area tested since address seemed to correlate with mutagen content. Bozeman was added as a control site to see if Anaconda was really representative of a negative population. Results of that sampling are presented in Figures 3 and 4. Monroe school was not nearly as high for that sampling as it had been before. Emerson, the new Butte School, was still higher than Anaconda's Lincoln. All of the samples listed in Figure 3 as less than −10 were toxic to the test bacteria. Dilution of the samples 1:2 or 1:4 in DMSO revealed activity as high as ten times background, but only on 3 samples. Subsequent analysis of control XAD-2 columns indicated that a malfunctioning batch of commercially prepared columns were obtained and used during the second sampling period. Only half as much of any control mutagen was retained by these columns as is normally retained by our own columns or other commercial batches.

A third sampling in Butte has now been accomplished under auspices of NIH. In February, 1982 three schools were sampled. Data indicate that the time students spend outside their houses correlates with mutagenesis. Since this time the mining operations in Butte have stopped. The open pit was in operation during the last sampling but is no longer. We were originally scheduled to follow 10 high and 10 low students daily for 14 days to establish any relationship of urine mutagen content with air quality. Now all we may do is establish the normal background variability of this test when run among children of these ages.

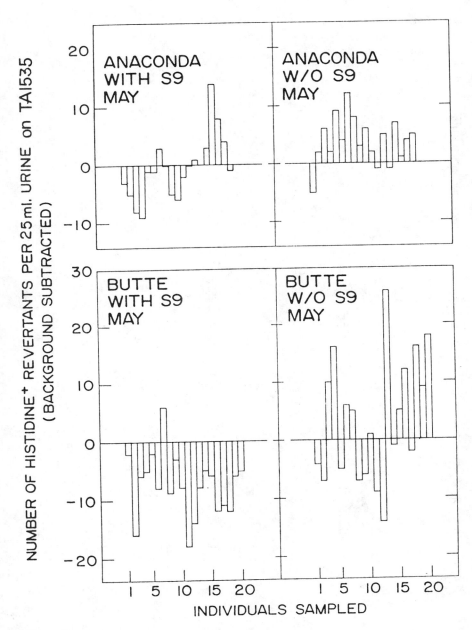

Figure 2
Mutagenic activity of XAD-2 urine extracts on TA1535 with and without Arochlor-induced S9 enzyme activation.

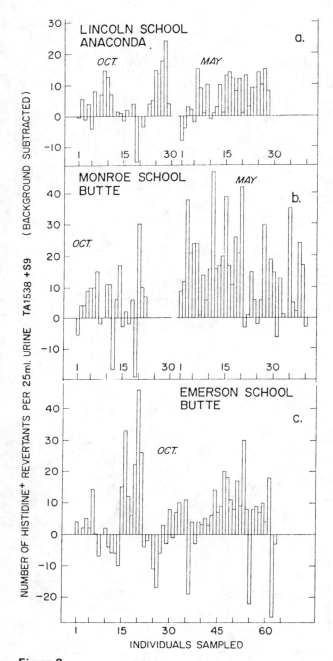

Figure 3
Mutagenic activity on TA 1538 of XAD-2 column extracts of urine samples from Anaconda and Butte assayed with Arochlor-induced rat S9.

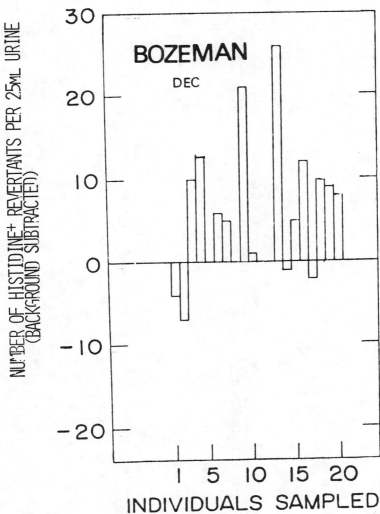

Figure 4

Mutagenic activity on TA1538 of XAD-2 column extracts of urine samples from Bozeman assayed with Arochlor-induced rat liver S9.

CONCLUSION

Table 1 summarizes the TA1538 Ames and filter data tabulated over the entire sampling period to date. Means range from 4.8 to 14.6 revertants/25 ml urine with those in Butte higher than anywhere else. Schools vary considerably from one sampling time to another. There may be a gross correlation with air filter mutagen levels. Diet is a possible variable. Another is exposure, since children

Table 1
Summary of Urine Mutagen Screening Data

City	School	Date	Number consented	Number tested	Average mutations 25 ml urine[a]	Air filter mutation levels Sampling Days[b]	
						(WS9)	(WOS9)
Butte	Monroe	5/78	47	40	14.6	24.0	17.0
	Monroe	10/78	29	22	8.7	15.8	10.1
	Emerson	10/78	74	68	7.8	15.8	10.1
Anaconda	Lincoln	5/78	43	28	6.3	2.1	9.2
	Lincoln	10/78	36	29	4.8	14.6	5.7
Bozeman	Longfellow	12/79	31	21	5.4	7.0	4.6

[a]On TA1538 + Arochlor-induced rat liver S9 (50 μl enz/plate). Two totally toxic < −10 samples from Monroe and 5 from Emerson and 1 from Lincoln have not been included in the averages. Three of these (1 from Monroe and 2 from Emerson) gave elevated levels (as high as 1280 muts/24 ml urine) upon dilution. The others did not.
[b]DMSO extracts of Hi-vol filters run at a sampling station close to school one day previous to sampling and on the day of sampling. Averages of the 2 days are presented.

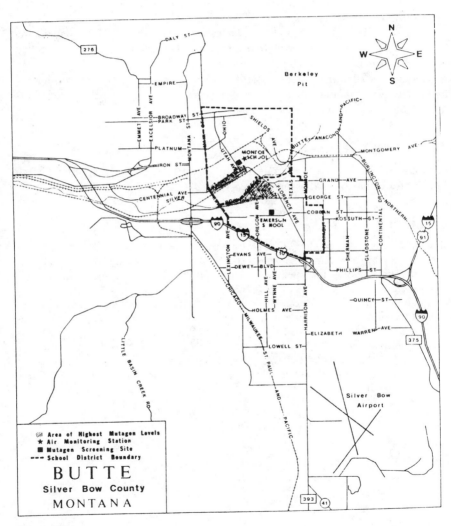

Figure 5
Map of Butte, Montana.

spend more time inside their homes during the winter although the air quality is worse while they roam outside in the summer although the air quality is better. Children also may be capable of more efficient excretion of toxicants.

Data from our questionnaires and interviews is only now being compiled. Only two items have correlated to this point with higher mutagen levels. They are length of time spent outdoors and address. Figure 5 shows a map of Butte and the shaded areas indicate where children were likely to have higher urine mutagen levels. Economic factors may play a role in this correlation since these same areas tend to be inhabited by out-of-work families. They are also near a main thoroughfare carrying heavy traffic.

A total of 303 individual samples have now been run. The tests and the techniques are sufficiently easy that any laboratory proficient in Ames testing could run them reproducibly. Improvements in bioassay should allow even more sensitive determinations. The technique appears to be promising for immediate effects monitoring and could be done to actually substantiate effects of environmental carcinogens.

ACKNOWLEDGMENTS

Portions of this work were supported by the 1976 and 1978 Montana State Legislature as part of MAPS and by PHS grant # CA26647 NCI.

REFERENCES

Newman, J.A.V., E. Archer, G. Saccomanno, M. Kuschner, O. Averbach, R.D. Grandahl, and J.C. Wilson. 1976. Histologic types of bronchogenic carcinoma among members of copper mining and smelting communities. *Ann. N.Y. Acad. Sci.* 271:260-269.

Yamasaki, E. and B.N. Ames. 1977. Concentration of mutagens from urine by adsorption with the non-polar resin XAD-2: Cigarette smokers have mutagenic urine. *Proc. Natl. Acad. Sci. U.S.A.* 74:3555.

Warren, G.R. 1981. Detection of genetically toxic metals by a microtiter microbial repair assay. In *Short-term bioassays in the analysis of complex mixtures II* (ed. M.D. Waters et al.), p. 101. Plenum Publishing, New York.

An Assessment of Fecal Mutagen Analysis for Predicting Genotoxic Exposure of Rats to Some Orally Administered Carcinogens

ROBERT D. COMBES, C. NICHOLAS EDWARDS, AND
JOHN M. WALTERS
Department of Biological Sciences
Portsmouth Polytechnic
King Henry I Street
Portsmouth PO1 2DY
Hants, England

Recently much interest has been directed towards the investigation of fecal material for genotoxicity. No doubt such work will receive added impetus from the recent assertion that a significant proportion of human cancer is due to dietary factors (Doll and Peto 1981).

There have been several studies demonstrating the presence of mutagens and clastogens in human fecal extracts. There is evidence that the activities of such extracts are dependent upon the nature of the diet. Attempts have been made to relate these observations to the epidemiology of colorectal cancer (Bruce et al. 1981; Venitt 1982).

In contrast there is a paucity of information concerning the genotoxic analysis of animal feces. Attempts to relate diet and presence of genotoxic compounds in animal feces have been undertaken for three carnivores and five herbivores by Stich et al. (1980). All the animals, maintained on their usual respective diets, produced fecal extracts capable of inducing clastogenic effects in cultured Chinese hamster ovary (CHO) cells. Variations in potency between species were apparent but there was no clear relationship with dietary intake. The presence of precursors of mutagens in animal feces was demonstrated by Rao et al. (1981). Feces from several species (especially rat, dog, and sheep) induced transitions when treated with nitrite and incubated with a *Salmonella* tester strain in a sealed container. These effects, which were also obtained using human stool samples, were attributed to the production of volatile nitrosamines.

It is also possible to test fecal extracts obtained from animals administered specific doses of potential genotoxicants in a host-mediated assay. Chow et al. (1980) were unable to detect any mutagenicity of ether extracts of feces from rats fed nitrate or nitrite in drinking water. This result implies a lack of intestinal nitrosation in this organism under the conditions used. In a recent report (Willems et al. 1981), it was found that either ether or methanol extracts of feces from rats given B[a]P per os were mutagenic to strains TA 98 and TA 100 with or without further metabolism.

This type of assay may be particularly useful for gaining further information concerning the potential hazard from ingested substances shown to be active using solely in vitro systems. We are currently applying fecal mutagenicity testing to the detection of the genotoxicity of several food colors. This paper describes the methodology used and the results obtained thus far with fecal extracts obtained from rats orally dosed with four known carcinogens, three of which caused the excretion of mutagenic urines.

MATERIALS AND METHODS

Bacteria

Salmonella typhimurium TA 98 and TA 100 (Ames et al. 1975) were kindly provided by Professor B. N. Ames.

Animals

Sprague-Dawley male rats maintained on Laboratory Animal Diet 41B (E. Dixon & Sons Ltd., Ware, Herts., U.K.) were used as a source of S9 fraction (5-6-week old animals) and for collection of urine and fecal samples (8-week old animals).

Chemicals

Trypan Blue was from Hopkin and Williams. Isoniazid, glucose-6-phosphate, (NADP) and glucuronidase-sulphatase type H_1 were from Sigma. The enzyme metabolized betanin, releasing glucose detected by paper chromatography. 2-Acetylaminofluorene (2-AAF) was from Aldrich. Safrole was purchased from Phase Separations Ltd (Deeside, Clwyd, Wales). All other chemicals were of pure grade. The chemicals, media, and positive controls for fluctuation assays are described in Haveland-Smith et al. (1979).

Preparation of S9 Mix

S9 mix was prepared as for Ames et al. (1975), and contained the 9000 g supernatant of a 25% w/v liver homogenate from rats preinduced with Aroclor (5 mg/kg bw).

Dosing and Sample Collection

Test compounds in solution were administered (< 4 ml/animal) by gavage to individual animals maintained in metabolism cages. Urines and feces were collected from 4 rats in each batch during the 24-hour period up to (controls) and for the 24 hours (treatments) following dosing.

Extraction of Feces

The respective samples were pooled, lyophilized, and extracted with peroxide-free diethyl ether followed by water. Ether extracts were obtained by adding ether (10 ml) to each dried sample and homogenizing with a spatula. After 3 hours in ether, the solvent was filtered off through Whatman No. 1 paper. The remaining solid was washed with a further 10 ml ether. The extraction was repeated with fresh ether (20 ml), the solid matter again washed with 10 ml ether. All extracts and washings from each sample were combined and the ether evaporated in an air stream. The resulting residues were suspended in dimethylsulfoxide (DMSO) (1.6 ml) and centrifuged at 2600 g for 15 minutes. The resulting clear solutions were transferred to sterile containers and stored at $-18°C$. These contained the extract from 5 g (wet weight) of the original feces per ml. Aqueous extracts were prepared by adding distilled water (16 ml) to the solid matter of each sample remaining after the ether extractions. These were pooled, centrifuged at 4000 g for 20 minutes, and the supernatant was membrane filtered and stored at $-80°C$ in sterile containers. The aqueous samples contained the extract from 0.5 g (wet weight) of the original feces per ml.

Mutation Assays

Fluctuation tests were conducted as previously published (Haveland-Smith et al. 1979). Bacteria were exposed in liquid medium containing low levels of histidine. After incubation the numbers of turbid tubes (due to continued divisions of prototrophic revertants) were counted in a set of 50 tubes and statistically compared with those arising in a corresponding solvent control series. Ames plate incorporation assays were undertaken according to the standard protocol (Ames et al. 1975) using 0.5 ml S9 mix/plate. The strains were characterized with several direct- and indirect-acting frame-shift and transition mutagens.

Spectrofluorimetry

Extracts were analyzed using an Aminco-Bowman spectrofluorimeter with a xenon lamp using 1 cm cells at the excitation and emission wavelength maxima for 2-AAF (Bowman 1979). The machine was calibrated with standard solutions of 2-AAF.

RESULTS

Preliminary studies with liquid fluctuation tests without activation (Table 1) showed that all samples of both ether and aqueous extracts from control and dosed animals did not induce significant increases in numbers of turbid tubes compared with each respective solvent control series. Urines collected from the

Table 1
Activities of Ether and Aqueous Extracts of Rat Feces in Fluctuation Tests without Activation

Agent (mg/kg bw)	Sample	Mutagenicity (N)[a]			
		TA 98		TA 100	
		Aqueous[b]	Ether[c]	Aqueous[b]	Ether[c]
Trypan Blue (800)	Control	12 14	14 7	27 19	11 13
	Treated	10 20	13 12	20 22	13 18
Isoniazid (400)	Control	7 16	10 12	17 19	36 27
	Treated	12 15	9 12	24 23	32 16
Safrole (400)	Control	13 12	8 8	22 26	35 21
	Treated	20 18	10 10	19 22	30 25
2-AAF (400)	Control	12 21	16 9	22 23	37 22
	Treated	10 25	18 6	21 26	33 22
Solvent controls		(21) (19)	(13) (10)	(20) (18)	(30) (23)

[a]Numbers of tubes turbid/50; (All values considered negative, with respect to solvent and sample controls). Results given for duplicate experiments;
[b]0.5 μl/tube-equivalent to 0.25 mg extracted feces;
[c]0.1 μl/tube-equivalent to 0.5 mg extracted feces. Isoniazid and 2-AAF urines were mutagenic to TA 100 and 98 respectively, without activation.

same animal groups in each case were likewise tested (data not presented). Those from 2-AAF-treated animals were mutagenic to TA-98 and samples from animals receiving isoniazid were active with TA-100. The other urines were inactive.

Lack of mutagenicity may be due to a necessity for further activation and due to the testing of insufficient quantities of extract. Both these factors were studied in the Ames test, initially using 50 μl/plate with and without S9 mix and (or) deconjugation of aqueous extracts. The data (Tables 2 and 3) show that ether extracts from one control group were mutagenic to TA 98. After dosing with 2-AAF this mutagenicity was greatly potentiated. Although aqueous extracts from the same control animals were inactive, weak mutagenicity was detected in extracts obtained after dosing. All activities were S9-dependent and unaffected by incubation with β-glucuronidase-sulphatase. All other control and treatment samples were nonmutagenic to the respective strains used (Tables 2 and 3). Urines from the same animals collected before and after dosing with either safrole or trypan blue were likewise subjected to Ames tests. Safrole urine was inactive, even following deconjugation, whereas trypan blue urine was mutagenic to TA 98 with S9 mix (data not presented).

Dose responses were obtained for aqueous and ether fecal extracts from control and 2-AAF-treated animals (Figs. 1 and 2). Ether extracts from dosed animals exhibited potent mutagenicity, which was dose-dependent. However, in common with the mutagenicity of control extracts, activity reached a maximum at 10 μl/plate. Aqueous extracts from control animals were mutagenic at and above 100 μl/plate. Those obtained after administration of 2-AAF were significantly more active, both extracts exhibiting a dose-response without saturation.

The possibility that much, if not all, of the activity observed may be due to unchanged 2-AAF in the extracts was studied by spectrofluorimetry (Fig. 3). The small peak in the control ether extract was due to DMSO and was also present with nonmutagenic control samples. However, there was evidence for substantial amounts of 2-AAF in the extracts from treated animals. Unfortunately fluorescence of controls precluded the detection of 2-AAF in aqueous extracts by this method. Dose-responses for 2-AAF in DMSO, directly plated, have been included (Figs. 1 and 2) to indicate the amounts of 2-AAF necessary to induce equivalent levels of mutagenicity to those observed with the fecal extracts.

DISCUSSION

The data with control fecal extracts show that in only one batch of animals, which was later administered 2-AAF, was mutagenicity detected. The cause of this activity is unknown but was not due to substances fluorescing at the excitation and emission wavelengths for 2-AAF itself. No evidence for alteration of transit times for intestinal material was seen in these animals. It is likely that mutagenicity was mainly due to nonpolar substances since ether extracts were more active than aqueous samples despite the greater quantity of fecal material

Table 2

Activities of Ether Extracts of Rat Feces in Ames Test

Agent (mg/kg bw)	Sample (50 µl/plate)[a,d]	Strain	Mean Revertants per Plate[b,c]	
			+S9	−S9
Trypan Blue (800)	Control	TA 98	70	66
	Treated		88	61
	(Bg)[c]		(59)	(45)
Isoniazid (400)	Control	TA 100	120	124
	Treated		138	121
	(Bg)		(182)	(189)
Safrole (400)	Control	TA 100	107	129
	Treated		131	139
	(Bg)		(182)	(189)
2-AAF (400)	Control	TA 98	<u>307</u>	60
	Treated		<u>2013</u>	64
	(Bg)		(59)	(45)

[a] 10 µl/plate also tested for inactive samples—all nonmutagenic.
[b] significant increases underlined
[c] background counts (DMSO). Preincubations of cells +S9 with extracts had no effect on results
[d] equivalent to 250 mg extracted feces

Table 3
Activities of Aqueous Extracts of Rat Feces in Ames Test

Agent (mg/kg bw)	Sample (50 μl/plate)[b]	Strain	Mean Revertants/Plate[a]			
			+S9 -e[c]	-S9 -e	+S9 +e	-S9 +e
Trypan Blue (800)	Control	TA 98	70	51	53	36
	Treated[d]		71	56	68	25
	(Bg)		(59)	(54)	(50)	(59)
Isoniazid (400)	Control	TA 100	210	268	159	194
	Treated		169	274	183	197
	(Bg)		(189)	(218)	(181)	(187)
Safrole (400)	Control	TA 100	188	307	183	185
	Treated		228	315	170	199
	(Bg)		(189)	(218)	(181)	(187)
2-AAF (400)	Control	TA 98	68	57	53	50
	Treated		112 (?)[e]	54	107	49
	(Bg)		(59)	(54)	(50)	(59)

Safrole urine was – (± S9, ± e). Trypan blue urine was + (+S9, ±e).
[a] significant increases underlined;
[b] equivalent to 25 mg extracted feces.
[c] β-glucuronidase/sulphatase (5 mg/plate, pH 7.4);
[d] background counts (water)
[e] (?) marginal effect

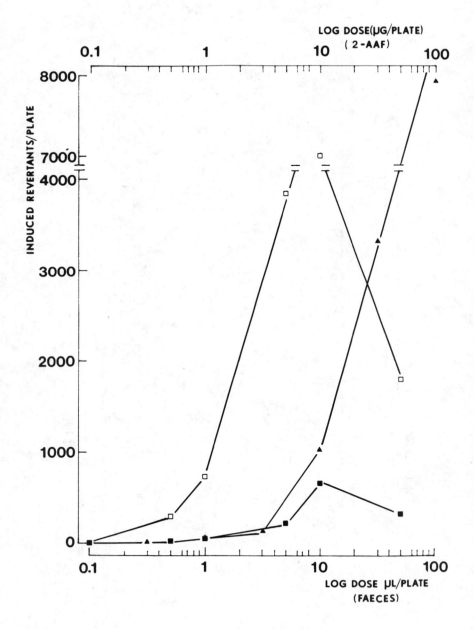

Figure 1
Dose-responses for 2-AAF (in DMSO) and for ether fecal extracts. (▲) 2-AAF; (□) feces from animals dosed with 2-AAF (400 mg/kg bw); (■) feces from animals before dosing.

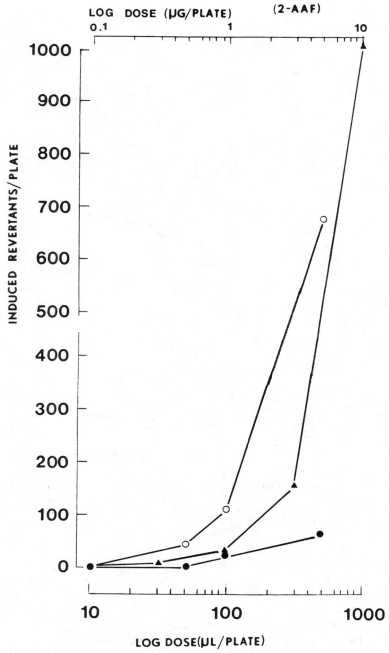

Figure 2
Dose-responses for 2-AAF (in DMSO) and for aqueous fecal extracts. (▲) 2-AAF; (○) feces from animals dosed with 2-AAF (400 mg/kg bw); (●) feces from animals before dosing.

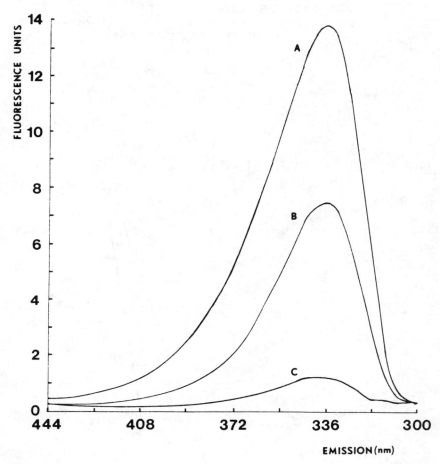

Figure 3
Spectrofluorimetry of 2-AAF and ether fecal extracts. (*A*) feces from animals dosed with 2-AAF (400 mg/kg bw) extracts were diluted 10^{-4} before analysis; (*B*) 2-AAF (in DMSO) at 0.33 μg ml^{-1}; (*C*) feces from animals before dosing or DMSO alone. Emissions were obtained in 1 cm cuvettes, using an excitation wavelength of 297 nm.

extracted. Aqueous and ether extracts from at least six further control groups of rats maintained under the same conditions have proved nonmutagenic. Inactivity of control fecal extracts from rats has been reported by others (Chow et al. 1980; Willems et al. 1981).

Oral dosing with 2-AAF caused a significant increase in mutagenicity of both ether and aqueous extracts. In view of the large oral dose (100 mg to each animal) and as activity was S9-dependent and seen with TA 98, it is likely to be due to the extraction of unaltered 2-AAF. This contention is supported by the fluorescence data which suggest a maximum of 6 mg 2-AAF/ml extract (Fig. 3). The optimum mutagenic response at 10 μl/plate (Fig. 1) is therefore equivalent to 60 μg/plate. Although such high counts are unreliable, the data (Fig. 1) suggest that the potency observed could have been due to unchanged 2-AAF in the ether extracts. Moreover, the dose-responses for 2-AAF and the ether extract were qualitatively similar. The likely amounts of 2-AAF and metabolites in aqueous extracts were not determined. The dose-responses (Fig. 2) and the data with deconjugation (Table 2) imply that activity may be due to 2-AAF, or more likely to a polar, nonconjugated metabolite thereof.

By reference to the dose-responses obtained with aqueous and ether extracts of treated samples at 100 and 1 μl/plate respectively (Figs. 1 and 2) and taking account of the tenfold less concentration of the aqueous extracts, it would seem that ether extracts were approximately seventy times more potent.

The predominant route for excretion of aromatic amines is in the urine in the form of N-hydroxy conjugates. In the present experiments, the same animals orally dosed with 2-AAF excreted mutagenic urine, the activity of which in fluctuation was independent of further metabolism. This result is not surprising in view of the large oral dose (see Commoner et al. 1974; Durston and Ames 1974). The possibility of fecal excretion of metabolites of 2-AAF is less understood. However, i.v. administered 2-aminofluorene (2-AF) appears to be excreted mainly as the glucuronide in rat bile which can be activated to mutagenicity by deconjugation (Conner et al. 1979). If it is assumed that significant intestinal absorption of 2-AAF was occurring in our experiments, then a comparison of the estimated dose per animal and the levels of 2-AAF in the ether extracts would infer that the methodology used was relatively efficient for the extraction of this compound at least.

The observed decrease in revertants induced by the ether extracts of feces from treated animals at 50 μl/plate may be due to many factors, especially at such high revertant counts. However, a similar decline in mutants was observed at the same dose with control extracts at lower revertant counts. The effect was absent when larger volumes of aqueous extracts were plated (Figs. 1 and 2). It is thus feasible that at 50 μl/plate of ether extract a substance was present which inhibited the detection of mutagenesis. This would seem the more likely since higher revertant numbers were induced by 2-AAF in DMSO. This phenomenon has been seen with ether extracts of feces from rats orally dosed with B[a]P (Willems et al. 1981) and may be due to the simultaneous extraction of unsaturated fatty acids. The latter, when present in ether extracts of human stools, are antimutagenic (Hayatsu et al. 1981).

Oral administration with large doses of isoniazid, safrole, or trypan blue failed to result in the isolation of either aqueous or ether fecal extracts with any mutagenicity irrespective of the presence or absence of further metabolism. All inactive ether extracts remained negative either when tested at 10 μl/plate or when preincubated with the respective tester strain for 30 minutes. The lower dose was used to reduce the possibility of antimutagenesis, as seen with 2-AAF extracts. Preincubation of fecal extracts has previously been observed to be potentiating (Lederman et al. 1980).

Lack of mutagenicity may also be due to a prevalence for in vivo detoxication or because the substance and its metabolites are mainly eliminated by other routes. Evidence for the latter situation in the present experiments is available for isoniazid and trypan blue since the same animals excreted mutagenic urines after separately receiving these compounds. However, we were unable to recover mutagenic urine after safrole dosing.

It should be emphasized that choice of tester strain in these experiments was based upon expected responses of the compounds and their metabolites, isoniazid and safrole urines inducing transitions (Miller and Stoltz 1978; H.S. Rosenkranz pers. comm.), the other agents inducing frame-shifts (see Hartman et al. 1978).

The results obtained suggest that fecal analysis alone is not always a reliable indicator of genotoxic exposure of rats to large oral doses of carcinogens. This is likely to be due to excretion of active substances via other routes or because insufficient levels of active compounds could be extracted by the methods used. A study of the mutagenicities of fecal extracts from rats given different, lower doses of 2-AAF should facilitate the further characterization of this system.

ACKNOWLEDGMENTS

We wish to thank Dr. R. Walker for advice, Dr. S. Venitt for the opportunity to see a manuscript before publication, and the MRC and SRC for financial support.

REFERENCES

Ames, B.N., J. McCann, and E. Yamasaki. 1975. Methods for detecting carcinogens and mutagens with the *Salmonella*/mammalian-microsome mutagenicity test. *Mutat. Res.* **31**:347.

Bowman, M.C. 1979. *Carcinogens and related substances. Analytical chemistry for toxicological research*, p. 66. Marcel Dekker Inc., New York.

Bruce, W.R. 1981. Properties of a mutagen isolated from feces. *Ban. Rep.* **7**:227.

Chow, C.K., C.J. Chen, and C. Gairola. 1980. Effect of nitrate and nitrite in drinking water on rats. *Toxicol. Lett.* **6**:199.

Commoner, B., A.J. Vithayathil, and M. Stoeckel. 1975. Detection of metabolic carcinogen intermediates in urine of carcinogen-fed rats by means of bacterial mutagenesis. *Nature* **249**:850.

Connor, T.H., G. Cantelli Forbi, P. Sitra, and M.S. Legator. 1979. Bile as a source of mutagenic metabolites produced *in vivo* and detected by *Salmonella typhimurium*. *Environ. Mutagen.* **1**:269.

Doll, R. and R. Peto. 1981. The causes of cancer: Qualitative estimates of avoidable risks of cancer in the United States today. *J. Natl. Cancer Inst.* **66**:1191.

Durston, W.E. and B.N. Ames. 1974. A simple method for the detection of mutagens in urine: Studies with the carcinogen 2-acetylaminofluorene. *Proc. Natl. Acad. Sci. U.S.A.* **71**:737.

Hartman, C.P., G.E. Fulk, and A.W. Andrews. 1978. Azo reduction of trypan blue to a known carcinogen by a cell-free extract of a human intestinal anaerobe. *Mutat. Res.* **58**:125.

Haveland-Smith, R.B., R.D. Combes, and B.A. Bridges. 1979. Methodology for the testing of food dyes for genotoxic activity: Experiments with Red 2G (C. I. 18050). *Mutat. Res.* **64**:241.

Hayatsu, H., S. Arimoto, K. Togawa, and M. Makita. 1981. Inhibitory effect of the ether extract of human feces on activities of mutagens: inhibition by oleic and linoleic acids. *Mutat. Res.* **81**:287.

Lederman, M., R. Van Tassell, S.E.H. West, M.F. Ehrich, and T.D. Wilkins. 1980. *In vitro* production of human fecal mutagen. *Mutat. Res.* **79**:115.

Miller, C.T. and D.R. Stoltz. 1978. Mutagenicity induced by lyophilization or storage of urine from isoniazid-treated rats. *Mutat. Res.* **56**:289.

Rao, B.G., I.A. Macdonald, and D.M. Hutchinson. 1981. Nitrite-induced volatile mutagens from normal human feces. *Cancer* **47**:889.

Stich, H.F., W. Stich, and A.B. Acton. 1980. Mutagenicity of fecal extracts from carnivorous and herbivorous animals. *Mutat. Res.* **78**:105.

Venitt, S. 1982. Mutagens in human faeces: Are they relevant to cancer of the large bowel? *Mutat. Res.* **98**:265.

Willems, M.I., G. Dubois, N. Verdugt, and J.K. Quirijns. 1981. Examination of rat faeces for mutagenic activity—evaluation of different extraction procedures. In *3rd Int. Conf. Environmental Mutagens, Abstracts,* p. 49.

COMMENTS

HEDDLE: Bob, how many of these are colon carcinogens?

COMBES: Isoniazid spontaneously hydrolyzes to hydrazine, but I think it is only dimethylhydrazine which is, in fact, a colon carcinogen (Deschner et al. 1979). In answer to your question, I suppose none of them is.

SEGA: How similar is the bacterial content of the rat compared to the human bacterial content?

COMBES: In the large bowel, there are quite a few similarities. In terms of quantitative comparisons, in the stomach and the proximal portion of the small intestine there are large amounts of differences. In fact, the rabbit is probably a closer model to man than the rat. We are currently looking at the rabbit gut microbial system. So there are lots of differences, quantitative at least, for the stomach end, and at the other end, the cecal content, of course, is qualitatively different.

WEINSTEIN: I am not sure what your interpretation of the significance is. Is this just material which has failed to be absorbed?

COMBES: We don't know. We wanted to characterize the system. What we have to do is proper dose responses with different oral doses, lower doses. We wanted to see whether our ether extractions and our aqueous extractions were actually taking out anything from the feces which would have genotoxic activity.

WEINSTEIN: Have you given any of these materials intraperitoneally?

COMBES: No, since we are interested in food additives.

THILLY: In the use of extractions of complex mixtures prior to mutation assay one wonders what was left behind or destroyed in the process. Would it make sense to take *S. typhimurium* with some additional drug-resistance markers and use it directly in the mixture along with the antibiotics to which it is resistant?

COMBES: We have thought of doing that, yes. And putting it into the gut?

THILLY: Yes, putting it into the gut or the sample.

COMBES: Well, people, in fact, have done this with germfree animals as well. They have put *S. typhimurium* in (Wheeler et al. 1975).

THILLY: Do germfree animals really represent what normally occurs in the intestine?

COMBES: Well, yes. I mean, then you would have to use drug resistance markers. Then you have to assume that they are going to survive and that there is going to be no further interaction. It is something we have thought of doing, yes.

SORSA: I think the only way you can avoid the toxic response in your biological samples is to analyze your sampling in several different dilutions, even though it is extremely tedious. We do it. For the mutagenic activity, every urine sample we analyze, in seven different concentrations, starting from the concentrate and ending up at 1/74th. You can very clearly see that, with the most diluted concentration, your mutagenic activity is rather low, because there are not enough mutagens. With the highest concentration, which is the full concentrate, your mutagenic activity goes down because of toxicity. We think it is reasonable to take this concentration which gives the highest amount of mutagenicity. There is variability between different samples. Some people have the highest urinary mutagenic activity with the least concentration, which is obviously due to the fact that there is this dilution coefficient in the urine. But it is very tedious.

EVANS: Can I ask one general question of everyone who has done this urine work? You are talking about toxic effects and mutations in the bacterium. Aren't we really concerned with mutations in mammalian cells? The first question to ask is, have you looked to see if your extracts cause mutations in human cells in culture or hamster cells in culture?

The second question is, what is toxic to the bacterium? The concentration gives high toxicity to bacteria but may be quite different in mammalian cells. So I think Marja's point about looking at a certain part of the peak of toxicity and saying, "Use that," may be quite wrong as far as mammalian cells are concerned.

Would someone care to answer those questions?

THILLY: I would like to address them directly. I think, with all due respect, that the multiple experiments you set for yourself will not give you a reasonable measurement of the mutagenicity even for the bacteria, and failure to independently assay the fraction of the bacteria killed will lead to a miscalculation of the amount of mutation induced in the population with complex mixtures and pure compounds. This is clearly outlined in the literature (Skopek et al.).

But as far as the bacterial-mammalian comparison, you are quite right. Ampicillin or other bacteriocidal drugs will kill *S. typhimurium* but not CHO cells or other cells.

The point here is that these bacterial assays are being used to measure mutation in bacteria. If they are being used inaccurately to measure mutation in bacteria, I think the second-level question of productivity for mammalian cell response has yet to be approached.

EISENSTADT: These are all very important questions in need of answers, but the bacterial assays do a little bit more than measure mutation in bacteria. My understanding is that what the bacterial assays measure is the capacity of some agent, something to which they are exposed, to damage DNA. You measure that indirectly as a mutation in the tester system. But, what this test really amounts to is a system which exposes DNA to agents that can, directly or following activation, introduce some kind of a lesion in the DNA. Then that gets translated by the bacteria to mutation and that alerts you that you had some DNA-damaging activity in the system. Whether that activity mutates human cells or mammalian cells is obviously an important point, but it is not quite so narrow as I think some people are implying; the implication being you are measuring some weird phenomenon in bacteria called mutation.

SORSA: Just a comment. I think we are still on an extremely qualitative basis in these test systems. The only thing we can do is to compare different types of exposure. We have to have the control, the unexposed group, to compare to the exposed group. But we should not go to any far-reaching risk extrapolations. It is a test of exposure, but nothing else.

COMBES: I think one of the problems is that when you go to mammalian cell systems, sometimes you go down in sensitivity as well. Although it might be more relevant, in a sense, you might miss a possible genotoxic effect which might be occurring more easily in vivo, which you are trying to pick up, having taken out the body substance and tested it in vitro.

SKINNER: Dave Amacher has just published a paper using urine in mouse lymphoma cells (Amacher 1981). I don't remember what he used as a test compound, but it can be applied in that way.

COMBES: Did it work?

SKINNER: It did.

EVANS: Well, Bert Baker has looked at tumor lymphocytes in CHO cells with Ames-positive smoking extracts. They are quite negative for SCE inducttion and negative for one other parameter. But they are very positive for producing aberrations of smokers' and nonsmokers' urine extracts in XAD-2. It is a very complicated story.

References

Amacher, D., G.N. Turner, and J.H. Ellis, Jr. 1981. Detection of mammalian cell mutagens in urine from carcinogen-dosed mice. *Mutat. Res.* **90**:79.

Deschner, E.E., F.C. Long, and A.P. Maskens. 1979. Relationship between dose, time and tumour yield in mouse dimethyl hydrazine induced colon tumorigenesis. *Cancer Lett.* **8**:23.

Skopek, T., H.L. Liber, J.J. Krolewski, and W.G. Thilly. 1978. Quantitative forward mutation assay in *Salmonella typhimurium* using 8-azaguanine resistance as a genetic marker. *Proc. Natl. Acad. Sci. U.S.A.* **75**:410.

Wheeler, L.A., J.H. Carter, F.B. Soderberg, and P. Goldman. 1975. Association of salmonella mutants with germfree rats: site specific model to detect carcinogens as mutagens. *Proc. Natl. Acad. Sci. U.S.A.* **72**:4607.

Mutagens in Nipple Aspirates of Breast Fluid

NICHOLAS L. PETRAKIS*, MARY EDITH DUPUY*, ROSE E. LEE*,
MICHAEL LYON*, CHRISTOPHER A. MAACK*, LARRY D. GRUENKEt,
AND JOHN C. CRAIGt
*Department of Epidemiology and International Health
tDepartment of Pharmaceutical Chemistry
University of California
San Francisco, California 94143

The adult nonlactating female breast is physiologically active and secretes small amounts of fluid into the ductal lumen. The secretions can be easily sampled with a breast aspiration device. We found that the breast alveoli actively secrete and concentrate many exogenous substances, including nicotine, fatty acids, technetium[99], immunoglobulins, and other substances (Petrakis 1977). The nonlactating breast appears to be unique among the secretory glands of the body in that many substances in its secretions are variably retained and metabolized by cells within the breast gland. We propose that by the processes of secretion, retention, and metabolism, carcinogenic substances can reach and affect the breast epithelia. It is now widely recognized that human milk fat and body fat contain a variety of pesticides and carcinogenic chlorinated hydrocarbons that enter the body through the food chain (Rogan et al. 1980). Several studies have shown that mutagenic substances are present in human foods, pharmaceuticals, hair dyes, and in other modern chemical products (Doll and Peto 1981). Since breast fluids closely resemble milk in composition, it is reasonable to propose that mutagenic chemical substances reaching human milk could also be found in the breast fluid of nonlactating women.

Earlier we reported on the detection of mutagens by the Ames *Salmonella* mutagenesis test and of cholesterol expoxides in nipple aspirates of breast fluid (Petrakis 1980, 1981a). This presentation will review these findings and present some new data on 1) mutagens in cystic breast fluid, and in colostrum samples from near-term pregnant women, and 2) studies of the induction of sister chromatid exchange (SCE) in lymphocytes by cholesterol beta-epoxide (β-epoxide) known to be present in breast secretions. Some of these findings were reported in Petrakis (1981).

COLLECTION OF BREAST FLUID

Breast fluid was aspirated from adult, nonlactating, nonpregnant women attending the Breast Screening Clinic at the University of California, San Francisco.

The technique of aspiration is a modification of that described by Sartorius et al. (1977), and employed previously by our group (Petrakis et al. 1975; 1981b). Approximately 20-30 μl of breast fluid were obtained from a single aspiration in each woman and were used for biochemical and mutagenesis tests. The fluids were stored at −35°F until tested.

Mutagenic Activity in Breast Fluids as Indicated by the Ames *Salmonella* Mutagenesis Test

Each fluid sample was diluted into 100 μl of 50% dimethylsulfoxide (DMSO) in distilled water. Quantities of fluid ranged from 5-50 μl and because of the small volume, permitted only one test per sample specimen. The entire sample was tested on a single plate after mixing with rat liver microsomal S-9 homogenate and one of the Ames histidine-dependent *Salmonella* tester strains. In most studies we used TA 1538 because of its broad spectrum of sensitivity to mutagens and its low rate of spontaneous revertant mutations (Petrakis et al. 1980).

The spontaneous revertant control rates were derived from the several hundred tests during the period of the study. A positive test for presumptive mutagens in breast fluid samples was arbitrarily set at at least 2 S.D. above the mean of the spontaneous revertant control values. The mean and 2 S.D. of the controls for TA 1538 were 26 ± 14 revertant colonies/plate.

We found that about 7% of breast fluids were "positive" with TA 1538 + S-9. Of these, 10 were at least two times the spontaneous revertant rates.

Ames tests of cyst fluid aspirated from women with clinical fibrocystic disease gave a similar proportion of positive fluids (Table 1). Recently we have studied colostrum obtained from pregnant women shortly before term. Fifty-five women were from the prenatal clinic at the University of California, San Francisco. Seven (12.7%) of 55 fluids from these "urban" women were positive tests with TA 1538 + S-9. In a sample of 44 fluids from pregnant farm workers, 7 (58.3%) gave highly positive at greater than three times the spontaneous revertant rate.

Table 1
Ames Test Results on Breast Secretions (TA 1538 with S-9)

Secretion	Number positive / Number women	Percent positive
Nipple aspirates	31/456	6.7
Cyst fluids	7/66	10.6
Colostrum		
UCSF[a]	7/55	12.7
Fresno; Salinas	26/44	59.1

[a]Samples taken from women at the University of California at San Francisco.

This finding appears to be due to exposures to pesticides used in rural farm areas. It is well recognized that pesticides are secreted into human milk (Rogan et al. 1980). These findings indicate that secretory products from the breasts of adult women may contain substances that produce positive tests for mutagenesis in the Ames system.

SISTER CHROMATID EXCHANGE IN LYMPHOCYTES BY 5β-6β-CHOLESTEROL EPOXIDE

Many earlier investigators (Bischoff 1969; Hieger 1949) have suggested that cholesterol metabolites have etiologic importance in carcinogenesis. Previously we (Petrakis et al. 1981) reported high concentrations of cholesterol in breast aspirates of many women that reached mean levels of 3554 mgm/dl in the premenopausal age group (40-49 years). Two cholesterol epoxides, 5α-6α-cholesterol epoxide (5,6α-epoxy-5αcholestan-3β-ol) and 5β-6β cholesterol epoxide (5,6β-epoxy-5β cholestan-3β-ol), were associated with the high levels of cholesterol (Petrakis et al. 1981a). Epoxide levels ranged from 0-31,500 μg/dl as shown in Table 2.

The origin of these substances found in many body tissues is in dispute; it is attributed by many workers to auto-oxidation of cholesterol by peroxides formed from polyunsaturated fatty acid esters of sterols, glycerol, phospholipids, etc. (Smith 1981). Recent studies report that they may be formed by epoxidases and metabolized by hydrases to 3α,5β,6β-triol (Smith and Kulig 1975; Kadis 1978). The possible biological significance of the cholesterol

Table 2

Cholesterol Epoxide Concentrations at Different Levels of Free and Total Cholesterol in Breast Fluid Aspirates for 37 Women

Cholesterol (mg/dl)	Cholesterol epoxide (α and β) (μg/dl)
< 99	
Free	0 in 6; 20, 90
Total	0 in 6; 85
100-199	
Free	0 in 1; 66, 422
Total	0 in 2; 16700
200-499	
Free	0 in 3; 3,140
Total	130, 600, 31500
> 500	
Free	1810, 2284
Total	0 in 2; 190, 315, 900, 950, 1260

Reprinted from Petrakis et al. (1981).

epoxides, irrespective of their mechanism of formation in the breast, is that cholesterol α-epoxide has been reported 1) to produce sarcomas and other tumors on injection (Bischoff 1969); 2) to be highly active in the embryo hamster transformation system (Kelsey and Pienta 1979); and, 3) to induce chromosome damage and DNA repair synthesis in human fibroblast cultures (Parsons and Goss 1978). Because the ductal system of the breast in many women is more or less continuously exposed to endogenously derived cholesterol epoxides, as well as to environmentally derived mutagens and toxic substances, we believe it possible that long exposure to these substances can lead to dysplastic and malignant change in breast epithelia. We found the highest proportion of dysplastic cells in nipple aspirates to be associated with the highest cholesterol levels in breast fluids (Petrakis 1981). Furthermore, the epoxides of cholesterol are negative in Ames tests with TA 1538, TA 98, and TA 100 (Maack et al. 1978).

During the past year, my colleagues Drs. John Craig (Craig et al. 1982) and Larry Gruenke have synthesized the α- and β-epoxide of cholesterol (5,6-α-epoxy-5α cholestan-3β-ol and 5,6β-epoxy-5β cholestan-3β-ol) and the triol (3β,5α,6β triol). We now have ample quantities of these substances, both deuterated and undeuterated, for our biological studies. I will present some preliminary findings on the activity of these substances in producing SCE in human lymphocyte cultures.

Toxicity tests on human lymphocyte cultures were made of α- and β-epoxides and triol dissolved in DMSO in concentrations ranging from 10^{-5} to 10^{-9} M. The optimal nontoxic concentration at 24 hours incubation was 10^{-7} M.

A 0.5 ml whole blood sample was obtained from 10 volunteer subjects and added to 5 ml of Roswell Park Memorial Institute tissue culture medium 1640 containing 15% fetal calf serum, 1% phytohemagglutinin, penicillin (100 units/ml), streptomycin (100 mg/ml), glutamine (2mM), and bromodeoxy-uridine (BrdU) (20 μM). Fifty μl α- and β-epoxides and triol dissolved in DMSO were added to each culture at 24 hours. Control cultures of lymphocytes from each woman contained only the culture medium and DMSO. The cultures were washed at 24 hours and replaced with fresh culture medium containing BrdU. The cells were harvested at 66 hours. The technique of preparation and counting of SCE was that described by Perry and Wolff (1974) and Hill and Wolff (1982).

SCEs were analyzed in 50 second-division metaphases for each (α- and β-epoxide and triol) compound and for controls in the blood samples from the 10 adult women and the data were expressed as the mean number of SCEs per cell ± SE. The results (Table 3) indicate that the β-epoxide significantly increased (35%; p < .001) the number of SCEs above the mean SCEs found in lymphocytes exposed to α-epoxide, triol, and controls. Studies are currently in progress to obtain dose-response curves for induction of SCEs by these substances, and to test breast fluids from normal women and from women with breast disease for SCE activity. Although these findings are statistically highly

Table 3
SCE in Lymphocyte Cultures Exposed to α- and β-Cholesterol Epoxide and Triol (N = 10 Women)

Compound	Number of metaphases	SCEs	Number of chromosomes	SCEs per chromosome (Mean ± SE)	SCEs per cell (Mean ± SE)	Increase (%)
Control	500	2825	23000	0.12 ± .01	5.65 ± .11	(−)
α-epoxide	500	2800	23000	0.12 ± .02	5.60 ± .11	(−)
β-epoxide	500	3807	23000	0.17 ± .02	7.61 ± .12[a]	(35)
Triol	500	2880	23000	0.12 ± .02	5.76 ± .11	(−)

[a]$P < 0.001$

significant and appear to indicate DNA-damaging effects caused by the β-epoxide, we cannot exclude the possibility that differential permeability of the lymphocytes to these compounds might lead to delays in mitoses and differences in SCEs. However, evidence from the studies indicating transforming activity, chromosome damage, and induction of cancer by cholesterol epoxides suggest that the finding is real.

DISCUSSION

Our studies of breast secretory activity carry the implication that the adult nonlactating breast glandular system acts as a "sink" wherein exogenous and endogenous chemical substances are secreted and can accumulate and remain for varying periods of time. The secretion of exogenous chemicals by the female breast is a dynamic process resulting from both active and passive transport by the epithelium in response to changing levels of estrogen, progesterone, prolactin, and possibly other hormones. Genetic factors have also been identified which influence the level of secretory activity (Petrakis 1977). The level of procarcinogen metabolizing activity of the breast epithelium will influence the activation of toxic, mutagenic, and carcinogenic chemical substances that reach the breast. Mammary tissue of rodents and human breast tumor lines have been found to possess inducible procarcinogen metabolizing activity (Maack et al. 1978). It might be expected that some of these chemical substances have cytopathologic effects on breast epithelium.

The several lines of evidence from our studies of breast fluid support the view that breast secretions often contain putative mutagenic substances. This evidence includes the presence of mutagens in a significant proportion of breast fluids, cyst aspirates, and colostrum samples, the presence of cholesterol epoxides in many breast fluids, and the induction of chromatid exchange in human lymphocyte cultures by β-epoxide. However, conclusive evidence is lacking as to whether any of these substances has an etiologic role in the development of atypical proliferative breast disease and breast cancer. "Proof" will have to be obtained by clinical and epidemiological studies of breast fluids from normal women, and from women with benign and malignant disease who are exposed to identifiable chemical substances under a variety of environmental conditions. The specific identification of mutagenic and DNA-damaging agents present in breast fluids would allow further epidemiological investigation of selected exposed populations. Such studies might permit the assessment of the risk of breast disease in these women and the removal of identifiable hazardous agents from the environment. If further studies of epoxides of cholesterol in breast fluid support the present evidence indicating mutagenic properties, a new approach to the etiology and prevention of breast disease might be developed.

ACKNOWLEDGMENT

We are grateful for the advice and assistance of Dr. Sheldon Wolff and Anna Hill in these studies. This research was supported in part by USPHS Grant PO1-CA-13556-11 from the National Cancer Institute, Bethesda, Maryland.

REFERENCES

Bischoff, F. 1969. Carcinogenic effects of steroids. *Adv. Lipid Res.* **7**:165.

Doll, R. and R. Peto. 1981. *The Causes of Cancer.* Oxford University Press, New York.

Hieger, I. 1949. Carcinogenic activity of lipoid substances. *Br. J. Cancer* **3**:123.

Hill, A. and S. Wolff. 1982. Increased induction of sister chromatid exchange by diethylstilbestrol in lymphocytes from pregnant and premenopausal women. *Cancer Res.* **42**:893.

Kadis, B. 1978. Steroid epoxides in biologic systems. A review. *J. Steroid Biochem.* **9**:75.

Kelsey, M.I. and R.J. Pienta. 1979. Transformation of hamster embryo cells by cholesterol-α-epoxide and lithocholic acid. *Cancer Lett.* **6**:143.

Maack, C.A., T.J. White, R.E. Lee, M. Lyon, and N.L. Petrakis. 1978. Metabolism of carcinogens to mutagens by mammary tissue extracts. *Proc. 69th Ann. Meeting Am. Assoc. Cancer Res.* **19**:229.

Parsons, P.G. and P. Goss. 1978. Chromosome damage and DNA repair induced in human fibroblasts by UV and cholesterol oxide. *Aust. J. Exp. Biol. Med. Sci.* **56**:287.

Perry, P. and S. Wolff. 1974. New Giemsa method for the differential staining of sister chromatids. *Nature* **251**:156.

Petrakis, N.L. 1977. Breast secretory activity in nonlactating women, postpartum breast involution and the epidemiology of breast cancer. *Natl. Cancer Inst. Monogr.* **47**:161.

_____. 1981. Epidemiologic studies of mutagenicity of breast fluids—relevance to breast cancer risk. *Ban. Rep.* **8**:243.

Petrakis, N.L., L.D. Gruenke, and J.C. Craig. 1981a. Cholesterol and cholesterol epoxides in nipple aspirates of human breast fluid. *Cancer Res.* **41**:2563.

Petrakis, N.L., C.A. Maack, R.E. Lee, and M. Lyon. 1980. Mutagenic activity in nipple aspirates of human breast fluid. (Letter). *Cancer Res.* **40**:188.

Petrakis, N.L., L. Mason, R. Lee, B. Sugimoto, S. Pawson, and F. Catchpool. 1975. Association of race, age, menopausal status and cerumen type with breast fluid secretion in nonlactating women, as determined by nipple aspiration. *J. Natl. Cancer Inst.* **54**:829.

Petrakis, N.L., V.L. Ernster, S.T. Sacks, E.B. King, R.J. Schweitzer, T.K. Hunt, and M.-C. King. 1981b. Epidemiology of breast fluid secretion: Association with breast cancer risk factors and cerumen type. *J. Natl. Cancer Inst.* **67**:277.

Rogan, W.J., A. Bagniewska, and T. Damstra. 1980. Pollutants in breast milk. *N. Engl. J. Med.* **302**:1450.

Sartorius, O.W., H.S. Smith, P. Morris, D. Benedict, and L. Friesen. 1977. Cytologic evaluation of breast fluid in the detection of breast disease. *J. Natl. Cancer Inst.* **59**:1073.

Smith, L.L. 1981. *Cholesterol Autoxidation.* Plenum Press, New York.

Smith, L.L. and M.J. Kulig. 1975. Sterol metabolism XXXIV. On the derivation of carcinogenic sterols from cholesterol. *Cancer Biochem. Biophys.* **1**:79.

COMMENTS

ALBERTINI: Have you looked at any aspirates from women in breast cancer families?

PETRAKIS: Oh, yes, we have—but not [many] chemical studies. We have cytology on them, and have found that the risk of dysplasia is higher in women who have a family first-degree history of breast cancer. There is about a twofold increased risk of dysplasia in women who come from a family with a first-degree relative with breast cancer.

ALBERTINI: In some of the striking families, you would wonder if the onset of the secretions would be moved down in age.

PETRAKIS: We haven't found any difference in whether you get fluid or not in families with or without breast cancer.

ALBERTINI: What about the presence of mutagens?

PETRAKIS: We have looked for mutagens but didn't find anything of significance relating to family history of breast cancer (Petrakis et al. 1982). At present our studies leave me with the feeling that almost anything women ingest or inhale will enter the breast secretions. We found several women taking a chlorinated phenothiazine, called Eskatrol. J. G. Jose has shown that these substances are strongly mutagenic by the Ames test. The two women taking it had very high levels of mutagenic activity in their breast fluid aspirates.

    I have concluded that many substances known to be mutagenic may reach the breast and could result in positive Ames tests if the timing of the aspiration was correct.

ALBERTINI: Reserpine?

PETRAKIS: We didn't find mutagenic activity with reserpine.

BRIDGES: A very naive question: Can you tell me what steps were taken to prevent microbial contamination coming up as a positive colony?

PETRAKIS: Well, DMSO kills most surface bacteria and the nipple is cleansed with antiseptic before aspiration is attempted. . . . Oh, you mean in the breast fluids?

BRIDGES: Yes.

PETRAKIS: We have cultured many breast fluids. They were always negative.

The nonlactating breast seems to be extremely resistant to infection. In the over 5,000 women we have aspirated we have never had a complication related to infection. In fact, we have not had complications of any form.

COMBES: There is a bacterium that comes off the nipple called *Lactobacillus bifidus* which protects against gastroenteritis. It affects the pH of the gut.

PETRAKIS: It is known that mother's milk contains antibody to specific bacteria in the colon of the mother, that the baby might also pick up. Mothers fed certain nonpathogenic *E. coli* late in pregnancy will develop and secrete into their milk antibodies specific for the *E. coli* bacterial strain fed to the mother.

LUTZ: What is the percentage of the β form of the epoxide in the total—

PETRAKIS: We have purified α and β epoxides. What I showed were the pure compounds as prepared by my collaborators, J. Craig and L. Gruenke (Craig et al. 1982). We tested the pure epoxides, the α versus the β, for their sister chromatid exchange.

LUTZ: So what is the percentage of each?

PETRAKIS: It is probably roughly 3:1 in the fluid. We haven't done enough work on this to be able to give you a firm answer on that.

LUTZ: And from the chemical synthesis—I presume with a peroxide, when you do the synthesis—what is the yield of each form?

PETRAKIS: I don't have that information available. I can get it for you, though.

LUTZ: In our search for endogenous DNA-binding agents we are pursuing this line of the cholesterol epoxides. We were not able to find any DNA binding in vivo from a mixture of α and β. I don't recall our fraction. The limit of detection for this negative evidence for DNA binding to liver DNA was on a binding index scale of 0.1. So it was not detectable at 0.1. Aflatoxin has something like 10,000. We do not find any DNA binding in vivo or in vitro. We find interaction with protein, but not with DNA in in vitro incubation, of this mixture of α and β. So I cannot say, if we had β form alone, what that would give. The limit of detection might be not as good if we had the pure β form.

But it seems to be a very inert epoxide. It reacts probably only with the very good nucleophiles you find on the proteins.

PETRAKIS: Remember we are talking about a substance that is going to be sitting in those breast fluids for a long time. It is probably turning over, but it is going to be there for a long time.

LUTZ: You are right, it is very important, not only the DNA binding capacity, but also the level of exposure.

BRIDGES: One comment about the Ames test, *Salmonella,* and *E. coli* tests, when you are near background level. Roy Foster found working with water samples that substances exist which can increase the number of spontaneous mutants detectable in a test, without being mutagenic (pers. comm.). The obvious one is histidine since it can be measured and ruled out. But there are substances which can spare the histidine requirement, to some extent, which are not histidine themselves.

    All these tests, whether you use the fluctuation test or the standard plate protocol, are designed so that your spontaneous mutant level, the number of colonies you count, is, in fact, determined largely by the amount of residual growth of the *Salmonella* in the tube or on the plate. So if there is more residual growth, you will get more mutants. Histidine, as I said, causes both more growth and more mutants in small quantities.

    One has to take into account the fact that there can be substances which have a similar effect which are not histidine, and which are not mutagens.

    In the Ames test there is absolutely nothing that can be done about it. There is no way of measuring the final level of growth in the agar. But in the fluctuation test, which is done all in liquid, in a tube, the total and viable bacterial numbers can be measured after residual growth. When Roy Foster did this with these water samples he thought some samples contained mutagenic activity but were just allowing a little bit more bacterial growth in the tube on the same amount of histidine that he put in the medium. So the mutant frequency was going up.

    I think Bruce Ames' figure—is it 2 or 2½ times?—quite empirically, has probably done it correctly. Any findings that are less than that, even if they give a dose response would just not be believeable, unless it was done in a fluctuation test and the possibility of extra residual growth was ruled out.

PETRAKIS: This is not a statistical question. The data I showed used two standard deviations as the cutoff point for a positive test, as is commonly used in epidemiologic studies. But, in the women I tested, about 1/3 had at least two times the activity and some had ten times more the activity. I worry about the borderline cases also.

    The details of how we determined the quantity of histidine in breast fluids was previously published (Petrakis et al. 1980). The mean level of histidine in 11 randomly selected breast fluid samples was 2.04 ± 2.86 nm per breast fluid sample since there are 100 nm of histidine in the top agar used in the Ames test. To increase the number of spontaneous revertants above the average normally found in control plates we would need at least 200 nm of histidine.

References

Craig, J.C., L.D. Gruenke, and N.L. Petrakis. 1982. Measurement of cholesterol epoxide formation and turnover in human breast fluid. Abstract. *International Symposium on the Synthetic application of isotopically labeled compounds.* Kansas City, Missouri. June 6-11, 1982.

Jose, J.G. 1979. Photomutagenesis by chlorinated phenothiazine tranquilizers. *Proc. Natl. Acad. Sci. U.S.A.* **76**:469.

Petrakis, N.L., V.L. Ernster, E.B. King, and S.T. Sacks. 1982. Epithelial dysplasia in nipple aspirates of breast fluid: Association with family history and other breast cancer risk factors. *J. Natl. Cancer Inst.* **68**:9.

Petrakis, N.L., C.A. Maack, R.E. Lee, and M. Lyon. 1980. Mutagenic activity in nipple aspirates of human breast fluid. (letter). *Cancer Res.* **40**:188.

General Discussion: Body Fluids

WEINSTEIN: I would like to raise a question related to sample collection. In collecting urine, of the three people who spoke, none of them collect 24-hour samples, as I understand it. That is worrisome because there would be daily fluctuations throughout the course of the day.

Do you always collect it at the same time of day?

SORSA: We collect in the evening, or if there are occupational exposures, after the work day, at the end of the week. The samples are stored deep-frozen, -20°. It is taken and frozen immediately.

WEINSTEIN: Do others add preservative?

EISENSTADT: It is hard to speak for others. We routinely freeze quickly on dry ice, and then store it frozen. We have looked at the stability of the frozen material over weeks and find that there is no shift up or down, whatever that means.

COMBES: You can also get effects if you store at room temperature, for example, isoniazid urine from animals. But this may be an effect that might occur with other exposures. So perhaps you ought to do both.

WEINSTEIN: And then for compensating for how dilute the urine might be, do you do creatinines in urine?

EISENSTADT: We don't routinely do them.

WEINSTEIN: How do you control for these interindividual variations? Do you simply assume that one sample is very concentrated and one is very dilute?

EISENSTADT: We have done it in the past, and I guess, like Guylyn [Warren], came to the conclusion that it didn't really add anything important. The variation in creatinine doesn't approach the variation in mutagenicity. I mean, it is not even close. That is not what we are looking at. We are not looking at concentrated versus dilute.

PETRAKIS: Wouldn't you expect that every one of these compounds would have its own renal clearance rate and that renal clearances of these substances would vary depending on what they are chemically.

WEINSTEIN: Yes, and how dilute the sample is, time of day. It would be surprising if these weren't important. The conventional way, which is tedious, is to get a 24-hour urine.

SORSA: Even in a hospital environment, it is almost impossible.

WEINSTEIN: Well, not in a controlled hospital environment. It is done frequently. The best way to do it is to instruct the patient, not the nurse or the doctor. The patient can be taught to collect 24-hour urines.

EISENSTADT: I quess I have the sense that urine volumes or creatinines are going to vary over about a range of 2 or so, but in the mutagenicity we are talking about orders of magnitude.

PETRAKIS: Also in fairness, it should be noted that if you are doing epidemiological studies, it will be almost impossible to obtain 24-hour urines from large numbers of people. You might collect morning urine samples. However, these will miss what may be excreted throughout the day.

WEINSTEIN: Well, of course, I am not sure that you are going to need hundreds of thousands of samples. The field might be at a stage where fairly well controlled cases would be sufficient.

HEDDLE: Is it true that people who are low-yielders have low yields repeatedly? For example, cigarette smokers are low-yielders. Is that reproducible or not? If it is reproducible, if a person is characteristically low and somebody else is characteristically high, can a retrospective study be done on people with lung cancer to determine whether one represents a low and the other a high risk?

SORSA: Well, we have to wait for 20 more years at least.

HEDDLE: No, no, that would be retrospective. I am suggesting that a study be done now.

But before that is worth doing, you have to know whether or not the low and the high represents a daily variation of an individual, or whether an individual can be classified as being typically low or high.

SORSA: Well, in our experience with this volunteer smoking study, it really seems to be that somebody is always high or always low. But it may also depend on the smoking habits. One can't tell whether it is an inborn metabolic capacity or whether it is just the habits of smoking, or how much you inhale?

EVANS: Our data show that, in fact, you do have high-yielders and low-yielders. But they stay high or low for 1-2 weeks, and then they can change. We don't have a consistent background. There are obviously factors that could cause change that we can't understand and we can't define. But over a short time period, some will give us a fairly high yield. This occurs over 2 or 3 weeks, and then they will change. Why, we don't know.

PETRAKIS: Have you looked at males versus females to see if there is a menstrual or endocrine pattern?

EVANS: In all these studies we are doing, we take a sample, run it in a column, take it out, store it, play with it and then we look at toxicity levels to decide what certain toxicity level is to be tested for a compound. We run this through and find that smokers give us a higher frequency of mutation than nonsmokers. We use the same compound, put it into human cells, and look for chromosome damage, and there is no difference between smokers and nonsmokers.

There is a different level of toxicity of the compound, of the extract. What is meat for bacterium isn't meat for human cells. Something very toxic to *E. coli* is harmless to Chinese hamster, and the other way around.

It is a very difficult area. I am not entirely convinced that all we are doing in looking at an Ames test-type system on smokers is really telling us very much about the mutagenicity of the product that is coming through into the smokers' cells themselves.

SORSA: We just recently completed a rather large study on volunteer smoking which involved the activity of many people. My colleagues, Dr. Kai Falck and Dr. Tuula Heinonen, have been responsible for the urine assays and studying the urine samples for extraction of thioethers, respectively.

This study was started because the Medical Board of Health in Finland was beginning to be concerned about the smoking habits of the Finns since many Finns are smoking low-tar cigarettes. We were interested in doing such a study because smoking is always a confounding factor in our studies on occupational exposures.

The subjects were 37 young males who had just entered the military service and were about the same ages. The military service in Finland is still rather disciplined so I think the information that we have obtained is rather reliable. Of course they had similar diets all the time. We had all the menus. There was a preselection of these volunteers, because they had taken a medical examination before they entered the army. We could select those persons who were without any previous occupational chemical exposures and also select for their personal smoking habits.

Three groups were selected: A, 12 high-tar cigarette smokers; group B, 15 light-tar cigarette smokers; group C, 10 nonsmokers.

The study design was a total of 51 days. We first wanted to have some kind of basic idea of what is the baseline in these 37 men for the parameters I will discuss. We took two samples after their first 6 days of service, when all the smoking groups had been smoking high-tar cigarettes. Then they had a 3-week period when group A smoked only high-tar cigarettes, and group B smoked low-tar cigarettes (all the cigarettes were free). Next group A smoked only low-tar cigarettes for a 3-week period while group B smoked only high-tar cigarettes. Urine and blood samples were taken four times during this 51-day study.

We know exactly how many cigarettes they smoked daily—a total of 24,140 cigarettes—both high-tar and low-tar cigarettes.

The chemical analysis of the cigarettes was as follows: Three times more tar, nicotine, and carbon monoxide levels, were in the high-tar cigarettes. The parameters which were studied with these four samples were carboxyhemoglobin—just to get an idea of the very recent exposure to carbon monoxide—urine samples; the amount of thioethers; and the urinary mutagenicity, both with the *Salmonella* TA98 strain (the sensitive strain for smoking effects) and also with the base-pair substitution strain, *E. coli* WPB 2 uvrA, just to get an idea of the individual differences in this type of very homogeneous environment. We then studied the sister chromatid exchanges in the blood samples. We calculated the urinary mutagenicity corrected for creatinine, which we call the mutagenic activity, with the fluctuation time test. It showed that group A and group B both had significantly higher urinary mutagenic activity than the group C. It also showed that there are huge variations between individuals. Individual results indicate that usually the same person either has high urinary mutagenic activity or low urinary mutagenic activity in all samples, and this may be related to the way of smoking; the type of puffs inhaled; how much you inhaled; and how long the cigarette was smoked. These are all difficult categories to control.

The individual values show that actually there seems to be no effect whether high-tar cigarettes or low-tar cigarettes were smoked. The amount of cigarettes and how much is smoked affects the mutagenic activity. The individual numbers of cigarettes smoked during this 51-day study period and the mean urinary mutagenic activities and individual activities of four analyses show that there is no baseline difference between the group A or group B people. The mean for the nonsmokers stays very low. There are always some individuals who are in the range of the nonsmoking persons. This may just be an effect of individual metabolic differences as was discussed previously.

In another study the consumption of cigarettes was increased in both groups—both group A and group B were smoking much more towards the end of the experiment, probably due to the fact that they got used to the system. Additionally, they found out that they could get as many cigarettes as they wanted to and probably also became accustomed to the rigors of military service, so that they could find the free time to smoke. The urinary mutagenic activities increased towards the end of the experiment, but there was no difference between the groups smoking low-tar or high-tar cigarettes.

Group B increased the number of cigarettes consumed, and probably also the mean mutagenic activity for group B was higher than in group A.

These details are unavailable but the other parameters which we were measured—the carboxyhemoglobin, the sister chromatid exchange, and the thioethers—pointed to the same direction. There really is not any difference in the brand of cigarettes or the amount of tar in the cigarette—it is the number of cigarettes smoked which increases genotoxic exposure.

SESSION III:
DNA DAMAGE AND REPAIR

Use of an In Vivo DNA Repair Assay as an Indicator of Genotoxic Exposure

JON C. MIRSALIS
Biochemical Genetics Department
SRI International
Menlo Park, California 94025

The measurement of chemically induced DNA repair as unscheduled DNA synthesis (UDS) has been shown to be an excellent indicator of the genotoxic and carcinogenic potential of chemicals. Various in vitro test systems utilizing several different cell lines for the measurement of UDS have been described (Trosko and Yager 1974; San and Stich 1975; Martin et al. 1978). A UDS assay employing primary cultures of rat hepatocytes offers the advantage of using a metabolically competent target cell (Williams 1977). These systems, however, often do not reflect the true genotoxicity present in the whole animal and even the in vitro hepatocyte UDS assay has been shown to be unresponsive to some classes of compounds such as nitroaromatics (Bermudez et al. 1979; Probst et al. 1981). The use of in vivo assays for the detection and study of genotoxic chemicals results in a more accurate profile of the effects of compounds observed in the whole animal since uptake, distribution, activation, detoxification, and elimination are all considered. In addition, in vivo systems allow the examination of genotoxicity in individual target organs.

The in vivo-in vitro hepatocyte DNA repair assay (Mirsalis and Butterworth 1980) measures chemically induced UDS in the liver following exposure of rats to chemicals. This in vivo system offers many advantages over similar in vitro assays that do not reflect the true metabolism of compounds in the intact animal.

METHODS

The procedures for this assay have been previously described (Mirsalis et al. 1982b). Fischer-344 rats are treated with chemicals by a suitable route of exposure, and at selected times after treatment, their livers are perfused with a collagenase solution. A single-cell suspension of hepatocytes is obtained by combing out the cells of the perfused liver into a petri dish containing collagenase solution. Cells are seeded into culture dishes containing coverslips and Williams Medium E supplemented with 10% fetal bovine serum, allowed to attach to the coverslips, and incubated with a solution of 10 μCi/ml ^{3}H-thymidine (^{3}H-dT) for 4 hours. Following overnight incubation (14-18 hr) in 0.25 mM unlabeled thymidine, cells are swelled, fixed, and washed; coverslips are mounted on microscope slides and dipped in Kodak NTB-2 photographic emulsion. After being exposed for 12-14 days, slides are developed and the cells are stained.

Quantitative autoradiographic grain counting is accomplished using a colony counter interfaced to a microscope via a TV camera; data are fed directly into a computer. Fifty morphologically unaltered cells from a randomly selected area of the slide are counted. The highest count from three nuclear-sized areas over the cytoplasm and adjacent to the nucleus is subtracted from the nuclear count to give the net grains/nucleus (NG). The percentage of cells in repair indicates the extent of damage throughout the liver and is calculated as those cells exhibiting > 5 NG.

DETECTION OF HEPATOCARCINOGENS

The in vivo-in vitro hepatocyte DNA repair assay has been shown to be responsive to genotoxic hepatocarcinogens from a variety of chemical classes (Table 1). Compounds that are not hepatocarcinogens, however, generally do not show genotoxic activity in the liver, even if they produce tumors in other tissues (e.g., benzo[a]pyrene, 7,12-dimethylbenz[a]anthracene). This indicates that measurement of UDS in the liver is suitable only for detection of genotoxic hepatocarcinogens—and perhaps direct-acting mutagens. Nongenotoxic carcinogens that may be acting as promoters (e.g., carbon tetrachloride) also produce negative responses.

Use of an in vivo system allows increased flexibility in the study of sex differences, chronic effects, different routes of exposure, effects of diet and environment, and other complex parameters absent in tissue culture assays. The major advantage of the in vivo UDS assay is its utility in studying a variety of different parameters of the genotoxicity of carcinogens.

TIME-COURSE OF DNA REPAIR

Uptake, distribution, metabolic activation, and detoxification are factors that affect a compound's ultimate genotoxicity. By measuring UDS at selected times after exposure of the rat to a chemical, the time-course of DNA repair can be determined. This offers some insight into the time required for a compound to produce genotoxic effects in the liver. A direct-acting mutagen such as methyl methanesulfonate (MMS) produces a peak response in UDS almost immediately after exposure, followed by a rapid decline in UDS (Fig. 1.). This reflects the fact that MMS does not require metabolic activation and indicates that its uptake by the liver is rapid. The fact that MMS is not a hepatocarcinogen suggests that repair of MMS-induced genotoxicity is both rapid and error-free. Con-

Table 1

Some Chemicals Studied in the In Vivo-In Vitro Hepatocyte DNA Repair Assay

Chemical	Hepato-carcinogenicity	UDS[b]
Nitrosamines		
NDMA	+	++++
NDEA[a]	+	++++
Aromatic amines		
2-AAF	+	++++
Benzidine	+	++++
2,4-Diaminotoluene	+	+++
2,6-Diaminotoluene	−	−
Nitroaromatics		
2,4-DNT	+	++++
2,6-DNT	+	++++
Nitrobenzene	?	−
2,4-Dinitroaniline	?	−
2,6-Dinitroaniline	?	−
Polycyclic aromatics		
7,12-Dimethylbenz[a]anthracene	−	−
Benzo[a]pyrene	−	−
Hepatotoxins		
Chloroform	−	−
CCl$_4$	±	−
Direct-acting		
MMS	−	++++
MNNG	−	±
Miscellaneous compounds		
1,2-Dimethylhydrazine	+	++++
AFB$_1$	+	++++
Azoxymethane	+	++++
Azaserine	±	+
Safrole	±	−

[a]Diethylnitrosamine
[b]−, < 0 NG; ±, 0-4.9 NG; +, 5-9.9 NG; ++, 10-14.9 NG (no chemicals in this category); +++, 15-19.9 NG; ++++, 720 NG.

versely, the hepatocarcinogen 2-acetylaminofluorene (2-AAF) does not show a peak response until 12 hours after treatment (Fig. 1). This is consistent with the requirement of metabolic activation for conversion of 2-AAF to a genotoxic form and may suggest a slower rate of uptake than that observed for MMS. The nongenotoxin carbon tetrachloride (CCl$_4$) induces no UDS at any time point examined.

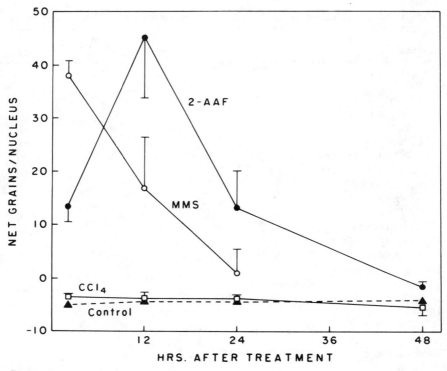

Figure 1
The time course of UDS. (o) MMS; (•) 2-AAF; (□) CCl_4. (– –▲– –) Controls which received corn oil. Livers were perfused at the times indicated after treatment of rats by gavage. S.E. shown represent the variation between animals; n = 2-4 for each point. Reprinted, with permission, from Mirsalis et al. (1982b).

DISTRIBUTION OF HEPATOCELLULAR GENOTOXICITY

In an in vitro test system, a monolayer of cells is uniformly exposed to a chemical. In the liver, however, cells may be exposed to different concentrations of a chemical depending on the reactivity of the compound, differential activation/deactivation in different lobes of the liver, the presence of hepatic necrosis and other factors. The percent of cells in repair, defined as those cells with > 5 NG, gives a rough indication of the extent of genotoxicity in the liver. A more complete representation of the distribution of genotoxicity may be obtained by examination of the frequency distribution of cells from an exposed animal (Fig. 2). Whereas most cells from control-treated rats show responses of less than 5 NG, cells from Aflatoxin B_1 (AFB_1)-treated rats are nearly all greater than 5 NG, with a mean of over 30 NG. This suggests that AFB_1 is widely distributed throughout the liver and that all cells are capable of metabolizing AFB_1 to a genotoxic form. The direct-acting mutagen N'-methyl-N'-nitro-N-

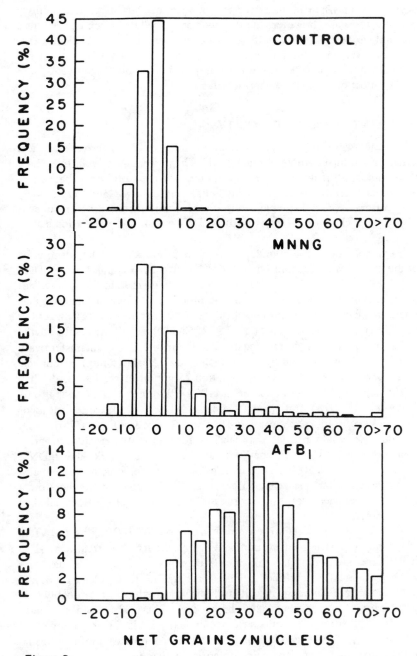

Figure 2

The distribution of NG for individual cells from rats treated with AFB_1, MNNG, or corn oil (control) 2 hr before sacrifice. Each treatment group represents 3 animals, 150 cells/animal. Reprinted, with permission, from Mirsalis et al. (1982b).

nitrosoguanidine (MNNG) is a highly reactive methylating agent that is not hepatocarcinogenic. When administered by i.p. injection, MNNG produces only a weak positive response, however, the distribution of cellular responses reveals a small percentage of cells (20%) with a high degree of UDS. This indicates that the highly reactive compound may actually come in contact with only a relatively small number of cells in the liver. Exposed cells show significant genotoxicity, but most of the liver remains unaffected.

MEASUREMENT OF HEPATOTOXICITY

Some chemicals may produce hepatotoxicity leading to liver necrosis regardless of whether or not the compound is genotoxic. The percentage of cells in S-phase in the liver indicates the degree of chemically induced hepatotoxicity that results in increased cell proliferation. Cells in S-phase are enlarged and very heavily labeled in autoradiographic preparations and are easily distinguished from cells in repair. Thus, a chemical's effects on both DNA repair and replication can be studied.

An example of a compound that is both genotoxic and hepatotoxic is dinitrotoluene (DNT). Technical-grade DNT (tgDNT) is a mixture of DNT isomers (76% 2,4-DNT; 19% 2,6-DNT; 5% other isomers) used in the synthesis of toluene diisocyanate, an intermediate in the production of polyurethane foams. tgDNT and its various isomers produce positive (Couch et al. 1981; Spanggord et al. 1982) and negative (Simmon et al. 1977; Chiu et al. 1978) results in the *Salmonella* mutagenesis assay and negative results in the Chinese hamster ovary cell mutagenesis assay (Abernethy and Couch 1982). Negative results were also seen in *Saccharomyces* mitotic recombination (Simmon et al. 1977), in vitro UDS in rat hepatocytes (Bermudez et al. 1979), dominant lethal tests in rats (Dougherty et al. 1978) and mice (Soares and Lock 1980), and coat-color mutations and sperm morphology assays in mice (Soares and Lock 1980). From these results, one would predict that DNT is not a potential carcinogen; however, in a bioassay in which Fischer-344 rats were fed 35 mg/kg/day tgDNT, 100% of the males and 50% of the females were diagnosed as having hepatocellular carcinomas after one year of treatment (CIIT 1978). DNT is therefore a potent hepatocarcinogen; yet most short-term genetic toxicology tests failed to predict this effect.

Treatment of male rats with a single oral dose of 100 mg/kg tgDNT in corn oil produced a striking elevation in UDS 12 hours after treatment (Fig. 3). Furthermore, a 50-fold elevation in the percentage of cells in S-phase was observed 48 hours after treatment. Therefore, tgDNT is a potent genotoxic agent and induces an increase in cell proliferation in the liver, probably due to hepatotoxicity. In addition, other hepatotoxins have also been shown to induce DNA replication 48 hours after treatment (Table 2), including CCl_4 which is not genotoxic (Fig. 1). Thus, this assay can provide useful information on the hepatotoxic and necrogenic potential of chemicals.

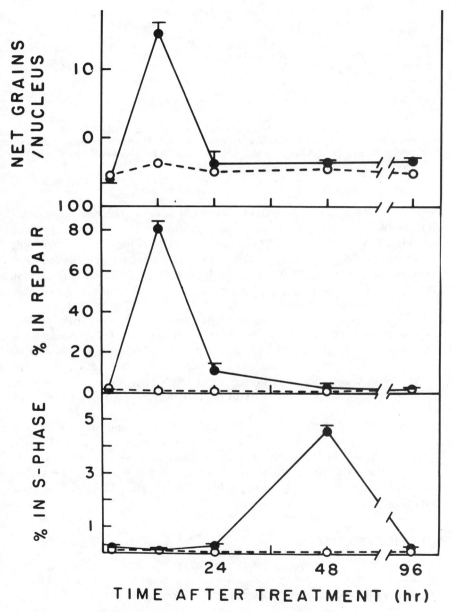

Figure 3
Induction of UDS and DNA replication following administration of 100 mg/kg tgDNT
(−●−) or corn oil (−○−). At times indicated after treatment, rat livers were perfused and
hepatocytes were isolated; 3-4 treated animals were used for each point. S.E. represent variation between animals. Reprinted, with permission, from Mirsalis and Butterworth (1982).

Table 2
Induction of DNA Replication Following Treatment with Hepatotoxins

Treatment	Dose (mg/kg)	% in S-phase
Control	—	0.08 ± 0.01
2-AAF	50	1.8 ± 0.5
tgDNT	100	4.6 ± 0.2
CCl₄	400	4.1 ± 1.0

All compounds were administered to rats by gavage in corn oil 48 hr prior to sacrifice; controls received corn oil. Variation shown represents the range of 2-3 animals for each group; 6000 cells scored/animal. (Adapted, with permission, from Mirsalis et al. (1982b).

SEX DIFFERENCES

Males and females frequently exhibit differences in rates of uptake, metabolism, and elimination of xenobiotics, resulting in differences in genotoxicity and carcinogenicity between sexes. This is evident for DNT where females show only one-half the incidence of hepatocellular carcinomas observed in males following chronic exposure to tgDNT (CIIT 1978). It is impossible to study differences in genotoxicity between sexes using in vitro assays, but with an in vivo system such a comparison can be easily made.

To determine whether there are differences in the genotoxicity of DNT between sexes, male and female rats were examined for UDS following treatment with tgDNT. Males showed a strong positive UDS response, but females exhibited only a relatively weak positive response (Table 3). Both sexes had strong positive responses following treatment with dimethylnitrosamine (NDMA), which confirms the ability of females to carry out high levels of DNA repair.

In addition, whereas males show approximately a 50-fold elevation in DNA replication 48 hours after treatment, females show only a 3- to 6-fold elevation (Table 3). These results appear to indicate that tgDNT is not only less genotoxic in females, but is also less hepatotoxic. Therefore, the use of such an in vivo system permits us to predict and study sex differences in hepatotoxicity, genotoxicity and, ultimately, carcinogenicity.

CHRONIC EXPOSURE

Short-term test systems usually measure genotoxicity following acute exposure, i.e., administration of a single, large dose of a compound. A more relevant approach would be to examine genotoxicity following continuous exposure to a much lower dose. To study the genotoxicity of DNT after chronic exposure, rats were maintained on a diet containing 0.1% tgDNT for up to 4 weeks. The observed level of UDS was quite low, with insignificant elevations in NG and only small, but significant, increases in the percent of cells in repair (Table 4).

Table 3

Induction of UDS in Male and Female Rats Following Treatment with tgDNT or NDMA

Chemical	Dose (mg/kg)	Sex (n)	NG ± S.E.	% in Repair
Control	—	M (4)	−4.2 ± 0.4	2 ± 2
		F (3)	−3.7 ± 0.7	2 ± 0
tgDNT	100	M (4)	15.1 ± 1.6	80 ± 4
		F (3)	4.6 ± 1.9	49 ± 12
NDMA	10	M (3)	54.9 ± 4.5	89 ± 5
		F (3)	43.0 ± 6.3	96 ± 1

Male and female Fischer-344 rats were treated with tgDNT in corn oil 12 hr prior to sacrifice or with NDMA in water 2 hr prior to sacrifice. Controls received corn oil. (n) number of rats treated. S.E. shown represent variation between animals.

Induction of DNA Replication in Male and Female Rats Following Treatment with tgDNT

Chemical	Sex (n)	% in S-phase ± S.E.
Control	M (5)	0.08 ± 0.01
	F (3)	0.16 ± 0.08
tgDNT	M (3)	4.57 ± 0.22
	F (3)	0.60 ± 0.05

Rats were treated with 100 mg/kg tgDNT or corn oil 48 hr prior to sacrifice. 3 slides scored/rat, 2000 cells/slide; (n) number of rats treated. Standard errors shown represent variation between animals. (Reprinted, with permission, from Mirsalis and Butterworth (1982).

This weak response may be due to the much lower administered dose than that used in acute studies as well as the rapid rate of DNT-induced DNA repair (Fig. 3). Evidence of hepatotoxicity was also observed following chronic exposure, with tgDNT-fed animals showing over a 20-fold elevation in DNA replication after 4 weeks of treatment (Table 4). Chronic exposure to tgDNT, therefore, produces genotoxicity and hepatotoxicity, both of which may contribute to its potent carcinogenicity.

EFFECTS OF GUT FLORA

Compounds that are metabolized to genotoxic intermediates outside the liver may still produce genotoxicity in hepatocytes, leading to tumor formation. In vitro assays that utilize liver-derived metabolic activation systems (e.g., S-9, intact hepatocytes) may not detect such genotoxic metabolites, but in vivo assays, which reflect the metabolism of the whole animal, will accurately predict the formation of such metabolites, regardless of their site of origin. DNT again

Table 4
DNA Repair, Replication and Growth Parameters of Rats Maintained on Feed Containing 0.1% tgDNT

Group	Number weeks on feed	Weekly % weight gain[a]	NG ± S.E.[b]	% in repair	S-phase
Control	1	24 ± 2	-4.1 ± 0.5	1 ± 1	0.7 ± 0.1
DNT	1	11 ± 2	-2.0 ± 0.7	12 ± 8[c]	1.2 ± 0.4
Control	2	18 ± 2	-5.2 ± 0.8	1 ± 1	0.4 ± 0.1
DNT	2	9 ± 3	-3.3 ± 0.5	10 ± 3[c]	1.9 ± 0.4[c]
Control	3	6 ± 3	—[d]	—	
DNT	3	8 ± 6	—[d]	—	
Control	4	5 ± 4	-3.2 ± 0.5	2 ± 2	0.03 ± 0.1
DNT	4	5 ± 3	-0.8 ± 0.8	17 ± 6[c]	2.0 ± 0.5[c]

[a]Initial body weights were 145 ± 9 g for the DNT group, 139 ± 4 g for the control group.
[b]Combined slide-to-slide and animal-to-animal variation, n = 3-6.
[c]Significant increase over control level (p < 0.01) by chi-square analysis.
[d]UDS was not measured on week 3. (Reprinted, with permission, from Mirsalis and Butterworth (1982).

serves as an excellent example: no UDS is observed in isolated hepatocyte cultures (Bermudez et al. 1979), but in the in vivo UDS assay DNT is clearly genotoxic (Fig. 3). A possible explanation for this discrepancy is that DNT may be converted to a genotoxic intermediate by reductive metabolism carried out by intestinal bacteria.

To study the role of gut flora in the metabolic activation of DNT, UDS was measured following treatment of germ-free (axenic) rats. Two groups of axenic male rats [CDR® (F-344)/CrlGN] were obtained from Charles River Breeding Laboratories. Before shipment, one group was associated with Charles River Altered Schaedler Flora (CRASF), a cocktail of eight anaerobic bacterial strains similar to the normal gut microflora of rats. Both groups were treated under sterile conditions with sterile tgDNT and examined for UDS.

Rats associated with CRASF exhibited a strong positive response following tgDNT treatment (Table 5; Mirsalis et al. 1982a), whereas axenic rats showed no significant increase in UDS. Treatment with NDMA produced a strong positive response, confirming the ability of axenic rats to carry out DNA repair.

Cultures of the cecal contents of each rat showed that four tgDNT-treated rats and one control rat from the axenic group were contaminated with individual bacterial contaminants (Table 5). These animals showed only low levels of bacteria relative to the CRASF-associated rats and none of these contaminated rats exhibited a degree of UDS comparable to that seen in the CRASF-associated rats. Therefore, it appears that DNT is genotoxic only in animals that possess the complete complement of gut flora. This, then, offers a plausible explanation for the lack of genotoxic activity observed for DNT in the in vitro test systems, which do not contain a reductive metabolic activation system. In the in vivo UDS system, genotoxicity is not only measured in the correct target tissue, but the metabolism by gut flora is accurately reflected as well.

SUMMARY

In vitro assays provide useful information on the genotoxicity of chemicals under certain conditions. These systems, however, do not take into account the sex, species, and tissue-specific effects involved in the complex processes of carcinogenesis. The in vivo-in vitro hepatocyte DNA repair assay is an example of a system that provides more relevant information about the genotoxicity of a specific target tissue, the liver, in the whole animal. This assay has been shown to be useful for the detection of a wide variety of genotoxic hepatocarcinogens, many of which are not detected by in vitro assays. In addition, such factors as the time-course of DNA repair and the distribution of cellular responses provide additional insight into the uptake, distribution, and activation of a chemical. This assay may also be used to measure increased hepatic cell proliferation and is a good indicator of chemically induced hepatotoxicity. Furthermore, other complex parameters, such as sex differences, effects of chronic exposure, and

Table 5
Induction of UDS in Germ-free (axenic) Rats

Bacterial Status	Chemical	Dose (mg/kg)	n	NG ± S.E.	% in repair
ASSOCIATED ANIMALS					
complete CRASF	Control	–	2	-4.2 ± 0.5	7 ± 2
complete CRASF	DNT	100	4	18.7 ± 1.3	87 ± 3
NON-ASSOCIATED ANIMALS					
Axenic	Control	–	1	-3.6	3
Bacillus leicheniformis + *Clostridium perfringens*	Control	–	1	-4.3	1
TOTAL CONTROL			2	-4.0 ± 0.4	2 ± 1
Axenic	DNT	100	2	-0.8 ± 0.2	14 ± 2
Streptococcus faecalis	DNT	100	2	-2.9 ± 0.7	34 ± 4
Clostridium perfringens	DNT	100	2	-3.5 ± 0.2	4 ± 2
TOTAL DNT		100	6	-0.5 ± 1.2	18 ± 6
Axenic	NDMA	10	1	43.7	97

Rats were treated with 100 mg/kg sterile tgDNT 12 hr before sacrifice. NDMA was administered 2 hr before sacrifice. Controls received corn oil. (n) number of treated animals. The bacterial content of each animal was determined from cecal contents at sacrifice. Axenic refers to animals whose cecal contents were found to be sterile. Bacteria found in the nonassociated group were low levels of adventitious contaminants. S.E, represent animal-to-animal variation. Reprinted, with permission, from Mirsalis et al. (1982a).

the effects of gut flora on genotoxicity have been studied using this whole-animal system. This assay is clearly useful for the study of genotoxicity in the liver. The development of similar assays in other key target tissues should provide valuable information on the genotoxicity of chemicals in the whole animal.

ACKNOWLEDGMENTS

I wish to thank my collaborators in all of these studies: Ms. Kim Tyson of SRI International and Dr. Byron Butterworth, Dr. Tom Hamm, Jr., Mr. Mike Sherrill, Ms. Patricia Mullins, and Mr. Steve Dennis of the Chemical Industry Institute of Toxicology. I also acknowledge the assistance of Ms. Erica Loh in the preparation of this manuscript.

REFERENCES

Abernethy, D.J. and D.B. Couch. 1982. Cytotoxicity and mutagenicity of dinitrotoluene in Chinese hamster ovary cells. *Mutat. Res.* **103**:53.

Bermudez, E., D. Tillery, and B.E. Butterworth. 1979. The effect of 2,4-diaminotoluene and isomers of dinitrotoluene on unscheduled DNA synthesis in primary rat hepatocytes. *Environ. Mutagen.* **1**:391.

Chemical Industry Institute of Toxicology (CIIT). 1978. A twenty-four month toxicology study in Fischer-344 rats given dinitrotoluene, 12 month report. Docket #327N8.

Chiu, C.W., L.H. Lee, C.Y. Wang, and G.T. Bryan. 1978. Mutagenicity of some commercially available nitro compounds for *Salmonella typhimurium*. *Mutat. Res.* **58**:11.

Couch, D.B., P.F. Allen, and D.J. Abernethy. 1981. The mutagenicity of dinitrotoluenes in *Salmonella typhimurium*. *Mutat. Res.* **90**:373.

Dougherty, R.W., G.S. Simon, F.I. Campbell, and J.F. Borzelleca. 1978. Failure of 2,4-dinitrotoluene to induce dominant lethal mutations in the rat. *Pharmacologist* **20**:155.

Martin, C.N., A.C. McDermid, and R.C. Garner. 1978. Testing of known carcinogens and noncarcinogens for their ability to induce unscheduled DNA synthesis in HeLa cells. *Cancer Res.* **38**:2621.

Mirsalis, J.C. and B.E. Butterworth. 1980. Detection of unscheduled DNA synthesis in hepatocytes isolated from rats treated with genotoxic agents: An *in vivo-in vitro* assay for potential carcinogens and mutagens. *Carcinogenesis* **1**:621.

Mirsalis, J.C. and B.E. Butterworth. 1982. Induction of unscheduled DNA synthesis in rat hepatocytes following *in vivo* treatment with dinitrotoluene. *Carcinogenesis* **3**:241.

Mirsalis, J.C., T.E. Hamm, Jr., J.M. Sherrill, and B.E. Butterworth. 1982a. Role of gut flora in the genotoxicity of dinitrotoluene. *Nature* **295**:322.

Mirsalis, J.C., C.K. Tyson, and B.E. Butterworth. 1982b. Induction of DNA repair in hepatocytes from rats treated *in vivo* with genotoxic agents. *Environ. Mutagen.* **4** (in press).

Probst, G.S., R.E. McMahon, L.E. Hill, C.Z. Thompson, J.K. Epp, and S.B. Neal. 1981. Chemically-induced unscheduled DNA synthesis in primary rat hepatocyte cultures: A comparison with bacterial mutagenicity using 218 compounds. *Environ. Mutagen.* **3**:11.

San, R.H.C. and H.F. Stich. 1975. DNA repair synthesis of cultured human cells as a rapid bioassay for chemical carcinogens. *Int. J. Cancer* **16**:284.

Simmon, V.F., S.L. Eckford, A.F. Griffin, R. Spanggord, and G.W. Newell. 1977. Munitions waste water treatments: Does chlorination or ozonation of individual components produce microbial mutagens? *Toxicol. Appl. Pharmacol.* **41**:197.

Soares, E.R. and L.F. Lock. 1980. Lack of an indication of mutagenic effects of dinitrotoluenes and diaminotoluenes in mice. *Environ. Mutagen.* **2**:111.

Spanggord, R.J., K.E. Mortelmans, A.F. Griffin, and V.F. Simmon. 1982. Mutagenicity in *Salmonella typhimurium* and structure-activity relationships of wastewater components emanating from the manufacture of trinitrotoluene. *Environ. Mutagen.* **4**:163.

Trosko, J.E. and J.D. Yager. 1974. A sensitive method to measure physical and chemical carcinogen-induced "unscheduled DNA synthesis" in rapidly dividing eukaryotic cells. *Exp. Cell Res.* **88**:47.

Williams, G.M. 1977. Detection of chemical carcinogens by unscheduled DNA synthesis in rat liver primary cell cultures. *Cancer Res.* **37**:1845.

COMMENTS

FURIHATA: Are there any false positive results?

MIRSALIS: It depends on what you mean by false positive. I might call MMS a false positive. Clearly, it is a mutagen, but it is not a hepatocarcinogen. If the question is if the assay will exclude compounds which are not hepato-carcinogenic, then we do have false positives. As far as compounds that are clearly not mutagenic, one of our major limitations right now is that we just haven't tested enough negative compounds. Most of the things we have looked at are carcinogens.

WEINSTEIN: Could you compare your assay to the Williams-type assay? Because I think polycyclic aromatics are positive there, and there are some differences.

MIRSALIS: There are several differences. Of course, the so-called Williams system is an in vitro assay, where hepatocytes are treated in culture, as opposed to the whole animal. In general, they will pick up some compounds that we will miss, for example, the polycyclics. There is some argument as to what the role is of metabolic activation of polycyclics by hepatocytes. When used in metabolic activation systems with mammalian cells, some people, like Bob Langenbach (Langenbach et al. 1981) don't see metabolism of DMBA and B[a]P, for example, while others, like Ed Bermudez (Bermudez et al. 1982) do.
　　　　　　On the other side of it, the Williams assay misses major classes of compounds, like nitroaromatics, some of the hydrazines, some azo-compounds, etc. These are all missed in the in vitro system, but are all detected in vivo.
　　　　　　So there are pros and cons. We are working with both systems right now, comparing them.

PEREIRA: How do you explain that you miss the polycyclics and the direct-acting alkylating agents, when in partial hepatectomized rats they are liver carcinogens?

MIRSALIS: I would argue that I don't think we miss B[a]P. In a 2-year bio-assay, if you run B[a]P or DMBA, they are not hepatocarcinogens.

PEREIRA: But if you partial hepatectomize the animal, they are.

MIRSALIS: Right, but we are not partial hepatectomizing our animals. It is very hard to say. Some studies suggest that there may be adducts formed in the liver, but that they are so rapidly repaired, or that detoxification of the compound is so great compared to the activation of the compound, that in a whole animal system, as opposed to a tissue culture system, we just don't see it.

PEREIRA: When you go to the methylating agents, MMS and MNNG, when administered to partial hepatectomized animals, are liver carcinogens. Do they bind and get O^6-methylguanine?

MIRSALIS: Yes, we pick up MMS and MNNG. It is interesting that we pick up the direct-acting agents and not the polycyclics.

THILLY: I have a cell physiology question that has always struck me in UDS experiments. Frequently, they are done in nondividing tissues. Those of us who study cell cycles know that thymidine kinase is a cell cycle-specific enzyme which doesn't start to arise until after the initiation of normal DNA replication. How come tritiated thymidine is taken into these? This paradox, no doubt, you have tried to resolve experimentally.

MIRSALIS: You mean why are nondividing cells taking up tritiated thymidine?

THILLY: Yes, into their DNA, under what the cell physiologists reported would be a condition where they wouldn't have cellular, as opposed to mitochondrial, thymidine kinase.

MIRSALIS: They are probably taking it up for DNA repair. I am not sure if that answers your question.

THILLY: No, it doesn't. You see, to take thymidine and put it into DNA, you need thymidine kinase in order to —

MIRSALIS: Right, and you are saying that there is no thymidine kinase there at the time of treatment.

THILLY: There is a wide literature that would lead us to believe that is true. Either that literature is false or we don't understand how you are getting tritiated thymidine to the DNA.

MIRSALIS: Well, we obviously are getting it into the DNA. So I would say there are probably low levels of thymidine kinase, or it is very rapidly inducible.

References

Bermudez, E., D.B. Couch, and D. Tillery. 1982. The use of primary rat hepatocytes to achieve metabolic activation of promutagens in the Chinese Hamster Ovary/Hypoxanthineguanine Phosphoribosyl Transferase Mutational Assay. *Environ. Mutagen.* **4**:55.

Langenbach, R., S. Nesnow, A. Tompa, R. Gingell, and C. Kuszynski. 1981. Lung and liver cell-mediated mutagenesis systems: Specificities in the activation of chemical carcinogens. *Carcinogenesis* **2**:851.

Chemically-induced DNA Repair in Rodent and Human Cells

**BYRON E. BUTTERWORTH, DAVID J. DOOLITTLE,
AND PETER K. WORKING**
Department of Genetic Toxicology
Chemical Industry Institute of Toxicology
Research Triangle Park, North Carolina 27709

**STEPHEN C. STROM, RANDY L. JIRTLE, AND
GEORGE MICHALOPOULOS**
Departments of Pathology and Radiology
Duke University Medical Center
Durham, North Carolina 27710

The ability to measure chemically-induced DNA damage in specific tissues is important in risk assessment because alteration of the DNA appears to be a key step in the initiation of carcinogenesis. Furthermore, mutations in germ cells are responsible for heritable genetic changes. Many techniques are available to assess the genotoxic activity of a chemical in a variety of bacterial or mammalian cell culture assays (for reviews, see Butterworth 1979; Hollstein et al. 1979). These techniques are referred to as short-term tests for potential carcinogenicity because of a demonstrated correlation between carcinogenicity and genotoxicity in these assays for many classes of chemicals. In vitro systems are, however, severely limited in their predictive abilities because they often do not reflect the important species, strain, sex, and organ specificities commonly observed in chemical carcinogenesis. Such specificity can be the result of differences in compound uptake, distribution, metabolism, DNA repair, detoxification, and excretion. For example, numerous carcinogens must be metabolized to active forms that react with the DNA. Yet, the cells and bacteria often used in assessment of mutagenic activity are generally not metabolically competent.

The exogenous metabolic activation system most commonly employed in such assays is a postmitochondrial supernatant (S-9) from a rat liver homogenate. This is a crude preparation which is biased toward activation at the expense of detoxification reactions. Further, important enzymes such as nitroreductases are either missing or nonfunctional in standard S-9 preparations. Thus, so-called correlations between short-term tests and carcinogenicity retain a high degree of subjectivity. In order to obtain more meaningful information, it is imperative to employ assays which examine the genotoxic activities of chemicals in whole animals and, where possible, human beings.

The determination of chemically-induced DNA repair as measured by unscheduled DNA synthesis (UDS) is becoming a valuable tool in the assessment of organ specific genotoxic activity. Covalently-bound DNA adducts can be removed by an enzymatic process of excision repair in which the strand of altered DNA is nicked, the damaged section removed, and the gap replaced as dictated by the correct pairing from the opposite DNA strand. If the repair process takes place in the presence of ^{3}H-TdR the incorporation of radioactivity can be quantitated by autoradiography and is a clear indication that the compound reached and altered the DNA.

Some direct-acting chemicals induce repair in cells in culture (San and Stich 1975) but the problem of the lack of appropriate metabolic activation remains. One of the most significant advances in this area was the demonstration that primary rat hepatocytes in culture would produce a UDS response to a wide variety of genotoxicants requiring metabolic activation (Williams 1976). A major advantage of this assay is that the target cell itself provides the metabolic activity.

A further advance was the development in this laboratory of an in vivo-in vitro hepatocyte DNA repair assay (Mirsalis and Butterworth 1980). Following treatment of the whole animal by an appropriate route of exposure, primary hepatocyte cultures are prepared and incubated with ^{3}H-TdR. If the DNA is damaged in the intact animal such that the cells are undergoing excision-repair, ^{3}H-TdR will be incorporated into the DNA of the freshly isolated cells in culture. The in vivo-in vitro rat hepatocyte DNA repair assay detects genotoxic hepatocarcinogens from a variety of chemical classes including nitroaromatics, aromatic amines, direct-acting agents, mycotoxins, nitrosamines, and azo compounds (Mirsalis et al. 1982a). One advantage of this assay is that factors such as uptake, metabolism, distribution, and excretion are inherently accounted for. The response of the cells in vivo can now be compared to the response in vitro.

A further advantage of measuring UDS as an endpoint is that most tissues have repair capability. Key questions regarding organ specificity and comparisons of animal models to the response in man can be addressed. Chemically-induced UDS has been measured in rat hepatocytes in vitro (Williams 1976; Probst et al. 1981); rat hepatocytes in vivo (Mirsalis and Butterworth 1980); human hepatocytes in vitro (Strom et al. 1982); mouse kidney in vivo (Brambilla et al. 1978); rat stomach in vivo (Furihata et al. 1981); rat trachea in vitro (Ishikawa et al. 1980); rat oral tissues in vitro (Ide et al. 1981b); rat lymphocytes in vivo (Skinner et al. 1980); human blood monocytes in vitro (Lake et al. 1980); human lymphocytes in vitro (Pero and Vopat 1981); human bone marrow cells in vitro (Lewensohn and Ringborg 1979); mouse spermatocytes in vivo (Lee and Suzuki 1979; Tanaka and Katoh 1979; Sega and Sotomayor 1980); and rabbit spermatocytes in vivo (Zbinden 1980). Currently, we are using or developing UDS assays for rat and human hepatocytes, trachea, bladder, and spermatocytes.

METHODS

Rat Hepatocytes

The methods for hepatocyte UDS assays have been described by Williams (1976) and Mirsalis and Butterworth (1980). Chemical exposure can be to the cells in culture or to the intact animal. The animal is anesthetized and the liver is perfused in situ with a solution of 100 units/ml of type 1 collagenase. The liver is dissociated and the hepatocytes are placed in culture, allowed to attach for 90 minutes and cultured with ^{3}H-TdR. Incorporation of label is determined by quantitative autoradiography. Autoradiography is superior to scintillation counting because one can clearly distinguish between cells in repair and cells in S-phase. This is important because some compounds induce cell division but not DNA repair (Mirsalis and Butterworth 1982). The net number of silver grains over the nucleus (NG) is calculated by subtracting the grain count from the highest of three adjacent nuclear sized areas over the cytoplasm from the grain count over the nucleus. Scoring is performed with a computerized grain counter interfaced via a television camera to the microscope.

The cytoplasmic background is particularly high in hepatocytes as compared to other cell types such as spermatocytes and may represent mitochondrial DNA synthesis. High cytoplasmic background counts are common in the in vitro assay where the cells must be incubated for an extended period with ^{3}H-TdR, and often presents a serious technical problem in scoring the slides (Mirsalis and Butterworth 1980). Compound related changes in the cytoplasmic background that are independent of nuclear counts are often evident (B.E. Butterworth, unpubl. results).

Human Hepatocytes

Fragments of normal human hepatic tissue are obtained from discarded surgical material removed during scheduled surgical procedures. For the case described here two benign tumors were removed from the liver of a 30-year old white female. A small portion of apparently healthy tissue removed with the tumor was placed into ice cold saline and transported to the laboratory. Catheters were inserted into the larger vessels on the cut surface and the tissue was perfused with a collagenase solution as described previously (Strom et al. 1982). The cell suspension contained 300,000 cells/ml with 80% of the cells viable; 0.4 ml was plated in each well of 8 well chamber/slides (Miles Laboratories) coated with rat tail collagen. Following attachment for 90 minutes, cells were incubated with the test chemical and ^{3}H-TdR for 18 hours. Autoradiography was performed and cells scored as described for rat hepatocytes.

Rat Spermatocytes

Testes can be obtained from the same treated animal used in the liver UDS assay so that a direct comparison can be made between the two tissues. To perform this assay the animal is anesthetized, the testes removed, rinsed in phosphate-buffered saline, and the tunica and blood vessels removed. Tubules are incubated for 10 minutes at 37° in 0.1% trypsin. Tubules are then rinsed, minced with scalpels, and incubated once again for 10 minutes at 37° in 0.1% trypsin. Fetal calf serum is added to 2.5% to stop the activity of the trypsin. Following vigorous pipetting to loosen the cells the tubules are allowed to settle. The cells are obtained from the supernatant by centrifugation. Cells can be exposed to the test chemical and ^{3}H-TdR in vivo or in vitro. The most promising approach appears to be in vivo exposure to the chemical followed by in vitro exposure to the ^{3}H-TdR in a manner analogous to the hepatocyte in vivo-in vitro assay. Incorporation of label is determined by quantitative autoradiography. No cytoplasmic background grains have been observed so that scoring of grains/nucleus is straightforward. Spermatogonia, spermatocytes, spermatids, and mature sperm are present in the preparation. Identification of different germ-cell stages of the rats is based on published morphological descriptions by LeBlond and Clermont (1952). Generally, only pachytene spermatocytes and early spermatids are scored, because of the ease of identification and the high numbers of each in the preparations.

In the in vivo study presented here, animals were untreated (control) or treated i.p. with 200 mg/kg ethylmethanesulfonate (EMS). One hour later each testis was injected with 50 μCi ^{3}H-TdR. Two hours later spermatocyte cultures were prepared and pachytene spermatocytes were purified by centrifugal elutriation (Grabske et al. 1975). Slides were prepared and exposed to photographic emulsion for 6 weeks. Following development, silver grains were quantitated by light microscopy.

In the in vitro study freshly prepared spermatocytes were exposed to 1% dimethylsulfoxide (DMSO) (control) or 1.0 mM EMS in the presence of 10 μCi/ml ^{3}H-TdR for 18 hours. Slides were prepared, fixed, washed, and exposed to photographic emulsion for 2 weeks. After staining with hematoxylin, pachytene spermatocytes were recognized by their morphology and scored as above.

RESULTS

Rat Hepatocytes

Most genotoxic hepatocarcinogens such as nitrosodimethylamine (NDMA) and 2-acetylaminofluorene (2-AAF) induce UDS both in vitro (Williams 1976) and in vivo (Mirsalis et al. 1982b). There are, however, interesting exceptions to this observation. For example, some polycyclic aromatic hydrocarbons which are not hepatocarcinogens such as benzo[a]pyrene (B[a]P), induce DNA repair in the in vitro hepatocyte assay (Probst et al. 1981) but not in the in vivo-in vitro hepatocyte system (Mirsalis et al. 1982a). The reasons for this are not presently understood, but may relate simply to delivery of the compound to the cells.

In contrast, the potent hepatocarcinogen technical grade dinitrotoluene (tech-DNT) induces a response in the in vivo assay but not in the in vitro assay because gut flora are obligatory in its metabolism to genotoxic products (Bermudez et al. 1979; Mirsalis et al. 1982b). Tech-DNT consists of approximately 80% 2,4-DNT and 20% 2,6-DNT. 2,6-DNT was shown to be at least an order of magnitude more potent than 2,4-DNT in inducing hepatocyte DNA repair and is presumed to be the active component in the mixture (Mirsalis and Butterworth 1982).

We have extended this work to determine if similar structure activity relationships and the obligatory role of gut flora in metabolic activation applies to other nitroaromatic compounds. 2-Nitrotoluene (2-NT) produces a dose-related increase in hepatic UDS at 12-hour post-treatment, while the 3-NT and 4-NT isomers do not (Table 1). Germ-free animals were obtained from Charles River Laboratories. Two weeks before shipment one group was separated and inoculated with Charles River Associated Flora (CRAF), a mixture of bacteria similar to the normal flora. As with the conventional animals, 2-NT produced a positive response in the hepatocytes from CRAF associated animals whereas no DNA repair was seen in hepatocytes isolated from 2-NT treated germ-free animals (Table 2). Cecal contents were taken at the time of liver perfusion

Table 1
Induction of Hepatic UDS in Conventional Fischer-344 Male Rats

Chemical	Dose (mg/kg)[a]	Time	Net grains	Percent in repair[b,c]
2-NT	200	12 hr	+++	+++
	500	12 hr	+++	+++
3-NT	100	12 hr	0	0
	500	12 hr	0	0
4-NT	100	12 hr	0	0
	500	12 hr	0	0
Corn oil	—	2 hr	0	0
NDMA	10	2 hr	+++	+++

[a]Chemicals were administered by oral gavage in corn oil (nitrotoluene) or water (NDMA) at 0.2 ml/100 g body wt.
[b]≥ 5 net grains/ nucleus is considered in repair

[c]
Score	Net grains	% in repair
0	< 5	0-10
±	≤ 5	10-20
+	5-10	20-50
++	10-20	50-80
+++	> 20	80-100

Table 2

The Influence of Gut Flora on the Hepatic Genotoxicity Produced by 2NT

Animal	Chemical	Dose (mg/kg)[a]	Time	Net grains	Percent in repair[b,c]
CRAF	2-NT	200	12 hr	+	++
	Corn oil	–	12 hr	0	0
Germ free	2-NT	200	12 hr	0	0
	Corn oil	–	12 hr	0	0
	NDMA	10	1 hr	+++	+++

[a]Chemicals were administered by oral gavage in corn oil NT or water NDMA at 0.2 ml/100 g body wt.
[b]≥ 5 net grains/nucleus is considered in repair.
[c]Scoring is as indicated in Table 1.

for quantitative bacteriology to confirm bacterial status. These data show that 2-NT but not 3-NT or 4-NT induces DNA repair in hepatocytes of treated animals and that metabolism by gut flora is obligatory for this activity. These experiments illustrate the value of the in vivo-in vitro hepatocyte DNA repair assay in predictive and mechanistic studies.

Human Hepatocytes

Given the often dramatic species differences in response to toxicants it becomes imperative to know the extent to which rodent models are relevant in predicting human health effects. The in vitro rat hepatocyte DNA repair assay has proven to be useful in detecting a wide variety of genotoxic agents (Williams 1976; Probst et al. 1981). These techniques are directly applicable to freshly isolated human hepatocytes obtained from surgical samples normally discarded during prescribed surgery. Results from the rat and human cell assays can be compared to evaluate the validity of the rat model. Compounds of interest can be examined for genotoxic activity directly in metabolically competent human cells. Aflatoxin B_1 (AFB_1) has been implicated in human liver cancer and is clearly a potent inducer of DNA repair in human hepatocyte cultures (Table 3). 2-AAF, B[a]P, and NDMA have all been shown to respond in the in vitro rat hepatocyte DNA repair assay (Probst et al. 1981) and also in the human hepatocyte assay (Table 3). Further comparative studies should provide information as to the extent and variability of the response within the human population and the relative response of the rat and human hepatocytes.

Rat Spermatocytes

Identification of a chemical as genotoxic not only raises concerns that it may be a potential carcinogen but that it also may have the potential of inducing heritable genetic damage in germ cells. Chemically induced UDS in germ cells

Table 3
Response in the In Vitro Human Hepatocyte DNA Repair Assay

Compound	Concentration (mM)[a]	Net grains	Percent in repair[b,c]
2-AAF	0.01	++	+++
2-AAF	0.001	++	++
AFB_1	0.01	+++	+++
AFB_1	0.001	++	+++
B[a]P	0.01	+	++
B[a]P	0.001	++	++
NDMA	1.0	+	++
NDMA	0.1	0	0
DMSO	1%	0	0
Untreated Control	—	0	0

[a]Chemicals were dissolved in DMSO and added to the medium. The DMSO concentration never exceeded 1%.
[b]$\geqslant 5$ net grains/nucleus is considered in repair.
[c]Scoring is as indicated in Table 1.

has been used as an indication that a compound can reach and damage spermatogenic cell DNA. One technique employed involves measurement of the premature appearance of label in sperm following injection of ^{3}H-TdR into the testes (Sega and Sotomayor 1980; Zbinden 1980). These techniques require many animals, take at least 5 weeks to complete the spermatogenic cycle, and do not give quantitative data of the cell stage affected.

When the genotoxicant EMS and ^{3}H-TdR are both administered to a rat, DNA repair can be detected by autoradiography as an increase in silver grains/nucleus in pachytene spermatocytes (Fig. 1, panels A and B). If > 5 grains/nucleus is taken as a tentative cutoff for cells in repair, 20% more of the cells were in repair in the treated than in the control animal. If spermatocytes are prepared and then incubated in vitro with EMS and ^{3}H-TdR, DNA repair is strongly induced in nearly all of the pachytene spermatocytes (Fig. 1, panels C and D), indicating that the cells were capable of repair. The observation that repair was induced in only a fraction of the cells in vivo is probably related to delivery of chemical and (or) ^{3}H-TdR to the cells in the intact animal. The most promising approach is to treat the whole animal with the test compound for realistic exposures, then culture the spermatocytes in the presence of ^{3}H-TdR for uniform labeling. This in vivo-in vitro technique presents the further advantage that effects in the testes can be assessed from the same animals used in the in vivo-in vitro hepatocyte DNA repair assay (Mirsalis and Butterworth 1980) and is under development in our laboratory.

Trachea

Lung cancer is currently the leading cause of mortality from cancer in men in the United States and the incidence in women is increasing. Carcinoma of the

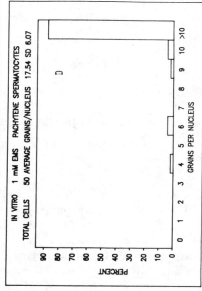

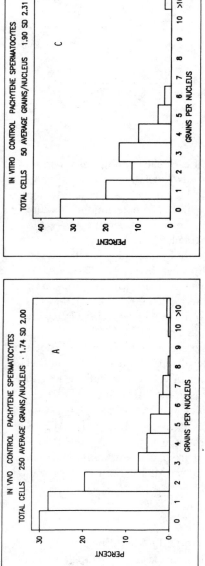

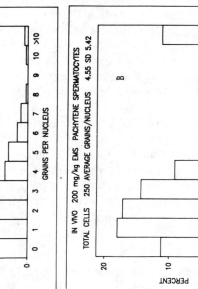

Figure 1

Induction of DNA repair in vivo and in vitro. Animals or spermatocyte cultures were exposed to EMS and ³H-dT as described in Methods. Quantitative autoradiography was performed to assess DNA repair. Histograms quantitate the silver grains/nucleus for pachytene spermatocytes in: (A) in vivo control; (B) in vivo treated; (C) in vitro control; (D) in vitro treated.

lung is usually bronchogenic with the most common site being the large bronchi at or near the lung hilus (National Cancer Institute 1980).

Rodent tracheas have been used extensively as model systems for studying human bronchogenic carcinoma because rodent tracheal epithelium closely resembles human bronchial epithelium (Kendrick et al. 1974; Becci et al. 1978). We are developing an assay for chemically induced DNA repair in rat trachea. Rats will be exposed to potential genotoxic agents in vivo (e.g., by inhalation, gavage, or i.p. injection), and DNA repair occurring in their tracheal cells will be assessed in vitro. Rat tracheal cells in organ culture are capable of repairing their DNA following in vitro chemical exposure (Ishikawa et al. 1980; Ide et al. 1981a), and the technology for placing viable, dissociated, rodent tracheal epithelial cells into culture has recently been developed (R. Wu, pers. comm.).

The major cell types in the tracheal epithelium are basal cells, mucus cells and ciliated cells. Basal cells and mucus cells can divide whereas ciliated cells appear to be terminally differentiated and incapable of cell division. Differential staining will be used to identify tracheal cell types during grain counting. This will enable us to make qualitative and quantitative comparisons between cell types in terms of chemically induced DNA repair.

Bladder

Cancer of the bladder will cause approximately 10,000 deaths in the United States this year (National Cancer Institute 1981). Certain environmental chemicals, particularly aromatic amines, have been suggested as playing an important role in the etiology of the disease. Cancer of the bladder usually originates in the epithelial layer, which is comprised of transitional cells. We believe that an in vivo-in vitro assay for UDS in transitional epithelial cells obtained from the bladder would provide a means to detect and study those chemicals which may produce genotoxicity in the bladder.

Available evidence indicates that for some bladder carcinogens initial metabolism and conjugation occurs in the liver, followed by clearance of the conjugated metabolites into the urine. The conjugates may then be chemically or enzymatically hydrolyzed in the bladder resulting in the generation of reactive intermediates which initiate the cancer. On the other hand, it has recently been demonstrated that bladder epithelium can itself metabolically activate several classes of carcinogens (Langenbach et al. 1981). By comparing in vivo and in vitro exposures in a transitional epithelium UDS assay it should be possible to examine the relative contribution of each of these metabolic pathways in the generation of chemically reactive intermediates within the bladder epithelium. The development of the assay should be facilitated by the recent development of techniques for culturing dissociated, viable rat bladder epithelial cells in vitro (Langenbach et al. 1981).

DISCUSSION

Historical data have shown that induction of DNA repair is useful as an indicator of genotoxic exposure in people and animals. A further strength of the approach is that induced repair can be assessed in many tissues including germ cells in the same treated animal, thus providing information as to target organ specificity. Development of these assays is currently the forefront of research in this area. Another focus of the research is to develop techniques to measure DNA repair in cultures of cells from human tissues. These studies should provide valuable information as to organ specificity and metabolic capability in human tissues, person-to-person variability, and the relevance of models employing rodent cells.

Limitations for these assays would be for compounds which are active by epigenetic mechanisms, or which do not induce excision repair. Because of organ specificity appropriate target organs must be assayed for induced DNA repair.

Although only modest doses of carcinogens are required to produce a response in the in vivo-in vitro hepatocyte DNA repair assay, sufficient repair to be detected may not be produced by low level chronic exposures (Mirsalis and Butterworth 1982). The only human tissue available for repair studies following in vivo exposure is peripheral blood lymphocytes, which may not show an effect if it is not a target tissue. A lymphocyte DNA repair assay may be of value in monitoring genotoxic effects following high accidental exposures or during clinical drug trials.

The fact that DNA repair is a vital defense mechanism to correct alterations of the DNA establishes chemically induced UDS as a valid indicator of genotoxic exposure. The ability to quantitate this repair in vitro and in vivo in specific tissues in man and animals makes this one of the most important tools in the hands of the genetic toxicologist.

ACKNOWLEDGMENTS

We thank Lynn Earle for excellent technical assistance.

REFERENCES

Becci, P.J., E.M. McDowell, and B.F. Trump. 1978. The respiratory epithelium. II. Hamster trachea, bronchus, and bronchioles. *J. Natl. Cancer Inst.* **61**:551.

Bermudez, E., B.E. Butterworth, and D. Tillery. 1979. The effect of 2,4-diaminotoluene and isomers of dinitrotoluene on unscheduled DNA synthesis in primary rat hepatocytes. *Environ. Mutagen.* **1**:391.

Brambilla, G., M. Cavanna, P. Carlo, R. Finollo, and S. Parodi. 1978. DNA repair synthesis in primary cultures of kidneys from BALB/c and C3H mice treated with dimethylnitrosamine. *Cancer Lett.* **5**:153.

Butterworth, B.E. (Ed.) 1979. Strategies for short-term testing for mutagens/carcinogens. CRC Press, West Palm Beach, Florida.

Furihata, C., S.S. Jin, and T. Matsushima. 1981. A short-term method for detecting organ specificity of potential carcinogens. In *Proceedings of the 3rd International Meeting on Environmental Mutagens*, p. 94, Tokyo.

Grabske, R.J., S. Lake, B.L. Gledhill, and M. Meistrich. 1975. Centrifugal elutriation: Separation of spermatogenic cells on the basis of sedimentation velocity. *J. Cell Physiol.* **86**:177.

Hollstein, M., J. McCann, F.A. Angelosanto, and W.W. Nichols. 1979. Short term tests for carcinogens and mutagens. *Mutat. Res.* **65**:133.

Ide, F., T. Ishikawa, and S. Takayama. 1981a. Detection of chemical carcinogens by assay of unscheduled DNA synthesis in rat tracheal epithelium in short-term organ culture. *J. Cancer Res. Clin. Oncol.* **102**:115.

Ide, F., T. Ishikawa, S. Takayama, and S. Umemura. 1981b. Autoradiographic demonstration of unscheduled DNA synthesis in oral tissues treated with chemical carcinogens in short-term organ culture. *J. Oral Path.* **10**:113.

Ishikawa, T., S. Takayama, and F. Ide. 1980. Autoradiographic demonstration of DNA repair synthesis in rat tracheal epithelium treated with chemical carcinogens *in vitro*. *Cancer Res.* **40**:2898.

Kendrick, J., P. Nettesheim, and A.S. Hammons. 1974. Tumor induction in tracheal grafts: A new experimental model for respiratory carcinogenesis studies. *J. Natl. Cancer Inst.* **52**:1317.

Lake, R.S., M.L. Kropko, S. McLachlan, M.R. Pezzutti, R.H. Shoemaker, and H.J. Igel. 1980. Chemical carcinogen induction of DNA-repair synthesis in human peripheral blood monocytes. *Mutat. Res.* **74**:357.

Langenbach, R., L. Malick, and S. Nesnow. 1981. Rat bladder cell-mediated mutagenesis of Chinese hamster V79 cells and metabolism of benzo(a)pyrene. *J. Natl. Cancer Inst.* **66**:913.

Lee, I.D. and J. Suzuki. 1979. Induction of unscheduled DNA synthesis in mouse germ cells following 1,2-dibromo-3-chloropropane (DBCP) exposure. *Mutat. Res.* **68**:169.

Lewensohn, R. and U. Ringborg. 1979. Induction of unscheduled DNA synthesis in human bone marrow cells by bifunctional alkylating agents. *Blood* **54**:1320.

LeBlond, C.P. and V. Clermont. 1952. Definition of the stages of the cycle of the seminiferous epithelium in the rat. *Ann. N.Y. Acad. Sci.* **55**:548.

Mirsalis, J.C. and B.E. Butterworth. 1980. Detection of unscheduled DNA synthesis in hepatocytes isolated from rats treated with genotoxic agents: An *in vivo-in vitro* assay for potential mutagens and carcinogens. *Carcinogenesis* **1**:621.

_____. 1982. Induction of unscheduled DNA synthesis in rat hepatocytes following *in vivo* treatment with dinitrotoluene. *Carcinogenesis* **3**:241.

Mirsalis, J.C., K.C. Tyson, and B.E. Butterworth. 1982a. The detection of genotoxic carcinogens in the *in vivo-in vitro* hepatocyte DNA repair assay. *Environ. Mutagen.* (in press).

Mirsalis, J.C., T.E. Hamm, Jr., M. Sherrill, and B.E. Butterworth. 1982b. The role of gut flora in the genotoxicity of dinitrotoluene. *Nature* **295**:322.

National Cancer Institute. 1980. *Research report: Cancer of the lung.* NIH Publication No. 81-526. Government Printing Office, Washington, D.C.

_____. 1981. *Research Report: Cancer of the bladder.* NIH Publication No. 81-722. Government Printing Office, Washington, D.C.

Pero, R.W. and C. Vopat. 1981. A human platelet-derived inhibitor of unscheduled DNA synthesis in resting lymphocytes. *Carcinogenesis* 2:1103.

Probst, G.S., R.E. McMahon, L.E. Hill, C.Z. Thompson, J.K. Epp, and S.B. Neal. 1981. Chemically-induced unscheduled DNA synthesis in primary rat hepatocyte cultures: A comparison with bacterial mutagenicity using 218 compounds. *Environ. Mutagen.* 3:11.

San, R.H.C. and H.F. Stich. 1975. DNA repair synthesis of cultured human cells as a rapid bioassay for chemical carcinogens. *Int. J. Cancer* 16:284.

Sega, G.A. and E. Sotomayor. 1980. Unscheduled DNA synthesis in mammalian germ cells—its potential use in mutagenicity testing. In *Chemical mutagens—Principles and methods for their detection* (ed. F.J. DeSerres and A. Hollaender), Vol. 17, p. 421.

Skinner, M.J., B. DeCastro, and J.F. Eyre. 1980. Detection of unscheduled DNA synthesis in rat lymphocytes treated *in vivo* with cyclophosphamide and triethylenemelamine. *Environ. Mutagen.* 2:277.

Strom, S.C., R.L. Jirtle, R.S. Jones, D.L. Novicki, M.R. Rosenberg, A. Novotny, G. Irons, J.R. McLain, and G. Michalopoulos. 1982. Isolation, culture and transplantation of human hepatocytes. *J. Natl. Cancer Inst.* 68:771.

Tanaka, N. and M. Katoh. 1979. Unscheduled DNA synthesis in the germ cells of male mice *in vivo*. *Jpn. J. Genet.* 54:405.

Williams, G.M. 1976. Carcinogen-induced DNA repair in primary rat liver cell cultures: A possible screen for chemical carcinogens. *Cancer Lett.* 1:231.

Zbinden, G. 1980. Unscheduled DNA synthesis in the testis, a secondary test for the evaluation of chemical mutagens. *Arch. Toxicol.* 46:139.

COMMENTS

SKOPEK: How many viable cells can be produced from the human liver biopsies?

BUTTERWORTH: It depends on the operation that is done. Sometimes we may get just a few cubic centimeters of liver. Other times we get a fairly large amount. The total number of cells ranges upward from 20 million with 90% viability as determined by trypan blue exclusion.

SEGA: Do you ever use any liver from a completely normal human liver?

BUTTERWORTH: I don't think we would ever get that. The data that you saw was from a patient with two very small benign tumors. She wasn't on any drugs. We believe it was a fairly normal, representative sample.

RAJEWSKY: One should not forget that UDS will only measure some rather specific types of DNA repair. It will not measure other types of DNA repair which can also occur, for instance, the type of repair that works on the O^6-alkyldeoxyguanosine, which is just a transferase of the alkyl group. The type of repair measurement that you use is perhaps more of a very general indicator of rather heavy damage to DNA which creates gaps.

BUTTERWORTH: That is correct.

RAJEWSKY: Maybe there are other types of repair which do not require formation of gaps, but still occur.

BUTTERWORTH: It is certainly a possibility, but the variety of chemicals that respond in this assay is quite broad.

RAJEWSKY: You cannot take the degree of UDS as a direct correlation for the degree of repair, because repair can occur simultaneously on very different types of lesions.

MIRSALIS: That is only partially true. Whether repair is long-patch or short-patch affects how much ^{3}H-TdR is incorporated. As far as missing certain types of repair, about the only thing that we will miss is direct demethylation. In fact, most compounds that produce simple methylated bases where direct demethylation occurs also induce some excision repair. For example, MMS and NDMA clearly induce UDS.

EISENSTADT: What proportion of the human hepatocytes are binucleate?

BUTTERWORTH: I would say 80 percent.

EISENSTADT: Is that deeply significant?

BUTTERWORTH: I don't think so. I would say that rat hepatocytes are about 5% binucleate. Both cell types appear to be equally capable of repair.

Unscheduled DNA Synthesis in Rat Lymphocytes Treated with Mutagens In Vivo

MICHAEL J. SKINNER AND CEINWEN A. SCHREINER
Mobil Oil Corporation
Toxicology Division
Princeton, New Jersey 08540

The detection and classification of genotoxic activity in vivo have become integral to our genetic toxicology screening program. Methods of reproducing genotoxic activity in vivo after it has been identified in vitro allows more precise assessments of dangers to the general population imposed by genotoxic agents.

There are cytogenetic techniques which detect and identify clastogenic events in vivo. Gene mutations and primary DNA damage should also be analyzed. We are currently developing techniques which detect unscheduled DNA synthesis (UDS) in various organs of the body. Solubility of chemicals greatly influences the bioavailability of a given agent; therefore, it is imperative that potential for activity be assessed in directly exposed target organs (e.g., respiratory system or alimentary canal) as well as in target cells that may be exposed after absorption and metabolism (e.g., lymphocytes). Analytical data on bioavailability are essential when designing studies to determine genotoxic activity at a specific target organ. One can initially address the question of absorption and then assess metabolites and specific site exposure. There is no reason to attempt to detect UDS in lymphocytes if the agent or its metabolites are never in the blood stream. Also, bioavailability data facilitates the establishment of critical sampling times for the testing protocol.

EXPERIMENTAL PROCEDURES

Male rats weighing approximately 200 ± 20 g are randomized into groups of five animals per test group. The rats are maintained in an air-conditioned room, $74 \pm 4°F$, with a 12/12 light cycle and a relative humidity range of 30-70%. They are given food and water ad libitum and acclimated for 5 days prior to treatment.

The vehicle for dosing and the route of administration depend on solubility, expected routes of human exposure, and bioavailability. Control animals are always dosed with vehicle by the same route of exposure as the test animals.

Peripheral blood is drawn by cardiac puncture into a heparinized syringe 2 hours after dosing. We have found that in most cases, if the test agent is absorbed, it will be in the blood within 2 hours but will not have cleared from the system. 0.5 ml of blood is maintained in culture with 7.5 ml McCoy's 5A-20% fetal calf serum and ^{3}H-thymidine (^{3}H-TdR; 20 Ci/mmol) at a final concentration of 6μCi/ml. The cultures were incubated at 37°C with 5% CO_2 for 4 hours. Cells are harvested, washed three times in Hanks' Balanced Salts Solution (HBSS) and fixed in Carnoy's fixative (Methanol:acetic acid, 3:1 v/v), which is chilled to 4°C. The fixative is changed three times; slides are made, randomly coded, and air-dried overnight.

Two different procedures are used for autoradiography: 1) A standard 2-week exposure procedure, and 2) a modification of the short-term scintillation enhancement method (Stanulis et al. 1979). These methods are described below:

1. Slides are washed in 95% ethanol and warmed to 47°C before they are dipped into NTB-3 (Kodak) photographic emulsion. They are drained, placed in a light-tight box and stored at 4°C for 2 weeks.
2. Slides are washed in 95% ethanol and warmed to 47°C before they are dipped into NTB-2 (Kodak) photographic emulsion (NTB-3 is too sensitive for this procedure). They are drained and air-dried for 1 hour; slides are dipped in Scinti Verse® (Fisher Scientific) for 2 minutes and air-dried for 20 minutes. Slides are then placed in a light-tight container with dessicant and stored at 4°C for 72 hours.

After the desired exposure period, the slides are developed by the procedures recommended by Kodak, stained with 2% Giemsa in Gurr's buffer pH 6.8 (Serile) and scored for cells undergoing UDS. A minimum of 300 cells are scored for each test animal. Cells are categorized into three groups: < background; > background but < 50 grains (indicative of UDS); and > 50 grains (S phase).

In this system, as in any UDS assay, it is critical to minimize background noise. The dipping emulsions are extremely sensitive and the level of UDS is very low; therefore, technical care is important for obtaining reliable data. Two means of reducing background are to thoroughly wash the cells and to use fresh dipping emulsion.

Results are analyzed by chi-square, comparing the data with concurrent negative control data (significance $p \leqslant 0.05$).

RESULTS

Analytical data from the blood of rats dosed orally with cyclophosphamide (CY) or 2-acetaminofluorene (2-AAF) shows that the test compounds were absorbed into the bloodstream by the 2-hour bleeding time (Figures 1 and 2). The blood was analyzed for the presence of parent compound, and the reduction in levels of the parent compound suggests that biotransformation had taken place.

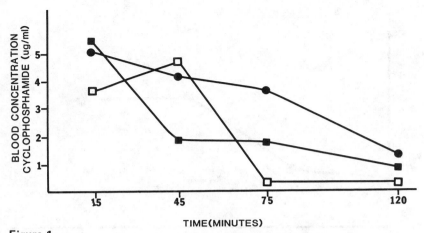

Figure 1

Cyclophosphamide concentration in the blood after oral dosing of 25 mg/kg (■); 50 mg/kg (●); 75 (□) mg/kg.

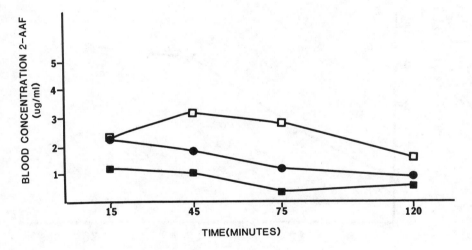

Figure 2

2-AAF concentration in the blood after oral dosing of 10 mg/kg (■); 20 mg/kg (●); 30 (□) mg/kg.

The 2-hour exposure of the rats to CY and 2-AAF induced a significant dose-related increase in UDS (Figure 3). CY is not as strong an inducer of UDS in this system as 2-AAF is.

Triethylenemelamine (TEM) also induced a significant dose-related response in rats dosed at 0.05 mg/kg, 0.5 mg/kg, and 5.0 mg/kg (Table 1).

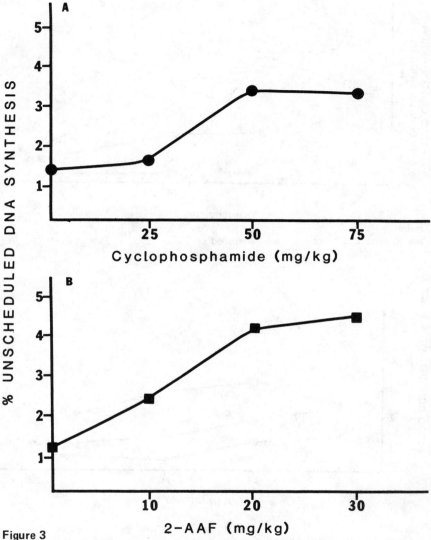

Figure 3
Detection of DNA damage caused by CY (A) and 2-AAF (B).

Methyl methanesulfonate (MMS), ethyl methanesulfonate (EMS), and N-methyl-N'-nitro-N-nitrosoguanidine (MNNG) were tested in a single-dose pilot study. EMS and MNNG induced UDS slightly, but not significantly, above control levels, but MMS gave a significantly higher level of UDS in the lymphocytes (Table 2).

Table 1

TEM: Number of Cells Classified as Undergoing UDS[a]

	Control[b]	0.05 mg/kg	0.5 mg/kg	5.0 mg/kg
$\overline{X}$	5.55	5.6	9.15	21.15
S.D.	1.2	1.4	5.5	3.0

[a]300 cells/animal
[b]distilled water

Table 2

Number of Cells Classified as Undergoing UDS[a]

	Control[b]	MNNG (250 mg/kg)	EMS (100 mg/kg)	MMS (100 mg/kg)
$\overline{X}$	8.0	9.2	9.6	15.5
S.D.	2.6	3.7	4.5	3.9

[a]300 cells/animal
[b]0.9% NaCl

DISCUSSION

The reliability of a qualitative UDS assay in rat lymphocytes depends on the elimination of variables in the testing protocol which impede generation of reproducible data. Bioavailability and absorption time of the test compound must be adequately characterized. Our results suggest that the CY and 2-AAF were metabolized before the blood was drawn for UDS detection. Since these two compounds must be biotransformed before activity can be detected, it was critical to permit these processes to take place in situ to maximize sensitivity.

Other factors which may adversely effect reproducibility are technical in nature. Autoradiography is a sensitive technique for detecting radioisotope incorporation into the cells; however, it also involves tedious work which must be carefully completed. Contaminated or partially exposed emulsions will compromise results by increasing background silver grain counts. Since the levels of UDS are extremely low in rat lymphocytes, it is necessary to minimize background count. Poor procedures in harvesting the cultures (i.e., incomplete washing of the cells) may also decrease the sensitivity by increasing background and masking UDS.

Hydroxyurea (HU), which inhibits scheduled DNA synthesis, is not used in the cultures. We have found that it is extremely difficult to stimulate the whole blood from rats to divide in culture; the percentage of S-phase cells is usually $\leq$ .3%. A level this low does not adversely effect the results. If studies of this type are used on human subjects exposed to chemicals in the workplace or to chemotherapeutic agents, HU may be necessary.

As was shown by the data on MMS, EMS, and MNNG (Table 2), rat lymphocytes are more sensitive to MMS damage than to EMS or MNNG. This is consistent with alkaline elution results published by Petzold and Swenberg (1978). The levels of UDS, although relative to each other (Petzold and Swenberg 1978), do not appear to be as high as might be expected. The low levels of detection may be a weakness of this particular system if repair enzymes are not rapidly induced in the lymphocytes. However, when TEM was given at increasing log doses, a significant dose-related response was seen (Skinner 1981). Therefore, it is necessary to give multiple doses over a wide range to verify genotoxic activity by eliciting a dose-response effect.

The detection and classification of genotoxic activity in vivo offers greater information about the relationship of different types of DNA damage. 2-AAF is not clastogenic in vivo but it does induce UDS, while CY is positive for both clastogenicity and UDS induction (Skinner et al. 1982). A positive response in a DNA perturbation assay such as UDS does not define a specific category of genetic damage but it does identify the necessity for evaluating potential hazard. The actual type of damage must be clearly defined in order for ultimate hazard to be assessed. In vivo testing provides a mechanism of definition by addressing probable target organ, biotransformation in situ and type of genotoxic activity.

ACKNOWLEDGMENTS

The authors would like to acknowledge the excellent technical assistance of S.E. Irwin and the editorial assistance of D.V. O'Leary.

REFERENCES

Petzold, G.L. and J.A. Swenberg. 1978. Detection of DNA damage induced *in vivo* following exposure of rats to carcinogens. *Cancer Res.* 38:1589.

Skinner, M.J. 1981. Detection of unscheduled DNA synthesis in rat lymphocytes treated with mutagens *in vivo*. *J. Environ. Pathol. Toxicol.* 5(3/4):785.

Skinner, M.J., C.A. Schreiner, and F.T. Davis. 1982. Unscheduled DNA synthesis detection and metaphase analysis in a common test system *in vivo-in vitro*. *J. Appl. Toxicol.* 2(3):172.

Stanulis, B.M., S. Sheldon, G.L. Grove, and V.J. Cristofalo. 1979. Scintillation fluid shortens exposure times in autoradiography. *J. Histochem. Cytochem.* 27(10):1303.

COMMENTS

KLIGERMAN: Do you see any way of making the system more sensitive? You use fairly potent mutagens.

SKINNER: No, I don't. That has been its major drawback. We have been a little disappointed about that. I think we are going to have to do some mechanistic studies to see what enzymes are induced—do some DNA elution, and get into the system. Just looking at autoradiography results doesn't tell us enough.

SEGA: Using the same dose of MMS as EMS, 100 mg/kg, with in vivo injections into mice, you get about four times as much alkylation with MMS in the DNA as you do with EMS. Therefore, your lower UDS response with EMS is probably just due to the fact that you have lower levels of alkylation in the DNA of the lymphocytes.

SKINNER: Right. We were hoping we would see that difference. When that study was in progress, I didn't know what slides I was reading. When I got the data, I was very pleased that the MMS was much more active.

MIRSALIS: Have you looked at any polycyclic aromatics?

SKINNER: No, not yet.

RAJEWSKY: The MMS is more sensitive in terms of expressing UDS in your system, but it doesn't mean that the EMS is less effective as a carcinogen or as a mutagen, because there is a completely different ratio of UDS-type repairable lesions created by MMS. MMS creates much less oxygen alkylation than EMS, but the oxygen alkylation is more dangerous, very likely, in terms of mutagenesis and carcinogenesis.

LUTZ: Generally, one should not use an UDS answer as equal to genotoxicity, because what repair of the damage is measured. One knows that the very persistent adducts are the most dangerous. I could imagine a compound which gives rise to a very persistent adduct which is perhaps not repaired at all, so you find no UDS, but it will be a very potent carcinogen.

SKINNER: That's correct.

BUTTERWORTH: What potent liver carcinogen is missed in Gary Williams' assay? What potent liver carcinogen is not detected either in vivo or in vitro in the liver assay?

LUTZ: Well, I could ask the question of the people who work—but I think the area where the UDS test fits is probably not for risk assessments, but more on the screening side. I would not make a quantitative statement out of the UDS measurement. Rather, I would say, this compound produces a damage that is repaired. But I think a quantification of the UDS in terms of carcinogenicity would be a very dangerous step.

SKINNER: That is why I said it is just an indicator and we have to look further. I won't try to use only UDS in risk assessment.

LUTZ: It was mostly the first presentation, where genotoxicity and UDS was almost taken synonymously.

MIRSALIS: But without genotoxicity, there would be no UDS. I agree that it is repair, and in some specific cases there may be adducts that are not repaired. We have not yet found a compound, at least in the liver, that fits that description. I am sure there are such compounds, but certainly compounds like MMS that are repaired very efficiently will show lots of UDS. I think you have to get away from plus-minus mentality and just try to interpret the results as you see them. In most cases you can come up with good explanations for them.

Unscheduled DNA Synthesis in Rat Stomach—Short-term Assay of Potential Stomach Carcinogens

CHIE FURIHATA AND TAIJIRO MATSUSHIMA
Department of Molecular Oncology
Institute of Medical Science
University of Tokyo
Tokyo 108, Japan

Stich et al. (1975) suggested that the organ specificity of a carcinogen could be determined by administration of the carcinogen in vivo and by measurement of unscheduled DNA synthesis (UDS) in vitro. They used autoradiography to measure UDS, but this method is time-consuming (Stich and Koropatnick 1977). Therefore, we tried to develop a rapid biochemical method for measuring UDS.

As stomach cancer is the most common type of cancer in Japan, we developed a short-term method for detecting potential stomach carcinogens. At least two kinds of potential stomach carcinogens are proposed: pyrolysis products of foods and food components and N-nitroso compounds. In Japan, broiling is a common cooking process and pyrolysis products of foods and food components were found to be mutagenic (Sugimura et al. 1977a,b; Sugimura 1982) and carcinogenic (Ishikawa et al. 1979; Hosaka et al. 1981; Matsukura et al. 1981). N-Nitroso compounds have been suggested to be formed from nitrogen containing compounds and nitrite in vivo (Sander et al. 1968). Four pyrolysis products and two nitroso compounds were examined by the present method.

MATERIALS AND METHODS

Materials

Dimethylnitrosamine (NDMA), N-methyl-N'-nitro-N-nitrosoguanidine (MNNG) and N-propyl-N'-nitro-N-nitrosoguanidine (PNNG) were purchased from Aldrich Chemical Co., 2-acetylaminofluorene (2-AAF) and ethyl methanesulfonate (EMS) were from Tokyo Kasei Kogyo Co. Ltd. and 4-nitroquinoline 1-oxide (4-NQO) was from Daiichi Kagaku Yakuhin Co. Ltd. Dinitrosocimetidine (DNCM) was kindly provided by Dr. H. F. Mower, University of Hawaii, Honolulu (Ichinotsubo et al. 1981); 2-hydroxy-3-nitroso-α-carboline (HNαC) was by Dr. M. Tsuda, National Cancer Center Research Institute, Tokyo (Tsuda et al. 1981); and methylnitrosourea (MNU) was by Dr. M. Nakadate, National

Institute of Hygienic Sciences, Tokyo. 3-Amino-1,4-dimethyl-5*H*-pyrido[4,3-*b*]indole (Trp-P-1), 3-amino-1-methyl-5*H*-pyrido[4,3-*b*]indole (Trp-P-2) (Sugimura et al. 1977a; Akimoto et al. 1977) and 2-amino-3-methylimidazo[4,5-*f*]quinoline (IQ) (Kasai et al. 1980, 1981) were obtained under the Research Resources Program for Cancer Research of the Ministry of Education, Science and Culture of Japan. The basic fraction from broiled sun-dried sardines was prepared by the method of Kasai et al. (Kasai et al. 1979). All these compounds except EMS and NDMA were made into solutions of dimethylsulfoxide (DMSO) and EMS and NDMA into solutions of distilled deionized water.

Methods

Male F344/Du Crj rats (Charles River Japan Inc.) of 7-8 weeks old were given test compounds as solutions in volumes of 0.5 ml by gastric tube. Control rats were given DMSO or distilled deionized water. The stomach was removed 2 hours later and opened along the greater curvature and washed with saline and L-15 medium containing 100 μg/ml streptomycin and 100 units/ml of penicillin G. The pyloric mucosa was separated from the serous membrane with scissors and cut into small pieces (1 $\times$ 1 $\times$ 1 mm) with a razor. Samples of 20 mg wet weight of tissue from each rat were cultured in 3 ml of L-15 medium containing 10 μCi/ml [^{3}H]thymidine ([methyl-^{3}H]thymidine 40-60 Ci/m mol or [methyl,1',2'-^{3}H]thymidine 70-100 Ci/m mol, the Radiochemical Centre), 10 mM hydroxyurea, 100 μg/ml of streptomycin and 100 units/ml of penicillin G at 37°C for 2 hours in an L-shaped glass tube with gentle shaking. With Trp-P-2, HNαC, and broiled fish extract, organ culture was performed for 3 hours, with IQ for 4 hours, with Trp-P-1 for 4.5 hours and with MNU for 5 hours. Hydroxyurea at 10 mM caused 93% inhibition of replicative DNA synthesis in the pyloric mucosa of rat stomach in organ culture (C. Furihata et al. in prep.). For examination of the cytotoxicity of carcinogens and mutagens, organ cultures in the absence of hydroxyurea were performed simultaneously and incorporation of [^{3}H]thymidine into DNA was determined. In the present study, doses of carcinogens and mutagens were used that did not inhibit incorporation of [^{3}H]thymidine into DNA more than 50%. For compounds that gave negative results, either the maximum dose for about 50% inhibition of DNA synthesis or 2/3 of the LD_{50} was used. After culture for 2 hours, tissues were further cultured at 37°C for 30 minutes by changing L-15 medium containing 1 mM unlabeled thymidine, 100 μg/ml streptomycin and 100 units/ml penicillin G, and then washed with phosphate-buffered saline and frozen on dry ice. The DNA fraction was extracted by a modification of the method of Schmidt, Thannhauser, and Schneider (Schmidt and Thannhauser 1945; Schneider 1957) as follows: The tissue was homogenized in 1 ml of cold 10% trichloroacetic acid (TCA) and the precipitate was washed with 10% TCA. The precipitate was homogenized in 1N KOH and incubated at 37°C overnight to hydrolyze RNA. Then DNA and protein were precipitated with 5% TCA; the precipitate was homogenized with 5% TCA and the DNA was hydrolyzed by incubation at 90°C for 10 minutes. The DNA in the supernatant was then determined with diphenylamine (Burton 1968). An aliquot of the supernatant was dissolved in ACS II and incorporation of [³H] thymidine into DNA was determined in a Beckman LS-355 liquid scintillation counter. Values were compared by Student's t-test, and levels of significance are expressed as p values.

RESULTS

Induction of UDS by Stomach Carcinogens

Figure 1 shows the induction of UDS in the pyloric mucosa of rat stomach by MNNG, which is a strong stomach carcinogen. Each value is for one rat and the horizontal line shows the mean value for each dose group. The effect of MNNG on UDS was dose-dependent with doses of 50-200 mg/kg body weight. At a dose of 200 mg/kg body weight, incorporation of [³H] thymidine into DNA was 7 times the control, as shown in Table 1. Preliminary studies showed that induction of UDS by MNNG was much higher in the pyloric mucosa than in the fundic mucosa of the stomach (data not shown). Moreover most chemical carcinogens induce tumors mainly in the pyloric mucosa of rat stomach (Sugimura and Kawachi 1973). Therefore, we examined only the pyloric mucosa in the present study. Table 1 also shows results on the effects of PNNG and 4-NQO. Both stomach carcinogens induced UDS. The induction was observed only with a high dose of PNNG, which is a weak stomach carcinogen (Sasajima et al. 1979), but with low doses of 4-NQO, although the incidence of gastric tumors with 4-NQO is very low (Takahashi and Sato 1969). Thus the carcinogenic potency of a compound may not be directly related with the extent of its induction of UDS.

UDS with Nongastric Carcinogens

Table 2 shows results on 2-AAF, EMS, MNU, and NDMA. Incorporation of [³H] thymidine into DNA was not increased at all by administration of 2-AAF, EMS, or NDMA. The results on NDMA are also shown in Figure 2; each value is for one rat and the horizontal line shows the mean value for each dose group. 2-AAF induces liver tumors (Weisburger & Weisburger 1958) and requires metabolic activation for conversion to its ultimate form. EMS is a direct-acting carcinogen and induces lung and kidney tumors when injected intraperitoneally (IARC 1974). MNU at a dose of 36 mg/kg body weight induced increase in the incorporation of [³H] thymidine into DNA in only 1 of 5 rats; at doses of 18 and 54 mg/kg body weight it induced no detectable increase. From these findings we conclude that MNU did not induce UDS in the pyloric mucosa, although it was reported to induce neurogenic tumors and squamous-cell

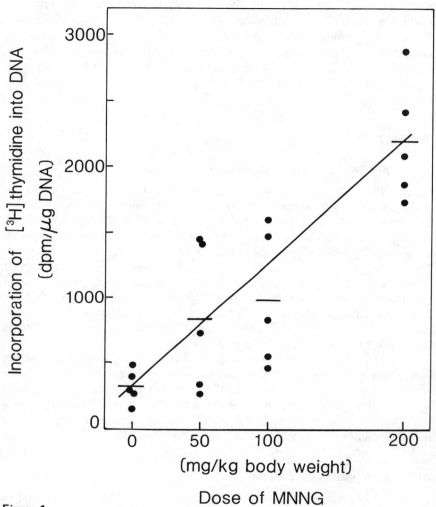

Figure 1
Induction of UDS in the pyloric mucosa of rat stomach by the administration of MNNG.

carcinomas of the forestomach (IARC 1978). NDMA induces liver and kidney tumors (IARC 1972) and requires metabolic activation to the ultimate carcinogen. No false positive results were obtained with these 4 chemicals.

Table 1
Induction of UDS by Stomach Carcinogens

Carcinogen	Dose (mg/kg body weight)	Number of rats	Incorporation of [^{3}H] thymidine into DNA (dpm/μg DNA)	Student's t-test
MNNG	0	5	318 ± 127[a]	
	50	5	834 ± 561	p > 0.05
	100	5	975 ± 519	p < 0.05
	200	5	2194 ± 460	p < 0.001
PNNG	0	4	340 ± 66	
	250	5	484 ± 144	p > 0.05
	500	5	466 ± 135	p > 0.2
	1000	5	856 ± 203	p < 0.01
4-NQO	0	5	93 ± 35	
	10	5	289 ± 101	p < 0.01
	20	5	352 ± 91	p < 0.01
	30	4	499 ± 114	p < 0.01

[a] mean ± standard deviation

Table 2
Incorporation of [^{3}H] Thymidine into DNA after Treatment with Nongastric Carcinogens

Carcinogen	Dose (mg/kg body weight)	Number of rats	Incorporation of [^{3}H] thymidine into DNA (dpm/μg)
2-AAF	0	5	147 ± 49^a
	400	5	141 ± 19
	500	5	174 ± 64
	600	5	173 ± 69
EMS	0	5	38 ± 10
	19	5	62 ± 17
	48	5	54 ± 25
	96	5	44 ± 11
MNU	0	5	75 ± 21
	18	5	74 ± 17
	36	5	117 ± 57
	54	5	82 ± 19
NDMA	0	5	138 ± 37
	11	5	169 ± 32
	16	5	117 ± 53
	21	5	110 ± 45

[a] mean ± standard deviation

UDS with Pyrolysis Products of Foods and Food Components

Table 3 shows results on IQ, Trp-P-1, Trp-P-2 and the basic fraction of broiled fish. IQ at doses of 77-230 mg/kg body weight did not increase the incorporation of [^{3}H] thymidine into DNA. IQ, isolated from the neutral fraction of an extract of broiled fish, is a potent mutagen (Kasai et al. 1980). Its carcinogenicity is now under study. Trp-P-1 at a dose of 169 mg/kg body weight induced increase in incorporation of [^{3}H] thymidine into DNA in 2 out of 5 rats, but at doses of 85 mg and 254 mg/kg body weight it did not cause any increase. Trp-P-2 at a dose of 44 mg/kg body weight also induced increase of [^{3}H] thymidine incorporation into DNA in 2 out of 5 rats, but at doses of 27 and 104 mg/kg body weight it had no effect. Trp-P-1 and Trp-P-2, isolated from a pyrolysate of tryptophan (Sugimura et al. 1977a), induced liver tumors but not stomach tumors by feeding (Hosaka et al. 1981; Matsukura et al. 1981). The basic fraction of an extract of broiled sardines at doses of 540 mg and 810 mg/kg body weight did not induce UDS. At a dose of 540 mg, the basic fraction was administered with NaCl at a dose of 250 mg/kg body weight. The basic fraction of an extract of broiled sardines is highly mutagenic (Kasai et al. 1979).

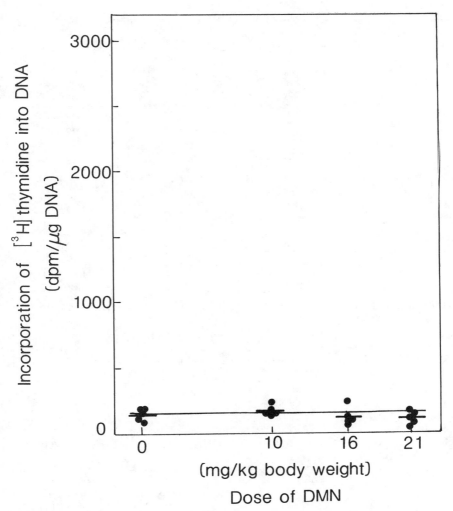

Figure 2
Incorporation of [³H] thymidine into DNA in the pyloric mucosa of rat stomach by the administration of NDMA.

UDS with Nitroso Compounds

Table 4 shows results on DNCM and HNαC. DNCM at doses of 500-2780 mg/kg body weight did not induce UDS. DNCM is mutagenic and is a nitroso derivative (Ichinotsubo et al. 1981) of cimetidine, which is used therapeutically for various disorders of the esophagus, stomach, and duodenum (Finkelstein 1978).

Table 3

Incorporation of [³H]Thymidine into DNA after Treatment with Pyrolysis Products of Food and Food Components

Pyrolysis product	Dose (mg/kg body weight)	Number of rats	Incorporation of [³H] thymidine into DNA (dpm/μg)
IQ	0	5	101 ± 12[a]
	77	5	72 ± 17
	154	5	98 ± 53
	230	5	68 ± 28
Trp-P-1	0	5	98 ± 39
	85	5	130 ± 34
	169	5	139 ± 97
	254	5	90 ± 49
Trp-P-2	0	5	70 ± 16
	27	5	65 ± 28
	44	5	147 ± 88
	104	5	38 ± 12
Broiled fish extract	0	5	49 ± 23
	540	5	48 ± 19
	0	5	23 ± 6
	810	5	27 ± 12

[a]mean ± standard deviation

Table 4

Incorporation of [³H]Thymidine into DNA after Treatment with Nitroso Compounds

Nitroso compound	Dose (mg/kg body weight)	Number of rats	Incorporation of [³H] thymidine into DNA (dpm/μg DNA)
DNCM	0	5	201 ± 127[a]
	500	4	107 ± 37
	1000	5	97 ± 22
	2780	5	228 ± 120
HNαC	0	5	108 ± 47
	50	5	86 ± 26
	150	5	119 ± 74
	300	5	131 ± 44

[a]mean ± standard deviation

of [^{3}H] thymidine into DNA in only 1 out of 5 rats, while at a dose of 50 mg or 300 mg/kg body weight it caused no increase. HNαC is a direct-acting mutagen and is a 2-hydroxy and 3-nitroso derivative (Tsuda et al. 1981) of 2-amino-α-carboline, which was isolated from a pyrolysate of soybean globulin (Yoshida et al. 1978).

DISCUSSION

In the present method, in vivo administration of a carcinogen was combined with in vitro determination of UDS. The organ-specific initiation activity of carcinogens was determined by a rapid method that took only 3 days. This method is useful for initial examination of the organ specificity of carcinogens.

Of the stomach carcinogens tested, MNNG and PNNG are direct-acting while 4-NQO requires metabolic activation to its ultimate form. Results showed that both types of carcinogen could be detected by the present method. Of the nongastric carcinogens tested, EMS and MNU are direct-acting carcinogens and 2-AAF and NDMA are procarcinogens. Neither type of carcinogen gave false positive results.

Of the pyrolysis products tested, IQ is an indirect-acting mutagen, while Trp-P-1 and Trp-P-2 are procarcinogens. All these chemicals gave negative results in the present test. The present results are consistent with the findings that Trp-P-1 and Trp-P-2 cause liver tumors. The basic fraction extracted from broiled sardines also gave negative results in the present study. Further studies are necessary on pyrolysis products of foods and food components.

The two nitroso compounds tested gave negative results in the present test. However, since Weisburger et al. (1980) suggested the formation of a stomach carcinogen from nitrite and a component of fish, further studies required on these compounds.

Many carcinogens are known to interact with DNA, causing its damage and repair. These effects of carcinogens are thought to be initiation actions. However, the precise relation between damage and repair of DNA by carcinogens and carcinogenesis is not yet clear. The interaction of carcinogens with DNA and damage of DNA by carcinogens and its repair were not found only in target organs; they were also found in some other organs (Rajewsky et al. 1976; Petzold and Swenberg 1978). These effects of carcinogens on DNA were prerequisites for carcinogenesis but alone they did not seem to be sufficient for carcinogenesis. As little is known about the relation between carcinogenesis and damage and repair of DNA, data on the relationship between UDS and target organs should be useful. The promoter effects of carcinogens on target organs must of course be tested by other methods.

ACKNOWLEDGMENTS

This work was supported in part by Grants-in Aid for Cancer Research from the Ministry of Education, Science and Culture and from the Ministry of Health and

Welfare of Japan, and by a grant from the Society for Promotion of Cancer Research. We thank Dr. H.F. Mower for giving dinitrosocimetidine, Dr. M. Tsuda for giving 2-hydroxy-3-nitroso-α-carboline and Dr. M. Nakadate for giving methylnitrosourea.

REFERENCES

Akimoto, H., A. Kawai, H. Nomura, M. Nagao, T. Kawachi, and T. Sugimura. 1977. Syntheses of potent mutagens in tryptophan pyrolysates. *Chem. Lett.* 1061.

Burton, K. 1968. Determination of DNA concentration with diphenylamine. *Methods Enzymol.* **XII-B**:163.

Finkelstein, W. 1978. Cimetidine. *New Engl. J. Med.* **299**:992.

Hosaka, S., T. Matsushima, I. Hirono, and T. Sugimura. 1981. Carcinogenic activity of 3-amino-1-methyl-5*H*-pyrido[4,3-*b*]indole (Trp-P-2), a pyrolysis product of tryptophan. *Cancer Lett.* **13**:23.

Ichinotsubo, D., E.A. Mackinnon, C. Liu, S. Rice, and H.F. Mower. 1981. Mutagenicity of nitrosated cimetidine. *Carcinogenesis* **2**:261.

International Agency for Research on Cancer. 1972. N-Nitrosodimethylamine. *IARC Monogr.* 1:95.

――――. 1974. Ethyl methanesulphonate. *IARC Monogr.* **7**:245.

――――. 1978. N-Nitroso-N-methylurea. *IARC Monogr.* **17**:227.

Ishikawa, T., S. Takayama, T. Kitagawa, T. Kawachi, M. Kinebuchi, N. Matsukura, E. Uchida, and T. Sugimura. 1979. In vivo experiments on tryptophan pyrolysis products. In *Naturally occurring carcinogens-mutagens and modulators of carcinogenesis* (eds. E.C. Miller et al.), p. 159. Japan Scientific Societies Press, Tokyo, Japan.

Kasai, H., S. Nishimura, M. Nagao, Y. Takahashi, and T. Sugimura. 1979. Fractionation of a mutagenic principle from broiled fish by high-pressure liquid chromatography. *Cancer Lett.* **7**:343.

Kasai, H., Z. Yamaizumi, S. Nishimura, K. Wakabayashi, M. Nagao, T. Sugimura, N.E. Spingarn, J.H. Weisburger, S. Yokoyama, and T. Miyazawa. 1981. A potent mutagen in broiled fish. Part 1. 2-Amino-3-methyl-3*H*-imidazo-[4,5-*f*]quinoline. *J. Chem. Soc. Perkin Trans.* I:2290.

Kasai, H., Z. Yamaizumi, K. Wakabayashi, M. Nagao, T. Sugimura, S. Yokoyama, T. Miyazawa, N.E. Spingarn, J.H. Weisburger, and S. Nishimura. 1980. Potent novel mutagens produced by broiling fish under normal conditions. *Proc. Jpn. Acad. Ser. B Phys. Biol. Sci.* **56**:278.

Matsukura, N., T. Kawachi, K. Morino, H. Ohgaki, T. Sugimura, and S. Takayama. 1981. Carcinogenicity in mice of mutagenic compounds from a tryptophan pyrolyzate. *Science* **213**:346.

Petzold, G.L. and J.A. Swenberg. 1978. Detection of DNA damage induced in vivo following exposure of rats to carcinogens. *Cancer Res.* **38**:1589.

Rajewsky, M.F., R. Goth, O.D. Laerum, H. Biessmann, and D.F. Hulser. 1976. Molecular and cellular mechanisms in nervous system-specific carcinogenesis by N-ethyl-N-nitrosourea. In *Fundamentals in cancer prevention* (ed. P.N. Magee et al.), p. 313. University of Tokyo Press, Tokyo, Japan.

Sander, J., F. Schweinsberg, and H.-P. Menz. 1968. Untersuchungen über die Entstehung cancerogener Nitrosamine im Magen. *Hoppe-Seyler's Z. Physiol. Chem.* **349**:1961.

Sasajima, K., T. Kawachi, N. Matsukura, T. Sano, and T. Sugimura. 1979. Intestinal metaplasia and adenocarcinoma induced in the stomach of rats by N-propyl-N'-nitro-N-nitrosoguanidine. *J. Cancer Res. Clin. Oncol.* **94**:201.

Schmidt, G., and S.J. Thannhauser. 1945. A method for the determination of desoxyribonucleic acid, ribonucleic acid, and phosphoproteins in animal tissues. *J. Biol. Chem.* **161**:83.

Schneider, W.C. 1957. Determination of nucleic acids in tissues by pentose analysis. *Methods Enzymol.* **III**:680.

Stich, H.F. and D.J. Koropatnick. 1977. The adaptation of short-term assays for carcinogens to the gastrointestinal system. In *Pathophysiology of carcinogenesis in digestive organs* (ed. E. Farber et al.), p. 121. University of Tokyo Press, Tokyo, Japan.

Stich, H.F., D. Kieser, B.A. Laishes, R.H.C. San, and P. Warren. 1975. DNA repair of human cells as a relevant, rapid, and economic assay for environmental carcinogens. *Gann Monogr.* **17**:3.

Sugimura, T. 1982. Tumor initiators and promoters associated with ordinary foods. In *Molecular interrelations of nutrition and cancer* (eds. M.S. Arnott et al.), p. 3. Raven Press, New York.

Sugimura, T. and T. Kawachi. 1973. Experimental stomach cancer. *Methods Cancer Res.* **7**:245.

Sugimura, T., T. Kawachi, M. Nagao, T. Yahagi, Y. Seino, T. Okamoto, K. Shudo, T. Kosuge, K. Tsuji, K. Wakabayashi, Y. Iitaka, and A. Itai. 1977a. Mutagenic principle(s) in tryptophan and phenylalanine pyrolysis products. *Proc. Jpn. Acad.* **53**:58.

Sugimura, T., M. Nagao, T. Kawachi, M. Honda, T. Yahagi, Y. Seino, S. Sato, N. Matsukura, T. Matsushima, A. Shirai, M. Sawamura, and H. Matsumoto. 1977b. Mutagen-carcinogens in food, with special reference to highly mutagenic pyrolytic products in broiled foods. *Cold Spring Harbor Conf. Cell Proliferation* **4**:1561.

Takahashi, M. and H. Sato. 1969. Effect of 4-nitroquinoline 1-oxide with alkylbenzenesulfonate on gastric carcinogenesis in rats. *Gann Monogr.* **8**:241.

Tsuda, M., M. Nagao, T. Hirayama, and T. Sugimura. 1981. Nitrite converts 2-amino-α-carboline, an indirect mutagen into 2-hydroxy-α-carboline, a non-mutagen, and 2-hydroxy-3-nitroso-α-carboline, a direct mutagen. *Mutat. Res.* **83**:61.

Weisburger, J.H., H. Marquardt, N. Hirota, H. Mori, and G.M. Williams. 1980. Induction of cancer of the glandular stomach in rats by an extract of nitrite-treated fish. *J. Natl. Cancer Inst.* **64**:163.

Weisburger, E.K. and J.H. Weisburger. 1958. Chemistry, carcinogenicity, and metabolism of 2-fluorenamine and related compounds. *Adv. Cancer Res.* **5**:331.

Yoshida, D., T. Matsumoto, R. Yoshimura, and T. Matsuzaki. 1978. Mutagenicity of amino-α-carbolines in pyrolysis products of soybean globulin. *Biochem. Biophys. Res. Commun.* **83**:915.

COMMENTS

PARODI: I have the feeling that it would be interesting to measure the number of adducts of MNU with stomach mucosa. If your observations are correct, we should find that MNU is not making adducts with stomach mucosa, which could be, perhaps, debated from the fact that there is no absorption because of the pH. These kinds of checks could be useful to see if in the negative cases there are really no adducts.

FURIHATA: Oral administration of MNU induces squamous cell carcinomas in the forestomach. This effect suggests that the absorption of MNU occurs at least in the forestomach. As a next step, it will be interesting to study the reason why MNU does not have this effect on the glandular stomach.

SWENBERG: In fact, when you do MNU (Ogiu et al. 1977) carcinogenesis, you do get some stomach tumors. They are much fewer than the squamous cell carcinomas of the forestomach, but there are a few gastric, and there are some sarcomas that also develop in deeper muscle. I don't think that is likely to be the case. The pH is going to be similar in the pyloric region and in the forestomach region because of the mixing that is going on there. I think it is much more likely that you have a lot less cell turnover in the pyloric.

FURIHATA: In the study on MNU, the incidence of tumors in the glandular stomach was very low. I think the incidence was not statistically significant. Therefore, I don't think that MNU induces tumors in the glandular mucosa of the stomach.

LUTZ: I think, if I remember the data correctly, the highest dose that you used with MNU was 54 mg/kg. That was a dose which was also negative with MNNG, because with MNNG you needed many hundreds of milligrams per kilogram. So it might still be that MNU dosage was too low, so that you did not get a positive answer, but it will definitely methylate the DNA in the stomach.

FURIHATA: Sometimes MNNG induced positive results at a dose of 50 mg/kg body weight. In our work MNNG induced positive results at doses of 100 and 200 mg/kg body weight.

LUTZ: Yes, but the dose of MNNG that you used to get the positive result was about ten times higher than the highest dose you used for MNU.

FURIHATA: In my preliminary study on MNU in the colon and bladder, MNU induced unscheduled DNA synthesis at a dose of 10 mg/kg body weight in the colon and at a dose of 5 mg/kg body weight in the bladder. There-fore, I think a dose of 54 mg/kg body weight was not too low for the stomach.

EVANS: Dr. Furihata, did I miss something? In your pyrolysis product IQ—

FURIHATA: IQ was negative.

EVANS: Not only negative. I thought, in fact, with an increasing concentra-tion of IQ, you get a decreasing count.

FURIHATA: At doses of 77 and 230 mg/kg body weight the mean count de-creased, but at a dose of 154 mg/kg body weight, the count was about the same as that of the control. The decrease was not statistically significant.

EVANS: I thought I saw dose decline continuously. Did this imply that you are measuring not merely incorporation in terms of repair, but your results are influenced considerably by proliferation rate in the stomach?

FURIHATA: In the pyloric mucosa cells are replaced in 10 days. Replication of DNA occurs in a significant number of cells at the proliferating zone at any time. But I used 10 mM hydroxyurea to inhibit more than 90% of the replicative DNA synthesis. I don't think that the results are influenced by the proliferation rate in the stomach.

EVANS: Well, yes, all right. In that case you wouldn't expect to get a de-creasing effect of the increasing IQ.

References

Ogiu, T., M. Nakadate, K. Furuta, A. Maekawa, and S. Odashima. 1977. Induc-tion of tumors of peripheral nervous system in female Donryu rats by continuous oral administration of 1-methyl-nitrosourea. *Gann* **68**:491.

Alkaline Elution In Vivo: Fluorometric Analysis in Rats
Quantitative Predictivity of Carcinogenicity,
as Compared with Other Short-term Tests

SILVIO PARODI, MAURIZIO TANINGHER, AND LEONARDO SANTI
Istituto Scientifico per lo Studio e la Cura dei Tumori
and Department of Oncology
University of Genoa
Viale Benedetto XV, 10—I-16132 Genoa, Italy

In their ultimate reactive form, many chemical carcinogens cause DNA damage and induce DNA repair synthesis. Consequently, DNA damage and (or) repair have been considered potentially useful for the detection of environmental carcinogens (Cox et al. 1973; Sarma et al. 1975; Stich et al. 1976). Alkaline elution, first performed by Kohn et al. (1976) on cultured cells, is a method which measures DNA damage as breaks or weak points in alkali, in single-stranded DNA. The velocity of elution of DNA through a Millipore filter, is inversely proportional to the single-strand length; furthermore, the rate of DNA eluted per time unit can be considered a direct complex function of the rate of DNA damage. The method has been subsequently adapted in vivo and proposed as a short-term test by different authors, including ourselves (Parodi et al. 1978; Petzold and Swenberg 1978; Schwarz et al. 1979). Several substances belonging to different chemical classes have been tested with it. Comparing the degree of positivity in the in vivo alkaline elution and the carcinogenicity in long-term assays for 57 chemicals, we have tried to establish the quantitative predictivity of the short-term test. Moreover, we compared the quantitative predictivity of the alkaline elution with that of two tests measuring two parameters that can be considered related to DNA damage: the mutagenesis Ames' test and the test of morphological transformation performed in vitro on Syrian hamster embryo cells (Pienta 1980). In addition, as a reference point for DNA damage we made correlation studies with the data on in vivo DNA adduct formation, as reported by Lutz (1979). Finally, because it could be suspected that the covalent binding of chemicals to macromolecules could be related, in some cases, with toxicity, we also considered the acute toxicity in a global work on quantitative correlations with the carcinogenic potency.

ALKALINE ELUTION: FLUOROMETRIC ANALYSIS IN RATS

The alkaline elution assay, as performed by our group, is essentially according to Kohn et al. (1976). There are, however, four major differences. Our technique is much faster because we elute at an elution rate of about 0.2 ml/min.; Kohn elutes at an elution rate of 0.03-0.06 ml/min. We work at pH 12.3; Kohn works at pH 12.1. We used mixed esters of cellulose filters, 5 μm in pore size; Kohn uses polyvinylchloride filters, 2 μm in pore size, and elutes against a standard with a fixed X-ray dosage. We elute without such an internal standard. The final effect of these four differences is that our method, while more rapid, is about five times less sensitive than the Kohn method, as it appears comparing elution rates of X-rays damaged DNA in ours (Brambilla et al. 1979) and Kohn conditions (Kohn et al. 1976).

The microfluorometric determination of the liver DNA in the eluted fractions is carried out as follows, according to the method of Kissane and Robins (1958). Since the alkaline elution has the advantage that almost all the non-DNA material of the cells can be removed before the analysis with alkali is begun, the steps for the recovery of the DNA from the eluted fractions are drastically simplified as follows, producing a net gain in time and a greater percentage of DNA recovery, which is about 85%. To each 1.0 ml aliquot of the collected fractions, 100 μl of bovine serum albumin (2 mg/ml) followed by 100 μl of 100% (wt./vol.) trichloroacetic acid are added. In order to suspend the DNA remaining on the filter, at the end of the elution the filter membrane is disrupted in 2.1 ml of eluant (the volume of the eluted fractions). In order to recover all the DNA from the elution apparatus, one more fraction is collected, after having removed the filter. The samples are then refrigerated for several hours and the precipitated DNA is pelleted by centrifuging at 1750 x g for 15 minutes at 4°C. Supernatants are carefully decanted, and 1.0 ml of 80% ethanol is layered on each pellet. After a second centrifugation the pellets are air-dried at 40°C, for 3 hours. Thirty μl of a 40% (wt./vol.) aqueous solution of 3,5-diaminobenzoic acid dihydrochloride are added to each sample and mixed. The tubes are then incubated for 30 minutes at 70°C. After the mixture has been cooled, 1.5 ml of 0.6 N perchloric acid are added to each tube. The fluorescence is read at 520 nm with an excitation wavelength of 420 nm (Brambilla et al. 1977; Parodi et al. 1978).

A similar method has been set up by Erickson et al. (1980). The fluorescent reaction with DNA was always according to Kissane and Robins (1958), however the precipitated DNA was trapped on 0.2 μm pore size Durapore filters. A different approach was followed by Cesarone et al. (1979). The eluted DNA, buffered to pH 7.0, was reacted with a fixed excess amount of Hoechst 33258, a dye binding specifically, but reversibly, with DNA. The DNA dosage was dependent on the increase in fluorescence quantum yield of the dye bound to DNA. The increase in quantum yield for bound dye was of about 50 times. The method appeared to be sensitive and reproducible. Schwarz et al.

(1979) were able to obtain enough sensitivity for their fractions using the diphenylamine method of Burton. In our laboratory we have experience with the methods of Parodi et al. (1978), Cesarone et al. (1979), and Erickson et al. (1980). All three methods can be considered suitable for the fluorometric dosage of the DNA.

CRITERIA ADOPTED FOR EVALUATING THE QUANTITATIVE PREDICTIVITY OF DIFFERENT SHORT TERM TESTS

Oncogenic potency was expressed with an Oncogenic Potency Index (OPI). Potency in inducing DNA fragmentation was expressed with a DNA Fragmentation Index (DFI). Mutagenic potency in the Ames' test was expressed with a Mutagenic Potency Index (MPI). Potency in inducing DNA adducts was expressed with a Covalent Binding Index (CBI). Potency in inducing morphological transformation in vitro in hamster embryo cells was expressed with a Transforming Potency Index (TPI). Finally, acute toxic potency was expressed with a Lethal Dose Index (LDI).

For lack of space we send the reader to previous publications (Parodi et al. 1981b,c; 1982a,b,c) for that which concerns the complex criteria of normalization adopted for OPI data. Special criteria of normalization were adopted also for DFI, CBI, MPI, TPI, and LDI data, always quoted in the above publications. For reasons of brevity, here we give only the formula for the calculation of the different potencies:

$$OPI = -\ln (1-I) / D\, t^n \tag{1}$$

according to Meselson and Russell (1977), where I is the cumulative single-risk incidence of tumors; t = the time of exposure (time unit = 2 years); D the dose (mmol/kg/day) equivalent to the total dose divided for a 2-year exposure; n was empirically set equal to 1, because no improvement of predictivity could be observed with higher exponents.

$$DFI = \frac{K_t - K_c}{\text{dosage in mmol/kg}} \times 1000 \tag{2}$$

where K_t is the elution rate constant per unit volume for treated samples and K_c the same for control samples.

$$CBI = \frac{\mu\text{mol chemical bound per mole nucleotides}}{\text{mmol/chemical administered/kg}} \tag{3}$$

$$MPI = \frac{\text{histidine revertants over controls per plate}}{\text{nmol of chemical per plate}} \tag{4}$$

$$TPI = \frac{(\mu g/ml \text{ of the lowest transforming dose})^{-1}}{\text{molecular weight}} \tag{5}$$

$$LDI = (LD_{50} \text{ in moles/kg})^{-1} \tag{6}$$

It is important to underline here that the parameters DFI and CBI in our investigation were both evaluated in vivo and in the liver cells. For that which concerns OPI we will only mention very briefly the criteria adopted for accepting positive and negative results:

Requirements for Accepting Positive Results

    a. only experiments lasting longer than 5 months were considered
    b. the dose giving the highest potency was selected
    c. the most responsive strain or species were selected
    d. at least 5 tumor-bearing animals in excess of controls
    e. $p < 0.05$ according to the χ^2 test.

Requirements for Accepting Negative Results

    a. only experiments lasting longer than 12 months were considered
    b. the dose given had important chronic toxic effects
    c. no other positive experiments existed at higher dosages in the same study
    d. at the end of the experiment more than 20 treated animals and 20 controls survived, (with substantially the same tumor frequency).

RESULTS AND DISCUSSION

Internal Consistence of Carcinogenicity Data

Normalization of potencies of carcinogenicity data is a serious problem, as it appears from the already quoted papers; the calculation of these potencies requires compromise solutions. One of the major complications is the following: Calculating the potencies from the animals with at least one tumor gives the same weight to all the different types of tumors. This is a serious compromise. However, it is absolutely impossible to find enough data for a single type of tumor, except for special classes of compounds (for instance: hydrazine derivatives and lung adenomas). Given all these problems, in order to test if the carcinogenicity data have some degree of internal consistence, we made the following type of comparison. For 56 compounds more than one value was available. We randomly generated two subsets of 56 single data. We repeatedly examined the correlation existing between these two subsets. We found an average correlation coefficient r equal to 0.77, p (that r = 0) < 0.001. For one typical comparison, the equation of the regression line between (Log OPI_1) and (Log OPI_2) was: $y = 0.31 + 0.76x$. Log OPI spaced between -1.5 and 5.0. In the central part of the regression line (Log OPI = 2.0) the belt zone including 90% of the values spanned in the following range: 2.0 ± 1.96. On an arithmetical scale, this is equivalent to a range 1/90-90 $\times$, for an interval of potencies greater than a millionfold. The above correlation coefficient and the above amplitude of the belt zone can be considered as the upper limit of correlation possible between carcinogenicity and any short-term test, at least for our approach to normalization of carcinogenic potencies. The reason is that no short-term test can be considered closer to carcinogenicity than carcinogenicity itself. The above result gives an idea of the degree of "noise" that affects our investigation (see Fig. 1).

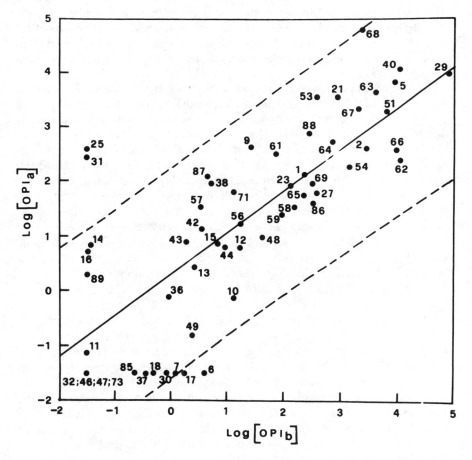

Figure 1
Internal correlation between couples of oncogenic potency data, for 56 compounds. The numbers refer to the compounds listed in Table 1. (————) regression line; (– – – –) 90% belt zone; $r = 0.77$; p (that $r = 0$) < 0.001; equation of the regression line: Log $(OPI_a) = 0.31 + 0.76$ (Log (OPI_b)).

Correlation Studies Amongst OPI, DFI, CBI, MPI, TPI, and LDI Data

In Table 1 the Logs of the averages of the different potency values for a series of 91 compounds are listed. For all 91 compounds we had OPI values. We had DFI values for 57 compounds, CBI values for 37 compounds, MPI values for 88 compounds, TPI values for 59 compounds, and LDI values for 76 compounds. While OPI, DFI, CBI, MPI, and LDI values were obtained from different literature sources, TPI values were obtained from the homogeneous data of Pienta (1980).

In Table 2 the correlations of the different short-term parameters with OPI are given. As it can be seen, the levels of correlation are rather similar for all the short-term parameters examined, and always around an $r = 0.4$. Correlations with OPI values were compared also for a more restricted number of compounds: the compounds tested at the same time for both short-term test "X" and "Y". Again, we obtained rather similar correlation values, reported in Table 3. There is perhaps a slight trend. CBI and TPI seem slightly better than DFI ($p < 0.15$).

In our opinion there are two important points that require a comment. First of all, LDI is better correlated than would be expected. Cytotoxic and mutagenic (carcinogenic) effects are generally regarded as the consequences of different lesions in macromolecules and as independent events (Suter et al. 1980). However, on the empirical statistical basis of our analysis, the significance of the correlation $r = 0.43$ ($p < 0.001$) is too high to be considered fortuitous. A possible explanation could be the following: Almost all our chemicals were capable of making adducts with macromolecules (albeit at very different levels), and their toxicity was mainly due to this fact. At the same time almost none of them probably had a significant toxicity related to other types of specific functional biological actions. In this situation, the capability of forming adducts with macromolecules will be correlated with the number of adducts in DNA, and this in turn could be correlated with carcinogenicity. It has been shown here that the amount of DNA adducts in vivo is significantly correlated with carcinogenicity; after the previous considerations, the above correlation of LDI with OPI should appear justified. The second point that requires a comment is in our opinion the fact that TPI, which should reflect an event closer to carcinogenesis than the other short-term parameters, is no better correlated than the other tests. While the closer correlation with carcinogenicity remains formally true, our data do not confirm this hypothesis.

Perhaps the explanation of this fact could be due to the following point: The transformation test utilizes (usually) the same activation system as the Ames test. It seems evident that the lack of predictivity related to problems of metabolic activation is dominant over the problem of the quality of the final end point of the short-term test considered.

In Table 4 we have examined the predictivity of some short-term tests for special classes of chemical compounds. This subdivision generates subsets of 10-15 compounds. A lot of caution must be exerted before accepting that 10-15 compounds can be considered representative of a chemical class. However, even

Table 1

Different Potencies for the Compounds Examined

Chemicals	Log (OPI)	Log (LDI)	Log (DFI)	Log (CBI)	Log (MPI)	Log (TPI)
Aromatic amines, etc.						
1) 2-Acetylaminofluorene	3.01[e,1]	2.29[b,4,5]	1.71[c,5-7]	2.03[e]	1.68[c,5,9,10]	1.65
2) N-Hydroxy-2-acetylaminofluorene	3.05[c,1]	3.66[a,4]	1.98[a,6]	2.59[e]	1.68[a,9]	2.38
3) 4-Acetylaminofluorene	−1.50[a,1]	3.03[a,4]	—	—	−0.331[c,9-11]	−1.50
4) Acridine orange	−1.50[a,3]		—	—	1.82[a,9]	5.42
5) 2-Aminoanthracene	3.92[b,12,13]		—	—	2.71[a,9]	3.29
6) 2-Naphthylamine	0.340[d,3,14]	2.66[b,4]	1.76[c,5,15,16]	0.208[b,8]	1.16[c,5,9,10]	3.16
7) 1-Naphthylamine	−0.240[b,3]	2.32[b,4,5]	1.49[b,5,15]	—	−0.146[d,5,9-11]	−1.50
8) 4-Aminobiphenyl	1.51[a,3]	2.53[a,4]	—	—	1.28[b,9,10]	1.53
9) 2′,3-Dimethyl-4-aminobiphenyl	2.35[b,1,2]		—	—	1.88[a,9]	3.37
10) Benzidine	3.07[d,3,12]	2.78[c,4,5]	1.95[d,5,6,15,16]	2.26[a,8]	−0.00612[c,5,9,17]	3.57
11) Aniline	−1.42[b,3,18]	2.31[d,4]	1.04[b,5,16]	0.568[a]	−4.00[c,5,9,11]	−1.50
12) 2,4-Diaminotoluene	1.06[b,3,19]	2.92[a,5]	1.11[a,5]	—	−0.598[b,5,9]	2.39
13) Auramine	0.246[c,3,20]	3.18[b,4,5]	2.75[a,5]	—	−2.52[a,5]	2.43
14) 2,4-Diaminoanisole	0.536[b,21,22]	3.18[a,4,5]	1.73[a,5]	—	1.56[a,5]	—
15) 4,4′-Oxydianiline	0.897[c,3,23,24]	2.65[d,4]	1.21[a,5]	—	0.111[a,5]	—
16) 4,4′-Methylenedianiline	0.442[b,3]	3.15[c,4]	2.15[a,5]	—	0.276[a,5]	—
17) Rhodamine B	−0.0730[b,f,3,25]	3.73[a,5]	0.50[a,5]	—	−1.30[b,5,10]	—
Alkyl halides, etc.						
18) Carbon tetrachloride	−0.777[e,1-3]	1.72[d,4]	0.50[a,7]	1.41[b,8]	−4.00[b,9,11]	—
19) Chloroform	0.00[a,3]	2.09[d,4]	0.50[a,6]	0.863[a,8]	—	—
20) Ethylene dibromide	1.57[a,3]	3.11[b,4]	—	2.26[a]	−1.22[a,9]	—
21) Nitrogen mustard	3.60[e,3]	4.97[c,4]	—	1.92[a]	0.114[a,9]	—
22) Uracil mustard	5.80[a,3]	4.71[b,4]	—	—	−0.398[a,9]	3.40
23) Cyclophosphamide	2.66[e,1,3,26]	3.47[e,4]	0.50[a,6]	1.79[a]	0.114[b,9,17]	0.747
24) Dimethylcarbamyl chloride	−1.50[a,3]	2.37[b,4]	—	—	−1.40[a,9]	3.03

143

Table 1 (*Continued*)

Chemicals	Log (OPI)	Log (LDI)	Log (DFI)	Log (CBI)	Log (MPI)	Log (TPI)
Polycyclic aromatics						
25) Benzo[a] pyrene	2.96[e,h,1,3,27]	3.70[a,4]	0.50[a,16]	1.18[e]	2.06[d,9-11,17]	2.40
26) Dibenz(a,h)anthracene	2.32[a,3]	—	—	0.447[b]	1.04[a,9]	2.75
27) 3-Methylcholanthrene	2.33[b,1]	—	—	—	1.76[a,9]	3.43
28) Benz[a] anthracene	-1.50[a,3]	—	—	—	1.06[b,9,11]	3.36
29) 7,12-Dimethylbenz[a] anthracene	4.36[e,1]	3.32[c,4]	0.50[b,6,16]	1.35[d]	1.18[b,9,10]	4.17
Esters, epoxides, carbamates, etc.						
30) Methyl methanesulfonate	-0.559[c,3]	2.88[b,4]	1.87[c,6,7,16]	2.60[c]	-0.030[c,9,11,17]	—
31) Ethyl methanesulfonate	2.50[e,1-3]	—	1.17[a,6]	1.79[a]	-0.810[b,9,17]	—
32) Myleran	-1.50[b,3]	3.87[b,4]	—	1.32[a]	-0.638[a,17]	—
33) β-Propiolactone	1.32[a,i,3]	2.32[a,4]	0.933[a,6]	—	0.620[c,9-11]	2.09
34) 1,3-Propane sultone	1.43[a,3]	2.96[a,4]	—	—	0.835[b,9,10]	3.24
35) 1,2,3,4-Diepoxybutane	3.15[a,3]	3.06[b,4]	—	—	-0.921[a,9]	2.88
36) Thiourea	-0.247[e,3]	2.42[c,4]	—	—	-4.00[a,9]	2.88
37) Ethyl carbamate	3.06[e,1,2]	1.69[d,4]	0.50[b,6,16]	1.61[e]	-4.00[a,9]	-1.45
38) Thioacetamide	1.70[b,j,3]	—	—	—	-4.00[b,9,11]	2.88
39) Acetamide	-1.26[a,3]	0.790[d,4]	—	—	-4.00[a,9]	1.77
Miscellaneous heterocycles						
40) 4-Nitroquinoline-1-oxide	4.18[c,28-30]	—	0.960[a,6]	—	3.38[c,9,10,17]	5.28
41) 2-(2-Furyl)-3-(5-nitro-2-furyl)-acrylamide	-1.50[a,1]	2.53[b,4]	—	—	4.20[b,9,10]	3.00
42) N-[4-(5-Nitro-2-furyl)-thiazolyl]-formamide	1.86[e,1,3]	—	—	—	4.22[a,9]	3.38
43) Phenobarbital	0.509[d,3]	3.08[e,4]	0.50[a,7]	—	-4.00[a,9]	-1.50
44) 3-Amino-1,2,4-triazole	0.722[c,3]	2.22[c,4]	—	—	-4.00[a,9]	1.92
45) Caffeine	-1.50[a,g,2]	3.13[e,4]	—	—	-4.00[a,9]	-1.50

Compound						
46) Saccharin	-1.50[d,3]				-4.00[a,11]	-1.50
47) Aminopyrine	-1.50[b,1,31]	2.91[e,4]	0.50[a,32]		-4.00[a,11]	–
Miscellaneous aliphatics and aromatics						
48) Ethionine	1.40[b,1,33]	1.88[b,4]		-0.347[b]	-4.00[a,9]	3.21
49) Safrole	0.0973[b,3]				-4.00[a,9]	5.21
50) 1-Naphthylisothiocyanate	-1.50[a,h,1]	2.92[b,4]			-4.00[a,9]	-1.50
51) Diethylstilbestrol	3.52[c,3]	3.77[b,4]		-0.301[b]	-4.00[a,9]	3.43
52) Cholesterol	0.143[a,3]				-4.00[a,11]	-1.50
Nitrosamines etc.						
53) Dimethylnitrosamine	3.13[e,1]	3.34[c,4]	2.63[d,6,7,32,34]	3.47[e]	-1.74[d,9-11,17]	2.87
54) Diethylnitrosamine	2.97[e,1,2]	2.67[a,4]	2.14[c,6,7,34]	2.14[d]	-2.00[a,9]	-0.787
55) Di-n-propylnitrosamine	2.47[a,3]	2.36[b,4]	1.80[a,34]		-1.10[a,9]	–
56) Di-n-butylnitrosamine	1.10[c,3]	2.12[b,4]	1.32[a,34]		-0.632[b,9,10]	–
57) N-Nitrosopyrrolidine	1.26[b,1,3]	2.05[a,4]	1.35[a,34]	2.25[a]	-1.70[a,9]	–
58) N-Nitrosomorpholine	2.12[d,1-3]	2.83[d,4]	1.61[a,7]	1.64[a]	-1.22[a,9]	–
59) N-Nitrosopiperidine	1.79[b,1]	3.08[c,4]		2.07[a]	-2.00[a,9]	3.06
60) Diphenylnitrosamine	-1.50[a,35]	1.93[b,4]			-4.00[a,9]	1.50
61) N-Methyl-N'-nitro-N-nitroso-guanidine	2.23[c,1-3]	2.56[b,4]	1.15[a,6]	3.00[a]	3.09[b,9,17]	3.47
62) Nitrosohexamethyleneimine	3.73[b,1,36]	2.58[a,4]	0.50[a,6]	2.06[b]		–
63) N-Nitrosomethylurea	3.43[d,1-3]	2.90[d,4]	1.80[c,6,34,37]	2.71[d]	0.481[b,9,10]	–
64) N-Nitrosoethylurea	2.73[c,1,2]	2.66[c,4]	1.27[a,6]	2.29[b,8]	0.0414[a,9]	2.37
65) Cycasin	2.16[b,1,3]	2.86[b,4]	2.47[a,38]		-4.00[a,9]	–
66) Methylazoxymethanol acetate	3.73[b,1,39]	4.12[a,4]	2.60[a,38]	3.64[a]	-1.68[a,10]	3.42
67) Streptozotocin	3.32[b,3]	3.16[b,4]	2.10[a,6]		3.29[a,9]	–
Fungal toxins						
68) Aflatoxin B$_1$	4.41[e,1-3]	4.61[d,4]	4.17[a,6]	4.09[e]	3.74[c,9,11,17]	3.49
69) Aflatoxin B$_2$	2.35[a,k,3]			2.75[a]	0.322[a,9]	2.80
70) Aflatoxin G$_1$	3.95[a,3]	6.91[a,4]		2.83[a]	2.06[a,9]	–

145

Table 1 (*Continued*)
Different Potencies for the Compounds Examined

Chemicals	Log (OPI)	Log (LDI)	Log (DFI)	Log (CBI)	Log (MPI)	Log (TPI)
Metal salts						
71) Cadmium chloride	1.81[c,3]	3.88[c,4]	—	—	-4.00[a,11]	3.79
72) Calcium chromate	-1.50[a,f,3]	2.77[a,4]	—	—	0.505[a,11]	4.49
73) Sodium nitrite	-1.50[e,1,1,40]	2.81[d,4]	0.50[a,32]	—	-2.29[c,9-11]	-1.50
Hydrazine derivatives						
74) Hydrazine	0.455[a,41,42]	2.50[c,4,42]	0.806[a,42]	—	-2.36[a,42]	2.11
75) 1,1-Dimethylhydrazine	0.680[a,41,42]	3.31[c,4]	0.50[a,42]		-3.06[a,42]	—
76) 1,2-Dimethylhydrazine	1.65[a,41,42]	2.88[c,4,42]	2.41[c,6,42,43]	3.40[c,8]	-2.35[b,9,42]	2.78
77) Carbamylhydrazine	-0.588[a,41,42]	2.74[e,4]	0.50[a,42]	—	-3.75[a,42]	—
78) Phenylhydrazine	0.582[a,41,42]	2.78[b,4,42]	1.54[a,42]	—	-1.79[a,42]	—
79) 1-Carbamyl-2-phenylhydrazine	-0.642[a,41,42]	3.24[b,4,42]	0.50[a,42]	—	-4.00[a,42]	—
80) Phenelzin	-0.857[a,41,42]	3.17[e,4]	0.50[a,42]	—	-1.08[a,42]	—
81) Procarbazine	3.00[a,41,42]	2.51[b,4,42]	1.61[a,42]	—	-3.59[b,9,42]	0.845
82) Isoniazid	-0.237[a,41,42]	2.82[e,4]	0.50[a,42]	—	-3.83[b,10,42]	—
83) Iproniazid	0.228[a,41,42]	2.51[e,4]	0.50[a,42]	—	-4.00[a,42]	—
84) Hydralazine	-0.928[a,41,42]	3.37[a,4]	0.50[a,42]	—	-1.30[a,42]	—
Azo dyes and diazo compounds						
85) 4-Aminoazobenzene	-0.955[b,3]	2.84[a,5]	1.51[a,5]	0.380[a]	-0.432[c,5,9,10]	1.30

86) o-Aminoazotoluene	2.14[c,1,3]	2.12[a,5]	0.50[a,5]	2.16[b]	1.03[c,5,9,10]	2.35
87) N,N-Dimethyl-4-aminoazobenzene	1.61[e,m,1-3]	2.92[d,4]	0.919[b,5,7]	0.831[e]	-0.986[c,5,9,11]	1.65
88) 3'-Methyl-4-dimethylamino-azobenzene	2.72[b,h,1]	—	0.50[a,5]	1.82[a]	0.190[c,5,9,10]	1.38
89) Ponceau MX	-0.168[d,3]	2.56[b,4,5]	0.50[a,5]		-1.16[a,5]	—
90) Azaserine	2.97[a,3]	3.15[d,4]	2.16[a,6]		4.08[a,9]	2.24
91) Azoxymethane	3.10[a,2]	3.44[a,4]	2.16[a,6]		—	—

[a] Log of 1 value;
[b,c,d,e] Log of the mean of 2, 3, 4, 5, values, respectively. Conditions slightly different from those established for evaluating oncogenic potency indexes;
[f] in one of the negative experiments there were 0/18 tumor-bearing animals;
[g] for one of the negative experiments the duration was of 242 days;
[h,j,m] for one of the positive experiments the duration was of 109, 120, and 140 days respectively;
[i,k,n] in one of the positive experiments 3/5 and 3/9 tumor-bearing animals respectively;
[l] for 3 of the negative experiments the duration was of 210-280 days.

References: [1,2]U.S. Public Health Service (1972-73) and (1978); [3]IARC (1972-81); [4]NIOSH (1978); [5]Parodi et al. (1981b); [6]Petzold and Swenberg (1978); [7]Schwarz et al. (1979); [8]the CBI values for Benzidine and Chloroform and additional values for 2-Naphthylamine, Carbon tetrachloride, Ethyl-nitrosourea and 1,2-Dimethylhydrazine were kindly given us directly from Dr. W.K. Lutz; all other values were obtained from Lutz (1979; [9]McCann et al. (1975); [10]Kawachi et al. (1980); [11]De Flora (1981); [12]Griswold et al. (1968); [13]Griswold et al. (1966); [14]Bonser et al. (1956); [15]Bolognesi et al. (1981); [16]Parodi et al. (1982b); [17]Painter and Howard (1978); [18]NCI (1978); [19]Cardy (1979); [20]Zeller et al. (1973); [21]Ward et al. (1979); [22]Evarts and Brown (1980); [23]Steinhoff (1977); [24]NCI (1980); [25]Umeda (1956); [26]Shimkin et al. (1966); [27]Vesselinovitch et al. (1975); [28,29,30]Mori (1962, 1963, 1964); [31]Taylor and Lijinsky (1975); [32]Parodi et al. (1980); [33]Ito et al. (1969); [34]Brambilla et al. (1981); [35]Argus and Hoch-Ligeti (1961); [36]Goodall et al. (1968); [37]Parodi et al. (1978); [38]Cavanna et al. (1979); [39]Laqueur and Matsumoto (1966); [40]Greenblatt and Lijinsky (1972); [41]Toth (1975); [42]Parodi et al. (1981a); [43]Brambilla et al. (1978). All TPI data were obtained from Pienta (1980).

The subdivision in chemical classes is essentially according to McCann et al. (1975).

The experiments in vivo concern only mice and rats.

Table 2
Degree of Predictivity of Different Short-term Parameters

Pair of parameters	Number of compounds	Correlation coefficient r	Probability that r = 0
(Log DFI) = f (Log OPI)	57	0.41	$p < 0.001$
(Log MPI) = f (Log OPI)	88	0.39	$p < 0.001$
(Log TPI) = f (Log OPI)	59	0.37	$p < 0.01$
(Log CBI) = f (Log OPI)	37	0.42	$p < 0.01$
(Log LDI) = f (Log OPI)	76	0.43	$p < 0.001$

Table 3
Degree of Predictivity of Different Short-term Parameters Compared for the Same Compounds

Pair of parameters	Number of compounds	Correlation coefficient r	Probability that r = 0
(Log DFI) = f (Log OPI)	54	0.43	$p < 0.001$
(Log MPI) = f (Log OPI)	54	0.53	$p < 0.001$
(Log DFI) = f (Log OPI)	29	0.24	N.S.[a]
(Log TPI) = f (Log OPI)	29	0.57	$p < 0.002$
(Log DFI) = f (Log OPI)	28	0.23	N.S.[a]
(Log CBI) = f (Log OPI)	28	0.55	$p < 0.01$
(Log DFI) = f (Log OPI)	54	0.46	$p < 0.001$
(Log LDI) = f (Log OPI)	54	0.26	$p < 0.05$

[a]N.S. = $p > 0.10$.

with this reservation in mind, it seemed interesting to investigate whether trends of higher or lower predictivity exist for a specific short-term test, with respect to a specific chemical class of compounds. The results, presented in Table 4, show that in many cases there are profound differences in predictivity between two short-term tests, for a given family of compounds. We reported in Table 4 only the differences reaching statistical significance or close to it. The number of cases is relatively relevant, suggesting that different short-term tests can be suitable for different classes of chemical compounds. Perhaps it is worthwhile to be underlined that the parameter more subjected to dramatic variations in predictivity, according to the chemical class of compounds tested, is toxicity. For instance, alkyl halides are predicted very well, but hydrazine derivatives show an inverse correlation. Perhaps toxicity, while as overall correlated with carcinogenicity, is a peculiar parameter, that allows for the largest discrepancies. The differences in predictivity do not reach complete statistical significance very often, but the general trend suggesting profound differences, in our

Table 4
Predictivity of Different Short-term Parameters for Specific Classes of Chemicals

Chemical class	Number of compounds	Correlation coefficient r	Probability that r = 0
(Log DFI) = f (Log OPI)			
Aromatic amines	12	0.33[a]	N.S.
Nitrosamines	13	0.24[b]	N.S.
Hydrazine derivatives	11	0.75[a,b]	p < 0.01
(Log MPI) = f (Log OPI)			
Aromatic amines	17	0.56[c]	p < 0.02
Alkyl halides	6	0.69[b]	N.S.
Hydrazine derivatives	11	−0.19[b,c]	N.S.
(Log TPI) = f (Log OPI)			
Esters, epoxides, carbamates	6	−0.27[d]	N.S.
Miscellaneous heterocycles	7	0.71[d]	p < 0.10
(Log LDI) = f (Log OPI)			
Aromatic amines	14	0.05[e]	N.S.
Alkyl halides	7	0.91[e,f]	p < 0.01
Hydrazine derivatives	11	−0.51[f,g]	N.S.
Nitrosamines	15	0.68[g]	p < 0.01

In each section the r values with the same index are significantly different with p values as follows: [a]p = 0.19; [b]p = 0.12; [c]p = 0.06; [d]p = 0.13; [e]p = 0.01; [f]p = 0.0006; [g]p = 0.002.

N.S. = p > 0.10.

opinion, clearly emerges, if one considers that the number of compounds per subclass is usually very small.

Increase in Predictivity using a Battery of Short-term Tests

The predicted carcinogenic potency (Y_R = Log OPI) can be correlated with two (or more) short-term parameters. For instance, let X_2 = Log DPI and X_3 = Log MPI. Let us examine the multiple regression equation $Y_R = a + \beta_2 X_2 + \beta_3 X_3$. (For this statistical analysis and for the symbols used see Snedecor and Cochran 1967). If r_{12} is the simple correlation between Log OPI and Log DFI and r_{13} is the simple correlation between Log OPI and Log MPI, it can be demonstrated that the multiple correlation r_{123} is linked to the simple correlations already mentioned and to r_{23} (correlation between Log MPI and Log DFI) by the following equation:

$$r_{123}^2 = r_{12} \cdot \frac{r_{12} - r_{13} \cdot r_{23}}{1 - r_{23}^2} + r_{13} \cdot \frac{r_{13} - r_{12} \cdot r_{23}}{1 - r_{23}^2} \cdot \qquad (7)$$

It can be seen that r_{123} is always better than r_{12} and r_{13}; at worst r_{123} will become equal to r_{12} and r_{13} when X_2 and X_3 have a complete correlation ($r_{23} = 1$).

It is interesting to underline the fact that the best improvement of r_{123} over r_{12} and r_{13} is obtained when X_2 and X_3, being both correlated with carcinogenicity, are poorly correlated between themselves. This fact indicates that X_2 and X_3 are looking at two different determining factors, so that there is an improvement in predictivity taking into consideration both factors. On the contrary, if X_2 and X_3 are strongly correlated between themselves it means that they are reflecting the action of a single factor, so that no improvement in predictivity can be obtained by the two parameters in association.

For the specific case of DFI and MPI r_{23} was = 0.33 and, as a consequence, a significant improvement of r_{123} over r_{12} and r_{13} could be obtained. For r_{12} (r_{DFI}) = 0.43 and for r_{13} (r_{MPI}) = 0.53, r_{123} was equal to 0.60. The different correlations between short-term parameters are shown in Table 5. This table shows that all the correlations between short-term parameters are around 0.3-0.4. The correlation is sufficiently low to make convenient the use of two (or more) short-term tests in a battery. The only couple that is not convenient for use in a battery of tests is DFI and CBI. Vice versa, CBI and TPI seem especially convenient for use in a battery of tests. Figure 2 shows the correlation between DFI and CBI.

CONCLUSIONS

An evaluation of the quantitative predictivity of different short-term tests is important for risk assessment. Preliminary to the evaluation of the quantitative potency in a given short-term test is the evaluation of carcinogenic potency. The evaluation of the carcinogenic potency is a difficult task and in order to make

Table 5
Correlation Between Couples of Different Short-term Parameters

Pair of parameters	Number of compounds	Correlation coefficient r	Probability that r = 0
(Log DFI) = f (Log MPI)	54	0.33	p < 0.02
(Log DFI) = f (Log TPI)	29	0.31	N.S.
(Log DFI) = f (Log CBI)	28	0.66	p < 0.001
(Log DFI) = f (Log LDI)	54	0.39	p < 0.01
(Log MPI) = f (Log TPI)	59	0.44	p < 0.001
(Log MPI) = f (Log CBI)	35	0.35	p < 0.05
(Log MPI) = f (Log LDI)	73	0.31	p < 0.1
(Log TPI) = f (Log CBI)	25	0.16	N.S.
(Log TPI) = f (Log LDI)	45	0.26	p < 0.10
(Log CBI) = f (Log LDI)	32	0.27	N.S.

[a]N.S. = p > 0.10.

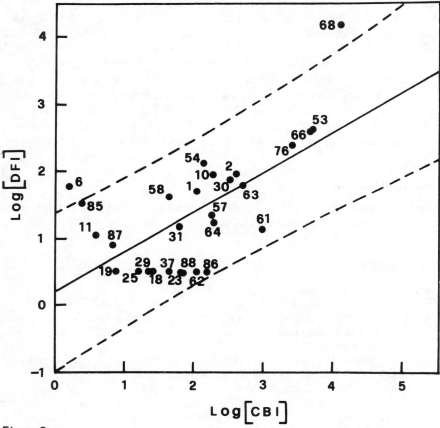

Figure 2
Correlation between DFI and CBI for 28 compounds. The numbers refer to compounds listed in Table 1. (————) regression line; (— — — —) 90% belt zone; r = 0.66; p (that r = 0) < 0.001; equation of the regression line: Log (DFI) = 0.23 + 0.59 Log (CBI).

computations compromise solutions have to be accepted. In order to evaluate whether carcinogenic potency was not completely without any objective fundamental, we evaluated the internal consistency of the carcinogenicity data. The correlation coefficient r was rather high (r = 0.77) with a 90% belt zone roughly 1/90-90X. This suggests that the computation of carcinogenic potency has some objective value.

We gave a short description of the fluorometric analysis of DNA alkaline elution. DFI showed a global predictivity similar to that of all the other short-term tests considered, and appeared strongly correlated with CBI. Obviously, if DNA fragmentation reflects closely the amount of adducts, it is a significant advantage for DFI that no labeling of DNA is required.

It was mathematically demonstrated that the use of a battery of short-term tests can improve the predictivity, especially if the tests are not strongly correlated among them.

There is a trend suggesting that different tests can have different predictivity for specific subclasses of chemical compounds. For instance, DFI appeared especially predictive for the class of hydrazine derivatives. Perhaps the specific short-term test most suitable for a given specific class of chemical compounds could be determined. Finally, TPI, even if theoretically closer to carcinogenicity, was not more predictive than the other short-term parameters.

ACKNOWLEDGMENT

This work was supported by the Italian National Council of Research (Progetto Finalizzato "Controllo della Crescita Neoplastica"; contract No. 81.01386.96).

REFERENCES

Argus, M.F. and C. Hoch-Ligeti. 1961. Comparative study of the carcinogenic activity of nitrosamines. *J. Natl. Cancer Inst.* **27**:695.

Bolognesi, C., C.F. Cesarone, and L. Santi. 1981. Evaluation of DNA damage by alkaline elution technique after *in vivo* treatment with aromatic amines. *Carcinogenesis* **2**:265.

Bonser, G.M., D.B. Clayson, J.W. Jull, and L.N. Pirah. 1956. The carcinogenic activity of 2-naphthylamine. *Br. J. Cancer* **10**:533.

Brambilla, G., M. Cavanna, A. Pino, and L. Robbiano. 1981. Quantitative correlation among DNA damaging potency of six *N*-nitrosocompounds and their potency in inducing tumor growth and bacterial mutations. *Carcinogenesis* **2**:425.

Brambilla, G., M. Cavanna, L. Sciabà, P. Carlo, S. Parodi, and M. Taningher. 1977. A procedure for the assay of DNA damage in mammalian cells by alkaline elution and microfluorometric DNA determination. *Ital. J. Biochem.* **26**:419.

Brambilla, G., M. Cavanna, S. Parodi, L. Sciabà, A. Pino, and L. Robbiano. 1978. DNA damage in liver, colon, stomach, lung and kidney of BALB/c mice treated with 1,2-dimethylhydrazine. *Int. J. Cancer* **22**:174.

Brambilla, G., M. Cavanna, P. Carlo, R. Finollo, L. Sciabà, S. Parodi, and C. Bolognesi. 1979. DNA damage and repair induced by diazoacetyl derivatives of amino acids with different mechanisms of cytotoxicity. Correlations with mutagenicity and carcinogenicity. *J. Cancer Res. Clin. Oncol.* **94**:383.

Cardy, R.H. 1979. Carcinogenicity and chronic toxicity of 2,4-toluenediamine in F344 rats. *J. Natl. Cancer Inst.* **62**:1107.

Cavanna, M., S. Parodi, M. Taningher, C. Bolognesi, L. Sciabà, and G. Brambilla. 1979. DNA fragmentation in some organs of rats and mice treated with cycasin. *Br. J. Cancer* **39**:383.

Cesarone, C.F., C. Bolognesi, and L. Santi. 1979. Improved microfluorometric DNA determination in biological material using 33258 Hoechst. *Anal. Biochem.* **100**:188.

Cox, R., I. Damianov, S.E. Abanobi, and D.S.R. Sarma. 1973. A method for measuring DNA damage and repair in the liver *in vivo. Cancer Res.* **33**: 2114.

De Flora, S. 1981. Study of 106 organic and inorganic compounds in the *Salmonella*/microsome test. *Carcinogenesis* 2:283.

Erickson, L.C., R. Osieka, N.A. Sharkey, and K.W. Kohn. 1980. Measurement of DNA damage in unlabeled mammalian cells analyzed by alkaline elution and a fluorometric DNA assay. *Anal. Biochem.* **106**:169.

Evarts, R.P. and C.A. Brown. 1980. 2,4-Diaminoanisole sulfate: Early effect on thyroid gland morphology and late effect on glandular tissue of Fischer 344 rats. *J. Natl. Cancer Inst.* **65**:197.

Goodall, C.M., W. Lijinsky, and L. Tomatis. 1968. Tumorigenicity of *N*-nitroso-hexamethyleneimine. *Cancer Res.* **28**:1217.

Greenblatt, M. and W. Lijinsky. 1972. Failure to induce tumors in Swiss mice after concurrent administration of amino acids and sodium nitrite. *J. Natl. Cancer Inst.* **48**:1389.

Griswold, D.P., Jr., A.E. Casey, E.K. Weisburger, and J.H. Weisburger. 1968. The carcinogenicity of multiple intragastric doses of aromatic and hetero-cyclic nitro or amino derivatives in young female Sprague-Dawley rats. *Cancer Res.* **28**:924.

Griswold, D.P., Jr., A.E. Casey, E.K. Weisburger, J.H. Weisburger, and F.M. Shabel Jr. 1966. On the carcinogenicity of a single intragastric dose of hy-drocarbons, nitrosamines, aromatic amines, dyes, coumarins, and miscel-laneous chemicals in female Sprague-Dawley rats. *Cancer Res.* **26**:619.

International Agency for Research on Cancer. 1972-81. *IARC Monogr. Eval. Carcinog. Risk Chem. Hum.,* vols. 1-26.

Ito, N., Y. Hiasa, Y. Konishi, and M. Marugami. 1969. The development of carcinoma in liver of rats treated with m-toluylenediamine and the syner-gistic and antagonistic effects with other chemicals. *Cancer Res.* **29**:1137.

Kawachi, T., T. Komatsu, T. Kada, M. Ishidate, M. Sasaki, T. Sugiyama, and Y. Tazima. 1980. Results of recent studies on the relevance of various short-term screening tests in Japan. In *The predictive value of short-term screen-ing tests in carcinogenicity evaluation* (ed. G.M. Williams et al.), p. 253. Elsevier, Amsterdam, The Netherlands.

Kissane, J.M. and E. Robins. 1958. The fluorometric measurement of deoxyribo-nucleic acid in animal tissues with special reference to the central nervous system. *J. Biol. Chem.* **233**:184.

Kohn, K.W., L.C. Erickson, R.A.G. Ewig, and C.A. Friedman. 1976. Frac-tionation of DNA from mammalian cells by alkaline elution. *Biochemistry* **15**:4629.

Laqueur, G.L. and H. Matsumoto. 1966. Neoplasms in female Fisher rats fol-lowing intraperitoneal injection of methylazoxymethanol. *J. Natl. Cancer Inst.* **37**:217.

Lutz, W.K. 1979. *In vivo* covalent binding of organic chemicals to DNA as a quantitative indicator in the process of chemical carcinogenesis. *Mutat. Res.* **65**:289.

McCann, J., E. Choi, E. Yamasaki, and B.N. Ames. 1975. Detection of carcinogens as mutagens in the *Salmonella*/microsome test: Assay of 300 chemicals. *Proc. Natl. Acad. Sci. U.S.A.* **72**:5135.

Meselson, M. and K. Russell. 1977. Comparison of carcinogenic and mutagenic potency. *Cold Spring Harbor Conf. Cell Proliferation* **4**:1473.

Mori, K. 1962. Induction of pulmonary tumors in rats by subcutaneous injections of 4-nitroquinoline 1-oxide. *Gann* **53**:303.

_____. 1963. Induction and transplantation of cancer of the lung in rats. *Gann* **54**:415.

_____. 1964. Induction of pulmonary and uterine tumors in rats by subcutaneous injections of 4-nitroquinoline 1-oxide. *Gann* **55**:277.

National Cancer Institute. 1978. Bioassay of aniline hydrochloride for possible carcinogenicity. *Natl. Cancer Inst. Carcinog. Tech. Rep. Ser.* **130**:1.

National Cancer Institute, National Institutes of Health, National Toxicology Program. 1980. Bioassay of 4,4'-oxydianiline for possible carcinogenicity. U.S. Department of Health and Human Services, DHHS Publication No. (NIH) 80-1761.

National Institute for Occupational Safety and Health. 1978. *Registry of toxic effects of chemical substances* (ed. R.J. Lewis). U.S. Government Printing Office, Washington, D.C., USA.

Painter, R. and R. Howard. 1978. A comparison of the HeLa DNA synthesis inhibition test and the Ames' test for screening of mutagenic carcinogens. *Mutat. Res.* **54**:113.

Parodi, S., M. Taningher, and L. Santi. 1981c. DNA fragmentation: Its predictivity as a short term test. In *Chemical Carcinogenesis* (ed. C. Nicolini). Plenum Press, New York (In press).

Parodi, S., M. Taningher, C. Balbi, and L. Santi. 1982a. Predictivity of the autoradiographic repair assay in rat liver cells, as compared with the Ames' test. *J. Toxicol. Environ. Health* (in press).

Parodi, S., M. Taningher, P. Boero, and L. Santi. 1982b. Quantitative correlations amongst alkaline DNA fragmentation, DNA covalent binding, mutagenicity in the Ames' test and carcinogenicity, for twenty-one different compounds. *Mutat. Res.* **93**:1.

Parodi, S., M. Taningher, M. Pala, G. Brambilla, and M. Cavanna. 1980. Detection by alkaline elution of rat liver DNA damage induced by simultaneous subacute administration of nitrite and aminopyrine. *J. Toxicol. Environ. Health* **6**:167.

Parodi, S., M. Taningher, P. Russo, M. Pala, M. Tamaro, and C. Monti-Bragadin. 1981b. DNA-damaging activity *in vivo* and bacterial mutagenicity of sixteen aromatic amines and azo-derivatives as related quantitatively to their carcinogenicity. *Carcinogenesis* **2**:1317.

Parodi, S., M. Taningher, L. Santi, M. Cavanna, L. Sciabà, A. Maura, and G. Brambilla. 1978. A practical procedure for testing DNA damage *in vivo*, proposed for the pre-screening of chemical carcinogens. *Mutat. Res.* **54**: 39.

Parodi, S., S. De Flora, M. Cavanna, A. Pino, L. Robbiano, C. Bennicelli, and G. Brambilla. 1981a. DNA-damaging activity *in vivo* and bacterial mutagenicity of sixteen hydrazine derivatives as related quantitatively to their carcinogenicity. *Cancer Res.* **41**:1469.

Petzold, G.L. and J.A. Swenberg. 1978. Detection of DNA damage induced *in vivo* following exposure of rats to carcinogens. *Cancer Res.* **38**:1589.

Pienta, R.J. 1980. Evaluation and relevance of the Syrian hamster embryo cell system. In *The predictive value of short-term screening tests in carcinogenicity evaluation* (ed. G.M. Williams et al.), p. 149. Elsevier, Amsterdam.

Sarma, D.S.R., S. Rajalakshmi, and E. Farber. 1975. Chemical carcinogenesis: Interactions of carcinogens with nucleic acids. In *Cancer* (ed. F.F. Becker), vol. 1, p. 235. Plenum Press, New York.

Schwarz, M., J. Hummel, K.E. Appel, R. Rickart, and W. Kunz. 1979. DNA damage induced *in vivo* evaluated with a non-radioactive alkaline elution technique. *Cancer Lett.* **6**:221.

Shimkin, M.B., J.H. Weisburger, E.K. Weisburger, N. Gubareff, and V. Suntzeff. 1966. Bioassay of 29 alkylating chemicals by the pulmonary-tumor response in strain A mice. *J. Natl. Cancer Inst.* **36**:915.

Snedecor, G.W. and W.G. Cochran. 1967. *Statistical methods.* The Iowa State University Press, Ames, Iowa.

Steinhoff, D. 1977. Cancerogene Wirkung von 4,4′-Diamino-diphenyläther bei Ratten. *Naturwissenschaften* **64**:394.

Stich, H.F., R.H.C. San, P.P.S. Lam, D.J. Koropatnick, L.W. Lo, and B.A. Laishes. 1976. DNA fragmentation and DNA repair as an *in vitro* and *in vivo* assay for chemical procarcinogens, carcinogens and carcinogenic nitrosation products. In *Screening tests in chemical carcinogenesis* (ed. R. Montesano et al.), *IARC Sci. Publ.* **12**:617.

Suter, W., J. Brennand, S. Mc Millan, and M. Fox. 1980. Relative mutagenicity of antineoplastic drugs and other alkylating agents in V79 Chinese hamster cells. Independence of cytotoxic and mutagenic responses. *Mutat. Res.* **73**:171.

Taylor, H.W. and W. Lijinsky. 1975. Tumor induction in rats by feeding aminopyrine or oxytetracycline with nitrite. *Int. J. Cancer* **16**:211.

Toth, B. 1975. Synthetic and naturally occurring hydrazines as possible cancer causative agents. *Cancer Res.* **35**:3693.

Umeda, M. 1956. Experimental study of xanthene dyes as carcinogenic agents. *Gann* **47**:51.

U.S. Public Health Service. 1972-73. *Survey of compounds which have been tested for carcinogenic activity.* U.S. Government Printing Office, Washington D.C.

Vesselinovitch, S.D., A.P. Kyriazis, N. Mihailovich, and K.V.N. Rao. 1975. Conditions modifying development of tumors in mice at various sites by benzo(*a*)pyrene. *Cancer Res.* **35**:2948.

Ward, J.M., S.F. Stinson, J.F. Hardisty, B.Y. Cockrell, and D.W. Hayden. 1979. Neoplasms and pigmentation of thyroid glands in F344 rats exposed to 2,4-diaminoanisole sulfate, a hair dye component. *J. Natl. Cancer Inst.* **62**:1067.

Zeller, H., H. Birnstel, K.O. Freisberg, P. Kirsch, and K.H. Hempel. 1973. Chronic toxicity of auramine. *Naturwissenschaften* **60**:523.

In Vivo Dosimetry by Means of Alkylated Hemoglobin—
A Tool in the Design of Tests for Genotoxic Effects

CARL JOHAN CALLEMAN
Department of Radiobiology
University of Stockholm
S-106 91 Stockholm
Sweden

A great number of compounds and mixtures of compounds are known to possess genotoxic activity. The majority of these compounds have been detected in experimental organisms, either in short-term tests such as the Ames test (Ames et al. 1973) or in cancer tests in rodents. The extrapolation to humans of data gained in such tests are, however, hampered by qualitative and quantitative differences between the experimental test systems and man, with regard to metabolism.

In order to provide data which may be extrapolated to risk in man and in order to avoid drawing erroneous conclusions from negative results, biological test systems need to be complemented with a dosimetry of ultimately reactive metabolites.

An important conclusion drawn by Miller and Miller (1974) from the work of their own and other research groups is that all cancer initiators and the majority of mutagens possess electrophilic reactivity in their ultimate forms. Electrophilic reagents, RX, react more or less indiscriminately with nucleophilic groups, Y_i, of proteins and nucleic acids according to

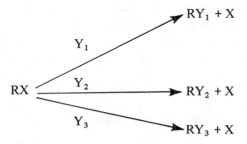

The lack in discrimination in the reactions of electrophiles with tissue nucleophiles has made it difficult to positively identify the targets, the reactions of which may be causative in mutagenesis and cancer initiation. For alkylating

157

agents, oxygen atoms of the DNA have been implicated to play such a role (Loveless 1969; Goth and Rajewsky 1974) in line with their characterization as weakly nucleophilic (Veleminsky et al. 1970; Ehrenberg and Hussain 1981; Hussain 1981) but the critical targets of many other important classes of genotoxic agents, such as aromatic amines (Neumann 1979) still remain to be identified.

It can, however, be shown that the degree of reaction of any nucleophile, Y_i, is proportional to the dose, D, (= time-integral of the concentration of free electrophilic reagent in the compartment under study) and consequently to the degree of modification of a critical target group according to equation 1,

$$\frac{1}{k_{Y_i}} \frac{|RY_i|}{|Y_i|} = D = \int_t |RX| dt = \frac{1}{k_{Y_{crit}}} \frac{|RY_{crit}|}{|Y_{crit}|} \tag{1}$$

where k_{y_i} is the second order reaction rate constant for the reaction between Y_i and RX (Ehrenberg 1974).

Thus, in principle any nucleophilic component, Y_i, in the cells can be used for dose monitoring provided that it and its reaction products have well established turnovers.

My aim is to demonstrate that hemoglobin is a suitable dose monitor in vivo, which may be used not only for risk estimation in man which has been discussed elsewhere (Calleman et al. 1978; Ehrenberg 1979) but also for the detection of new genotoxic compounds and for the design of biological tests for genotoxic effects.

RESULTS

Hemoglobin as a Dose Monitor for Genotoxic Compounds

Hemoglobin has been chosen for dosimetry in experimental animals and man mainly because of its availability in large quantities, its appropriate life-span and well-known turnover in vivo (Osterman-Golkar et al. 1976). In the cases studied to date (Table 1), alkylated hemoglobins have been found to be stable throughout the life-span of the erythrocytes independently of the chemical nature of the group being introduced. Since old erythrocytes (at the mean age of 6 weeks, 9 weeks, and 18 weeks in mice, rats, and humans, respectively) are constantly disappearing from the circulation of the blood this means that following an acute exposure to an electrophilic reagent the degree of alkylation will decrease linearly to the value zero at the time corresponding to the average life-span of the erythrocytes (Fig. 1). After an exposure which is longer than the life-span of the erythrocytes, chronic exposure to electrophilic reagents will therefore result in the attainment of a steady-state level. This model for the accumulation of alkylated hemoglobins has been verified in experiments with mice (Osterman-Golkar et al. 1976; Segerbäck et al. 1978).

Table 1
Estimated Life-spans In Vivo of Alkylated Hemoglobins Produced by Reactions with Different Compounds

Compound	Species	Estimated life-span (days)	References
Methyl Methanesulfonate	Mouse	43	(Segerbäck et al. 1978)
Dimethylnitrosamine	Mouse	39	(Osterman-Golkar et al. 1976)
Ethylene Oxide	Mouse	43	(Osterman-Golkar et al. 1976)
Benzo(*a*)pyrene	Mouse	40	
Benzo(*a*)pyrene	Mouse	43	(Löfroth G. et al., unpubl. results)
Chloroform	Rat	70	(Pereira and Chang 1982)

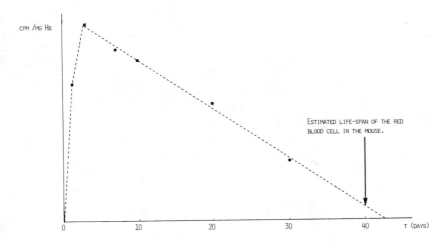

Figure 1
The degree of alkylation of mouse hemoglobin following a single injection of radiolabeled benzo(*a*)pyrene at T = 0.

On this basis the steady-state level in humans exposed to ethylene oxide was assumed to be nine times the weekly increment in the degree of alkylation after more than 18 weeks of chronic or intermittent exposure with approximately the same weekly dose increment (Calleman et al. 1978).

Important results gained with the use of hemoglobin as a dose monitor are the following:

1. Some 35 radiolabeled compounds have been shown to bind covalently to hemoglobin in animal experiments (for a compilation see Calleman 1982). These compounds include representatives of the most important classes of mutagens and cancer initiators currently known (Wieland and Neumann 1978; Pereira and Chang 1980). So far, no compound believed to derive its genotoxicity from its electrophilic reactivity has failed to yield covalent reaction products with hemoglobin in vivo.

2. It has been shown that many electrophilic reagents are relatively uniformly distributed in the bodies of experimental animals, and that the dose in other organs can be approximated by the dose determined in the red blood cells (Table 2). Important exceptions to this have, however, also been noted. Dimethylnitrosamine and vinyl chloride, both of which are metabolized mainly in the liver to reactive metabolites of very short life-spans produce lower doses in distal organs than in the liver and the red blood cells. Methyl bromide produces a lower dose in the liver than in the red blood cells which possibly may be related to its high lipophilicity.

Table 2

The Relationship Between Hemoglobin Dose (= 1) and Dose in the DNA Compartment of the Liver and the Testis for a Few Compounds at Low Doses

Compound	Hemoglobin dose	DNA-Dose		References
		liver	testis	
Methylmethanesulfonate	1 (Def.)	0.7-1.0	1	(Segerbäck et al. 1978)
Ethylmethanesulfonate	1	1.0-1.5	—	(Murthy M.S.S., in prep.)
Methyl Bromide	1	0.01	—	(Djalali-Behzad et al. 1981)
Dichlorvos	1	1		(Segerbäck and Ehrenberg 1981)
Dimethylnitrosamine	1	30[a]	<0.1	(Osterman-Golkar et al. 1976)
Ethylene	1	1.3-1.8	0.5-0.6	(Segerbäck D., in prep.)
Ethylene Oxide	1	1.3-1.8	0.5-0.6	(Segerbäck D., in prep.)
Vinyl Chloride	1	1	0.1[b]	(Osterman-Golkar et al. 1977)
Benzyl Chloride	1	1	2	(Walles 1981)

[a] Estimated from the amount of binding to DNA and hemoglobin assuming the same reactivity of DNA and hemoglobin, respectively.
[b] Estimated from the degree of binding to protein.

161

These aberrant cases emphasize the need for investigations in experimental animals of the distribution between the dose in the DNA compartment in different organs and the dose in the red blood cells. In relevant cases, such as ethylene oxide, risk estimations in humans may then be based on these relationships (Ehrenberg 1979).

3. Hemoglobin alkylation has been used to determine in vivo doses in persons occupationally exposed to ethylene oxide (Calleman et al. 1978) and propylene oxide (Osterman-Golkar et al., unpubl. results).

4. A risk assessment for human exposure to ethylene oxide has been proposed (Calleman et al. 1978; Ehrenberg 1979; Ehrenberg and Hussain 1981). This risk estimation has been confirmed by limited epidemiological studies (Hogstedt et al. 1979). Risk estimation based on hemoglobin alkylation takes into account all the processes affecting the in vivo fate of a genotoxic compound such as its rates of uptake, activation, and deactivation through chemical and enzymatic reactions and excretion. The need to base risk estimations on the in vivo dose rather than the exposure dose (= time-integral of the concentration in the outer environment; cf. Ehrenberg et al. 1982) is apparent from the fact that the above-mentioned rates may be very different in different animal species as well as in different individuals.

5. Measurements of hemoglobin alkylation have demonstrated the occurrence of "spontaneous" methylations in unexposed animals and man (Bailey et al. 1981).

6. It has been shown by a determination of 2-hydroxyethylated and 2-hydroxypropylated amino acids in hemoglobin that the alkenes ethylene (Ehrenberg et al. 1977) and propylene (K. Svensson and S. Osterman-Golkar, unpubl. results), are metabolized in mice to the corresponding epoxides, ethylene oxide and propylene oxide, both of which are well-established mutagens and carcinogens (Ehrenberg and Hussain 1981). Thus, by means of hemoglobin alkylation new genotoxic compounds (Ehrenberg and Osterman-Golkar 1980) such as ethylene and propylene have been detected, compounds which to the best of our knowledge have escaped detection in other test systems for genotoxic activity.

DISCUSSION

Dose Monitoring in Cancer Tests

Figure 2 shows the relationship between the in vivo dose and the exposure dose in rats exposed to ethylene and ethylene oxide. The filled squares representing the in vivo doses of ethylene oxide have been determined from the degree of 2-hydroxyethylation in histidine in hemoglobin in rats undergoing a long-term cancer test performed at the Carnegie-Mellon Institute (Snellings et al. 1981). The levels of increase and significant increase in the leukemia frequency in this test are indicated. As can be seen from Figure 2, the in vivo dose depended almost linearly upon the exposure dose of ethylene oxide with a slight tendency

WEEKLY IN VIVO DOSE OF ETHYLENE OXIDE

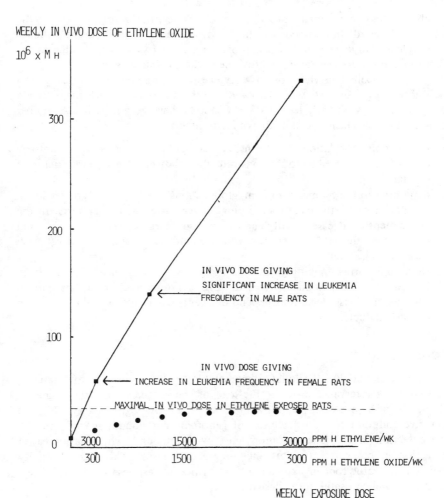

WEEKLY EXPOSURE DOSE

Figure 2
The relationship between exposure dose and in vivo dose in rats chronically exposed to ethylene oxide (■) and ethylene (●).

to a decreased slope. The filled dots represent calculations of the in vivo doses of ethylene oxide in rats exposed to ethylene. These calculations (cf. Osterman-Golkar and Ehrenberg 1981) were based on an inhalation study of ethylene in rats performed by Andersen et al. (1980) showing that the metabolic conversion of ethylene is a saturable process. The rate of uptake of ethylene followed Michaelis-Menten kinetics: $-\dfrac{d|\text{Ethylene}|}{dt} = V_{max}\dfrac{|\text{Ethylene}|}{K_m + |\text{Ethylene}|}$ asymptotically reaching a maximal rate, V_{max}, as the concentration of ethylene in the air was increased. In consequence, the in vivo doses of the primary metabolite of ethylene, ethylene oxide, will approach a maximal value as indicated in the figure, when the concentration of ethylene in the air is increased.

This maximal in vivo dose of ethylene oxide is below that which gives a significant response in the rats exposed to ethylene oxide, which may explain why cancer tests with ethylene (CIIT 1980) so far have been negative. A number of important conclusions can be drawn from these studies:

1. Hemoglobin alkylation is a sensitive end-point for detecting genotoxic compounds. It is able to demonstrate the relatively weak genotoxicity of ethylene.
2. Dose monitoring by means of hemoglobin alkylation may be of great value in the design of costly large-scale biological tests such as cancer tests. In the above-mentioned case of ethylene the in vivo doses determined can be used to estimate the size of the test which would demonstrate an effect with statistical significance.
3. Dose monitoring in biological tests may serve to explain deviations from linearity of dose-response curves at high exposure concentrations and to what extent these deviations depend on events from uptake to the formation of DNA adducts.

CONCLUSIONS

The chemical stability of alkylated hemoglobins makes it possible to determine in vivo doses integrated over the life-span of the red cells from the degree of alkylation of hemoglobin in both animals and man. Hemoglobin alkylation is a sensitive endpoint for the detection of genotoxic compounds as evidenced by the detection of ethylene and propylene, two compounds which have escaped detection in other test systems, including a cancer test. This high sensitivity makes it a useful tool in explaining the absence of response in biological test systems as well as useful in the design of these tests.

ACKNOWLEDGMENTS

Financial support from the Swedish Work Environment Fund, the Swedish Natural Science Research Council and the Swedish Cancer Society is gratefully acknowledged.

REFERENCES

Ames, B.N., W.E. Durston, E. Yamasaki, and F.D. Lee. 1973. Carcinogens are mutagens: A simple test system combining liver homogenates for activation and bacteria for detection. *Proc. Natl. Acad. Sci. U.S.A.* **70**:2281.

Andersen, M.E., M.L. Gargas, R.A. Jones, and L.J. Jenkins, Jr. 1980. Determination of the kinetic constants for metabolism of inhaled toxicants in vivo using gas uptake measurements. *Toxicol. Appl. Pharmacol.* 54:100.

Bailey, E., T.A. Connors, P.B. Farmer, S.M. Gorf, and J. Rickard. 1981. Methylation of Cysteine in Hemoglobin following exposure to methylating agents. *Cancer Res.* 41:2514.

Chemical Industry Institute of Toxicology (CIIT). 1980. Docket No. 12000, October 15. Research Triangle Park, N.C.

Calleman, C.J. 1982. Monitoring and risk assessment by means of alkyl groups in hemoglobin. In *The biological monitoring of exposures to industrial chemicals,* p. 331. Hemisphere, Washington, D.C. (In press).

Calleman, C.J., L. Ehrenberg, B. Jansson, S. Osterman-Golkar, D. Segerbäck, K. Svensson, and C.A. Wachtmeister. 1978. Monitoring and risk assessment by means of alkyl groups in hemoglobin in persons occupationally exposed to ethylene oxide. *J. Environ. Pathol. Toxicol.* 2:427.

Djalali-Behzad, G., S. Hussain, S. Osterman-Golkar, and D. Segerbäck. 1981. Estimation of genetic risks of alkylating agents. VI. Exposure of mice and bacteria to methyl bromide. *Mutat. Res.* 84:1.

Ehrenberg, L. 1974. Genetic toxicity of environmental chemicals. *Acta Biol. Iugosl., Ser. F., Genetika* 6:367.

———. 1979. Risk assessment of ethylene oxide and other compounds. *Ban. Rep.* 1:157.

Ehrenberg, L. and S. Osterman-Golkar. 1980. Alkylation of macromolecules for detecting mutagenic agents. *Teratog. Carcinog. Mutagen.* 1:105.

Ehrenberg, L. and S. Hussain. 1981. Genetic toxicity of some important epoxides. *Mutat. Res.* 86:1.

Ehrenberg, L., E. Moustacchi, and S. Osterman-Golkar. 1982. Dosimetry of genotoxic agents and dose-response relationships of their effects. Working document for ICPEMC. Committee 4.

Ehrenberg, L., S. Osterman-Golkar, D. Segerbäck, K. Svensson, and C.J. Calleman. 1977. Evaluation of genetic risks of alkylating agents. III. Alkylation of haemoglobin after metabolic conversion of ethene to ethene oxide in vivo. *Mutat. Res.* 45:175.

Goth, R. and M.F. Rajewsky. 1974. Persistence of O^6-ethylguanine in rat-brain DNA: Correlation with nervous system-specific carcinogenesis by ethylnitrosourea. *Proc. Natl. Acad. Sci. U.S.A.* 71:639.

Hogstedt, C., N. Malmqvist, and B. Wadman. 1979. Leukemia in workers exposed to ethylene oxide. *J. Am. Med. Assoc.* 241:1132.

Hussain, S.S. 1981. *Mutagenic action of radiation and chemicals: Parameters affecting the response of test systems,* Thesis, University of Stockholm, Stockholm, Sweden.

Loveless, A. 1969. Possible evidence of O-6-alkylation of deoxyguanosine to the mutagenicity of nitrosamines and nitrosamides. *Nature* 223:106.

Miller, E.C. and J.A. Miller. 1974. Biochemical mechanisms of chemical carcinogenesis. In *Molecular biology of cancer* (ed. H. Busch), p. 377. Academic Press, New York.

Neumann, H.-G. 1979. Pharmacokinetic parameters influencing tissue specificity in chemical carcinogens. *Arch. Toxicol. Suppl.* 2:229.

Osterman-Golkar, S. and L. Ehrenberg. 1981. Covalent binding of reactive intermediates to hemoglobin as an approach for determining the metabolic activation of chemicals—ethylene. At *Symposium on the metabolism and pharmacokinetics of environmental chemicals in man.* June 7-12. Sarasota, Florida. (In press).

Osterman-Golkar, S., D. Hultmark, D. Segerbäck, C.J. Calleman, R. Göthe, L. Ehrenberg, and C.A. Wachtmeister. 1977. Alkylation of DNA and proteins in mice exposed to vinyl chloride. *Biochem. Biophys. Res. Commun.* **76**:259.

Osterman-Golkar, S., L. Ehrenberg, D. Segerbäck, and I. Hällström. 1976. Evaluation of genetic risks of alkylating agents. II. Haemoglobin as a dose monitor. *Mutat. Res.* **34**:1.

Pereira, M.A. and L.W. Chang. 1980. Binding of chemical carcinogens and mutagens to rat hemoglobin. *Chem.-Biol. Interact.* **33**:301.

———. 1982. Binding of chloroform to mouse and rat hemoglobin. *Chem.-Biol. Interact.* **39**:89.

Segerbäck, D. and L. Ehrenberg. 1981. Alkylating properties of dichlorvos (DDVP). *Acta Pharmacol. Toxicol.* **49** (Suppl. V):56.

Segerbäck, D., C.J. Calleman, L. Ehrenberg, G. Löfroth, and S. Osterman-Golkar. 1978. Evaluation of genetic risks of alkylating agents. IV. Quantitative determination of alkylated amino acids in haemoglobin as a measure of the dose after treatment of mice with methyl methanesulfonate. *Mutat. Res.* **100**:71.

Snellings, W.M., C.S. Weil, R.R. Maronpot. 1982. Final report of ethylene oxide two-year inhalation study on rats. Bush Run Research Center, Pittsburgh, Pennsylvania.

Veleminsky, J., S. Osterman-Golkar, and L. Ehrenberg. 1970. Reaction rates and biological action of N-methyl- and N-ethyl-N-nitrosourea. *Mutat. Res.* **10**:169.

Walles, S. 1981. Reaction of benzyl chloride with hemoglobin and DNA in various organs of mice. *Toxicol. Lett.* **9**:379.

Wieland, E. and H.-G. Neumann. 1978. Methemoglobin formation and binding to blood constituents as indicators for the formation, availability and reactivity of activated metabolites derived from *trans*-4-aminostilbene and related aromatic amines. *Arch. Toxicol.* **40**:17.

COMMENTS

SEGA: Is histidine the only amino acid in hemoglobin that is alkylated by ethylene oxide?

CALLEMAN: No, the N-terminal valine residues, and cysteine and to lesser extents probably a few others are alkylated by ethylene oxide.

WEINSTEIN: If you are going to generalize your last remarks, don't we have to see data showing that the relative potency of a series of carcinogens correlates with the extent of hemoglobin adduct formation? Otherwise you are limited to cases where you are comparing a precursor and its reactive metabolite.

CALLEMAN: Carcinogenic potencies of precursors and their metabolites represent unusually clear-cut cases where predictions can be based on hemoglobin dose monitoring, provided that the distribution in the body of the compounds are properly taken into account. For monofunctional alkylating agents you can also make predictions between diverse compounds based on their rates of reaction with nucleophiles of low nucleophilic strength, such as oxygen atoms. (cf. Ehrenberg 1979; Ehrenberg and Osterman-Golkar 1980; Ehrenberg and Hussain 1981). For other classes of carcinogens, such as aromatic amines, I don't think we're ready for risk estimation since too little is known about the selectivity of their reactions with nucleophiles or the chemical nature of their critical target groups in the DNA.

SWENBERG: What is your direct evidence for metabolism of ethylene to ethylene oxide? John Dent spent a full year trying unsuccessfully to show metabolism.

CALLEMAN: How was that done? The evidence I showed from mice is quite clear. Since this conversion has been shown in many other organisms I assume it is ubiquitous.

SWENBERG: Dent only looked at rats. Perhaps there is a species difference.

CALLEMAN: Andersen and coworkers demonstrated that metabolism does take place, but they do not identify any metabolite in the rat. Ortiz de Montellano showed hydroxyethylations of the heme group of cytochrome P-450 in rats.

SWENBERG: With ethylene or ethylene oxide?

CALLEMAN: With ethylene oxide I think there is no doubt that we will get hydroxyethylation of nucleophilic groups. Did John Dent try to determine hydroxyethylations of nucleophilic target groups?

SWENBERG: Yes, they used some extensive radioisotopic methods with some very, very high specific activity and up to 10,000 ppm, as I recall, of ethylene, and just never were able to show any covalent binding.

CALLEMAN: It would be interesting to know which nucleophile was used for trapping hydroxyethylations.

SWENBERG: I don't know the exact details but he spent at least a year doing it and based it on the work in your laboratory.

CALLEMAN: I'd like to point out one interesting thing, namely, that if you want to demonstrate hydroxyethylations and don't have a sufficiently high specific activity of ethylene, it does not help very much to increase the exposure concentration to very high values since the rate of conversion of ethylene reaches a maximal value. This is also interesting in terms of risk estimation, because if you have some compound whose metabolic conversion obeys Michaelis-Mentens kinetics and make linear extrapolations from very high doses where you find a positive response, or like in the case of ethylene an upper confidence limit for the response, then you will tend to underestimate the risk at low-exposure concentrations where the ratio between in vivo dose and exposure dose is much higher.

The Occurrence of S-methylcysteine in the Hemoglobin
of Normal Untreated Animals

PETER B. FARMER
MRC Toxicology Unit
MRC Laboratories
Carshalton, Surrey SM5 4EF
England

It has been proposed that genotoxic exposure to alkylating agents could be determined by measurements of the degree of alkylation of hemoglobin in the exposed species (Osterman-Golkar et al. 1976; Ehrenberg and Osterman-Golkar 1980). Such measurements have been carried out using radiochemical methods in animals for a wide variety of alkylating agents (Ehrenberg et al. 1977; Segerbäck et al. 1978; Pereira and Chang 1981, 1982) and the results indicate that exposure to a large number of carcinogens and mutagens could be monitored by these techniques.

For human exposure to alkylating agents radiochemical techniques cannot be used, therefore there is a need for the development of sensitive analytical techniques for monitoring protein alkylation by nonradioactive compounds. Such methods would also be of value in the interpretation of some of the radiochemical experimental results. Thus, for example, some of the radioincorporation into hemoglobin following exposure to a labeled alkylating agent could arise from biosynthetic processes rather than alkylation reactions. The availability of a specific analytical method for alkylated material would enable the relative extents of these possible routes for radiolabeling to be determined.

We have developed methods using high resolution gas chromatography-mass spectrometry (GC/MS) for the determination of the alkylated amino acids formed in protein chains following exposure to alkylating agents (Bailey et al. 1980; Farmer et al. 1980, 1982). Our initial work was carried out on the formation of S-methylcysteine following exposure of animals to methylating agents. During this investigation we observed the presence of this alkylated amino acid in the globin from normal untreated animals (Bailey et al. 1981). The amount of the compound was species-dependent, ranging from several hundred nmol/g in avian species to 6 nmol/g in hamsters. In view of these results it was important to check that the presence of S-methylcysteine in our final analyzed extract was really due to its presence in the globin molecule and was not caused by the work-up procedure or an impurity in the globin. It was also of interest to determine if background levels of other alkylated amino acids, such as the closely related compound S-ethylcysteine, are present in globin. The existence of these would diminish the value of the hemoglobin alkylation technique for the detection of exposure to low levels of alkylating agents. This paper describes our preliminary work in these areas.

METHODS

Globin was isolated from blood samples essentially by the method of Anson and Mirsky (1930). Normally the protein was isolated by precipitation with 1% HCl in acetone from the lysate of washed erythrocytes. In some experiments the lysate was dialysed against water prior to globin precipitation, and in others the final isolated product was eluted from a column of Sephadex G25 (28 × 2 cm) in 10% acetic acid.

S-Methylcysteine in the globin was determined by previously described methods (Bailey et al. 1980; Farmer et al. 1980). In brief the method was as follows: The protein was hydrolyzed in 6 M HCl in vacuo and the alkylated amino acid separated from the resulting mixture by ion exchange chromatography (AG 50W H-form). The amino acid was then derivatized (n-butyl ester, N-heptafluorobutyryl) and subjected to capillary gas chromatography-chemical ionization mass spectrometry. Quantitation was carried out by selected ion monitoring using either a deuterated or an enantiomeric internal standard.

Enzymic hydrolysis was also carried out on some protein samples. For this, the globin (10 mg) and the internal standard S-trideuteromethylcysteine (1 μg) were dissolved in *tris* buffer (1 ml, 0.1 M, pH 7.25) and were treated with protease (Sigma Chemical Co., Type XIV, 12 units) at 37° for 16 hours. The pH was then adjusted to 8.1 by addition of NaOH solution (60 μl, 1M) and leucine aminopeptidase was added (Sigma Chemical Co., Type III-CP, 12.5 units). After another 4 hours at 37° the product was chromatographed in the same way as that used for the acid hydrolysate. Assays with ninhydrin (Moore and Stein 1954) were used to check that the duration of the enzymic hydrolyses was sufficient for maximum amino acid production.

S-Ethylcysteine was determined in a similar fashion to S-methylcysteine using S-trideuteromethylcysteine as an internal standard. In this case a larger fraction size from the ion exchange separation had to be used in order to incorporate both compounds. Hydrolyzed globin was applied to a column of AG 50W-X4 (12.5 × 0.6 cm) and the amino acids were eluted with 1 M HCl. The fraction 15-30 ml was evaporated to dryness under a stream of nitrogen, and the product converted to its n-butyl ester N-heptafluorobutyryl derivative. Chemical ionization (isobutane) selected ion monitoring was carried out on the MH$^+$ ions (402 and 391) of the compounds as they eluted from a capillary column (20 m × 0.3 mm) coated with Chirasil Val. Quantitation was carried out by reference to calibration lines established for known mixtures of S-ethylcysteine and S-trideuteromethylcysteine added to hydrolyzed rat globin and taken through the assay procedure.

RESULTS

This further investigation into the presence of S-methylcysteine in globin is being carried out on the blood from the four species: rat, mouse, chicken, and human. The levels of the alkylated amino acid in these globins are shown in Figure 1 (Bailey et al. 1981). These values were obtained by the acidic hydrolysis procedure on undialyzed globin. Dialysis of the globin prior to its hydrolysis has not so far shown any appreciable losses of S-methylcysteine (Table 1). Thus a sample of rat globin (59 nmol S-methylcysteine/g globin) after dialysis had 56 nmol S-methylcysteine/g globin. Similarly a sample of chicken globin (205 nmol/g) was dialyzed and was then shown to contain 191 nmol S-methylcysteine/g globin. The dialysate contained ≤ 15 nmol/g.

Gel chromatography on Sephadex G25 did remove some S-methylcysteine from the globin product. The S-methylcysteine content of samples of rat globin decreased by up to 50% on Sephadex chromatography. However, the column fractions which would contain low molecular weight material such as S-methylglutathione showed no detectable S-methylcysteine. An even greater loss of S-methycysteine has been seen after gel filtration of a chicken globin sample.

When the acidic hydrolysis stage in the isolation procedure of S-methylcysteine from globin was replaced by enzymic hydrolysis, the amounts of S-methylcysteine detected in the products again showed marked variation between species (Table 1). Additionally, the absolute amounts seen were similar to those observed following acid hydrolysis.

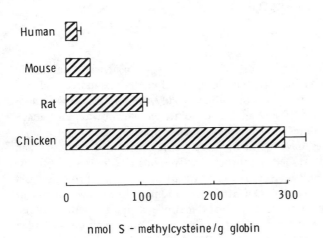

nmol S - methylcysteine / g globin

Figure 1
The amount of S-methylcysteine isolated from rat, mouse, chicken and human globin. Standard deviations are shown for humans (n = 7), rat (n = 20), and chicken (n = 6) (Bailey et al. 1981). For the mouse most S-methylcysteine determinations were carried out on globin isolated from pooled blood samples (3-5 animals/sample). The result shown is the mean of 11 such samples.

Table 1

S-Methylcysteine Isolated from Globin after Various Treatments (nmol/g globin)

	Undialyzed acid hydrolysis	Dialyzed acid hydrolysis	Undialyzed enzyme hydrolysis
Rat 1	70	—	64
Rat 2	70	—	69
Rat 3	64	—	53
Rat 4	59	56	—
Human	13	—	16
Chicken 1	360	—	601
Chicken 2	205	191	—

A fuller description of the treatments carried out on the globin is given in the Methods Section. The results now being obtained for the rat are lower than those previously demonstrated (Fig. 1). It is possible that this difference may have been caused by the fact that the strain of rat used has recently been rederived.

The quantitation of S-ethylcysteine using S-trideuteromethylcysteine as internal standard gave a satisfactorily linear calibration line (correlation coefficient 0.998). A little extra contamination was seen on the GC/MS traces owing to the use of a larger fraction from the ion exchange separation which resulted in a greater content of interfering material in the final derivatized extract. At the present stage of development of this assay, the limit of sensitivity when 10 mg of globin is used as 5 nmol/g globin. No S-ethylcysteine could be detected above this limit in mouse or rat globin (10 mg).

DISCUSSION

Although this is a preliminary report of our work with insufficient animals in some of the experiments for statistical interpretation, the results do indicate that S-methylcysteine (or a compound which readily decomposes to S-methylcysteine) is a constituent of globin. Some points should, however, be made about the analysis procedures used and their possible drawbacks.

First, the analyses have normally been carried out on a crude protein precipitate from lysed washed erythrocytes. The possible presence of low molecular weight contaminants or of minor quantities of other proteins apart from globin cannot be excluded. Our attempts to remove low molecular weight material by dialysis have not led to loss of S-methylcysteine from the protein. However, we have a preliminary indication that chromatography on Sephadex G25 results in considerable losses of alkylated amino acid from the protein fraction. The reason for this is unknown.

Secondly, it is known that the presence of methionine in a protein hydrolysis (6 M HCl) solution can cause S-methylation of cysteine (Calleman et al. 1979). In our acidic hydrolysis solutions we have included mercaptoethanol which is reported to prevent this methylation process. The effectiveness of this procedure was checked by the addition of a 50-fold excess of methionine to a hydrolysis mixture, which resulted in only a 150% increase in S-methylcysteine content. This would indicate that the background levels of S-methylcysteine are not due to reaction products of cysteine and methionine. Additionally, enzymic hydrolysis of the protein with protease and leucine aminopeptidase has also yielded S-methylcysteine in approximately the same quantities as acid hydrolysis.

S-Methylcysteine sulphoxide is widely distributed in nature being particularly abundant in vegetables such as broccoli, cauliflower, and turnips (Morris and Thompson 1956). As the sulphoxide is unstable under acid conditions (Synge and Wood 1956) it is conceivable that it is the sulphoxide that is present, at least in part, in the globin samples studied. S-Methylcysteine has been isolated from human urine (Tominaga et al. 1963) and in this case it was believed that this was formed from the sulphoxide in the isolation procedure. Further investigation of the nature of the globin constituent is thus necessary although this would present considerable experimental difficulties owing to the instability of the sulphoxide.

Another possibility for the source of the S-methylcysteine is that it is formed by methylation of cysteine by dietary constituents. Alkylating agents including dimethylnitrosamine (Kann et al. 1977) have been found in laboratory diets although it seems unlikely that dietary differences could result in the observed species dependence. A marked variation in the levels of reduced glutathione in the erythrocytes of the species studied could also result in different degrees of protection against the effects of absorbed alkylating agents. However this also seems unlikely on the basis of both the reported levels of glutathione in the literature (Long 1961) and our unpublished determinations using high performance liquid chromatographic separation of the S-carboxymethyl N-dinitrophenyl derivative of reduced glutathione (Reed et al. 1980). Whatever the origin of the S-methylcysteine, its presence means that exposure to low levels of methylating agents cannot be determined by monitoring its formation in hemoglobin. S-Ethylcysteine however may well be a suitable compound for monitoring ethylation as its background levels are much lower.

N-3'-(2-Hydroxyethyl)histidine is also being studied owing to its involvement in monitoring procedures for exposure to ethylene, ethylene oxide, and vinyl chloride (Ehrenberg et al. 1977; Osterman-Golkar et al. 1977; Calleman et al. 1978). Background levels (0-3 nmol/g globin) of this alkylated amino acid are also being observed. Once again the possibility of an experimental artifact should be considered very carefully as the amount of material detected in these analyses is only in the low nanogram range. However the in vivo formation of hydroxyethylhistidine is a conceivable process, as it could arise for example from interactions with ethylene oxide derived from the metabolism of ethylene. Ethylene is a commonly found urban pollutant and has also been detected in the exhaled gases of humans (Chandra and Spencer 1963). Additionally, its production by rat liver extracts has been demonstrated (Lieberman and Hochstein

1966; Fu et al. 1979). The homologue of hydroxyethylhistidine, N-3'-(2-hydroxypropyl) histidine, which could be used in the monitoring of exposure to propylene or propylene oxide, appears to be much less abundant than hydroxyethylhistidine in untreated species.

In conclusion, it appears that the existence of background levels of some alkylated amino acids in globin may restrict the applicability of hemoglobin alkylation technique for the monitoring of exposure to certain genotoxic agents. However, the possibility that some of these background levels may be due to previous unsuspected exposure to electrophilic species should be examined further as this may be very relevant to the identification and quantitation of environmental genotoxic hazards.

ACKNOWLEDGMENTS

The author wishes to acknowledge the assistance of Ms. S.M. Gorf, J.H. Lamb, and Dr. E. Bailey in the provision of blood samples, the performance of the hemoglobin analyses and the preparation of this manuscript.

REFERENCES

Anson, M.L. and A.E. Mirsky. 1930. Protein coagulation and its reversal. The preparation of insoluble globin, soluble globin and heme. *J. Gen. Physiol.* 13:469.

Bailey, E., P.B. Farmer, and J.H. Lamb. 1980. The enantiomer as internal standard for the quantitation of the alkylated amino acid S-methyl-L-cysteine in haemoglobin by gas chromatography-chemical ionisation mass spectrometry with single ion detection. *J. Chromatogr.* 200:145.

Bailey, E., T.A. Connors, P.B. Farmer, S.M. Gorf, and J. Rickard. 1981. Methylation of cysteine in hemoglobin following exposure to methylating agents. *Cancer Res.* 41:2514.

Calleman, C.J., L. Ehrenberg, S. Osterman-Golkar, and D. Segerbäck. 1979. Formation of S-alkylcysteines as artifacts in acid protein hydrolysis, in the absence and in the presence of 2-mercaptoethanol. *Acta Chem. Scand.* B33:488.

Calleman, C.J., L. Ehrenberg, B. Jansson, S. Osterman-Golkar, D. Segerbäck, K. Svensson, and C.A. Wachtmeister. 1978. Monitoring and risk assessment by means of alkyl groups in hemoglobin in persons occupationally exposed to ethylene oxide. *J. Environ. Pathol. Toxicol.* 2:427.

Chandra, G.R. and M. Spencer. 1963. A micro apparatus for absorption of ethylene and its use in determination of ethylene in exhaled gases from human subjects. *Biochem. Biophys. Acta* 69:423.

Ehrenberg, L. and S. Osterman-Golkar. 1980. Alkylation of macromolecules for detecting mutagenic agents. *Teratog. Carcinog. Mutagen.* 1:105.

Ehrenberg, L., S. Osterman-Golkar, D. Segerbäck, K. Svensson, and C.J. Calleman. 1977. Evaluation of genetic risks of alkylating agents. III. Alkylation of haemoglobin after metabolic conversion of ethene to ethene oxide *in vivo. Mutat. Res.* **45**:175.

Farmer, P.B., S.M. Gorf, and E. Bailey. 1982. Determination of hydroxypropyl-histidine in haemoglobin as a measure of exposure to propylene oxide using high resolution gas chromatography-mass spectrometry. *Biomed. Mass Spectrom.* **9**:69.

Farmer, P.B., E. Bailey, J.H. Lamb, and T.A. Connors. 1980. Approach to the quantitation of alkylated amino acids in haemoglobin by gas chromatography-mass spectrometry. *Biomed. Mass Spectrom.* **7**:41.

Fu, P.C., V. Zic, and K. Ozimy. 1979. Studies of ethylene-forming system in rat liver extract. *Biochem. Biophys. Acta* **585**:427.

Kahn, J., B. Spiegelhalder, G. Eisenbrand, and R. Preussmann. 1977. Occurrence of volatile N-nitrosamines in animal diets. *Z. Krebsforsch.* **90**:321.

Lieberman, M. and P. Hochstein. 1966. Ethylene formation in liver microsomes. *Science* **152**:213.

Long, C. ed. 1961. *Biochemists handbook,* p. 845. E. and F.N. Spon, London.

Moore, S. and W.H. Stein. 1954. A modified ninhydrin reagent for the photometric determination of amino acids and related compounds. *J. Biol. Chem.* **211**:907.

Morris, C.J. and J.F. Thompson. 1956. The identification of (+) S-methyl-L-cysteine sulfoxide in plants. *J. Am. Chem. Soc.* **78**:1605.

Osterman-Golkar, S., L. Ehrenberg, D. Segerbäck, and I. Hallstrom. 1976. Evaluation of genetic risks of alkylating agents. II. Haemoglobin as a dose monitor. *Mutat. Res.* **34**:1.

Osterman-Golkar, S., D. Hultmark, D. Segerbäck, C.J. Calleman, R. Gothe, L. Ehrenberg, and C.A. Wachtmeister. 1977. Alkylation of DNA and proteins in mice exposed to vinyl chloride. *Biochem. Biophys. Res. Commun.* **76**:259.

Pereira, M.A. and L.W. Chang. 1981. Binding of chemical carcinogens and mutagens to rat hemoglobin. *Chem.-Biol. Interact.* **33**:301.

———. 1982. Binding of chloroform to mouse and rat hemoglobin. *Chem.-Biol. Interact.* **39**:89.

Reed, D.J., J.R. Babson, P.W. Beatty, A.E. Brodie, W.W. Ellis, and D.W. Potter. 1980. High-performance liquid chromatography analysis of glutathione, glutathione disulfide and related thiols and disulfides. *Anal. Biochem.* **106**:55.

Segerbäck, D., C.J. Calleman, L. Ehrenberg, G. Löfroth, and S. Osterman-Golkar. 1978. Evaluation of genetic risks of alkylating agents. IV. Quantitative determination of alkylated amino acids in haemoglobin as a measure of the dose after treatment of mice with methyl methanesulfonate. *Mutat. Res.* **49**:71.

Synge, R.L.M. and J.C. Wood. 1956. (+)-(S-Methyl-L-cysteine S-oxide) in cabbage. *Biochem. J.* **64**:252.

Tominaga, F., S. Kobayashi, I. Muta, H. Takei, and M. Ichinose. 1963. On the isolation and identification of S-methylcysteine from human urine. *J. Biochem.* **54**:220.

Hemoglobin Binding as a Dose Monitor for Chemical Carcinogens

MICHAEL A. PEREIRA AND LINA W. CHANG
U.S. Environmental Protection Agency
Health Effects Research Laboratory
Cincinnati, Ohio 45268

The estimation from animal bioassay data of the carcinogenic risk to humans resulting from an exposure to a chemical carcinogen requires knowledge of the dose and pharmacokinetics of the chemical in humans. Most of the other assays described at this conference such as mutagenicity of body fluids, unscheduled DNA synthesis, alkaline elution, cytogenetics, sister chromatid exchange, micronuclei, and lymphocyte mutagenesis are indicators of an exposure to a genotoxic agent or tumor initiator but are not dose monitors. They do not give either the identity of the genotoxic agent or a quantitative estimate of the dose received. The indicators of exposure to genotoxic agents can only be used to distinguish between exposed and nonexposed populations. This is because 1) the background level of the indicator varies among individuals and 2) the background level for a given individual is usually not known. The systemic dose in an individual can be obtained by quantitation of the carcinogen and metabolites in such body fluids as blood, urine, sputum, semen, milk, and feces or can indirectly be estimated by hemoglobin binding. Since the background level of hemoglobin binding is zero, hemoglobin binding can be used to demonstrate an individual's exposure and dose. The presence of a benzo[a]pyrene or chloroform adduct in human hemoglobin demonstrates that the individual has been exposed to the carcinogen. The presence of methyhistidine demonstrates exposure not to a particular carcinogen but to the class consisting of direct and indirect methylating agents. Therefore, hemoglobin binding has the potential to be an individual dose and exposure monitor.

A pharmacokinetic model is outlined in Figure 1 that relates hemoglobin binding to the systemic dose and by way of target organ dose to carcinogenic potency (Pereira et al. 1981). Two types of tumor initiators have been identified; direct-acting initiators are electrophiles capable of binding macromolecules, while indirect-acting initiators require metabolic activation to an electrophile (RM) before they can bind. If RM is not detoxified, it can bind to protein, lipid, RNA or DNA. The binding of RM to lipid and protein can result in cytotoxicity. The binding to DNA results in the formation of altered bases

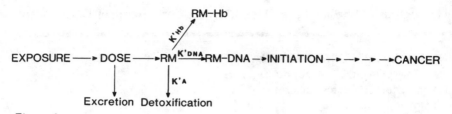

Figure 1
Pharmacokinetic model of hemoglobin alkylation. The pharmacokinetic model of hemoglobin alkylation by environmental chemicals can be used to estimate the concentration of RM of carcinogens. The estimated concentration of RM can then be used to derive estimates of dose, exposure, and target organ macromolecular binding. RM-DNA, RM-lipid, and RM-protein represents the reactive metabolite covalently bound to DNA, lipid and protein, respectively.

that can be repaired, be innocuous, or result in mispairing of the altered base during DNA replication. The alteration in the genome is then fixed and transmitted onto daughter cells. A critical alteration in the genome can result in a transformed cell and thus initiate the carcinogenic progression.

Of major importance in the extrapolation of animal data to estimate human response, is a knowledge of species variation in metabolism to RM and in detoxification. Estimates of human metabolism can be obtained from accidently exposed people by measurement of metabolites in body fluids, feces, and expired air. A major drawback for measuring the presence of metabolites in humans is that the analyses must be performed immediately, that is, prior to the excretion of the metabolites. The binding of RM to two different macromolecules in blood (i.e., hemoglobin and lymphocyte DNA) have been proposed as surrogates for estimating RM (Fig. 1). Neither type of binding directly results in carcinogenesis but is rather an indirect measure of RM, and can be used to estimate species variation in metabolism, systemic and target organ dose and carcinogenic activity (Fig. 1).

Chloroform has been demonstrated to induce hepatocellular carcinomas in mice and kidney epithelial tumors in male rats (National Cancer Institute 1976). Chemicals can increase the incidence of cancer by two distinct mechanisms: genetic and epigenetic (Boutwell 1974, 1978). Chloroform does not appear to be a genotoxic carcinogen. Chloroform was nonmutagenic in the Ames *Salmonella*/microsome assay (Simmon et al. 1977), *Escherichia coli* K 12 for basepair substitution (Uehleke et al. 1976); and 8-azaguanine locus in Chinese hamster lung fibroblast in cell culture (Sturrock 1977). The amount of covalent binding of chloroform to rat liver and kidney DNA was very minimal compared to other carcinogens (Reitz et al. 1980; Pereira et al. 1982). The negative evidence for chloroform mutagenicity and the low level of DNA binding has resulted in the proposal that chloroform is an epigenetic carcinogen. Therefore, the indicators of exposure that are based on mutagenesis including binding to lymphocyte DNA are not appropriate for demonstrating human exposure to epigenetic carcinogens such as chloroform.

Humans are exposed to ppb levels of chloroform in their drinking water (Symons et al. 1975). The chloroform is produced during chlorination by the reaction of chlorine with the organics including the fulvic and humic acids present in the water. Chloroform has been demonstrated to bind covalently to protein (Reynolds and Yee 1967; Ilett et al. 1973; Hill et al. 1975; Uehleke and Werner 1975). Therefore, the binding to hemoglobin of epigenetic carcinogens such as chloroform which covalently bind protein might be an appropriate dose monitor.

METHODS AND EXPERIMENTAL PROCEDURES

Determination of In Vivo Binding to Hemoglobin

The procedure for determination of the in vivo binding of carcinogens has been previously described by Pereira and Chang (1981, 1982). Briefly, blood was obtained by cardiac puncture under light ether anesthesia. The red blood cells were isolated, washed, and hemolyzed. The globin was precipitated with acetone containing 1% HCl (-20°C) and washed with acetone. The extent of radio-labeled carcinogen bound to the globin was determined by combustion in a Packard Oxidizer followed by liquid scintillation counting.

Determination of In Vitro Binding of Chloroform -^{14}C to Hemoglobin

Microsomes (109,000xg for 1 hour) were isolated from Aroclor 1254 induced Sprague-Dawley rats (Charles River, Portage, MI). The microsomes were suspended in 0.05M KPO_4 (pH 7.5). The incubation medium contained 1g rat hemoglobin, 4.4mCi chloroform -[14]C (ICN Pharmaceuticals, Inc., Irvine, CA; 10 mCi/mmole; Lot No.: 981492). 0.4 mmole $MgCl_2$, 1 mmole KPO_4 (pH 7.0), 0.128 mmole glucose-6-phosphate, 40 mg NADPH, 500 units glucose-6-phosphate dehydrogenase and 1 mg microsomal proteins in a total volume of 40 ml. The incubation was performed under oxygen at 37°C for 2 hours. An aliquot (0.5 ml) of the incubation was diluted with 1 ml water and extracted 3 times with 1 ml toluene. The globin was precipitated with acetone containing 1% HCl (-20°C). The extent of chloroform -^{14}C bound to the globin was determined after combustion by liquid scintillation counting.

RESULTS AND DISCUSSION

Stability of Hemoglobin Binding and Dose-response Relationships

The major advantages of hemoglobin binding as a dose monitor are easy access from humans in large quantities (5 ml blood yields 0.5-0.75 g hemoglobin), and

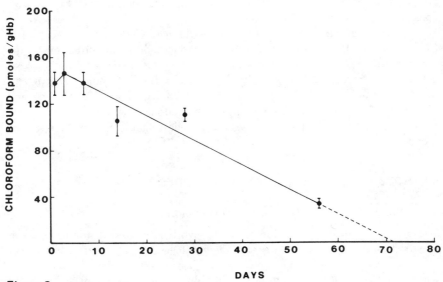

Figure 2
Time course of the binding of chloroform-[^{14}C] to rat hemoglobin. Rats were administered orally 20 μmol/kg chloroform-[^{14}C] and sacrificed at various times later. Results are averages from 6 rats ± S.E. Data redrawn, with permission, from Pereira and Chang 1982.

circulating hemoglobin is stable with a lifetime of 45-68 days in rats and 120 days in humans. The lifetime of hemoglobin containing covalently bound metabolites of chloroform shown in Figure 2, was determined in rats to be approximately 70 days which is (approximately the lifetime of rat erythrocytes (Pereira and Chang 1982). Over 70% of the radioactivity in these preparations was determined to be amino acid adducts and not to represent metabolic incorporation into de novo synthesis of hemoglobin. The shape of the lifetime curve indicates that hemoglobin of all ages contained bound metabolite of chloroform. This is similar to the data reported by Ehrenberg (1979) for the binding of dimethyl-nitrosamine (NDMA) in mice. We have also shown that the extent of chloroform binding to hemoglobin resulting from 10 daily doses was equal to the binding that resulted from a single dose equivalent to the total of the 10 daily doses. Therefore, the binding of a carcinogen to hemoglobin can be used to integrate the dose obtained during the long lifetime of circulating hemoglobin so that low chronic human exposures might be detected even if the daily exposure was below detection. This is a major advantage that hemoglobin binding has over measurement of metabolites in body fluids and lymphocyte DNA binding (DNA repair could prevent the accumulation of adducts to bases).

A list of the chemical carcinogens that we have shown to bind hemoglobin when administered orally, is presented in Table 1. Other carcinogens and

Table 1
Carcinogens that Covalently Bind Hemoglobin In Vivo[a]

Chemical	Binding Index[b]
A. Direct-Acting Carcinogens	
1. Methyl Methanesulphonate	3320
2. N-Nitroso-Methylurea	302
3. N-Nitroso-Ethylurea	217
4. N-Methyl-N'-Nitro-N-Nitrosoguanidine	180
B. Indirect-Acting Carcinogens	
1. Dimethylnitrosamine	697
2. Benzo[a]pyrene	181
3. 3-Methylcholanthrene	165
4. Diethylnitrosamine	153
5. Benzene	144
6. 2-AAF	108
7. 7,12-Dimethylbenz[a]anthracene	99.8
8. Aflatoxin B_1	39.4
9. Aniline	24.8
10. Benzidine	22.5
11. Chloroform	14.2
12. Carbon Tetrachloride	10.2

[a]Pereira and Chang (1981).
[b]Binding Index: pmole bound/g hemoglobin/μmole/kg bw

mutagens that bind hemoglobin include ethylene oxide (Ehrenberg et al. 1974; Osterman-Golkar et al. 1976) and vinyl chloride (Osterman-Golkar et al. 1977). Both direct- and indirect-acting carcinogens bound hemoglobin. The direct-acting carcinogens in general bound hemoglobin to a greater extent than indirect acting carcinogens. Much of the RM of indirect-acting carcinogens formed presumably in the liver was probably detoxified, hydrolyzed, and bound to other macromolecules before it reached the circulating erythrocytes to bind hemoglobin. This might explain why weak carcinogens that are direct-acting alkylating agents bind hemoglobin to a greater extent than potent indirect-acting carcinogens. There was no relationship among the indirect-acting carcinogens between the extent of binding to hemoglobin and carcinogenic potency. For example, aflatoxin B_1, the most potent carcinogen tested, was eighth in the order of binding among the indirect-acting carcinogens and benzo[a]pyrene bound to a greater extent than 7,12-dimethylbenz[a]anthracene. Differences in the stability of the RMs of the carcinogens and in their ability to reach circulating erythrocytes as well as differences in the relationship of the RMs to carcinogenic potency are reasons why hemoglobin binding can not be used to rank carcinogens according to their potency.

The dose-response relationship of the binding of chloroform and 2-acetylaminofluorene (2-AAF) to hemoglobin is shown in Figure 3. The extent of

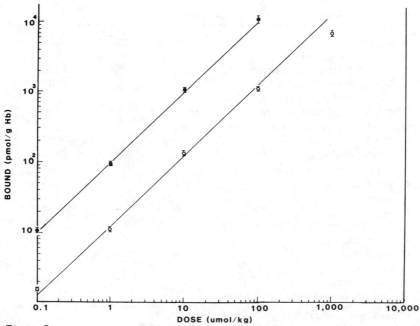

Figure 3

Dose dependency of 2-AAF-[^{14}C] and chloroform-[^{14}C] bound to rat hemoglobin. Rats were administered orally various doses of 2-AAF-[^{14}C] chloroform-[^{14}C] and sacrificed 24 hr. later. Results are averages from 6 rats ± S.E. The line with a slope of 1 that best fits the data is shown. (Data redrawn, with permission, from Pereira and Chang (1982) and Pereira et al. 1981). (●) 2-AAF; (○) chloroform.

chloroform and 2-AAF binding to hemoglobin was linearly dependent upon dose between 0.1 and 100 μmoles/kg (Pereira et al. 1981; Pereira and Chang 1982). Above 100 μmole/kg chloroform the binding increased at a reduced rate. The binding of 2-AAF to liver DNA was also related to dose in a linear manner from 1-100 μmole/kg (Pereira et al. 1981). However, the binding of 2-AAF to hemoglobin and DNA did not support, as proposed in Figure 1, a quantitative relationship between macromolecular binding and carcinogenic potency. The lack of a quantitative relationship between target organ dose measured either as DNA binding or indirectly as hemoglobin binding might be due to the many confounding steps in the neoplastic progression between initiation and cancer.

ISOLATION OF CARCINOGEN ADDUCTS TO HEMOGLOBIN

Isolation of Hemoglobin Containing Carcinogen Adducts

Chloroform-^{14}C was bound to hemoglobin in vitro in a microsomal incubation as described in Methods and Experimental Procedures. The aqueous phase, after three toluene extractions, was saturated with carbon monoxide, centrifuged, and 0.5ml-1ml was applied to a Bio-Rex 70 column (0.9 × 11 cm; Bio-Rad Laboratories, Richmond, CA). The column was eluted sequentially with Developer 4 (pH 6.91, 50 mm Na^+) and high phosphate buffer (pH 6.37, 392 mM Na^+) as described by Cole et al. 1978. A high specific activity peak I (4.4 × 10^5 dpm/protein) was eluted early in Developer 4 (Fig. 4a). A much lower specific activity peak II (8.9 × 10^4 dpm/mg protein) was eluted in the high phosphate buffer. Peak I represented a 4.9-fold enrichment of the hemoglobin-containing chloroform-derived adducts. Over 60% of the chloroform-^{14}C present in peak I remained bound to the globin after precipitation with acetone containing 1% HCl ($-20°C$) and washing with acetone. Hemoglobin from rats not treated with carcinogens contained very little, if any, of the peak eluted by Developer 4 (Fig. 4b). The separation of hemoglobin containing bound carcinogen by Bio-Rex 70 chromatography should prove very useful as a first step in the purification of the carcinogen adducts to the amino acids in hemoglobin.

Isolation of Chloroform Adduct

Peak I of the Bio-Rex 70 elution containing the in vitro bound chloroform-^{14}C was reduced with sodium borohydrate and desalted by dialysis. The globin was

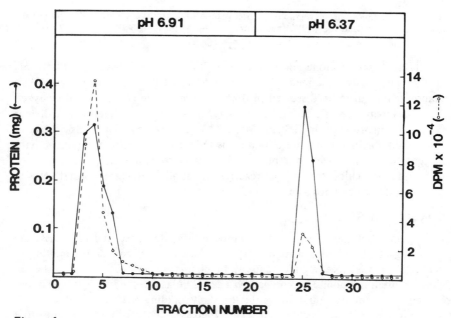

Figure 4a
Bio-Rex 70 column chromatography elution profile of chloroform-[^{14}C] bound in vitro to hemoglobin. Protein was determined by the Bradford (1976) procedure. (•———•) protein (mg); (o— — — —o) DPM × 10^{-4}.

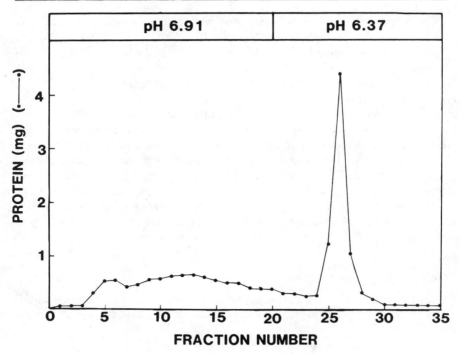

Figure 4b
Bio-Rex 70 column chromatography elution profile of control hemoglobin. Protein was determined by the Bradford (1976) procedure. (●———●) protein (mg); (o— — — —o) DPM ×
10^{-4}.

precipitated with acetone containing 1% HCl (-20°C) and hydrolyzed with 6N HCl as previously described (Pereira and Chang 1982). The hydrolysate was evaporated to dryness, dissolved in 0.2N sodium citrate (pH 2.2) and analyzed in a Beckman 118 BL amino acid analyzer equipped with a stream divider (Beckman Instrument Inc., Palo Alto, CA). The radioactivity was eluted in four peaks distinct from the amino acids (Fig. 5). Therefore, the radioactivity present in the hemoglobin represented adducts to amino acids of a reactive metabolite of chloroform. The reactive metabolite of chloroform has been demonstrated to be phosgene (Pohl 1979).

CONCLUSIONS

Hemoglobin binding of chemical carcinogens was demonstrated to possess the following attributes of a dose monitor: a) the extent of carcinogen binding to hemoglobin was dose related, and b) in circulating erythrocytes, the hemoglobin containing bound carcinogen was as stable as normal hemoglobin so that hemoglobin binding could integrate the daily binding that occurred over the long lifetime of the erythrocyte. This increases the sensitivity of the dose monitor to chronic low-level doses such as those that result from environmental exposure.

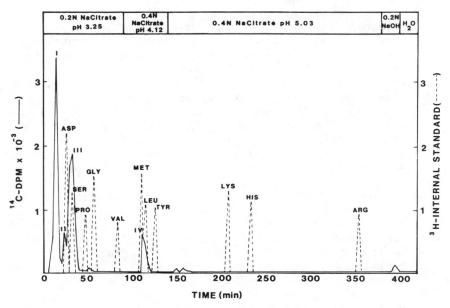

Figure 5
Amino acid analysis of the acid hydrolyzate of chloroform-[^{14}C] bound in vitro to hemoglobin. Amino acid analysis was performed with a Beckman 118 BL. Tritium-labeled amino acids were added as internal standards to the hydrolysate. (———) ^{14}C-DPM $\times$ 10^{-3}; (– – – –) ^{3}H-internal standard.

A procedure is being developed for the isolation of carcinogen adducts in hemoglobin that employs as its initial step the isolation of the hemoglobin containing the bound carcinogen.

REFERENCES

Boutwell, R.K. 1974. Function and mechanism of promoters of carcinogenesis. *CRC Crit. Rev. Toxicol.* 2:419.

———. 1978. Biochemical mechanism of tumor promotion. In *Carcinogenesis Mechanisms of Tumor Promotion and Cocarcinogenesis* (eds., T.J. Slaga et al.), vol. 2, p. 49. Raven Press, New York.

Bradford, M. 1976. A rapid and sensitive method for the quantitation of microgram quantities of protein utilizing the principle of protein-dye binding. *Anal. Biochem.* 72:248.

Cole, R.A., J.S. Soeldner, P.J. Dunn, and H.F. Bunn. 1978. A rapid method for the determination of glycosylated hemoglobins using high pressure liquid chromatography. *Metabolism* 27:289.

Ehrenberg, L. 1979. Risk assessment of ethylene oxide and other compounds. *Banbury Rep.* 1:157.

Ehrenberg, L., K.D. Hiesche, S. Osterman-Golkar, and I. Wennberg. 1974. Evaluation of genetic risks of alkylating agents: Tissue doses in the mouse from air contaminated with ethylene oxide. *Mutat. Res.* **24**:83.

Farmer, P.B., E. Bailey, J.H. Lamb, and T.A. Connors. 1980. Approach to the quantitation of alkylated amino acids in hemoglobin by gas chromatography mass spectrometry. *Biomed. Mass Spectrom.* **7**:41.

Hill, R.M., T.L. Clemens, D.L. Liu, E.S. Vesell, and W.D. Johnson. 1975. Genetic control of chloroform toxicity in mice. *Science* **190**:159.

Ilett, K.F., W.D. Reid, I.G. Sipes, and G. Krishna. 1973. Chloroform toxicity in mice: Correlation of renal and hepatic necrosis and covalent binding of metabolites to tissue macromolecules. *Exper. Molec. Pathol.* **19**:215.

National Cancer Institute. 1976. Carcinogenesis bioassay of chloroform. Nat. Tech. Inf. Service No. PB 264018/AS. Government Printing Office, Washington, D.C.

Osterman-Golkar, S., L. Ehrenberg, D. Segerbäck, and I. Hallstrom. 1976. Evaluation of the genetic risks of alkylating agents. II. Hemoglobin as a dose monitor. *Mutat. Res.* **34**:1.

Osterman-Golkar, S., D. Hultmark, D. Segerbäck, D.J. Calleman, R. Goethc, L. Ehrenberg, and C.A. Wachmeister. 1977. Alkylation of DNA and protein in mice exposed to vinyl chloride. *Biochem. Biophys. Res. Commun.* **76**:259.

Pereira, M.A. and L.W. Chang. 1981. Binding of chemical carcinogens and mutagens to rat hemoglobin. *Chem. Biol. Interact.* **33**:301.

———. 1982. Binding of chloroform to mouse and rat hemoglobin. *Chem. Biol. Interact.* **39**:89.

Pereira, M.A., L.-H.C. Lin, and L.W. Chang. 1981. Dose-dependency of 2-acetylaminofluorene binding to liver DNA and hemoglobin in mice and rats. *Toxicol. Appl. Pharmacol.* **60**:472.

Pereira, M.A., L.-H.C. Lin, J.M. Lippitt, and S.L. Herren. 1982. Tri halomethanes as initiators and promoters of carcinogenesis. *Environ. Health Perspect.* (in press).

Pohl, L.R. 1979. Biochemical toxicology of chloroform. In *Reviews in biochemical toxicology* (eds., E. Hodgson et al.). Vol. 1, p. 79. Elsevier/North-Holland, New York.

Reitz, R.H., J.F. Quast, W.T. Scott, P.G. Watanabe, and P.J. Gehring. 1980. Pharmacokinetics and macromolecular effects of chloroform in rats and mice: Implications to carcinogenic risk estimation. In *Water chlorination: Environmental impact and health effects* (eds., R.L. Jolley, W.A. Brungs, R.E. Cumming, and V.A. Jacobs), vol. 3, p. 983. Ann Arbor Science Publishers, Inc., Ann Arbor.

Reynolds, E.S. and A.G. Yee. 1976. Liver parenchymal cell injury. V. Relationships between patterns of chloromethane-^{14}C incorporation into constituents of liver in vivo and cellular injury. *Lab. Invest.* **16**:591.

Segerbäck, D., C.J. Calleman, L. Ehrenberg, G. Lofroth, and S. Osterman-Golkar. 1978. Evaluation of genetic risk of alkylating agents. IV. Quantitative determination of alkylated amino acids in hemoglobin as a measure of the dose after treatment of mice with methyl methanesulfonate. *Mutat. Res.* **49**:71.

Simmon, V.F., K. Kauhaven, and R.G. Tardiff. 1977. Mutagenic activity of chemicals found in drinking water. *Dev. Toxicol.* 2:249.

Sturrock, J. 1977. Mitosis in mammalian cells during exposure to anesthetics. *Br. J. Anesth.* **49**:207.

Symons, J.M., T.A. Bellar, J.K. Carswell, J. DeMarco, K.L. Kropp, R.G. Robeck, D.R. Seeger, C.J. Slocum, B.L. Smith, and A.A. Stevens. 1975. National organics reconnaissance survey for halogenated organics. *J. Am. Water Works Assoc.* **67**:634. (update in *J. Am. Water Works Assoc.* **67**:708).

Uehleke, H. and Th. Werner. 1975. A comparative study on the irreversible binding of Tabeled halothane, trichlorofluoromethane, chloroform, and carbon tetrachloride to hepatic protein and lipids in vitro and in vivo. *Arch. Toxicol.* **34**:289.

Uehleke, H., H. Greim, M. Kraemer, and T. Werner. 1976. Covalent binding of haloalkanes to liver constituents, but absence of mutagenicity on bacteria in a metabolizing test system. *Mutat. Res.* **38**:114.

SESSION IV:
DNA ADDUCTS

The Covalent Binding Index—DNA Binding in Vivo as a Quantitative Indicator for Genotoxicity

WERNER K. LUTZ
Institute of Toxicology
Swiss Federal Institute of Technology and University of Zurich
CH-8603 Schwerzenbach, Switzerland

The probability for the induction of a tumor by a genotoxic carcinogen is a function primarily of *dose* and *potency* of the carcinogen. A number of contributions to this conference deal with the monitoring of genotoxic *exposure*. My presentation, however, deals with the evaluation of *carcinogenic potency* of a genotoxic compound. In principle, only long-term bioassays on carcinogenicity can provide a respective estimate, but because of the numerous problems connected with this assay, short-term tests have to be evaluated for this purpose. With mutagenicity tests on isolated cells or microorganisms, it has been shown repeatedly that a quantitative correlation between mutagenicity for bacteria and carcinogenicity might be found within selected groups of compounds (Meselson and Russell 1977; Lee and Guttenplan 1981), in a number of cases not even there (Andrews and Lijinsky 1980; Brambilla et al. 1981; Parodi et al. 1981a, 1981b; Ashby et al. 1982), and definitely not with carcinogens of entirely different structure (Bartsch et al. 1980; Parodi et al. 1982).

The limiting factor seems to be the complexity of activation or inactivation processes which are heavily-distorted in in vitro assays (Ashby and Styles 1978; Glatt et al. 1979). Only intact organisms, preferably with circulating blood, can exhibit human-like toxicokinetics and be used for quantitative purposes. One possibility, the study of the covalent binding of chemicals in vivo to DNA, is discussed in this chapter.

THE COVALENT BINDING INDEX

The expression of a DNA binding per unit dose was introduced by Lutz and Schlatter (1977a) for a comparison of the DNA-binding activities of various chemicals tested in a number of laboratories under widely different conditions. The unit chosen for the Covalent Binding Index (CBI) is

$$CBI = \frac{\mu\text{mol chemical bound} / \text{mol DNA nucleotide}}{\text{mmol chemical applied} / \text{kg body weight}} \qquad (1)$$

More conspicuously, these units mean that a compound which exhibits a CBI of 1 will generate one DNA adduct per 10^6 nucleotides after a theoretical dose of 1 mmol/kg.

CBIs for more than 80 compounds have been compiled from the literature and from our laboratory in a review (Lutz 1979). The values measured span about 5 orders of magnitude, ranging from more than 10^4 (aflatoxin B_1; Lutz et al. 1980) down to about one (e.g., benzene; see Lutz and Schlatter 1977b). There can be no doubt that there are compounds that bind to DNA on an even lower level. However, it will be shown below that a long-term bioassay on carcinogenicity with such a compound will not give a positive result unless this compound exhibits some type of cocarcinogenic or promoting activity in addition to its minute genotoxicity.

ROUGH CLASSIFICATION WITH RESPECT TO GENOTOXIC POTENCY

CBIs have been shown to very roughly reflect the genotoxic potency of a chemical in long-term bioassays. CBIs of the order of 10^3-10^4 are found with potent carcinogens, of around 10^2 for moderate carcinogens, and a range of 1-10 for weak carcinogens. A plot of the CBI in the target organ as a function of the respective tumorigenic potency derived from long-term bioassays on carcinogenicity has been set up with 13 chemicals where both values were published (Lutz 1982). The correlation coefficient of 0.74 calculated from a linear regression analysis was surprisingly high and about equal to the internal consistency of carcinogenic potencies calculated from replicate bioassay data of the same compound (Parodi et al., this volume). Therefore it seems that the determination of a CBI should allow a rough classification of compounds into categories of varying genotoxic potency in vivo. This correlation holds not only for related chemicals that are metabolized via the same activation and inactivation reactions but seems to include all chemical classes.

With carcinogens that require metabolic activation, it has been shown that liver DNA is always damaged to a high extent, even if the liver is not a target for tumor development. Thanks to this empirical fact it is possible with a new test compound to study the DNA binding in liver first and obtain valuable information about a possible formation in vivo of reactive metabolites able to penetrate to and react with DNA.

The accuracy of this general correlation of CBI vs. genotoxic carcinogenic potency is limited by the fact that neither persistence nor mutagenicity of the various DNA adducts formed by one single carcinogen are taken into account.

A number of applications will be given in a later section where these limitations do not interfere.

EXPERIMENTAL

The low amount of DNA-carcinogen adducts formed in vivo calls for the use of radiolabeled test compound, at least until the new methods adopted from molecular biology, using the phosphorylation of nucleotide-carcinogen adducts with [^{32}P]phosphate of highest specific activity will be more generally introduced.

A DNA-binding experiment in vivo includes the following steps:

Protocol for a DNA-Binding Assay in Vivo

1. Administration of radiolabeled test compound to an animal, preferably by the route of human exposure
2. Waiting period of hours, sometimes days, to allow for absorption, distribution, and metabolism of the test compound (→ maximum level of DNA binding), possibly for DNA repair[1]
3. Isolation of DNA and determination of the specific activity[2]
4. Degradation of DNA to nucleosides or bases and analysis for the type of radioactivity (covalent binding; biosynthetic incorporation)[3]
5. Calculation of CBI (i.e., DNA binding per unit dose).[4]

[1] Two time points are interesting: first, the time of maximum binding, and second, a much later time point, when the efficiency of the repair processes can be studied. For the time of maximum binding, pharmacokinetic data should help to define the time when most of the dose administered has been metabolized to a putative reactive metabolite. If no data are available, a time-dependent study might have to be performed. For the assessment of DNA repair, much later time points (weeks to months) can, in addition, be taken.

[2] DNA should be purified to constant specific activity, i.e., to a state of purity where an additional purification step does not alter the specific activity anymore. This would indicate that neither contaminations with protein or RNA nor noncovalently bound chemical is the reason for the radioactivity measured on the DNA.

[3] It is highly probable that DNA will be radiolabeled if the labeled atom of the test compound can enter the pool of precursors of nucleic acid biosynthesis, e.g., as formaldehyde, carbon dioxide, acetate, tritiated water (HTO). It is important to deduct this fraction from the total radioactivity because only the remaining radioactivity can be due to DNA adducts. With new test compounds, we now regularly measure the specific activity of expired HTO or the amount of [^{14}C]O_2 expired from animals treated with a [^{3}H]- or [^{14}C]-labeled compound, respectively, using metabolism cages where the humidity of the expired air is frozen out and the carbon dioxide is trapped. The fraction of the radioactivity dose expired in the form of these breakdown products can then be put into relation to the specific activity of the DNA. A comparison with the respective data obtained from control experiments with radiolabeled compounds that are degraded intracellularly to HTO or [^{14}C]O_2 but are not expected to bind to DNA (methylamine, ethanol) will allow a rough estimate for whether the radioactivity measured on the DNA after administration of the

Limit of Detection; Exclusion of DNA Binding in Vivo

Most new chemicals to be tested do not bind to DNA in vivo. In order to establish this important information, it is of prime importance to determine, in CBI units, on what level such a negative finding can be made. A statement of 'zero' obviously is not acceptable. We must answer the question: How many dpm in the scintillation vial containing X mg DNA would be needed to be detectable in our system? The confidence limits are given by the level of significance chosen (2 or 3 standard deviations) and by the *variability of the background DNA counts*. Any contaminations in the background DNA samples will increase the amount of radioactivity required in the DNA sample from the treated animals to be considered significant. For a start, DNA from at least four untreated animals should be isolated in parallel with the isolation of DNA from the treated animals.

The limit of detection, and with this, the relevance of a negative finding for a CBI, is also dependent on the amount of radioactivity administered. A few hundred μCi [^{14}C] or mCi [^{3}H] amounts of radiolabel are required per rat for a limit of detection of about CBI <0.1 in liver DNA, if the limit of detection for a significant DNA radioactivity can be shown to be a few cpm.

Contaminations with Radiolabels

In order to avoid contaminations, a few important precautions are suggested:

1. One laboratory unit should be specifically designed for the work with low level radioactivities encountered on the DNA of the final purification stages. No material should be kept or handled there which contains more than some 1000 dpm. No equipment is to be used for the low-count sample which was used also for the high radioactivity samples (pipettors, lab coat).
2. Disposable or strictly personal and dedicated glassware is used if it comes into contact with DNA. No dishwasher is used for this type of material.
3. The handling of highly labeled samples must be avoided on the same day when DNA is processed.

---

test compound could be due to the biosynthetic incorporation of the radiolabeled breakdown products. This analysis of the data might help to attribute a minute DNA radioactivity to a biosynthetic incorporation in cases where the specific activity of the DNA is too low for a chromatographic analysis of the nucleosides or bases.

[4] A conversion of the experimental data to the CBI is easily done, when taking into account that 1 mol deoxynucleotides weighs 309 g. The experimentally available data for the DNA binding in the units of dpm/mg DNA and for the dose administered in the units of dpm/kg body weight, can be processed according to

$$\text{CBI} = (\text{dpm/mg DNA})/(\text{dpm/kg b.w.}) \cdot 3.09 \cdot 10^8 \qquad (2)$$

APPLICATIONS

A list of useful applications for a DNA-binding assay in vivo is compiled in Table 1. A classification is attempted according to the knowledge about the compound concerned with resect to bacterial mutagenicity tests and long-term bioassays on carcinogenicity. A quick look at the number of examples given in each column reveals the main field for suitable applications of a DNA-binding study in vivo. It includes, above all, those situations where long-term data were positive or not available. The results of a bacterial mutagenicity test system seem not to be crucial for whether or not a DNA-binding assay is indicated.

In *situation A* (positive/positive), the questions to be answered are mainly for risk assessments from unavoidable exposures to carcinogens, for instance about extrapolations of long-term data to the lower doses of human exposure (A1) or about the extent of formation in vivo from noncarcinogenic precursors (A2). In this type of study, the DNA binding must be determined in the organ that was the target in the long-term bioassay. A3 and A4 describe more mechanistic studies on the effect of pretreatments of animals on the DNA damage set by a subsequent administration of a genotoxic compound (A3) or on the efficiency of DNA repair.

Dose-Response Relationships (Example A1)

A CBI can be determined at low dose levels down to the limit of detection in order to find out whether the dose-response relationship to be used for an extrapolation of the bioassay data to lower doses is linear or not. Most studies performed along these lines have revealed a linear relationship between the dose of a genotoxic compound and the level of DNA damage at the lowest dose levels. This supports the hypothesis that there is *no threshold* dose for this type of carcinogen. In higher dose ranges, however, a number of studies have revealed nonlinearities, for instance for the binding of benzo[a]pyrene to rat liver DNA (Lutz et al. 1978), or for the level of O^6-methylguanine in hamster liver DNA after administration of dimethylnitrosamine (NDMA) (Stumpf et al. 1979). The former effect was attributed to an enzyme induction in the higher dose range, the latter finding can be explained by an overloading of the respective DNA repair capacities at the higher levels of methylation. In addition, saturation effects at high dose levels are to be expected. Although these findings do not interfere with the understanding of first order kinetics at the very lowest doses encountered in human situations, it is important to collect data also in the intermediate dose range. They will be needed for biologically sound extrapolation of long-term bioassay data obtained at even higher dose levels to lower doses.

Table 1
Situations Appropriate for a DNA-binding Assay in Vivo

Bacterial Mutagenicity Test	Bioassay Carcinogenesis	Reason for Performing DNA-binding Assay	Examples	Reference
Positive	Positive	A1 Dose-response in vivo A2 Extent of formation in vivo	Benzo[a]pyrene Dimethylamine + Nitrite	Lutz et al. 1978 Meier-Bratschi et al. 1982a
		A3 Modulation of DNA damage A4 Assessment of DNA repair	Aminopyrine + Nitrite Enzyme Inducers —	Meier-Bratschi et al., unpubl. Viviani et al. 1980
	No Data	B1 Prediction of carcinogenic potency B2 Assay with minimum amount of compound available	Gyromitrin Aflatoxin M_1 Aflatoxin D_1 Aflatoxin Resid.	Meier-Bratschi et al. 1981b Lutz et al. 1980 Schroeder et al. unpubl. Jaggi et al. 1980
		B3 In vitro = in vivo? B4 Extent of formation	Cs tear gas: no Primary amine + Nitrite	Däniken et al. 1981a Huber et al. 1982
	Negative	—	—	
No Data	No Data	C1 Predication of carcinogenic potency	Cholesterol-epoxide	Caviezel et al. 1982
Negative	Positive	D Lack of DNA binding in vivo?	Clofibrate/Fenofibrate Saccharin	Däniken et al. 1981b Lutz & Schlatter 1977a
	No Data	E Genotoxicity in vivo? Long-term bioassay indicated?	—	
	Negative	No apparent problem		

Extent of Formation In Vivo from a Noncarcinogen (Example A2)

If the genotoxic compound can be formed in vivo from noncarcinogenic precursors, the extent of formation can be estimated on the basis of DNA binding. The formation of the methylating agent and hepatocarcinogen NDMA from the dietary constituent dimethylamine (DMA) and nitrite under acidic conditions is well known in vitro but there is a lack of quantitative data about the extent of this reaction in the stomach of mammalians. After oral administration of [^{14}C]DMA and potassium nitrite to rats and mice, a methylation of liver DNA could be detected. On comparison with the methylating power of NDMA the conditions used (Meier-Bratschi et al. 1982a). It is important to note that these experimental conditions gave rise to intraintestinal concentrations of amine and nitrite that were both higher (by a factor of about one thousand) than is expected in humans after a meal. An extrapolation to the lower concentrations must therefore be made, possibly on the basis of the well-known data on in vitro nitrosation rates.

Modulations of the DNA Damage (Example A3)

The influence of pretreatments of an animal on the CBI can be investigated. A variety of compounds alter the enzyme activities involved in the activation/ inactivation pathways of genotoxic carcinogens. Such a modulation can result in a higher or lower extent of DNA binding of ubiquitous or unavoidable genotoxic agents. DNA binding studies might be a useful tool in the quantitative evaluation of such influences. We have studied the influence of a number of *enzyme inducers* on the binding of benzo[a]pyrene to rat liver DNA. By concomitant determination of the enzyme activities in the same liver samples we were able to attribute to each activity a role in the modulation of the concentration of DNA-binding benzo[a]pyrene metabolites (Viviani et al. 1980). This type of approach might well be expanded to studies on the influence of fasting, of alcohol, and other important dietary factors.

The *situations of the type B* arise if a positive response in a bacterial mutagenicity test has to be translated to an in vivo situation in order to get some information about the potency in an intact mammalian system.

Known Type of DNA Damage → Accurate Prediction of Potency (Example B1)

A compound which gives rise to a type of DNA damage already known from another carcinogen can be evaluated relatively precisely, on the basis of a CBI, for its mutagenic and carcinogenic potency. Gyromitrin is the toxic principal in the false morel mushroom species *Gyromitra esculenta*. The metabolism of gyromitrin can yield a number of methylating agents. Our analysis of liver DNA of rats that had been given p.o. [³H] methyl-labeled gyromitrin revealed a binding index of about ten (Meier-Bratschi et al. 198ab). This value is many hundred times below the methylating power of the strong hepatocarcinogen NDMA and the risk from consumption of this mushroom can now be compared with the DNA damage arising from the intake of other methylating agents, such as NDMA in food.

Assay with Minimum Amount of Compound Related to a Known Carcinogen (Example B2)

If not enough material for a long-term bioassay on carcinogenicity is available, a CBI can be determined using minimal amounts of chemical. Aflatoxins (AF) form a large group of related compounds but only the most abundant have been studied thoroughly for carcinogenicity. Metabolites occurring only in minor quantities, such as *aflatoxin M_1* (Lutz et al. 1980), or *aflatoxin residues* bound to macromolecules (relay study, Jaggi et al. 1980) are much less well studied. Since a contamination of food with aflatoxins will never be completely avoided, technologies for a decontamination will be required. Of practical use is the degradation of aflatoxin B_1 with ammonia. Among the products is *aflatoxin D_1*, for which no carcinogenicity data are available. In our latest studies, we have investigated the extent of binding of [^{14}C] AFD_1 to liver DNA in the rat. We could not detect any radioactivity on the DNA (T. Schroeder et al., unpubl. results). The limit of detection was at 300 times below the CBI for AFB_1. AFB_1 can therefore really be called a highly decontaminated product.

Discrepancy between In Vivo and In Vitro Metabolism (Example B3)

A compound which is slightly positive in one of the in vitro tests for point mutations but is found to be negative in a DNA binding assay in vivo might represent one of those compounds which are genotoxic only in vitro. *o-Chlorobenzylidene malononitrile (CS)* is a lacrimating riot control agent. Its acute toxicity has been studied in various animal species but a possible carcinogenic effect was merely discussed. CS, as an activated olefin, is chemically reactive, preferentially towards SH-groups of proteins. For an evaluation of a potential carcinogenicity, it would be important to find out whether CS can also interact with DNA. We found a minute mutagenicity with strain TA 100 in the Ames test, but no DNA binding in rat liver although chromatin protein carried a detectable radioactivity (Däniken et al. 1981a). In view of these results, and taking into account the rare and low exposure of man, it is concluded that CS will not create a risk for the induction of point mutations or of a carcinogenic process mediated by DNA binding.

Extent of Formation In Vivo of Chemically Unstable Compound (Example B4)

For chemically reactive compounds that are formed only in vivo, a DNA-binding study at the site of formation will allow the investigator to determine whether the compound formed is stable enough to penetrate to and react with DNA of the nearest cells. This situation is currently under investigation with the *nitrosation of primary amines* in the stomach. The reaction products are known to decay spontaneously to alkylating agents. We have preliminary evidence for the methylation of nitrogen 7 of guanine in the stomach and the small intestine of rats that had been given p.o. [^{14}C]methylamine and potassium nitrite (Huber et al. 1982). Although these findings are based upon concentrations of the reactants that are much higher than is expected in humans, the role of primary amines in the induction of gastrointestinal tumors will have to be more carefully assessed.

Situation C: Prediction of Genotoxicity

Some chemicals cannot be adequately tested for bacterial mutagenicity because of insolubility in water or because of bactericidal properties. For such cases, a DNA-binding assay can provide information about a putative genotoxicity in vivo. Our data collected with *cholesterol-5,6-oxide*, a naturally occurring metabolite of cholesterol, indicate that this epoxide is not reactive towards DNA in vivo (CBI for liver DNA after intravenous administration <0.1) or in vitro (Caviezel et al. 1982), so that the role of cholesterol in carcinogenesis cannot be based upon a genotoxicity of the 5,6-epoxide.

Situation D: Exclude DNA Binding as Mechanism for Tumor Induction

Probably the most important application of the DNA-binding assay in vivo for the industry arises in situations where a long-term assay on carcinogenicity was clearly positive with TD$_{50}$ values of <1 mmol/kg/d, but where additional information, such as mutagenicity data, renders a genotoxic mode of action unlikely.

Here, it is important to verify the lack of DNA binding of the compound also in a mammalian organism. A negative binding assay (limit of detection at least as low as CBI <0.1 for compounds with an estimated daily intake by the general human population on the order of milligrams) would suggest that the tumorigenicity was probably due to a mode of action not related to DNA binding. Such a finding might allow the consideration of a threshold dose and it will be most important to find out whether the processes responsible for the induction of tumors in the bioassay with animals are found also in humans. One published example is given below.

Clofibrate (ethyl-α-p-chlorophenoxyisobutyrate), a widely used hypolipidemic drug, is known to induce hyperplastic and neoplastic changes in rat liver. It is also known to be a potent inducer of peroxisomes in rodents, and it was postulated that peroxisome proliferators might represent a novel class of carcinogens. An assay for DNA binding in rat liver was negative (Däniken et al. 1981b), so that the hepatocarcinogenicity of clofibrate cannot be based upon an initiating, DNA damaging, mode of action, but must be due to other mechanisms. An extrapolation of any carcinogenicity data from rats to man must therefore await a comparison over a wide dose range of a number of biochemical effects of this drug, viz, induction of microsomes and peroxisomes, hypertrophy and hyperplasia of the liver.

IMPROVEMENTS FOR THE GENERAL CORRELATION OF CBI v. CARCINOGENIC POTENCY

Time-Dependent DNA Binding

If DNA damage is properly repaired before the DNA is replicated, no mutation will be induced. DNA repair is therefore an important modulator for the number of mutations produced. The determination of DNA binding not only near the time of the maximum level but also after a few weeks, would therefore allow for the persistence of the DNA damage to be assessed.

Mutagenicity of DNA-Carcinogen Adduct

It is well known from the study of methylating agents that different adducts can widely differ with respect to their biological consequences and a determination of the total level of DNA adducts can distort a quantitative correlation. Thanks to the wide total span of carcinogenic potencies of about 7 orders of magnitude, this error still does not exclude the use of a CBI for placing a compound into a category of genotoxic potency. It is, however, obvious that it would be helpful to know for each adduct its efficiency for eliciting a mutation. Such data slowly become available from mutagenicity tests where, in addition to the scoring of mutations, the underlying DNA damage is determined in parallel.

LIMITATIONS

DNA binding is a very early event in the process of chemical carcinogenesis and one must always keep in mind that a wide variety of modulatory influences ultimately determines the probability for neoplastic growth. Therefore, only carcinogens which act predominantly by DNA binding are amenable to such analysis, and epigenetic carcinogens, such as solid state carcinogens, hormones, immunosuppressors, promoters, or cocarcinogens cannot be studied. Also, it is not possible to evaluate compounds that undergo noncovalent interactions with DNA, that bind exclusively to protein, or that damage DNA indirectly, via the generation of reactive oxygen species in coupled oxidation-reduction reactions. The relative importance of these latter mechanisms in chemical carcinogenesis has, however, not yet been shown conclusively.

The importance of the cell-specific modulatory influences is such as to exclude a prediction of organotropism from a DNA-binding experiment, if the DNA is isolated after a tissue has been homogenized. This situation might be much improved if a DNA damage in a single cell will become amenable to analysis.

FUTURE TRENDS

New Methods for DNA-binding Assays

Use of Unlabeled Test Compound and Postlabeling Techniques

As one of the experimental limitations for a DNA binding assay in vivo, there was a need for radiolabeled compounds. Methodologies adopted from recent advances in molecular genetics, e.g., the kinase-mediated 5'-labeling of nucleotide-3'-monophosphate-carcinogen adducts with $[^{32}P]$ ATP of highest specific activity will result in a phosphorylation to yield a radiolabeled nucleotide-3',5'-diphosphate-carcinogen adduct so that also unlabeled compound will be amenable to a DNA-binding assay in vivo (Randerath et al. 1981).

Use of Monoclonal Antibodies to Carcinogen-DNA Adducts

These new techniques discussed in detail in other chapters will ultimately allow the determination of DNA damage in one single cell after treatment of an animal with a reasonably low dose of a genotoxic carcinogen. This will allow insight into the problem of cellular specificities and organotropism besides its use in exposure monitoring, but this technique will not replace radiolabeled test compounds in the study of *new* chemicals because they will be more easily available.

Additional Determinations in a DNA-binding Assay

Mitogenic Activity of the Test Compound

A number of nongenotoxic carcinogens stimulate cell division and it is conceivable that carcinogens with a demonstrated genotoxicity could, in addition, exhibit this type of mitogenic activity. With the use of radiolabeled precursors for DNA synthesis it should be possible to study this effect and receive additional information on a biological activity of the test compound possibly also related to carcinogenicity.

Protein Binding

Covalent binding of a chemical to a critical protein could also be an important step in chemical mutagenesis. Although the importance of this interaction relative to DNA binding is unknown, it would be feasible to include protein binding in DNA-binding experiments. Such data might also be valuable in an evaluation of the role of protein binding in cytotoxicity and in the regenerative processes elicited thereafter. These might be important modulators in chemical carcinogenesis.

Future Trends in DNA-binding Assays for Risk Assessment

Endogenous and Unavoidable DNA Damage

Degradation of compounds via chemically reactive metabolites is not only a feature of xenobiotics but must be expected to occur also with endogenous substrates; a wide variety of natural food constituents, and with genotoxic products formed in the gastrointestinal tract from ubiquitous nongenotoxic precursors. The binding of such compounds to tissue macromolecules might represent the basis for unavoidable nuclei acid- or protein damage, and possibly contribute to what is called 'spontaneous' tumor incidence (Lutz 1982).

Numerous classes of unavoidable, reactive chemicals and metabolites will have to be considered, such as aldehydes, epoxides, quinones, oxygen radicals, aromatic amines, and diverse amines nitrosated in the stomach. Data on the interaction of estrone with liver DNA of rats treated with this hormone is tritiated form by Jaggi et al. (1978). In addition, preliminary evidence is now available that formaldehyde, generated in vivo from methanol, crosslinks DNA with chromatin protein in the liver of rats (U. Minini and W. K. Lutz, unpubl. results). In the light of such findings, the covalent interactions of xenobiotics with tissue macromolecules will have to be put into relation to an unavoidable background damage set by endogenous and ubiquitous compounds.

REFERENCES

Andrews, A. W. and W. Lijinsky. 1980. The mutagenicity of 45 nitrosamines in *Salmonella typhimurium. Teratog. Carcinog. Mutagen.* **1**:295.

Ashby, J. and J. A. Styles. 1978. Does carcinogenic potency correlate with mutagenic potency in the Ames assay? *Nature* **271**:452.

Ashby, J., P. A. Lefevre, J. A. Styles, J. Charlesworth, and D. Paton. 1982. Comparisons between carcinogenic potency and mutagenic potency to *Salmonella* in a series of derivatives of 4-dimethlyaminoazobenzene (DAB). *Mutat. Res.* **93**:67

Bartsch, H., C. Malaveille, A.-M. Camus, G. Martel-Planche, G. Brun, A. Hautefeuille, N. Sabadie, A. Barbin, T. Kuroki, C. Drevon, C. Piccoli, and R. Montesano. 1980. Validation and comparative studies in 180 chemicals with *S. typhimurium* strains and V79 chinese hamster cells in the presence of various metabolizing systems. *Mutat. Res.* 76:1.

Brambilla, G., M. Cavanna, A. Pino, and L. Robbiano. 1981. Quantitative correlation among DNA damaging potency of six N-nitroso compounds and their potency in inducing tumor growth and bacterial mutation. *Carcinogenesis* 2:425.

Caviezel, M., W. K. Lutz, and C. Schlatter. 1982. Cholesterol-5α-epoxide (ChE) does not interact covalently with DNA in vivo. *Experientia (Basel)* 38: 753.

Däniken, A. von, U. Friederich, W. K. Lutz, and C. Schlatter. 1981a. Tests for mutagenicity in *Salmonella* and covalent binding to DNA or protein in the rat of the riot control agent o-chlorobenzylidene malononitrile (CS). *Arch. Toxicol.* 49:15.

Däniken, A. von, W. K. Lutz, and C. Schlatter. 1981b. Lack of covalent binding to rat liver DNA of the hypolipidemic drugs clofibrate and fenofibrate. *Toxicol. Lett. (Amst.)* 7:311.

Glatt, H. R., H. Schwind, F. Zajdela, A. Croisy, P. C. Jacquignon, and F. Oesch. 1979. Mutagenicity of 43 structurally related heterocyclic compounds and its relationship to their carcinogenicity. *Mutat. Res.* 66:307.

Huber, K. W., W. K. Lutz, and C. Schlatter. 1982. Methylation of DNA by N-methylnitrosamine formed in vivo from methylamine and nitrite. *Experientia (Basel)* 38:755.

Jaggi, W., W. K. Lutz, and C. Schlatter. 1978. Covalent binding of ethinylestradiol and estrone to rat liver DNA *in vivo. Chem. Biol. Interact.* 23: 13.

Jaggi, W., W. K. Lutz, J. Lüthy, U. Zweifel, and C. Schlatter. 1980. *In vivo* covalent binding of aflatoxin metabolites isolated from animal tissue to rat liver DNA. *Food Cosmet. Toxicol.* 18:257.

Lee, S. Y. and J. B. Guttenplan. 1981. A correlation between mutagenic and carcinogenic potencies in a diverse group of N-nitrosamines: Determination of mutagenic activities of weakly mutagenic N-nitrosamines. *Carcinogenesis* 2:1339.

Lutz, W. K. 1979. *In vivo* covalent binding of organic chemicals to DNA as a quantitative indicator in the process of chemical carcinogenesis. *Mutat. Res.* 65:289.

_____. 1982. Constitutive and carcinogen-derived DNA binding as a basis for the assessment of potency of chemical carcinogens. In *Biological Reactive Intermediates 2: Chemical Mechanisms and Biological Effects* (ed. R. Snyder et al.) p. 1349. Plenum Press, New York.

Lutz, W. K. and C. Schlatter. 1977a. Saccharin does not bind to DNA of liver or bladder in the rat. *Chem. Biol. Interact.* 19:253.

_____. 1977b. Mechanism of the carcinogenic action of benzene: Irreversible binding to rat liver DNA. *Chem. Biol. Interact.* 18:241.

Lutz, W. K., A. Viviani, and C. Schlatter. 1978. Nonlinear dose-response relationship of the binding of the carcinogen benzo(a)pyrene to rat liver DNA *in vivo. Cancer Res.* 38:575.

Lutz, W. K., W. Jaggi, J. Lüthy, P. Sagelsdorff, and C. Shlatter. 1980. *In vivo* covalent binding of aflatoxin B_1 and aflatoxin M_1 to liver DNA of rat, mouse and pig. *Chem. Biol. Interact.* **32**:249.

Meier-Bratschi, A., W. K. Lutz, and C. Schlatter. 1982a. Methylation of liver DNA of rat and mouse by dimethylnitrosamine formed in vivo from dimethylamine and nitrite. *Food Chem. Toxicol.* (in press).

———. 1982a,b. Methylation by gyromitrin of DNA in the rat. *Food Chem. Toxicol.* (in press).

Meselson, M. and K. Russell. 1977. Comparisons of carcinogenic and mutagenic potency. *Cold Spring Harbor Conf. Cell Proliferation* **C**:1473.

Parodi, S., S. de Flora, M. Cavanna, A. Pino, L. Robbiano, C. Bennicelli, and G. Brambilla. 1981a. DNA-damaging activity *in vivo* and bacterial mutagenicity of sixteen hydrazine derivatives as related to their carcinogenicity. *Cancer Res.* **41**:1469.

Parodi, S., M. Taningher, P. Russo, M. Pala, M. Tamaro, and C. Monti-Bragadin. 1981b. DNA-damaging activity *in vivo* and bacterial mutagenicity of sixteen aromatic amines and azo derivatives, as related quantitatively to their carcinogenicity. *Carcinogenesis* **2**:1317.

Parodi, S., M. Taningher, P. Boero, and L. Santi. 1982. Quantitative correlations amongst alkaline DNA fragmentation, DNA covalent binding, mutagenicity in the Ames test and carcinogenicity, for 21 compounds. *Mutat. Res.* **93**:1.

Randerath, K., M. V. Reddy, and R. C. Gupta. 1981. ^{32}P-Labeling test for DNA damage. *Proc. Natl. Acad. Sci.* **78**:6126.

Stumpf, R., G. P. Margison, R. Montesano, and A. E. Pegg. 1979. Formation and loss of alkylated purines from DNA of hamster liver after administration of dimethylnitrosamine. *Cancer Res.* **39**:50.

Viviani, A., A. von Däniken, C. Schlatter, and W. K. Lutz. 1980. Effect of selected induction of microsomal and nuclear aryl hydrocarbon monooxygenase and epoxide hydrolase as well as cytoplasmic glutathione S-epoxide transferase on the covalent binding of the carcinogen benzo(a)pyrine to rat liver DNA *in vivo*. *J. Cancer Res. Clin. Oncol.* **98**:139.

COMMENTS

KLIGERMAN: Do you find much variation in the DNA methylation in the intestinal tract or stomach of control animals that have not been given any known carcinogens?

LUTZ: We cannot determine the level of a background methylation of the DNA of control animals because it is a radiolabel that we are looking for in the methylated DNA bases. The treated animals receive the methylating precursor with a radiolabeled methyl group. The control animals receive no radioactivity.

FURIHATA: Did you give methylamine and nitrite by oral gavage?

LUTZ: Yes, first an aqueous solution of methylamine hydrochloride, and about 20 seconds later a solution of potassium nitrite.

FURIHATA: And the methylnitrosamine produced from the reaction of methylamine and nitrite resulted in a methylation of DNA?

LUTZ: Yes.

FURIHATA: Do you have any data on the carcinogenicity of methylnitrosamine in the stomach?

LUTZ: No. Since methylnitrosamine decays so very rapidly, it will be exceedingly difficult to synthesize and administer it to animals in pure form.

FURIHATA: What is your feeling about its carcinogenicity?

LUTZ: Since methylnitrosamine is a very reactive methylating agent of the methyldiazonium ion type, there can be no doubt about its carcinogenic potential. The main question rather is quantitative in nature, for instance whether a critical level of DNA methylation can ever be reached from the probably minute amounts formed in vivo, taking into account the extremely short half-life.

WEINSTEIN: Why did you measure 7-methylguanine rather than O^6-methylguanine?

LUTZ: The O^6-methylguanine was not detectable because of the very low level of DNA methylation. But, assuming that the ultimate methylating agent is the methyldiazonium ion, the ratio of 7- to O^6-methylation is expected to be the same as the one you find with dimethylnitrosamine, i.e., about 10 to 1.

BUTTERWORTH: What were the target organs for benzo[a]pyrene and beta-naphthylamine that you mentioned in your correlation of covalent binding to DNA versus carcinogenic potency?

LUTZ: With benzo[a]pyrene, it was forestomach; with beta-naphthylamine it was liver.

Cell-specific DNA Alkylation and Repair:

Application of New Fluorometric Techniques to Detect Adducts

JAMES A. SWENBERG AND MARY A. BEDELL
Department of Pathology
Chemical Industry Institute of Toxicology
Research Triangle Park, North Carolina 27709

Many chemical carcinogens are tissue or organ specific in their induction of neoplasms. The mechanisms responsible for this tropism include the absorption and distribution of the compound, the sites of metabolic activation and detoxification, DNA binding, repair, and replication, and the route of excretion. In addition to a compound's organotropism, however, many chemicals selectively induce tumors in specific cell populations within the target organ. We have been investigating DNA alkylation, repair, and replication as potential mechanisms responsible for cell specificity in hepatocarcinogenesis.

Our initial investigations of cell-specific hepatocarcinogenesis utilized pulse exposure of rats to ^{14}C-1,2-dimethylhydrazine (SDMH) (Lewis and Swenberg 1980) and demonstrated that liver nonparenchymal cells were deficient in repair of O^6-methylguanine (O^6-MeGua). Subsequently, new methods of analysis which did not require radiolabeled carcinogen were sought in order to examine the effect of continuous carcinogen exposure on DNA alkylation. Two methods for nonradiometric quantitation of specific types of DNA damage were considered, radioimmunoassay (RIA) and high performance liquid chromatography (HPLC) in conjunction with high sensitivity optical detection. Each method offers distinct advantages and disadvantages. RIA is extremely sensitive and is amenable to evaluating large numbers of samples. The principal disadvantage is the requirement for high affinity, high specificity antisera for each DNA adduct. Multiple adducts can be resolved and quantitated in a single run with HPLC, but these methods frequently lack the exquisite sensitivity of RIA. By utilizing highly sensitive fluorescence detectors with HPLC separations, Herron and Shank (1979) were able to quantitate O^6-MeGua and 7-methylguanine (7-MeGua) in DNA from rats acutely exposed to methylating agents. The reported limits of detection were approximately 1 pmol and 40 pmol respectively for O^6-MeGua and 7-MeGua. We have modified their procedures and can detect as little as 300 fmol of O^6-MeGua and 10 pmol of 7-MeGua. These methods have been utilized to determine the concentrations of O^6-MeGua and 7-MeGua in target and nontarget liver cell DNA following exposure of rats and mice to hepatocarcinogens (Bedell et al. 1982; Lindamood et al. 1982; Swenberg et al. 1982b). Our experiences with HPLC/fluorescence quantitation of DNA alkylation products are summarized below.

MATERIALS AND METHODS

Animal Exposures

Animals were exposed to carcinogen ad libitum in their drinking water for intervals of up to 1 month. Male Fischer 344 rats were given SDMH at a concentration of 30 ppm in their water. The intake of this solution resulted in daily doses ranging from 1.7 mg/kg-2.4 mg/kg (Bedell et al. 1982). Male and female C3H mice were given dimethylnitrosamine (NDMA) at 10 ppm in their drinking water which resulted in dosages of 1.7 mg/kg-2.3 mg/kg/day (Lindamood et al. 1982). Fresh solutions of carcinogens were prepared in deionized water every 5 days for SDMH and every 8 days for NDMA. In both cases stability of the carcinogens was greater than 95% during the dosing intervals.

Cell Separations

On the day of sacrifice, rats and mice were anesthetized with Nembutal and their livers perfused in situ with collagenase using methods modified from those of Berry and Friend (1969). Mice were additionally given an i.p. injection of heparin prior to perfusion. The mixed liver cell suspension was then separated into purified populations of hepatocytes and nonparenchymal cells by elutriation centrifugation (Lewis and Swenberg 1980, 1982) or differential centrifugation (Lindamood et al. 1982). The purity of each cell population was greater than 95% for elutriation centrifugation and greater than 90% for differential centrifugation. Separated cells were then quickly frozen in liquid nitrogen and stored at $-80°C$.

DNA Isolations

DNA was isolated using hydroxyapatite chromatography by the method of Beland et al. (1979). Hepatocyte homogenates were extracted with a chloroform-phenol-isoamylalcohol solution prior to hydroxyapatite chromatography while nonparenchymal cell (NPC) homogenates were applied directly to the column without extraction. DNA was then dialyzed against water, concentrated on Amicon YM-10 ultrafiltration membranes (Amicon Corp., Lexington, MA) and stored at $-30°C$.

HIGH PERFORMANCE LIQUID CHROMATOGRAPHY

DNA samples were hydrolyzed in 0.1 N HCl at 80°C for 30 minutes to release purine bases. Hydrolysates were applied directly to the HPLC column. Strong cation exchange HPLC columns (Partisil-10 SCX, 25 × 0.45 cm, Whatman, Clifton, NJ) with Whatman column survival kits were employed in each of three analytical methods for the determination of unmodified and methylated purine bases. In all methods the equipment consisted of the following: A Rheodyne 7125 loop injector (Berkeley, CA), two Waters Solvent Delivery Systems (Milford, MA) controlled by a Chromatograph Control Module (CCM) (Laboratory Data Control, Riviera Beach, FL), a Waters 440 dual wavelength absorbance detector (Milford, MA), and a Perkin-Elmer 650-10 fluorescence spectrophotometer equipped with an 18 μl flow cell (Norwalk, CT). The HPLC columns were eluted with ammonium phosphate buffer ($NH_4H_2PO_4$) with or without methanol that had been adjusted to pH 2.0 with phosphoric acid (all reagents were HPLC grade, Fisher Scientific, Pittsburgh, PA). UV-absorbing and fluorescing peak areas were determined by electronic integration with the CCM. Sample concentrations of unmodified and methylated purine bases were calculated from linear regressions of standard curves. Standard solutions of 7-MeGua (Vega Biochemicals, Tucson, AZ), O^6MeGua (synthesized by Dr. D. Swenson), and guanine and adenine (Sigma Chemical Co., St. Louis, MO) were prepared in 0.01 or 0.1 N HCl.

The determination of methylated purines in DNA from hepatocytes and NPC of rats exposed to SDMH was accomplished using HPLC Method I (Bedell et al. 1982). Small aliquots (100 or 200 μl) of hydrolyzed DNA (50 to 150 μg and 100 to 300 μg respectively for NPC and hepatocytes) were injected unto two Partisil 10-SCX columns in series and eluted isocratically at 1.5 ml/min with 75 mM $NH_4H_2PO_4$ (pH 2.0) to determine concentrations of guanine, adenine, and 7-MeGua. The mean retention times were as follows: guanine (9.0 min), adenine (16.5 min), and 7-MeGua (25.4 min). Effluent was monitored for 254 nm absorbance at a very high sensitivity (0.01 absorbance units full scale) for the determination of 7-MeGua and at 280 nm with a low sensitivity (2.0 units full scale) for guanine and adenine. A larger aliquot of hydrolyzed DNA was used to determine O^6-MeGua concentrations in rats exposed to SDMH. 200 or 500 μl of the hydrolysate containing from 200 to 600 μg of NPC DNA or 400 to 1200 μg of hepatocyte DNA was injected unto a single Partisil 10-SCX column eluted with 50 mM $NH_4H_2PO_4$ plus 12% methanol, pH 2.0, at 1.5 ml/min. Column effluent was monitored for fluorescence at 295 nm excitation and 370 nm emission wavelengths. These conditions were optimal for the quantitation of O^6MeGua only (eluting at an average retention time of 17.7 minutes) and did not fully resolve other peaks.

A second HPLC method was utilized to determine methylated bases in DNA of hepatocytes and NPC from male and female mice exposed to NDMA (Lindamood et al. 1982). This second method was adapted from the first and used fluorescence detection for both 7-MeGua and O^6MeGua. Two Partisil 10-SCX columns in series were eluted with Buffer A (37.5 mM $NH_4H_2PO_4$, at pH 2.0) for the first 10 minutes after sample injection. A linear gradient of 0-40% Buffer B (75 mM $NH_4H_2PO_4$ plus 20% methanol, at pH 2.0) was then generated from 10 minutes-40 minutes. The flow rate was 1.5 ml/min for the first 28.5 minutes, and then it changed to 3 ml/min. The following elution order was obtained: guanine (11.0 min), adenine (16.5 min), 7-MeGua (23.0 min), and O^6-MeGua (34.5 min). Effluent was monitored for both 254 nm absorbance (2.0 units full scale) and fluorescence at 295 nm excitation and 370 nm emission for quantitation of unmodified and methylated purines respectively.

A third HPLC method was modified from the first two methods and utilizes fluorescence detection for both 7-MeGua and O^6-MeGua in two separate injections. Chromatographic conditions are similar to those of Method 1 with 7-MeGua determined after isocratic elution of $NH_4H_2PO_4$ at pH 2.0 without methanol, whereas O^6-MeGua was determined after isocratic elution of the same buffer plus 7% methanol. Guanine and adenine concentrations were determined from 254 nm absorbing peaks as in Method 2.

Miscellaneous Methods

In all experiments, DNA concentrations were determined in aliquots of the samples prior to hydrolysis by either modified diphenylamine (Burton 1956) or diaminobenzoic acid (Hinegardner 1971) procedures.

Calibration of the standard solutions of methylguanines was achieved by injecting hydrolyzed DNA that had been alkylated with ^{3}H-N-methyl-N-nitroso-urea (^{3}H-MNU). Calf thymus DNA (Sigma Chemical Corp., St. Louis, MO) was incubated for 30 minutes with ^{3}H-MNU (New England Nuclear, Boston, MA., specific activity 1.4 Ci/mMol), precipitated with ethanol and dialyzed against water. Neutral thermal hydrolysis was performed according to Beranek et al. (1980) to release 3- and 7-methylpurines and the residual DNA (containing O^6-MeGua was acid hydrolyzed as described above. Although 3-methyladenine (3MeAde) does not fluoresce under the analytical conditions used (see Results Section below) and as such does not interfere with the fluorescence quantitation of O^6-MeGua, the radioactive peak of 3-MeAde was not fully resolved from O^6-MeGua. The two-step hydrolysis procedure was thus necessary for accurate quantitation of O^6-MeGua by radioactivity. The ^{3}H-methylated DNA contained approximately 6 pmol O^6-MeGua and 60 pmol 7-MeGua/mg DNA with the specific activity of the methyl groups being 3108 dpm/pmol. HPLC method 2 was utilized to separate the purine bases. Fluorescing peak areas of O^6-MeGua and 7-MeGua were determined as described above. 1 min fractions of column effluent were collected, 15 ml of aqueous counting scintillant (ACS, Amersham, Arlington Heights, IL) were added, and radioactivity was determined in a Searle Mark III liquid scintillation system (Analytic Inc., Des Plaines, IL). The counting efficiencies of these samples were 37%. Linear regressions of disintegrations per minute dpm versus area of methylguanine peaks were obtained.

Other methylated purines (1-MeGua, 1-MeAde, 3-MeGua, 3-MeAde and 7-MeAde) were obtained from Vega Biochemicals (Tucson, AZ) and prepared for chromatography in the same fashion as for O^6-MeGua and 7-MeGua. The

The elution order and retention times of these products were determined using HPLC method 2. The relative UV-absorbance and fluorescence of each of these peaks were determined under the same conditions as for 7-MeGua and O^6-MeGua (at 254 nm absorbance, 0.01 units full scale; 295 nm excitation and 370 nm emission, 12 nm slits).

Fluorescence spectra for 0.1 mM solutions of each of the methylated and unmodified bases in 10 mM $NH_4H_2PO_4$ at pH 2.0 were determined using an Aminco SPF-500 fluorescence spectrophotometer (Silver Springs, MD). The pH dependence of O^6-MeGua fluorescence was determined by adjusting the pH of a 0.1 mM solution of O^6-MeGua in 10 mM $NH_4H_2PO_4$ with either phosphoric acid or ammonium hydroxide and reading the fluorescence intensity at 303 nm excitation and 363 nm emission in the Aminco spectrophotometer.

RESULTS

The methods presented here (adapted from Herron and Shank 1979) utilized optical quantitation of methylated purines in DNA from liver cell populations after continuous exposure of rodents to methylating carcinogens. In developing these methods several criteria had to be satisfied. First, the chromatographic conditions had to separate the methylated purines and unmodified purines under conditions where concentrations of the latter were up to a million times that of methylated purines. Secondly, optimal pH and concentrations of the mobile phase were required for both chromatographic separation and fluorescence of the methylated bases. Accurate and reproducible integration of the peaks had to be achieved with a linear detector response throughout the range of sample concentrations. Finally, the analysis needed to be reproducible and relatively rapid.

Three HPLC modes were evaluated. The use of strong cation exchange columns satisfied all of the criteria mentioned above. Both reverse phase and ion-pair reverse phase chromatography were evaluated. They did not allow adequate resolution under conditions where fluorescence of the methylated products was optimal. As seen in Figure 1, fluorescence of O^6-MeGua is strongly pH-dependent. Maximum fluorescence occurs in the pH range of 2-3 with decreasing fluorescence intensity at pH levels outside of this range. Cation exchange columns are thus more suitable than reverse phase columns for the analysis of O^6-MeGua by fluorescence. A disadvantage of the cation exchange columns was that optimal conditions had to be determined for the mobile phase concentration and flow rate on each set of columns. In our experience, buffer concentrations of 35-75 mM at a constant pH of 2.0 with flow rates of 1-3 ml/min gave optimal separations and fluorescence detection on four sets of Whatman cation exchange columns. These columns have given good separations for up to 600 samples or approximately 45 liters of mobile phase. The relatively small methanol gradient (up to 5-15 percent) generated in the latter part of the sample run gave earlier elution and a sharper peak for

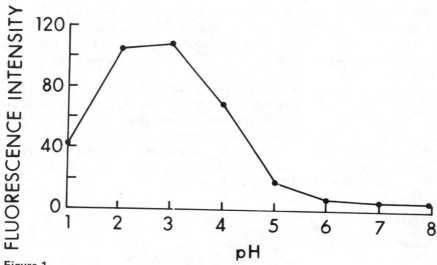

Figure 1

Dependence of O^6-MeGua fluorescence on pH. The pH of a 0.1 mM solution of O^6-MeGua in 10 mM $NH_4H_2PO_4$ was adjusted with either phosphoric acid or ammonium hydroxide. The fluorescence intensity was then determined at 303 nm excitation and 363 nm emission. Each point represents the mean of 2 or 3 determinations.

O^6MeGua. This gradient caused baseline changes that could affect electronic integration of small peaks; however, the integrator used was able to correct for this baseline change.

Figure 2 illustrates the separation of 5 methylated purine bases and the two unmodified purines using the conditions of HPLC method 2. The only methylated purines which could not be resolved under' these conditions were 1-MeGua and 3-MeGua. These coeluted with guanine and adenine, respectively. The peak areas relative to guanine for 254 nm absorbance and fluorescence were determined following injection of equimolar amounts of each of the products in Figure 2 (Table 1). It is noteworthy that the fluorescence of O^6-MeGua is approximately 26 times that of guanine, while the next highest fluorescing product, 7-MeGua, has peak areas only slightly greater (by a factor of 1.3) than guanine. As mentioned (in the Materials and Methods section) O^6-MeGua and 3-MeAde were not fully resolved under these conditions and radiometric analysis required that neutral thermal hydrolysis prior to acid hydrolysis be used. The lack of 3-MeAde fluorescence allowed accurate quantitation of O^6-MeGua by fluorescence under conditions where these products were not fully separated. In contrast to fluorescence intensity, UB-absorbing peak areas were nearly identical for 5 of the purines. The exceptions were 3-MeAde, which had approximately 60% that of guanine, and O^6-MeGua, which had less than 1% that of guanine.

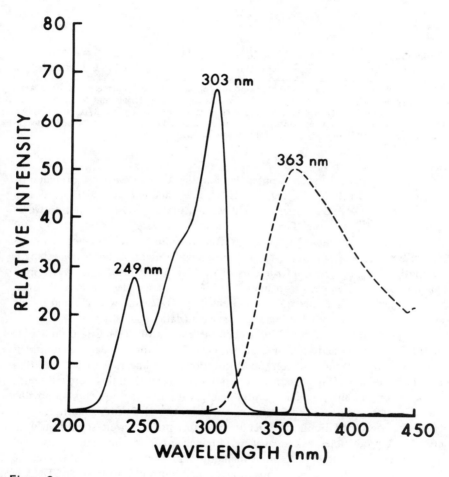

Figure 2

Separation of guanine, adenine, and 5 methylated purine bases by strong cation exchange chromatography. HPLC method 2 (see Materials and Methods) was used to separate equimolar amounts of each of these products. (upper panel) absorbance at 254 nm (with a detector sensitivity of 0.2 units full scale); (lower panel) fluorescence intensity at 295 nm excitation and 370 nm emission. (↓) positions of bases that do not absorb or fluoresce under these conditions.

Table 1

Relative Absorbance and Fluorescence Peak Areas of Some Unmodified and Methylated Purine Bases

Purine base[a]	254 nm absorbance	Fluorescence
Guanine	1.0	1.0
Adenine	1.1	0.08
7-MeGua	1.2	1.4
7-MeAde	1.1	0.01
1-MeAde	0.8	0.08
O^6-MeGua	<0.01	25.6
3-MeAde	0.6	<0.01

[a]Solutions containing equimolar amounts of each of the 7 purine bases listed were chromatographically separated using HPLC method 2 (see Materials and Methods). Peak areas for 254 nm absorbance and fluorescence (295 nm excitation and 370 nm emission) were determined by electronic integration and normalized to the areas of guanine peaks. Each value represents the mean of 3 or 4 determinations.

The fluorescence spectra obtained from the Aminco spectrophotometer of O^6MeGua in 10 mM $NH_4H_2PO_4$ at pH 2 is presented in Figure 3. The emission maximum for O^6MeGua when excited with white light (i.e., excitation wavelength of 0) is 363 nm. Two excitation maxima were observed, at 249 and 303 nm. These maxima are slightly different from those obtained with the Perkin-Elmer spectrophotometer and probably reflect different wavelength calibrations of the two instruments. We therefore utilized the wavelength maxima obtained on each instrument for analysis of samples. Spectra obtained at pH 7 demonstrated that only the intensity of O^6MeGua was affected by pH and a shift in wavelength maxima did not occur (data not shown).

Calibration curves for fluorescence peak areas of O^6MeGua using HPLC method 2 are presented in Figure 4. Peak areas for standard amounts of O^6MeGua in the range of 2-22 pmol/injection were linear, with a correlation coefficient of 0.99 (Fig. 4, Panel A). Amounts smaller than this were either not apparent or did not integrate reproducibly with this method. By optimizing the chromatographic conditions in methods 1 and 3 for quantitation of O^6MeGua only, the limits of detection for this product were significantly increased. Reproducible peak areas were obtained for O^6MeGua amounts as small as 0.3 pmol, compared to 2 pmol in method 2. The linearity of fluorescence peak area vs. peak dpm from injection of acid hydrolyzed ^{3}H-methyl-DNA is shown in Panel B of Figure 4. The data generated had a correlation coefficient of 0.98. Similar curves have been obtained for both UV-absorbance and fluorescence peak areas for 7-MeGua (data not shown) with the limits of detection being 70 and 10 pmol/injection, respectively.

Figures 5 and 6 depict fluorescence chromatograms obtained after injecting hydrolyzed DNA from liver NPC and parenchymal cell populations of rats and mice, respectively. HPLC method 1 was used to obtain the chromato-

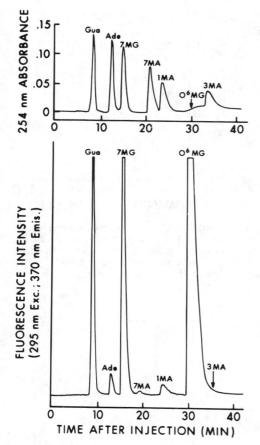

Figure 3
Excitation and emission spectra of O⁶MeGua. The emission spectra (- - -) were determined by exciting a solution of O⁶-MeGua at pH 2.0 with white light (i.e., all excitation wavelengths) and the excitation spectra (——) were obtained at the emission maxima of O⁶-MeGua (363 nm). Wavelengthmaxima of emission and excitation are indicated in the figure.

grams in Figure 5 that demonstrate O⁶MeGua peaks in NPC cell DNA (Panel A) and in hepatocyte DNA (Panel B) of a rat exposed to SDMH for 8 days (Bedell et al. 1982). Elution of O⁶MeGua occurred at an average retention time of 17.7 minutes under these conditions while guanine, adenine, and 7-MeGua were not resolved but were quantitated in a separate run with different conditions. In Figure 6, HPLC method 2 was used to quantitate both 7-MeGua and O⁶MeGua in mouse liver cell DNA by fluorescence (Lindamood et al. 1982). 7-MeGua eluted at a retention time of 23 minutes while O⁶MeGua eluted at

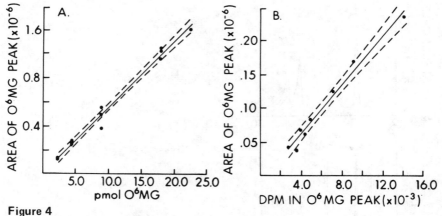

Figure 4
Calibration curves for standard solutions of O⁶MeGua. (——) linear regression curves for fluorescence peak areas versus pmol O⁶MeGua (Panel A) and peak areas versus dpm of ³H-labeled O⁶MeGua (Panel B) (- - -) the 95% confidence intervals of the regression curves. The specific activity of O⁶MeGua was 3108 dpm/pmol. Chromatographic conditions were the same as HPLC method 2 (see Materials and Methods). Correlation coefficients were greater than 0.98 for both curves.

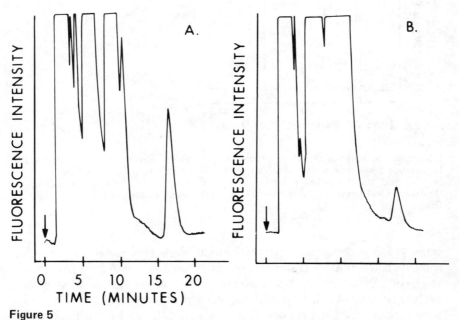

Figure 5
Fluorescence chromatograms of acid hydrolyzed DNA from NPC (Panel A) and hepatocytes (Panel B) of a rat exposed to SDMH at 30 ppm in the drinking water for 8 days. O⁶MeGua elutes between 15 and 20 minutes while guanine, adenine, and 7-MeGua elute earlier and were quantitated by 254 nm absorbance in a separate analysis. (Reprinted, with permission, from Bedell et al. 1982.)

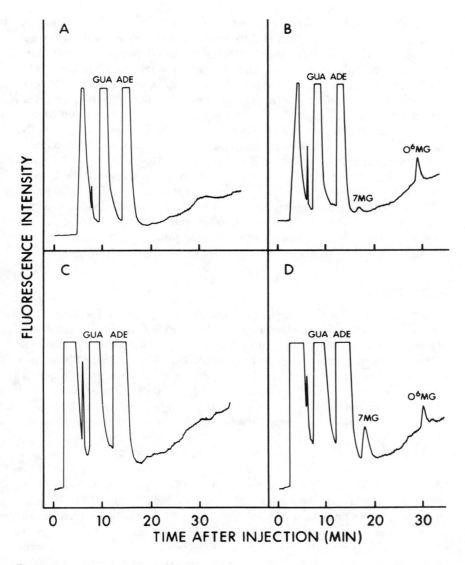

Figure 6
Fluorescence chromatograms of DNA from NPC (Panels A and B) and hepatocytes (Panels C and D) from control mice (Panels A and C) or mice exposed for 16 days to 10 ppm NDMA in the drinking water (Panels B and D). (Reprinted, with permission, from Lindamood et al. 1982.)

34.5 minutes. No peaks eluting after adenine (at 16.5 minutes retention) were observed in NPC (Panel A) or hepatocyte (Panel B) DNA of control mice. Peaks of 7-MeGua and O^6MeGua in NPC DNA (Panel C) and hepatocyte DNA (Panel D) from a mouse exposed to NDMA for 16 days are shown.

Figures 7 and 8 illustrate data obtained from the analysis of DNA in liver cell populations of rats and mice chronically exposed to two methylating carcinogens. SDMH (at 30 ppm) and NDMA (at 10 ppm) administered in drinking water primarily induces vascular tumors in the livers of rats (Druckrey 1970; Bedell et al. 1982) and mice (Grasso and Hardy 1975), respectively. NPC DNA had amounts of 7-MeGua similar to hepatocyte DNA in both rats (Figure 7A) and mice (Lindamood et al. 1982). Distinct differences were observed for the cumulative concentrations of O^6MeGua between hepatocytes and NPC DNA. Hepatocytes of both species had significantly lower amounts of O^6MeGua in their DNA than did NPC (Figure 7B and Figure 8). The ratios of O^6MeGua to 7-MeGua were much higher in NPC DNA than in hepatocyte DNA in both rats (Figure 7C) and mice. These data indicate that rat NPC cells repair O^6MeGua more rapidly during the first 4 days of exposure than during the second week of exposure. This is consistent with studies showing decreased activity of the O^6-alkylguanine alkyl acceptor protein in NPC cells of rats similarily exposed to SDMH (Swenberg et al. 1982a).

DISCUSSION

Fluorescence HPLC is a highly sensitive, reproducible technique for determining the concentrations of 7-MeGua and O^6MeGua in DNA from cells and tissues exposed to methylating agents. It is less useful for such determinations following exposures to ethylating agents because the overall extent of alkylation is lower, resulting in problems with the limits of detection. Moreover, fluorescence HPLC is unable to quantify the alkylated pyrimidines, due to their lack of fluorescence. It is possible that future modifications, such as post-column derivatization, will permit fluorescence detection of these and additional DNA adducts.

Fluorescence HPLC data have demonstrated distinct differences in repair of O^6MeGua by hepatocytes and NPC. The accumulation of O^6MeGua in the NPC coupled with greatly increased de novo DNA synthesis in NPC (Bedell et al. 1982; Lewis and Swenberg 1982) provides a high likelihood for GC $\rightarrow$ AT transitions due to mispairing of O^6MeGua. Since O^6MeGua is the predominant promutagenic lesion following alkylation of DNA by SN_1 methylating agents such as SDMH, these data strongly support O^6MeGua as the primary promutagenic lesion responsible for initiation of hemangiosarcomas of the liver following exposure to these agents. In contrast, O^6MeGua does not appear to be as important in initiating hepatocytes. Other, more slowly repaired lesions such as O^2- and O^4-alkylthymine and O^2-alkylcytosine may play important roles in hepatocyte initiation. Support for this hypothesis is also provided by the greater

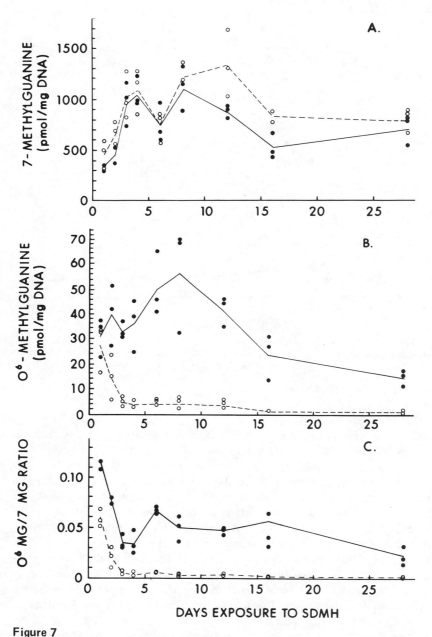

Figure 7
Normalized concentrations of 7-MeGua (Panel A), O⁶-MeGua (Panel B) and the O⁶-MeGua/
7-MeGua alkylation ratios (Panel C) in DNA from NPC (●——●) and hepatocytes (○- - -○)
of rats exposed to 30 ppm SDMH in their drinking water. Each data point represents one
animal while the curve connects the means of three animals at each time point. (Reprinted,
with permission, from Bedell et al. 1982.)

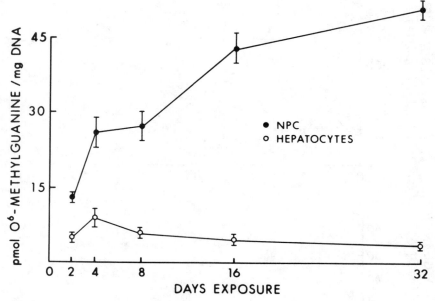

Figure 8
Normalized concentrations of O⁶MeGua in DNA of NPC (●——●) and hepatocytes (○——○) from mice exposed to 10 ppm NDMA in their drinking water. Each point represents the mean ± S.E.M. for 4-8 analyses of DNA pooled from liver cells of three animals. (Reprinted, with permission, from Lindamood et al. 1982.)

specificity of methylating agents for NPC and of ethylating agents for hepatocytes (Bedell et al. 1982; Swenberg et al. 1982a,b).

Nonradiometric HPLC detection methods have provided a relatively inexpensive means to determine DNA alkylation during continuous carcinogen administration. Such studies provide valuable information on the kinetics of DNA alkylation under conditions that simulate carcinogen bioassays. As can be appreciated from the data presented, cumulative concentration curves have vastly different shapes depending upon the amount of DNA repair. It is clear, however, that DNA adducts eventually attain steady state levels, where formation equals removal or loss. In the case of hepatocytes, the steady state actually achieved for O⁶MeGua is much lower than would be predicted by pulse exposure to methylating agents. This is primarily due to the enhanced activity of the O⁶-alkylguanine alkyl acceptor protein (Swenberg et al. 1982a). NPC from SDMH-exposed rats appear to have attained steady-state concentrations of O⁶MeGua by the second week of administration (Bedell et al. 1982), while NPC from mice were still accumulating O⁶MeGua after 4 weeks of NDMA exposure (Lindamood et al. 1982). Steady-state kinetics have also been demonstrated using radioisotopes and RIA. Croy and Wogan (1981) showed that steady-state levels of DNA adducts were achieved within the first 5 days of exposure to daily doses of 25 μg of aflatoxin B_1. Poirier et al. (1982) demonstrated that DNA adducts of 2-acetylaminofluorene reached maximal concentrations after administration for 4 weeks and maintained those levels for an additional 4 weeks. Data on steady-state kinetics of DNA alkylation together with comparable data on cell replication should provide greater insight into the mechanisms of chemical carcinogenesis. Coupling these data to specific cell types of known susceptibility or lack thereof will hopefully provide information on the relative importance of different DNA adducts in the initiation and promotion of chemical carcinogenesis.

ACKNOWLEDGMENTS

The authors wish to acknowledge the invaluable contributions of Dr. Charles Lindamood III, Dr. James G. Lewis, and Ms.Kathryn C. Billings to the work presented here.

REFERENCES

Bedell, M.A., J.G. Lewis, K.C. Billings, and J.A. Swenberg. 1982. Cell specificity in hepatocarcinogenesis: O^6-methylguanine preferentially accumulates in target cell DNA during continuous exposure of rats to 1,2-dimethylhydrazine. *Cancer Res.* 42:3079.

Beland, F.A., K.L. Dooley, and D.A. Casciano. 1979. Rapid isolation of carcinogen-bound DNA and RNA by hydroxyapatite chromatography. *J. Chromatogr.* 174:177.

Beranek, D.T., C.C. Weis, and D.H. Swenson. 1980. A comprehensive quantitative analysis of methylated and ethylated DNA using high pressure liquid chromatography. *Carcinogenesis* 1:595.

Berry, M.N. and D.S. Friend. 1969. High yield preparation of isolated rat liver parenchymal cells. *J. Cell Biol.* 43:506.

Burton, K. 1956. The conditions and mechanism of the diphenylamine reaction for the colorimetric estimation of deoxyribonucleic acid. *Biochem. J.* 62:315.

Croy, R.G. and G.N. Wogan. 1981. Temporal patterns of covalent DNA adducts in rat liver after single and multiple doses of aflatoxin B1. *Cancer Res.* 41:197.

Druckrey, H. 1970. Production of colonic carcinomas by 1,2-dialkylhydrazines and azoxyalkanes. In *Carcinoma of the Colon and Antecedent Epithelium* (ed. W.J. Brudette) p. 267. Charles C. Thomas, Springfield, IL.

Grasso, P. and J. Hardy. 1975. Strain difference in natural incidence of response to carcinogens. In *Mouse Hepatic Neoplasia.* (ed. W.H. Butler), p. 111. Elsevier Publishing Corp., Amsterdam.

Herron, D.B. and R.C. Shank. 1979. Quantitative high-pressure liquid chromatographic analysis of methylated purines in DNA of rats treated with chemical carcinogens. *Anal. Biochem.* 100:58.

Hindegardner, R.J. 1971. An improved fluorometric assay for DNA. *Anal. Biochem.* **39**:197.

Lewis, J.G. and J.A. Swenberg. 1980. Differential repair of O^6-methylguanine in DNA of rat hepatocytes and nonparenchymal cells. *Nature* **288**:185.

Lewis, J.G. and J.A. Swenberg. 1982. Effect of 1,2-dimethylhydrazine and diethylnitrosamine on cell replication and unscheduled DNA synthesis in target and nontarget cell populations in rat liver following chronic administration. *Cancer Res.* **42**:89.

Lindamood, III, C., M.A. Bedell, K.C. Billings, and J.A. Swenberg. 1982. Alkylation and de novo synthesis of liver cell DNA from C3H mice during continuous dimethylnitrosamine exposure. *Cancer Res.* **42**:4153.

Poirier, M.C., B'A. True, and B.A. Laishes. 1982. Formation and removal of (guan-8-yl)-DNA-2-acetylaminofluorene adducts in liver and kidney of male rats given dietary 2-acetylaminofluorene. *Cancer Res.* **42**:1317.

Swenberg, J.A., M.A. Bedell, K.C. Billings, D.R. Umbenhauer, and A.E. Pegg. 1982a. Cell specific differences in O6-alkylguanine-DNA repair activity during continuous exposure to carcinogen. *Proc. Natl. Acad. Sci. USA* **79**:5499.

Swenberg, J.A., D.E. Rickert, B.L. Baranyi, and J.I. Goodman. 1982b. Cell specificity in DNA binding and repair of chemical carcinogens. *Environ. Health Perspect.* (in press).

Aflatoxin-DNA Adducts:

Detection in Urine as a Dosimeter of Exposure

PAUL R. DONAHUE, JOHN M. ESSIGMANN, AND G.N. WOGAN
Department of Nutrition and Food Science
Massachusetts Institute of Technology
Cambridge, Massachusetts 02139

We have had a long-standing interest in the possible role of aflatoxins as etiologic agents for primary hepatocellular carcinoma in exposed populations. The work reported in this paper derives in large part from experiences gained during an epidemiologic survey which we conducted in Thailand (Shank et al. 1972). In that survey, we attempted to determine whether a relationship existed between aflatoxin exposure and liver cancer incidence by direct measurement of aflatoxin ingestion and liver cancer incidence in the same population groups in different regions of the country. Quantification of aflatoxin intake was accomplished by a diet-survey technique in which teams of sample collectors visited statistically selected homes several times each year and systematically collected aliquots of all food items (to the extent possible) as they were prepared for consumption. At the same time, total food intake by each member of the household was measured, as were body weights. Food samples were frozen, shipped to a central laboratory and analyzed chemically for aflatoxin levels. The combined data were then used to estimate aflatoxin intake on a per-family basis. A number of epidemiologic studies of essentially similar design and approach were subsequently carried out in Africa, which collectively produced strong incidence of a statistically positive association between aflatoxin intake and liver cancer incidence (Shank 1977). The results will be discussed further in a subsequent section of this report.

This kind of approach to quantifying exposure of humans to carcinogens suffers from several very serious limitations. It is logistically difficult, expensive, and inevitably leads to underestimation of exposure, since certain items of the diet escape collection and analysis (e.g., those items consumed outside the home at times other than mealtime). Additionally, exposure levels can rarely be calculated on an individual basis, but must be averaged among members of the family or other units. All of these complicating factors emphasize the need for approaches that would permit measurement on individuals and provide more direct, quantitative information on persons at different levels of risks by virtue of quantitatively different exposures.

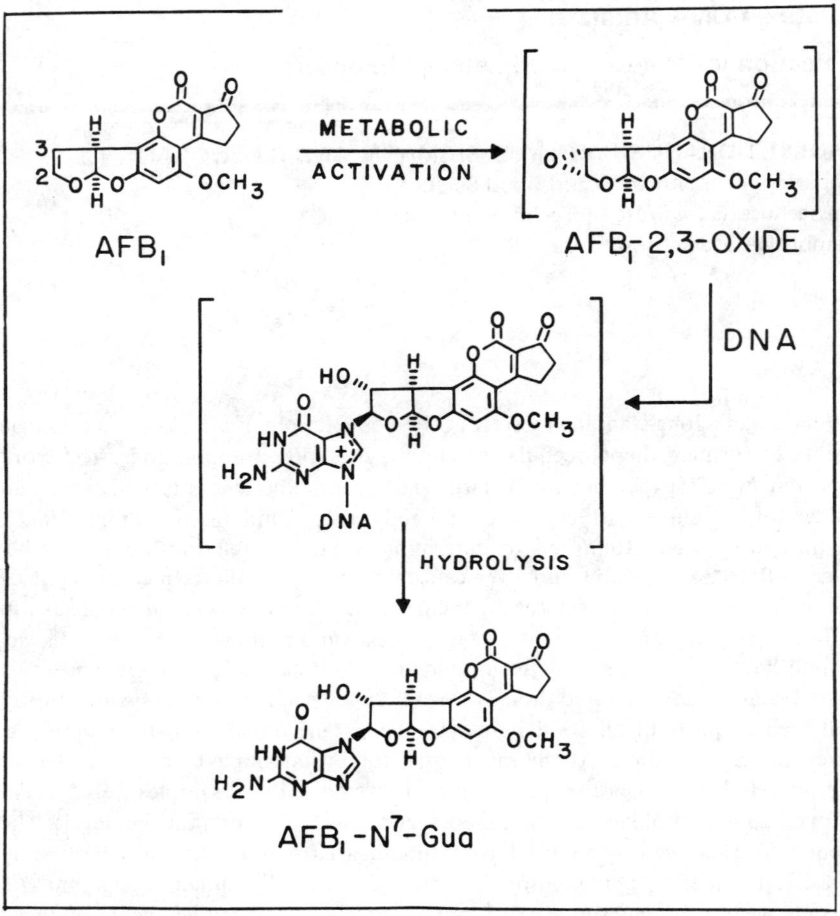

Figure 1
Formation of AFB_1-N^7-Gua through AFB_1 epoxidation.

Our results of recent years on the activation and macromolecular binding of aflatoxin B_1 (AFB_1) in animals have provided sensitive experimental tools that we currently are attempting to use in evaluating human risk.

Figure 1 summarizes the main features of activation of AFB_1 on which the method is based. The primary activation step is formation of the 2,3 epoxide which, when formed in the presence of DNA, undergoes nucleophilic attack preponderantly by the N^7 atom of guanine (N^7-Gua). Formation of this adduct in DNA induces a strong positive charge in the imidazole ring, labilizing the glycosidic bond, and facilitating depurination of the AFB_1-N^7-Gua adduct.

This major adduct is excreted in the urine of rats dosed with AFB_1 and our work has been directed toward its detection and quantification.

Our initial experiments were successful in developing a method for detection of the adduct in the urine of rats dosed with comparatively large amounts of AFB_1 (Bennett et al. 1981). Following extraction and chromatographic purification, by high pressure liquid chromatography (HPLC), the adduct was detected by absorbance at 365 nm (the AFB_1 maximum), and its identity was established by comparison of its properties with a sample of authentic adduct. Using this method of detection and quantification, we then conducted a series of studies in rats to determine the relationships between adduct excretion and levels of adduct in liver DNA. These studies established that a relatively constant proportion (about 35%) of the maximum amount of adduct formed in liver DNA is excreted in urine during the 48 hours after dosing. This relationship appeared to hold over a wide range of AFB_1 doses, from 0.125 mg/kg–1.0 mg/kg. Although these results were encouraging, the sensitivity of the method was inadequate to detect levels of the adducts anticipated to occur in human populations, even in regions where AFB_1 intakes are very high (see below). Consequently, we directed our recent efforts to increasing the sensitivity of the method to a point where it should be applicable for field studies in human populations.

METHODS, EXPERIMENTAL PROCEDURES, AND RESULTS

In its present form, the method includes the following successive stages: (1) extraction and purification of the AFB_1-N^7-Gua adduct; (2) methylation of the adduct with [^{3}H]-dimethyl sulfate of very high specific activity; (3) liberation of [^{3}H]-9-methylguanine (MeGua); and (4) purification and quantification of [^{3}H]-9-MeGua.

The desired increase in sensitivity is accomplished by the introduction of tritiated methyl groups of very high specific activity by methylation with radioactive dimethyl sulfate. This reaction was based on observations made in the course of structure elucidation of the AFB_1-N^7-Gua adduct (Essigmann et al. 1977) that when the adduct is treated with dimethyl sulfate, the primary product is the 9-methyl derivative. Consequently, when the AFB_1 moiety is removed, 9-MeGua is released, and it is this compound that we are quantifying as the marker of AFB_1-N^7-Gua levels.

The major steps of the method are outlined in Figure 2. Human urine samples were randomly collected from laboratory personnel. Urine aliquots (10 ml) were spiked with known quantities of authentic AFB_1-N^7-Gua prepared by the method of Essigmann et al. (1977). These samples were then subjected to extraction with two volumes of isopropanol at 4°C overnight. The isopropanol extractable material was separated from precipitated protein and other solids by centrifugation at 800 $\times$ g for 10 minutes at 4°C. The super-

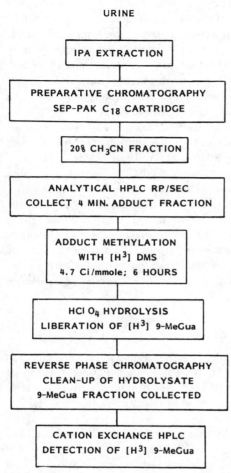

Figure 2
Outline of method for analysis of AFB_1-N^7-Gua in urine.

natants were evaporated under vacuum to approximately 3 ml and then reconstituted to a final volume of 10 ml by the serial addition of 5 ml 0.1 M HCl, 1 ml 0.5 M KAc (pH 5.3), 0.5 ml 1.0 M KOH, 0.5 ml methanol, and H_2O.

The samples were submitted to an initial cleanup step by preparative chromatography on C_{18} Sep-Pak cartridges using a slightly modified method described by Bennett et al. (1981). The sample, 20 ml 5% methanol, and 10 ml 10% methanol were sequentially passed through the cartridge to remove polar constituents of urine. Retained components, including AFB_1-N^7-Gua, were then eluted from the Sep-Pak with 10 ml 20% acetonitrile. The recovery of spiked adduct in these fractions was essentially quantitative (more than 95% in all cases). Non-polar constituents remaining in the cartridge were discarded. The 20% acetonitrile fraction of each sample was evaporated under vacuum to 0.3 ml, reconstituted to 1.0 ml with H_2O and 0.5 ml 0.1 molar HCl, and finally neutralized to pH 5. The AFB_1-N^7-Gua-containing fractions were then subjected to further cleanup on an Aquapore RP 300 analytical column (Brownlee Labs) by HPLC (Fig. 3). Elution of AFB_1-N^7-Gua from this column was accomplished by running a gradient from 10%-40% acetonitrile at 1 ml/min in a period of 30 minutes at 32°C. A fraction was collected which eluted 2 minutes on either side of the retention time of authentic adduct (13 min). The fraction was rotary evaporated and lyophilized to dryness.

The samples were subjected to methylation with tritiated dimethyl sulfate, which can be obtained at a specific activity of up to 1-5 Ci/mmole. A method similar to that of Jones and Robbins (1963) was used, which has been shown to result in the selective methylation of the imidazole ring nitrogen atoms of guanine. To the dry sample derived from urine was added 1 mCi [³H]-dimethyl sulfate in 50 μl N, N-dimethylacetamide, and the mixture was stirred for 6 hours at ambient temperature. After methylation, the samples were treated with perchloric acid to liberate the 9 MeGua moiety (Essigmann et al. 1977).

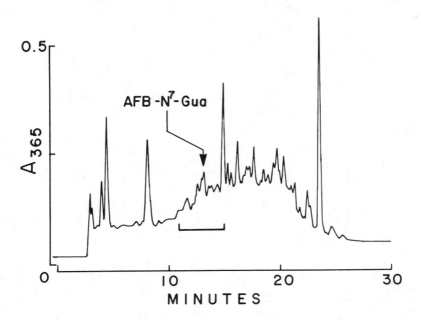

Figure 3
Chromatogram showing partial purification of AFB_1-N^7-Gua prior to methylation by reverse phase-size exclusion HPLC.

The methylated guanine samples were purified by reverse-phase chromatography before being submitted to base analysis by cation-exchange HPLC. This cleanup step essentially removed all the interfering material and radioactivity that was not associated with 9 MeGua. The perchloric acid hydrolysate was initially injected onto a C_{18} Zorbax column (DuPont), and elution of [^{3}H]-9 MeGua was accomplished by running a gradient from 0-40% acetonitrile in a period of 20 minutes at ambient temperature and a 1.0 ml/min flow rate. The fraction was collected that eluted 3 minutes on either side of the retention time of 9 MeGua (14 minutes), and it was evaporated under vacuum to less than 0.1 ml.

Analytical determination of [^{3}H]-9 MeGua was peformed by cation-exchange chromatography (Fig. 4). Samples were injected on a 25 × 0.22 cm Durrum DC-4A column eluted isocratically with 0.1 M ammonium formate (pH 4.5) at 0.25 ml/min at 51°C. Fractions were collected during the chromatogram for tritium counting. A radioactive peak co-eluting with an authentic 9-MeGua spike provides strong evidence that AFB_1-N^7-Gua was initially present in the urine.

DISCUSSION

As discussed earlier, a method in which the determinative step was based on absorbance of the AFB_1-N^7-Gua would be of inadequate sensitivity for use in

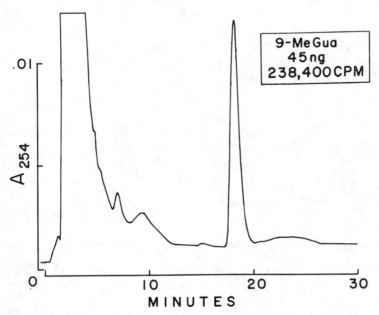

Figure 4
Chromatogram showing isolation of 9 MeGua (9 MeGua; 45 ng; 238,400 CPM) by cation-exchange HPLC.

studies in human populations owing to the low levels of the adduct expected to occur in urine even under conditions of maximal known exposure. Fortunately, in the case of the aflatoxin-liver cancer question, sufficient data from human populations are in hand to enable us to set some precise objectives with respect to limits of detection. This information is devised from the epidemiologic surveys mentioned earlier, the results of which are summarized in Table 1.

This represents a composite of a series of studies that were done in Thailand and Africa in regions where liver cancer occurs at differing incidence among men over 15 years of age. As can be seen, the incidence gradient extends over approximately one order of magnitude, and are rank-ordered according to estimated aflatoxin intakes, expressed in terms of nanograms per kilogram body weight. Intake values range from 4 to 222 ng/kg body weight, or in individuals weighing 50 kg, from 200 to 11,100 ng/day.

If we make the assumption that the same proportion of ingested aflatoxin is bound to liver DNA in the human as in the rat (namely, 1% of the ingested dose), and that the same proportion (30%) of adduct formed in DNA is excreted in 1 liter of urine over a 24-hour period, the expected concentration of AFB_1-N^7-Gua in urine would be 0.6-33.3 picograms/ml. These values then set reasonable limits of detection for the eventual application of the method in field studies. Based on the data already shown in Figure 4, it is clear that this objective is attainable and within reach. The chromatogram in Figure 4 shows a 9-MeGua peak containing substantial absorbance at 254 nm and also more than 238,000 cpm of $[^3H]$. This peak was produced by 45 ng of adduct carried through the entire analytical procedure and methylated with undiluted ^3H-dimethyl sulfate. Extrapolating this information to the anticipated urinary levels of adduct indicated in Table 1, it will be possible to detect in a 100 ml sample of urine containing 100 picograms of adduct (i.e., a concentration of 1 pg/ml), a peak of radioactivity containing about 1000 cpm. On this basis, we anticipate that the method described herein should be adequate for application even in populations experiencing low level aflatoxin exposure.

However, further validation studies will be required before the method is routinely applicable in field surveys. We are currently establishing the characteristics of the isolation and determinative steps over the full range of expected adduct levels. Its applicability will then be tested in rats fed different levels of aflatoxin to determine the relationship of excretion to DNA binding over the appropriate dose range. In addition, we are now in the process of establishing collaborative arrangements with epidemiologists carrying out studies in various areas in which aflatoxin exposure levels are expected to vary over wide ranges, with the objective of evaluating the adequacy of the method under carefully controlled conditions.

Several additional points concerning this approach are noteworthy. First, the method in its present form should be applicable, with appropriate minor modifications, to the detection of N^7-Gua adducts formed by any carcinogen. It could, therefore, be useful in a more general sense for detecting DNA damage and depurination even under circumstances in which the nature of the damaging

Table 1
Summary of Data Relating AFB_1 Intake, Liver Cancer Incidence and Expected AFB_1-N^7-Gua Levels in Human Urine

Location	Liver Cancer Incidence[a]	Average Dietary Aflatoxin Intake		Expected Adduct Concentration[c] in Urine
		ng/kg body wt./day	ng/day[b]	pg/ml
Kenya (high altitude)	3.1	4.0	200	0.6
Thailand (Songkhla)	2.0	6.5	325	1.0
Swaziland (Highveld)	7.1	7.0	350	1.0
Kenya (medium altitude)	10.8	7.0	350	1.0
Kenya (low altitude)	12.9	12.5	625	1.9
Swaziland (Middleveld)	14.8	11.5	575	1.7
Swaziland (Lebombo)	18.7	17.5	875	2.6
Thailand (Ratburi)	6.0	61.0	3050	9.2
Swaziland (Lowveld)	26.7	48.5	2425	7.2
Mozambique	35.0	222.0	11,100	33.3

[a]Cases/100,000/yr. in men > 15 yr
[b]per 50 kg body weight
[c]assumes binding of 1% of total AFB_1 ingested to DNA; excretion of 30% of DNA adduct in 1 liter urine in 24 hr.

agent is unknown. Similarly, in principle, the approach should be useful in quantification of total levels of DNA modification in tissue samples obtained at autopsy or surgical biopsy, under which circumstances methylated derivatives other than 9 MeGua can be produced and analyzed thus broadening the capacity of the method to detect other forms of DNA damage. We are currently undertaking evaluation of this and related approaches in the characterization of DNA modification by the polynuclear aromatic hydrocarbon fluoranthene, an important product of combustion.

ACKNOWLEDGMENT

Financial support for these studies has been provided by grant no. 5PO1-ES00597 from the National Institute of Environmental Health Sciences.

REFERENCES

Bennett, R.A., J. Essigmann, and G. Wogan. 1981. Excretion of an aflatoxin-guanine adduct in the urine of aflatoxin B_1-treated rats. *Cancer Res.* **41**: 650.

Essigmann, J.M., R. Croy, A. Nadzan, W. Busby, V. Reinhold, G. Buchi, and G. Wogan. 1977. Structural identification of the major DNA adduct formed by aflatoxin B_1 in vitro. *Proc. Natl. Acad. Sci.* **74**:1870.

Jones, J.W. and R. Robbins. 1963. Purine nucleosides. III. Methylation studies of certain naturally occurring purine nucleosides. *J. Am. Chem. Soc.* **85**: 193.

Shank, R.C. 1977. Epidemiology of aflatoxin carcinogenesis. In *Environmental Cancer* (eds. H.F. Kraybill and M.A. Mehlman). p. 291. J. Wiley and Sons, New York.

Shank, R.C., J. Gordon, G. Wogan, A. Nondasuta, and B. Subhamani. 1972. Dietary aflatoxins and human liver cancer. III. Field survey of rural Thai families for ingested aflatoxins. *Food Cosmet. Toxicol.* **10**:71.

Quantitative Relationships Between DNA Adduct Formation and Biological Effects

GAIL THEALL, I. BERNARD WEINSTEIN, AND DEZIDER GRUNBERGER
Division of Environmental Sciences and Cancer Center
Institute of Cancer Research
Columbia University College of Physicians and Surgeons
New York, New York 10032

STEPHEN NESNOW
Carcinogenesis and Metabolism Branch
U.S. Environmental Protection Agency
Health Effects Research Laboratories
Research Triangle Park, North Carolina 27711

GEORGE HATCH
Northrop Environmental Sciences
Research Triangle Park, North Carolina 27709

The primary objective of several recently developed short-term carcinogenicity bioassays is the identification of chemicals that might pose a risk to humans. Although identification of carcinogens by use of in vitro bioassays has been established, the use of quantitative data from these bioassays in human risk extrapolation has not yet been attempted. We reasoned that the "parallelogram" approach of Sobels (1977) for prediction of risk of mutation in humans could be applied towards predicting the risk of cancer. In this approach, knowledge of the relationship between oncogenic responses in rodents and rodent cells in culture and responses in human cells in culture might allow an estimation of oncogenic response in humans. Many of the problems encountered in dosimetry such as absorption, distribution, metabolic activation, degradation, and excretion (Committee 17 1975) could be avoided if dose is expressed in terms of cellular reaction products rather than as the concentration of the parent compound. Since DNA damage and, in many cases, the formation of stable carcinogen-DNA products are considered critical events in the initiation of the carcinogenic process (Miller 1978), it seems likely that assays for the level of carcinogen-DNA adducts in target cells might provide a measure of the biologically effective dose of certain carcinogens.

This study represents our initial effort using benzo[a]pyrene (B[a]P) as a model carcinogen in various in vitro bioassays. B[a]P was chosen since considerable information exists on the metabolism and types of DNA adducts formed from this compound in a variety of mammalian systems (Gelboin and

Ts'o 1978). In this study, we investigated three short-term assay systems: The *Salmonella typhimurium* mutagenicity assay (STT) developed by Ames et al. (1975), the chemical enhancement of SA-7 viral transformation in Syrian hamster embryo cells (Casto 1973) and the C3H 10T1/2 CL8 oncogenic transformation system (Reznikoff et al. 1973).

METHODS

Salmonella typhimurium Test

S. typhimurium strain TA100 was incubated in liquid suspension with various doses of $[^3H]B[a]P$ and a Sencar mouse liver S9 activation system for 60 minutes at 37°C. The STT was performed as described by Ames et al. (1975) except that the reaction volume was 50-100-fold larger. Aliquots were removed for the determination of mutagenicity. DNA was isolated from the remaining cells and the $[^3H]B[a]P$-DNA binding levels were determined by quantitation of DNA by UV spectroscopy and covalently bound $B[a]P$ by levels of radioactivity (Theall et al. 1981).

Chemical Enhancement of Viral Transformation of Syrian Hamster Embryo Cells

Cultures treated with either the indicated concentrations of $B[a]P$, $[^3H]B[a]P$ or the solvent were incubated for 18 hours. The cultures treated with $[^3H]B[a]P$ were used for $B[a]P$-DNA binding studies. SA-7 virus was absorbed for 3 hours to the remaining two mass cultures. These cultures were split and seeded into dishes and scored 4 weeks later for morphologically transformed foci to determine the enhancement ratios resulting from the pretreatment with $B[a]P$ (Casto 1973).

C3H 10T1/2 CL8 Oncogenic Transformation

Cells were seeded at a density of 1000 cells/dish and treated 24 hours later with the indicated concentration of $B[a]P$. Antibiotics were omitted from the medium during carcinogen treatment. The carcinogen was removed 24 hours later, the media was changed, and antibiotics were added. Transformation was scored as type II and type III foci/10^3 survivors at 6 weeks (Gold et al. 1980). In parallel studies, cells were seeded at 7.5×10^4 cells/dish and on reaching log phase they were treated with $[^3H]B[a]P$ for 24 hours, harvested, and used for $B[a]P$-DNA-binding studies as described above.

RESULTS AND DISCUSSION

Salmonella typhimurium Test

In the STT the biological response of mutagenesis is measured in terms of His$^+$ revertants. Since mutations are the result of DNA damage, the quantitative relationship of B[a]P-DNA adduct formation to His[+] revertants was investigated. After performing the experiments with strain TA100 under the same conditions several times, two types of relationships emerged. These were designated type 1 and type 2. Figure 1 shows a composite of four experiments that gave the type 1 result. The relationship between B[a]P-DNA binding and revertants is linear. In the type 1 response the slope of the line indicates that there are about 70 His[+] revertants per plate formed for every molecule of B[a]P bound per 10^6 nucleotides of cellular DNA.

The type 2 relationship of B[a]P-DNA binding to revertants is seen in the four experiments displayed in Figure 2. The response is also linear; however, in contrast to the type 1 response, about 1400 revertants per plate are generated for every B[a]P molecule bound per 10^6 nucleotides. Thus the type 2 response is 20-fold higher than the type 1 response seen in Figure 1. We were unable to pinpoint the specific variable that determined whether in a given experiment we obtained a type 1 or type 2 response. In all experiments, the same stock of TA100 was used, and the medium, cell density, culture age, and protocol were the same. We believe that the most likely variable is the extent of induction of an error-prone repair function of the *muc* gene (Shanabruch and Walker 1980) carried by the plasmid pKM101 in strain TA100. Other

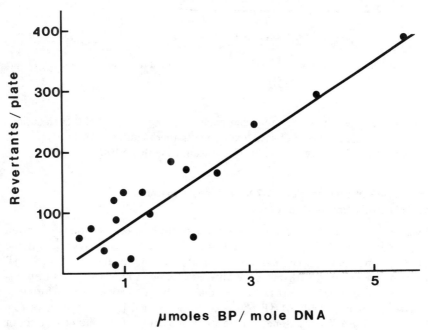

Figure 1
His[+] revertants as a function of B[a]P-DNA binding in the type 1 response of *S. typhimurium* strain TA100.

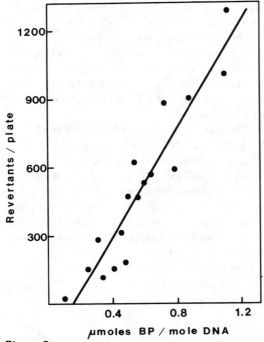

Figure 2
His[+] revertants as a function of B[a]P-DNA binding in the type 2 response of *S. typhimurium* strain TA100.

explanations have, however, not been excluded. Regardless of the explanation, our results indicate that between experiments there can be a 20-fold variation in this type of dosimetry in studies of mutagenesis with TA100.

Carcinogen Enhancement of SA-7 Viral Transformation in Syrian Hamster Embryo Cultures

It was also of interest to examine the relationship between cell transformation and B[a]P-DNA binding levels in a mammalian cell culture systems. In the Syrian hamster embryo cell system, preincubation of the cultures with B[a]P just prior to infection with the SA-7 virus results in an enhanced level of transformation. The enhancement ratio (i.e., the number of foci [adjusted for cytotoxicity]) obtained with B[a]P plus SA-7 virus divided by the number of foci obtained with SA-7 virus alone, is used as an indicator of the biologic effect of B[a]P.

Figure 3 shows that the B[a]P-DNA binding levels were linearly related to the dose of B[a]P (0.1-1 μg/ml) in Syrian hamster embryo cultures. On the

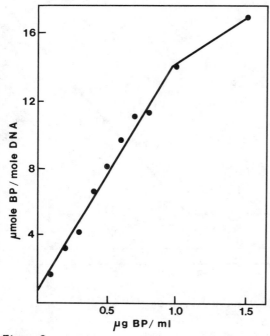

Figure 3
The relationship of B[a]P-DNA binding to B[a]P dose in Syrian hamster embryo cells.

other hand, when the enhancement ratio for transformation was plotted as a function of the B[a]P-DNA binding levels (Fig. 4) a biphasic curve was obtained. There was no increase in the enhancement ratio until a binding level of 8-10 μmoles B[a]P/mole DNA was reached. After this "threshold" level was achieved, there was a linear increase in the enhancement ratio as a function of the extent of B[a]P-DNA binding. This biphasic response was seen in three separate experiments and the enhancement ratios at specific binding levels were also reproducible.

Transformation of C3H 10T1/2 CL8 Cells

Similar experiments were done in C3H 10T1/2 CL8 cultures because of the simplicity and sensitivity of this transformation assay (Reznikoff et al. 1973). We observed a linear relationship between B[a]P dose (.05-1.2 μg/ml) and B[a]P-DNA binding levels (Fig. 5). When transformation, expressed as the number of type II and type III foci per 10^3 survivors, was plotted as a function of B[a]P-DNA binding, a linear relationship was also observed (Fig. 6). In contrast to the results obtained with the SA-7 system described above, we found no evidence

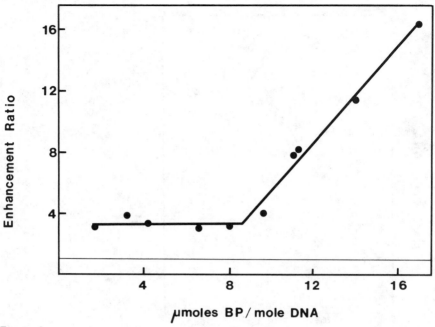

Figure 4
The relationship of B[a]P enhancement of SA-7 viral induced transformation of Syrian hamster embryo cells to B[a]P-DNA binding.

for a "threshold" effect. Thus, there appears to be a direct and reproducible relationship between the B[a]P dose, the level of B[a]P-DNA adducts, and the frequency of cell transformation in the C3H 10T1/2 CL8 transformation system. We must stress, however, that the B[a]P-DNA binding levels were determined at a higher cell density than that used in the transformation assays because of the insufficient amounts of DNA available under the latter conditions. It remains to be determined to what extent this influences the results obtained.

We also characterized the types of B[a]P-DNA adducts formed in the C3H 10T1/2 CL8 cultures used in the B[a]P-DNA binding studies. Throughout the concentration range of BP studied, the HPLC profiles were similar to those previously described (Brown et al. 1979). There was one predominant B[a]P-DNA adduct which cochromatographed with the 7R enantiomer of BPDE I bound to the exocyclic amino group of deoxyguanosine. This adduct is also the predominant B[a]P-DNA adduct in human bronchial explants (Jeffrey et al. 1977), human cell cultures (MacNicoll et al. 1980; Theall et al. 1981), and in other mammalian cells (Ivanovic et al. 1978) exposed to B[a]P.

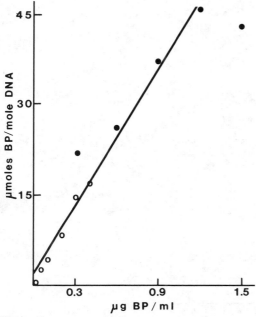

Figure 5
B[a]P-DNA binding in C3H 10T1/2 CL8 cells as a function of B[a]P dose. Open (o) and closed circles (•) represent the results obtained in two different experiments.

SUMMARY

It is of interest to compare the relationship of adduct formation to mutation response with that of adduct formation to transformation response in the systems described above. We have calculated these values from the slopes of the lines in the linear portion of the curves in Figures 1, 2, 4, and 6. The mutation response to B[a]P adducts in the STT (TA100) (expressed as the number of revertants generated per surviving cell for each B[a]P molecule bound per 10^6 nucleotides) in the type 1 and type 2 response are 5.8×10^{-7} and 1.2×10^{-5} units respectively. The transformation response to B[a]P adducts (expressed in terms of the number of transformed foci for each B[a]P molecule bound per 10^6 nucleotides) in the C3H 10T1/2 CL8 oncogenic transformation system and in the SA-7 viral transformation of Syrian hamster embryo cells were 4.0×10^{-4} and 6.9×10^{-4} units respectively. It would appear that B[a]P adducts are less efficient in inducing mutations in bacteria than they are in inducing or enhancing cell transformation in mammalian cells. Similar studies on the induction of mutations in mammalian cells will be of interest in the future.

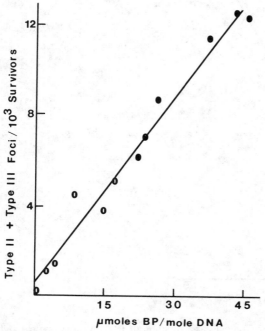

Figure 6
The relationship of transformation of C3H 10T1/2 CL8 cells to B[a]P-DNA binding. (o) and (•) represent the results obtained in two different experiments.

The use of chemical enhancement of viral transformation in Syrian hamster embryo cells for dosimetry studies is complicated by our finding that there is a biphasic response because of the existence of an apparent "threshold" level of adducts below which no enhancement occurs. The mechanisms by which B[a]P and other carcinogens enhance virus-induced cell transformation is not known, but our results suggest that it requires a critical level of DNA damage before the effect becomes manifest. On the other hand, in the C3H 10T1/2 CL8 oncogenic transformation system the relationship of DNA adducts to transformation is linear over a wide range of B[a]P concentration and there is no evidence of a "threshold." The C3H 10T1/2 CL8 system is also simplified by the fact that only one major type of B[a]P-DNA adduct is formed (Brown et al. 1979). In contrast, the Syrian hamster embryo cells give a much more compex profile of B[a]P-DNA adducts (Ivanovic et al. 1978). Thus the 10T1/2 CL8 oncogenic transformation system has emerged as a very useful assay for the study of carcinogen dosimetry. We believe that the determination of the adduct response relationship in the C3H 10T1/2 CL8 oncogenic transformation system can serve as the first of the three factors needed in a "parallelogram" approach (Sobels 1977) to human cancer risk estimation.

ACKNOWLEDGMENTS

We would like to thank R. M. Santella of Columbia University for valuable discussions; P. Mamay, C. Cristensen, and G. Curtis of Northrop Environmental Sciences, and H. Garland of the EPA, Health Effects Lab for technical assistance. This investigation was supported by Grant CR 807282 from the U.S. Environmental Protection Agency.

REFERENCES

Ames, B.N., J. McCann, and E. Yamasaki. 1975. Methods for selecting carcinogens with the *Salmonella*/mammalian microsome mutagenicity test. *Mutat. Res.* **31**:347.

Brown, H.S., A.M. Jeffrey, and I.B. Weinstein. 1979. Formation of DNA adducts in C3H10T1/2CL8 mouse embryo fibroblasts incubated with benzo[a]-pyrene or dihydrodiol oxide derivatives. *Cancer Res.* **39**:1673.

Casto, B.C. 1973. Enhancement of adenovirus transformation by treatment of hamster cells with ultraviolet radiation, DNA base analogs and dibenz[a]-anthracene. *Cancer Res.* **33**:402.

Committee 17. 1975. Environmental mutagenic hazards. *Science* **187**:503.

Gelboin, H.V. and P.O.P. Ts'o. 1979. Polycyclic hydrocarbons and cancer: Molecular, and cell biology. Vol. 2. Academic Press, New York.

Gold, A., S. Nesnow, M. Moore, H. Garland, G. Curtis, B. Howard, D. Graham, and E. Eisenstadt. 1980. Mutagenesis and morphological transformation of mammalian cells by a non-bay-region polycyclic cyclopenta(CD)-pyrene and its 3,4 oxide. *Cancer Res.* **40**:4482.

Ivanovic, V., N.E. Geacintov, H. Yamasaki, and I.B. Weinstein. 1978. DNA and RNA adducts formed in hamster embryo cell cultures exposed to benzo-[a]pyrene. *Biochemistry* **17**:1597.

Jeffrey, A.M., I.B. Weinstein, K.W. Jenette, K. Grzeskowiak, K. Nakanishi, R.G. Harvey, H. Autrup, and C. Harris. 1977. Structures of benzo[a]-pyrene-nucleic acid adducts formed in human and bovine bronchial explants. *Nature* **269**:348.

MacNicoll, A.D., G.C. Easty, A.M. Neville, P.L. Grover, and P. Sims. 1980. Metabolism and activation of carcinogenic polycyclic hydrocarbons by human mammary cells. *Biochem. Biophys. Res. Comm.* **95**:1599.

Miller, E.C. 1978. Some current perspectives on chemical carcinogenesis in humans and experimental animals: Presidential address. *Cancer Res.* **38**:1479.

Reznikoff, C.A., J.S. Bertram, D.W. Brankow, and C. Heidelberger. 1973. Quantitative and qualitative studies of cloned C3H mouse embryo cells sensitive to postconfluence inhibition of cell division. *Cancer Res.* **33**:3239.

Shanabruch, W.G. and G.C. Walker. 1980. Localization of the plasmid (pKM101) gene(s) involved in $recA^+lexA^+$-dependent mutagenesis. *Mol. Gen. Genet.* **179**:289.

Sobels, F.H. 1977. Some problems associated with the testing for environmental mutagens and a prospective for studies in "comparative mutagenesis." *Mutat. Res.* **46**:245.

Theall, G., M. Eisinger, and G. Grunberger. 1981. Metabolism of benzo[a]-pyrene and DNA adduct formation in cultured human epidermal keratinocytes. *Carcinogenesis* 2:581.

Yamasaki, H., P. Pulkrabek, D. Grunberger, and I.B. Weinstein. 1977. Differential excision from DNA of the C-8 and N^2 guanosine adducts of N-acetyl-2-aminofluorene by single strand specific endonucleases. *Cancer Res.* **37**: 3756.

COMMENTS

THILLY: If you look at the number of mutations you get from every benzo[a]pyrene adduct in *Salmonella*, it looks like 1 out of 3 adducts can lead to a reversion, assuming one base pair at risk. For the transformation, it looks like about 1 in 2000. Have you compared those systems?

THEALL: The two systems are difficult to compare. The frequency of mutation in the STT is 10^{-5} to 10^{-6} and the frequency of transformation is 10^{-2} to 10^{-3}. Therefore, the mechanisms of transformation and mutagenesis are probably different. Our results show B[a]P adducts are less efficient in inducing mutations in bacteria than they are in inducing or enhancing transformation. Also, it cannot be assumed that one base pair is at risk even in *Salmonella*.

SWENBERG: Do you think the variability in the Syrian hamster embryo system was due to different batches of cells with different activating potential?

THEALL: Yes. The experiments were done over a period of 1 year and the cells were from different batches. Since the enhancement ratio to binding was relatively constant, the variability was probably due to the activation systems.

HSIE: Maybe you could obtain more data with the 10T1/2 than with the SHE cells, because 10T1/2 is a homogeneous cell type. There could be variation between batches. The cultures are heterogeneous cell types to begin with, therefore I think it is worthwhile taking this into consideration.

WATERS: I think it would be better to draw that comparison if you were working strictly with a SHE transformation system. If you froze the cells down it would be much less variable, instead of repetitive isolation.

BRIDGES: What was the difference between the two types of *Salmonella* experiments that gave you these different slopes? Could you please explain?

THEALL: Although we have no direct experimental evidence, we believe that there is an induction of the *muc* gene carried on the plasmid pKM101 which was inserted into the strain to increase its sensitivity to various carcinogens. Since when high numbers of revertants are obtained we observe only low levels of binding, an induction of the error prone function attributed to the *muc* gene is very likely. It would also explain the wide variation of dose response reported in the literature. We have tried without success to determine the factor involved in this "induction."

BRIDGES: What strain did you use?

THEALL: Strain TA100.

BRIDGES: Is this a liquid incubation assay?

THEALL: Yes. The preincubation is for 60 minutes at 37°C.

BRIDGES: And at the end of that time you do your DNA binding?

THEALL: Yes. I preincubate a mass reaction mixture, remove aliquots for the mutagenicity assay and then isolate the DNA from the rest of the cells. The experiments that we have done indicate that there is essentially no further metabolism with the B[a]P after the preincubation period. The binding levels should correlate.

BRIDGES: When you plate for mutants, are the microsomes in the plate?

THEALL: Yes.

BRIDGES: Why? What possible reason is there to have them in the plate?

THEALL: To the extent possible, we wished to use the mutagenicity protocol outlined by Ames and colleagues. Aliquots were removed before the cells were centrifuged and DNA extracted. We felt that it was important to plate the number of cells treated and not the number that were centrifuged and possibly clumped after resuspension.

BRIDGES: It seems to me you should get rid of the microsomes before you plate, otherwise you have no control over what might be going on after you have taken your sample for measuring binding. You've still got potentiality for further binding while the cells are on the plate.

THEALL: We recognized this problem. However, it has been shown by Bartsch and colleagues (1975) with vinyl chloride that the activity of the S9 preparation is very low after incubation at 37°C for 60 minutes. Our studies of B[a]P metabolites in each of the experiments that were performed indicated that at the highest dose less than 3% of the B[a]P remained.

BRIDGES: There is one thing about microsomes: Some people find that they last for 6 hours and some people find that they last for 30 minutes. This could be the root of your problem, i.e., in some experiments mutants occur on the plate long after you have taken your sample to DNA binding. You've got more mutants coming on the plate, because your microsomes are still active.

WEINSTEIN: Well, maybe the final comment is that even in simple model systems the conversion of adducts to biological effect is dependent upon a large number of variables, the system used, and how it's done, as well as the endpoint which was measured.

References

Bartsch, H., C. Malaveille, and R. Montesano. 1975. Human, rat and mouse liver-mediated mutagenicity of vinyl chloride in *S. typhimurium* strains. *Int. J. Cancer* 5:429.

Adducts in Human DNA Following Acute Dimethylnitrosamine Poisoning

DEBORAH C. HERRON* AND RONALD C. SHANK
Department of Community and Environmental Medicine
University of California Southern Occupational Health Center
University of California
Irvine, California 92717

This study of adducts in human DNA involves dimethylnitrosamine (NDMA) which is a potent and specific hepatic carcinogen in a broad range of animal species (Magee et al. 1976). However, no human cancers have been associated with exposure to NDMA despite widespread occurence of this compound in the environment as well as evidence of in vivo formation in the human stomach and bladder (IARC 1978).

It is understandable that NDMA is used extensively to investigate the mechanism by which certain chemicals produce cancer, since a single administration of this agent to rats is capable of producing both liver (Craddock 1973) and kidney tumors (Magee and Barnes 1962). Biochemical evidence suggests that NDMA exerts its carcinogenicity by metabolic activation to a methylating agent which forms adducts with various cellular macromolecules including DNA (reviewed in Magee et al. 1976). In particular, the formation of the promutagenic lesion O^6-methylguanine (O^6-MeGua) in DNA is suspected to be a critical step in NDMA carcinogenesis (reviewed in Lawley 1976; Singer 1979). This sequence of events is consistent with the somatic mutation theory of cancer which hypothesizes that direct interaction of an electrophile with DNA potentially can lead to a mutation and subsequent cellular transformation (Miller and Miller 1966).

While substantial in vivo evidence from animal studies has been accumulated to link the presence of O^6-MeGua with NDMA-induced liver and kidney tumors, little is known about methylated base formation in human tissue. Montesano and Magee (1970) used human liver slices in vitro to study the metabolic activation of NDMA and measured subsequent DNA methylation at the N-7 position of guanine. Recently, however, an unusual homicide case presented the opportunity to examine DNA from a victim of acute NDMA poisoning (Herron and Shank 1980). Previous development of a highly sensitive technique

*Present address: Atlantic Richfield Company, 515 South Flower Street, Los Angeles, California 90071

for the optical detection of alkylated bases in DNA enabled us to measure both 7-MeGua and O^6-MeGua in liver DNA (Herron and Shank 1979). The presence of these adducts in DNA suggests that humans indeed activate NDMA to the methylating agent in vivo. The data presented here compare 7-MeGua and O^6-MeGua production in human liver tissue with that found in rats treated with NDMA. Data from rats treated with two additional methylating agents, 1,2-dimethylhydrazine (SDMH) and N-methyl-N-nitrosourea (MNU) has also been included for comparative purposes.

EXPERIMENTAL PROCEDURES

Chemicals

MNU was obtained from ICN Pharmaceuticals (Irvine, California). SDMH as the dihydrochloride and 7-MeGua were obtained from Sigma Chemical Company (St. Louis, MO). O^6-MeGua was prepared according to the procedure of Balsiger and Montgomery (1960).

Tissue: Animal Study

Young, 125 g male Sprague-Dawley rats from Charles River Breeding Laboratories (North Wilmington, MA) were injected intraperitoneally (i.p.) with 163 mg SDMH/kg b.w., 80 mg MNU/kg b. w. or 0.01N HCl as a control. Animals were killed by decapitation 6 hours after administration of SDMH and 1 hour after administration of MNU. Liver, kidney, and colon were rapidly removed, washed in ice-cold isotonic saline solution and stored at −80 C. We are grateful to Dr. Mathuros Ruchirawat, Mahidol University, Bangkok, Thailand for supplying liver DNA from Fischer rats treated with 25 mg NDMA/kg b.w. and killed 6 hours after injection.

Tissue, Human Study

There were five victims of NDMA poisoning; two of the victims died. At first, illness was attributed to infectious causes, however, later it was determined that a toxic substance had been involved (Cooper and Kimbrough 1980). Further investigations and the trial established that the toxin ingested by the victims was NDMA, and the suspect was convicted of two counts of first degree murder and three counts of poisoning.

Frozen samples of human liver, kidney, and heart tissue were obtained from the Center for Disease Control (Atlanta, GA) and Poison Lab (Denver, CO). All samples had been encoded to insure that there was no knowledge of the cause of death.

DNA Isolation and Hydrolysis

In both the rat and human experiments, DNA was isolated by a modified Kirby (1957) phenolic extraction method (Swann and Magee 1968). Severe tissue autolysis prohibited DNA isolation from five of the human samples. DNA was hydrolyzed in 0.1 N HCl (5mg/ml) for 30 minutes at 70°C, conditions which release purines as free bases and pyrimidines as oligonucleotides. Hydrolysates were filtered through a 0.65 μm filter, Millipore Corporation (Bedford, MA).

Analysis of DNA Methylation

Optical detection of methylated bases in DNA was accomplished by the method of Herron and Shank (1979). Fractionation of DNA hydrolysates was carried out using a Partisil-10 strong cation-exchange column (25 cm $\times$ 4.6 mm I.D.; Whatman, Inc., Clifton, NJ) and 0.05 M ammonium phosphate (pH 2.0) at 2.0 ml/min. Fluorescence of the eluting bases was measured with a Farrand LCF-100 variable wavelength detector using a 286-nm excitation wavelength and 366 nm emission interference filter. Electronic integration of peaks (after calibration with standard solutions of authentic guanine 7-MeGua and O^6-MeGua) was used for quantitation of methylated purines.

RESULTS

Liver DNA isolated from control animals did not contain detectable amounts of 7-MeGua or O^6-MeGua. Chromatographic elution profiles of liver DNA hydrolysate from a control rat and from a rat treated with 25 mg NDMA/kg b.w. are compared in Figure 1. In both chromatograms the pyrimidines, present as oligonucleotides, elute rapidly as three to four merged peaks. These are followed by the two major purines, guanine and adenine. However, it is apparent that DNA of the NDMA-treated rat contains two additional peaks, representing 7-MeGua and O^6-MeGua. Similar results were obtained with DNA hydrolysates from rats treated with SDMH or MNU (chromatograms not shown). Table 1 summarizes the quantitative analyses of methylated bases in DNA of rats treated with these three compounds.

We were able to isolate DNA from only seven of the 12 human tissue specimens since five were severely autolyzed, thus preventing nucleic acid extraction. Two DNA samples contained detectable levels of 7-MeGua and O^6-MeGua. A representative chromatogram is pictured in Figure 2. It was confirmed that these two DNA samples were isolated from two different liver specimens from the same victim of NDMA poisoning. Although there had been an autopsy on a second victim of NDMA poisoning, no tissue specimens were available for analysis. The remaining specimens had been collected from a victim of methylbromide poisoning and two cases of Reyes Syndrome. Quantitation of methylated purine content is presented in Table 2.

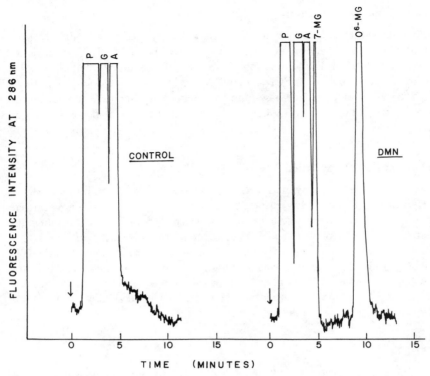

Figure 1

Chromatographic elution profiles of DNA hydrolysate from control rat and rat treated with 25 mg NDMA/kg body wt; P, pyrimidine oligonucleotides; G, guanine; A, adenine.

Table 1

Methylated Bases in Rat DNA

Treatment	Methylated bases (μmol/mol guanine)	
	7-MeGua	O^6-MeGua
Control (0.1N HCl)	—[a]	—
NDMA (25 mg/kg)	10800	926
MNU (80 mg/kg)	2300	242
SDMH (163 mg/kg)	2510	214

[a]None detected

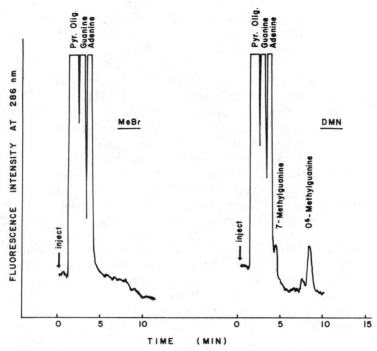

Figure 2
Elution profiles of liver DNA hydrolysate from victims of methylbromide (MeBr) and
dimethylnitrosamine (NDMA) poisoning; (Pyr. Olig.) pyrimidine oligonucleotides. (Redrawn from original). Reprinted, with permission, from Herron and Shank (1980).

Table 2
Methylated Purines in Victim of NDMA Poisoning

	Methylated purines (μmol/mol guanine)	
	7-MeGua	O^6-MeGua
Specimen #1	1363	273
Specimen #2	1373	317

DISCUSSION

The results of these studies provide in vivo evidence that humans, like rodents,
activate NDMA metabolically to a strong methylating agent capable of forming
adducts at both the N-7 and O^6 position of guanine. The presence of these

methylated purines in human DNA also characterizes the sequence of events following similar methylating carcinogens: SDMH, and MNU. This emphasizes a possible common mechanism of action for such agents (i.e., formation of potentially mutagenic methylated bases in DNA).

The in vitro metabolic activation of NDMA by rat and human liver slices was compared in a report by Montesano and Magee (1970). They observed the level of 7-MeGua in human DNA (0.13%) to be lower than in the rat (0.19%). Due to the analytical method used in this experiment, O^6-MeGua was not observed. These data suggest that humans form the active methylating agent at 68% of the rate at which the rat metabolizes NDMA.

During the homicide investigation, it was important to determine whether the alkylation levels were consistent with administration of a lethal dose of NDMA. Based on 7-MeGua levels of 1300 μmol/mol guanine in liver DNA at the time of death (5 days after oral ingestion of NDMA), and assuming a half-life for 7-MeGua of 48 hours (Herron and Shank 1981), we extrapolated the initial, maximum level of alkylation to be between 6000-8000 μmol/mol guanine. This level of 7-MeGua is attained in rat liver DNA upon administration of 25-30 mg NDMA/kg (Montesano and Magee 1970) a dose which nears the LD_{50} level of 27-41 mg NDMA/kg b.w. (Heath and Magee 1962). Since the data from Montesano and Magee suggest that man exhibits 68% of the rodent's rate of NDMA metabolism, we assumed that the amount of NDMA administered was $\geqslant$ 25-30 mg/kg. Thus, the results obtained in this study are compatible with the ingestion by the victim of an acutely toxic, and probably lethal, dose of NDMA.

Extrapolation to initial O^6-MeGua levels in the human liver also can be approximated. Pegg (1977) has shown that at high doses of NDMA, the ration of O^6-MeGua : 7-MeGua approaches 0.10. Therefore, the initial O^6-MeGua levels in the victim's liver were between 600-800 μmol/mol guanine. Moreover, since O^6-MeGua levels 5 days after ingestion of NDMA were still 300 μmol/guanine; one can speculate that the human has a removal rate for O^6-MeGua of approximately 120 hours. It should be cautioned, however, that recent evidence in both rat and human cells indicate that O^6-MeGua repair is markedly dose-dependent. Therefore, this seemingly inefficient removal of O^6-MeGua in the human liver specimen may reflect the lethal dose of NDMA received by the victim. The hypothesis that DNA alkylation plays a critical role in carcinogenesis has not yet been proven; however, there does appear to be a correlation between methylated base levels in an organ and distribution of tumors in rats treated with NDMA (Pegg and Nicoll 1976). Since human liver DNA also contains comparable levels of methylated bases after NDMA administration, it is tempting to speculate that man may be equally as susceptible as the rat to the carcinogenic action of this agent.

A tragic epilogue to the homicide described here may result in an answer to this question. Of the three surviving victims, one was a girl who was 2½ years at the time of the incident. A recent report (Cooper and Kimbrough 1980) indicates that the patient shows continuous hepatosplenomegaly and elevated liver enzymes in addition to chronic active hepatitis, indicative of possible extreme and permanent liver damage.

REFERENCES

Balsiger, R.W. and J.A. Montgomery. 1960. Synthesis of potential anticancer agents. XXV. Preparation of 6-alkoxy-2-aminopurines. *J. Org. Chem.* **25**:1573.

Cooper, S.W. and R.D. Kimbrough. 1980. Acute dimethylnitrosamine outbreak. *J. Forensic Sci.* **25**:874.

Craddock, V.M. 1973. Induction of liver tumors in rats by a single treatment with nitrosocompound given after partial hepatectomy. *Nature* **245**:386.

Heath, D.F. and P.N. Magee. 1962. Toxic properties of dialkylnitrosamines and some related compounds. *Br. J. Ind. Med.* **19**:276.

Herron, D.C. and R.C. Shank. 1979. Quantitative high-pressure liquid chromatographic analysis of methylated purines in DNA of rats treated with chemical carcinogens. *Anal. Biochem.* **100**:58.

_____. 1980. Methylated purines in human liver DNA after probable dimethylnitrosamine poisoning. *Cancer Res.* **40**:3116.

_____. 1981. In vivo kinetics of O^6-methylguanine and 7-methylguanine formation and persistence in DNA of rats treated with symmetrical dimethylhydrazine. *Cancer Res.* **41**:3967.

IARC. 1978. Dimethylnitrosamine. *IARC Monogr.* **17**:125.

Kirby, K.S. 1957. New method for the isolation of deoxyribonucleic acids: Evidence on the nature of bonds between deoxyribonucleic acid and protein. *Biochem. J.* **66**:495.

Lawley, P.D. 1976. Carcinogenesis by alkylating agents. *ACS Monogr.* **173**:84.

Magee, P.N. and J.M. Barnes. 1962. Induction of kidney tumors in the rat with dimethylnitrosamine. *J. Path. Bact.* **84**:19.

Magee, P.N., R. Montesano, and R. Preussmann. 1976. N-nitroso compounds and related carcinogens. *ACS Monogr.* **173**:491.

Miller, E.C. and J.A. Miller. 1966. Mechanisms of chemical carcinogenesis: Nature of proximate carcinogens and interactions with macromolecules. *Pharmacol. Rev.* **18**:805.

Montesano, R. and P.N. Magee. 1970. Metabolism of dimethylnitrosamine by human liver slices *in vitro*. *Nature* **228**:173.

Pegg, A. 1977. Alkylation of rat liver DNA by dimethylnitrosamine: Effect of dosage on O^6-methylguanine levels. *J. Natl. Cancer Inst.* **58**:681.

Pegg, A. and J.W. Nicoll. 1976. Nitrosamine carcinogenesis: The importance of the persistence in DNA of alkylated bases in the organotropism of tumour induction. *IARC Sci. Publ.* **12**:571.

Singer, B. 1979. N-nitroso alkylating agents: Formation and persistence of alkyl derivatives in mammalian nucleic acids as contributing factors in carcinogenesis. *J. Natl. Cancer Inst.* **62**:1329.

Swann, P.F. and P.N. Magee. 1968. Nitrosamine-induced carcinogenesis. The alkylation of nucleic acids of the rat by N-methyl-N-nitrosourea, dimethylnitrosamine, dimethylsulphate, and methylmethanesulphonate. *Biochem. J.* **110**:39.

COMMENTS

HERRON: In reference to my speculation on the sensitivity of man compared to the rat, I would like to add that Dr. Swenberg (pers. comm.) has informed me of a study in R. Montesano's laboratory. Apparently, the promutagenic lesion O^6-MeGua is removed from human liver slices at a rate 10 times faster than it is in the rat. Therefore, it might be that humans are much *less* sensitive to the carcinogenic action of NDMA.

SWENBERG: First, that factor of tenfold in the O^6-MeGua repair rate would not likely occur in cases of extreme amounts of DNA alkylation because it is a saturable process. So it may not be relevant at the high dose of NDMA that occurred here.

One of the things I noticed, though, was if you did an O^6:N^7 ratio, you get a value of about 0.2, which is higher than theoretical, and indicates that you are losing 7-Me-Gua faster than O^6-MeGua. So it's entirely possible that in man, the 7-methylglycosylase reaction is a much more important enzymatic system than it is in the rat.

HERRON: I should emphasize that we used 48 hours as a figure for the 7-MeGua half-life because it appeared to be a fairly good average based on the literature. I also should emphasize that there's a limited amount of manipulation one can do with these data, since there was only one case yielding one point on the curve. Naturally, in view of the uniqueness of this type of opportunity, we did quite a bit more speculating than we would normally.

SWENBERG: Yes, in addition, you probably can assume that such high levels of O^6-MeGua are not repaired significantly since you probably really overwhelmed the system; 7-MeGua values suggest that this also may be a factor in man. I don't know of any other data.

MIRSALIS: Has anyone tried to make any actual dose calculations on how much NDMA was administered? If the accused confessed that he dumped a bottle in, and if they know, for example, that he dumped a 25-g bottle of NDMA into a 2-quart pitcher of lemonade and each person drank a 12-ounce glass—I figure out in a 70 kg person that would be 75 mg/kg, which is over twice the estimated LD_{50}. Any comments?

HERRON: That is probably a reasonable estimate. However, unfortunately the bottle of NDMA was not recovered. The only estimates we have made of actual dosage have been based on DNA alkylation levels.

New Methods for the Detection of DNA Damage to Human Cellular DNA by Environmental Carcinogens and Anti-tumor Drugs

KWOK MING LO, WILLIAM A. FRANKLIN,
JUDITH A. LIPPKE, WILLIAM D. HENNER, AND
WILLIAM A. HASELTINE
Laboratory of Biochemical Pharmacology
Sidney Farber Cancer Institute
Boston, Massachusetts 02115

The case just reported by Herron and Shank (this volume) is a very bizarre case of poisoning. Someone said in the previous discussion "It could only happen in America." A similar case was reported in Germany; it was not the same chemical, but it's not an American phenomenon.

Unfortunately, I come from an institute in which poisoning of humans is done on a routine scale. That is, of course, in a cancer clinic. I think it's a relevant point to make at this juncture, because many of the technologies that we have developed were motivated by a desire to analyze what happens to people exposed to toxic substances. If you're going to look at the high-dose range of exposure to a variety of potent carcinogens and mutagens, cancer patients provide a reasonably controlled environment.

In the area of antitumor therapy carcinogenesis, one would like to have some measure of the actual damage inflicted on DNA; that is, methods that are specific as well as sensitive. We have addressed the issues of specificity and sensitivity (Grunberg and Haseltine 1980; Lippke et al. 1981; Lippke and Haseltine 1982).

The first method described was developed for use with human cells. It provides the type of information regarding DNA damage that can be correlated with the ultimate genetic effects. The example of ultraviolet light will be presented. However, the method is general for analysis of specific sites of DNA damage. Knowing the sites of DNA damage really doesn't reveal the nature of the chemical alterations. However, such information may provide a starting point for chemistry.

The second method described is a post-labeling technique that was designed for sensitivity (Franklin and Haseltine 1982). This method provides information regarding specificity as well.

The first method uses DNA sequencing techniques to determine sites of DNA damage.

The basic methodology is a very simple one: A defined fragment of DNA which is generated by restriction enzyme cleavage is labeled at one terminus, and then exposed to DNA damaging agents.

The analysis of the sites of DNA alteration may proceed along several different lines. If the agent breaks the DNA, the analysis is straightforward. The basis of the method is to treat the DNA fragment with a series of agents that break the DNA. These may be the DNA sequencing reagents of Maxam and Gilbert (Maxam and Gilbert 1980). For example, breakage at G produces a series of fragments of defined length; breakage at T will produce another series of fragments.

Polyacrylamide gel analysis of the products separates molecules in the DNA products on the basis of the length of the nucleotide chain.

By way of illustration, treatment of such a DNA fragment with several reagents is shown in Figure 1. On this system, the shorter the DNA fragment, the faster the DNA migration. In one lane, the DNA was treated with a reagent that broke it at every single nucleotide. Such a reagent produces a ladder of bands. On a separate lane, we analyze the products reactions done so as to break the DNA at G residues, (and to a lesser extent at As). A clear pattern of products is observed since that is the basis of the Maxam-Gilbert DNA sequencing techniques (Maxam and Gilbert 1980).

Using this method, one can rapidly compare the cleavage specificity of a number of different reagents. We first developed this method for use with neocarzinostatin (D'Andrea and Haseltine 1978). Neocarzinostatin is a small peptide that carries a prosthetic group. It breaks the DNA at very specific places which become mostly Ts and As.

The effect of bleomycin is also shown in Figure 1. Bleomycin cleaves DNA at positions different from that of neocarzinostatin. This method provides a very easy and direct way for looking at the sites of DNA modification. Figure 1 also illustrates DNA cleavage at all sites is an interesting reaction. In this case ferrous ion was added to the DNA solution. If the solution is oxygenated superoxide, hydroxyl-free radicals are produced. We have noted that similar cleavage patterns are observed in reactions in which other reagents are used to generate hydroxyl-free radicals.

Figure 2 shows exposure of DNA to ionizing radiation which produces a similar cleavage pattern. There is a general advantage to this methodology since one observes every product. Information regarding independent events can be obtained. For example, the breaks created by ionizing radiation turn out to be complex. The break is not a simple nick that can be ligated. We find that the breaks result in the removal of a base and that sugar fragments are left at the 3' terminus.

Several years ago we described the effects of benzo[a] pyrene (B[a] p) diol epoxide on DNA (Haseltine et al. 1980). The major adduct is known to occur

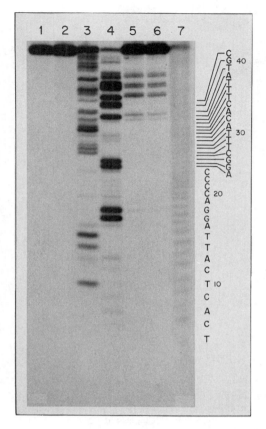

Figure 1

A fragment of the *lac* p-o gene of *E. coli* was labeled with ^{32}P at the 5' terminus. The DNA was layered onto a 20% polyacrylamide-urea gel and subjected to electrophoresis. The DNA was denatured prior to layering in the gel. (*Lane 1*) Unmodified DNA-migrates as a single, homogeneous species; (*Lane 2*) treatment with the experimental anti-tumor agent neocarzinostatin in the absence of the necessary thiol cofactor. No breaks apparent in the DNA. (*Lane 3*) treatment with neocarzinostatin (for conditions, see D'Andrea and Haseltine 1978 in the presence of a thiol cofactor. (*Lane 4*) treatment with the methylating agent dimethylsulfate followed by treatment at high temperature, followed by treatment with hot alkali. The procedure, developed by Maxam and Gilbert (1980), creates breaks at sites of G (predominant) and A (minor). Breaks at specific sites in the DNA are apparent. Comparison of the electrophoretic mobility of these fragments with those produced by the DNA sequencing reagents permits identification of the cleavage sites as As and Ts. (*Lane 5,6*) treatment of the DNA with bleomycin (D'Andrea and Haseltine 1978) without (*Lane 5*) or with (*Lane 6*) a third reagent. A pattern of DNA cleavage different from that of either dimethylsulfate or neocarzinostatin is observed. (*Lane 7*) treatment with 1 mm FeSO$_4$ shows uniform cleavage at all bases of the DNA.

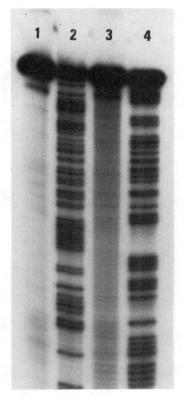

Figure 2

Gamma radiation fragments DNA at each nucleotide site. pMCI plasmid DNA was cleaved with BotE II restriction endonuclease and labeled at 3′ termini in a reaction with large fragments of *E. coli* DNA polymerase and ^{32}P deoxynucleoside triphosphates. Labeled DNA fragments were then cleaved with Hine II restriction endonuclease to generate singly-end-labeled DNA. The 320-base-long fragment was resolved by polyacrylamide gel electrophoresis. After elution from the gel and alcohol precipitation, the DNA was resuspended in distilled H_2O. Samples were radiated (level 3) with 5000 rads (^{60}Co source 3.5 krads/min), treated with the G+A (*Lane 2*) or the C+T (*Lane 4*) DNA sequencing reactions of Maxam and Gilbert (1980) received no further treatment (*Lane 1*). Samples were then electrophoresed on an 8% polyacrylamide gel and one autoradiogram of the gels prepared.

at N7G. Such adducts don't lead to DNA breakage. However, if a defined fragment of DNA is trained with B[a]P diol epoxide followed by a mild alkaline hydrolysis, a specific pattern of DNA breaks is observed. Breaks occur at a position of Gs and to a lesser extent at As and Ts. This experiment indicates that some interesting chemistry occurs in these reactions. The base specificity of the cleavage reaction implies that the epoxide is not reacting with the phosphodiester backbone (Haseltine et al. 1980).

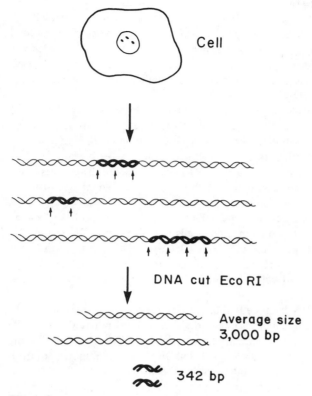

Cell

DNA cut Eco RI

Average size
3,000 bp

342 bp

Figure 3
Schematic diagram of the DNA extraction procedure. Total cellular DNA is extracted from
cells and then digested with the restriction endonuclease Eco RI. This treatment yields the
highly reiterated 342 base pair alphoid sequence.

LIMITATIONS OF THE METHOD?

The first limitation is that the test is certainly not done in vitro for use with
intact human cells. The requirement for this method is that the cells must be-
have as if they are purified fragments of DNA. Human cells are ideally suited for
this purpose.

Figure 3 shows that there exists a sequence in all human cells called the
alphoid sequence. This sequence is thought to be associated with centrometric
structures which exist in tandem arrays. Cleavage of cellular DNA with the restriction enzyme, Eco RI, results in production of a 342 base pair long alphoid

fragment. Most of the cellular DNA is cut by the same enzyme into very large pieces. Alphoid DNA can be taken from any tissue or any tumor tissue. It is readily purified from total DNA by gel electrophoresis. One percent of the total DNA can be obtained as a pure DNA fragment. In contrast to DNA which has been cloned, this is a piece of DNA that is embedded in 300,000 different locations in the human cellular genome. Nonetheless a readable DNA sequence is obtained from bulk alphoid DNA.

Alphoid DNA provides a probe that permits treatment of populations of human cells as if they were pure clones of DNA. You could imagine, for example, using circulating lymphocytes from a person exposed to a toxic agent.

We have done experiments of the following sort: A petri dish of cells is exposed for 2 hours to sunlight and the DNA was extracted. The damage deposited in the cellular DNA was apparent. This method is sufficiently sensitive to pick up reasonable levels of exposure.

A direct comparison between the effects of UV-light exposure of naked DNA to the same DNA sequence in intact cells was also done. DNA from untreated cells is purified and analyzed for evidence of UV damage. DNA was purified from cells that had been exposed to UV-light before DNA extraction. Naked DNA from untreated cells was purified and then exposed to ultraviolet light.

Figure 4 shows DNA from cells that were not exposed to ultraviolet light and unmodified by subsequent treatments that break UV-irradiated DNA. The treatments are digested with an enzyme that cuts DNA at sites of cyclobutane pyrimidine dimers, with hot alkali ($90°$, 0.1MNaOH), or a combination of both treatments. DNA extracted from cells that has been exposed to UV light but not treated with alkali or enzyme is unbroken. Treatment of the same DNA preparation with an enzyme that cuts at sites of cyclobutane pyrimidine dimers produces a series of discrete bands. These bands reflect breaks at sites of potential dimers.

A comparison of the effects on cellular DNA with a similar dose given to naked DNA is shown in Figure 5a. The data shows that the relative distribution of UV-induced damage is the same under the two conditions (formation of pyrimidine dimers). However, the effective dose is diminished by a factor of two in the case of exposure of intact cells.

Such experiments can be done, not only with radiation, but also with chemicals. Quite a bit of specific information came from this particular experiment. At moderately low doses—50 joules/m^2—alkali labile lesions are picked up. This kind of damage can be seen at 15 joules/m^2, the limit of sensitivity to this method.

In Figure 5b the dose-response curve for the alkali labile lesion is illustrated. At high doses it doesn't saturate as does that of the cyclobutane dimers, therefore, a different chemistry from that of dimers is involved.

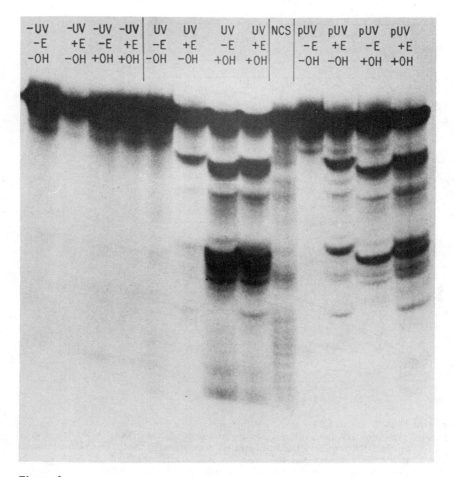

Figure 4

Comparison of UV-light-induced damage to cellular DNA and purified DNA. (*Lanes 1-4*) no irradiation; (*Lanes 5-8*) irradiation of purified DNA at 5000J/M²; (*Lane 9*) NCS sequencing reaction; (*Lane 10-13*) irradiation of intact cells at 5000J/M² : (*Lanes 1, 5, 10*) control—no treatment; (*Lanes 3, 7, 12*) treated with 1 M piperidine at 90°C for 20 min; (*Lanes 4, 8, 13*) treated with M. Luteus pyrimidine dimer endonuclease followed by treatment with 1M piperidine at 90°C for 20 min.

From this starting point, we have deduced the chemistry of the alkali labile lesions. It is considerably different from that of cyclobutane dimers. A reasonable case can be made that this lesion is a premutagenic event. Using this methodology, i.e., analysis of specific breaks, one can often pick up events

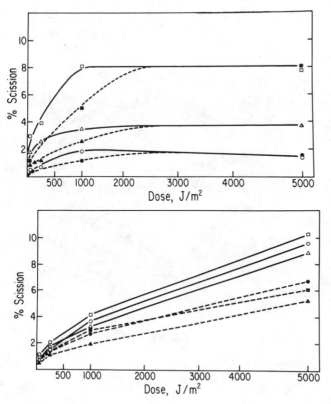

Figure 5

Dose-response for UV light-induced damage to human alphoid DNA. (■, ●) intact cells, (□, ○) purified DNA were irradiated and treated as described in the text. (A) Dose-response of pyrimidine dimer damage was determined by treatment of the irradiated DNA with the M. luteus pyrimidine dimer-specific endonuclease followed by resolution of the scission products on urea containing polyacrylamide sequencing gels. Scission occurred at the sequences indicated by the asterisks: (0–0) G–T*T–G (37′–34′); (△–△) G–T*T*C–A (51′–47′); and G–T*T*C–A (79′–74′). The position of the sequences within the alphoid DNA are indicated by the numbers in parentheses. The percentage of scission is the fraction of the input molecules broken at the sequences indicated. (B) Dose-response for alkali-labile lesions was determined by treatment of the irradiated DNA with 1M piperidine at 90°C for 20 min prior to layering on the gel. The percentage of scission was computed as in A. Alkali-induced scission occurred at the sequences indicated by asterisks: (0–0) A–C–T–C̆–T–G (46′–41′); (△–△), G–T–T–C–A (51′–47′); and G–T–T–T–T–C̆–A (79′–74′).

that may have been missed using other methods. Other methods can be used for analysis of the DNA of cancer patients, provided, of course, that exposure to the antitumor agents is sufficiently high.

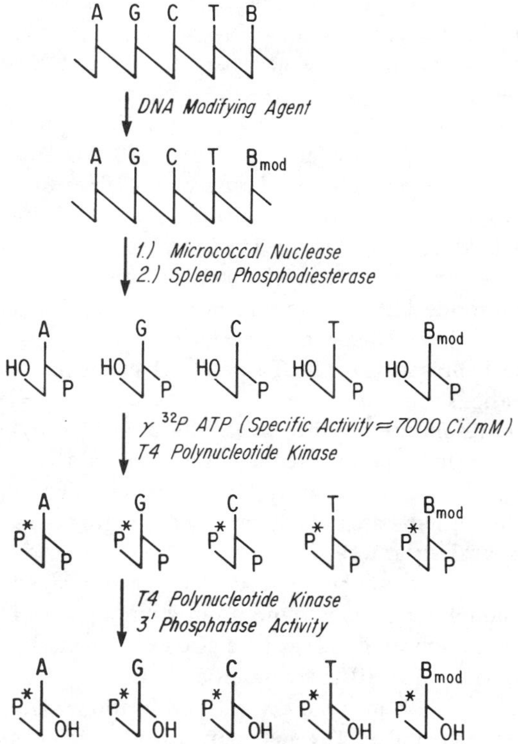

Figure 6

DNA post-labeling method. DNA (in vitro or in vivo) is modified covalently by some modifying agent. Following damage to the DNA, the DNA is isolated and digested enzymatically via the action of micrococcal nuclease and spleen phosphodiesterase to yield 3'deoxynucleoside monophosphates. The 3' deoxynucleotides are then post-labeled by the action of a mutant T4 polynucleotide kinase lacking 3' phosphatase activity and high specific activity γ ^{32}P 5' ATP. Following the 5' kinasing, the 3'5' deoxynucleoside diphosphates are treated with T4 polynucleotide kinase containing a 3' phosphatase activity. The resultant 5' labeled deoxynucleotides are then resolved by HPLC or TLC.

SENSITIVITY OF THE ASSAY

One base in 2,000 can be detected. Modification of DNA to this extent requires a reasonably high dose of the agent. However, some clinical exposures should result in sufficient modification of the target cells for this analysis.

This method is primarily an analytical tool; it is a tool to provide insight into the spectrum of lesions that might be responsible for some of the lethal and mutagenic effects of DNA modification.

We have also been interested in developing methods that are sensitive. A general approach to the problem of detection of stable DNA adducts has been developed which is based on the notion that it is necessary to detect DNA modifications in tissues of people who have been exposed to chemicals. The method that we developed is called the post-labeling method. A similar method was developed independently by Randerath and colleagues (1981). This is illustrated in Figure 6. The DNA is exposed to a mixture of nucleases that yield 3' deoxynucleoside monophosphates. This is an important feature of the method as the T4 polynucleotide kinase used in the next step requires a 3' phosphoryl moiety. The nucleotides are incubated with high specific activity gamma[32] p ATP and T4 polynucleotide kinase, resulting in an addition of a radioactive phosphate to the nucleotides. Under the correct conditions, one has a representative mix of labeled phosphates. The T4 polynucleotide kinase is insensitive to most base modifications.

To improve the subsequent resolution of the bases, the 3' phosphoryl groups are removed by incubation with T4 polynucleotide kinase under a very special set of reaction conditions. We normally use high pressure liquid chromatography systems to resolve modified bases. Thin layer chromatographic systems are suitable for this sort of analysis as well.

Using this method an extraordinary number of counts can be obtained from a very small sample, i.e., a couple of billion counts from a microgram of DNA. The separation methods that have been developed are also extraordinarily powerful. It is certainly feasible to detect 1 in 10^6 base modifications.

We have used this procedure to look at a variety of modifications. The method uniformly labels all four nucleotides. The radioactive profile mimics the optical density scan of the separate nucleotides. In normal human DNA the 5-methyl dCMP is detected as a discrete peak. This is a nice way of measuring the extent of DNA methylation.

The precursors to the alkaline labile lesions and cyclobutane pyrimidine dimers are among the compounds analyzed. For this analysis the DNA is digested to nucleosides by inclusion of a phosphatase in the nuclease. The only compound labeled under these conditions will be unusual nucleotides. This assay is a sensitive one as nothing else will be labeled. It may be useful for analysis of another kind of problem. Suppose one is interested in the major DNA adducts consequent to an exposure to a complex chemical mix, i.e., cigarette smoke. Cellular DNA could be purified, digested, and labeled. The major peaks of modified bases could be resolved and chemical analysis could be done on the modified nucleotides. I think that the post labeling technique is a useful method to add to the growing repertoire of techniques already existing for analysis of the biological damage that is sustained by human cells when exposed to complex mixes. The amount of material needed for such studies is quite small. One can contemplate doing a full analysis with as little as 5,000 cells.

REFERENCES

D'Andrea, A. D. and W. A. Haseltine. 1978. Sequence specific cleavage of DNA by the antitumor antibiotics neocarzinostatin and bleomycin(antitumor drugs/DNA sequencing/DNA damage). *Proc. Natl. Acad. Sci. U.S.A.* 75: 3608.

Franklin, W. and W. A. Haseltine. 1982. The use of post-labelling methods to detect and characterize infrequent base modifications in DNA. In *DNA repair: A laboratory manual of research procedures* (eds. P. C. Hanawalt and E. C. Friedberg), Vol. II. Marcel Dekker, Inc., New York. (In press).

Grunberg, S. M. and W. A. Haseltine. 1980. Use of an indicator sequence of human DNA to study DNA damage by methylbis(2-chloreothyl) amine. *Proc. Natl. Acad. Sci. U.S.A.* 77:6546.

Haseltine, W. A., K. M. Lo, A. D. D'Andrea. 1980. Preferred sites of strand scission in DNA modified by anti-diol epoxide of benzo(a)pyrene. *Science* 209:929.

Lippke, J. and W. A. Haseltine. (1982). The human alphoid sequence as an indicator of DNA damage in intact cells. In *DNA Repair: A laboratory manual of research procedures* (eds. P. C. Hanawalt and E. C. Friedberg), Vol. II, Marcel Dekker, Inc., New York. (In press).

Lippke, J. A., L. K. Gordon, D. E. Brash, and W. A. Haseltine. 1981. Distribution of ultraviolet light induced damage in a defined sequence of human DNA: Detection of alkaline sensitive lesions at pyrimidine-nucleoside-cytidine sequences. *Proc. Natl. Acad. Sci. U.S.A.* 78:3388.

Maxam, A. M. and W. Gilbert. 1980. Sequence end-labelled DNA with base-specific chemical cleavages. In *Methods of Enzymology* (eds. L. Grossman and K. Moldave), Vol. 65, 499, Academic Press, New York.

Randerath, K., M. V. Reddy, and R. C. Gupta. 1981. ^{32}P-labelling test for DNA damage. *Proc. Natl. Acad. Sci. U.S.A.* 78:6126.

High-affinity Monoclonal Antibodies Specific for DNA Components Structurally Modified by Alkylating Agents

JÜRGEN ADAMKIEWICZ, WOLFGANG DROSDZIOK, WILFRIED
EBERHARDT, URSULA LANGENBERG, AND MANFRED F. RAJEWSKY
Institut für Zellbiologie (Tumorforschung)
Universität Essen (GH)
Hufelandstrasse 55, D-4300 Essen 1
Federal Republic of Germany

Chemical mutagens, as well as most chemical carcinogens and cancer chemo-therapeutic agents, cause structural alterations of DNA in the chromatin of target cells (Grover 1979; Rajewsky 1980). The ultimate reactive derivatives generated from these agents in vivo either enzymatically or via nonenzymatic decomposition, are mostly electrophilic and react with electron-pairs in nucleo-philic atoms of cellular macromolecules (Miller and Miller 1979). During the past years, considerable progress has been made with regard to the precise characterization of DNA components structurally modified by chemical carcino-gens and mutagens (Grover 1979; Grunberger and Weinstein 1979; Pullman et al. 1980; Rajewsky 1980). Notably, the modifications effected in DNA by alkylating N-nitroso compounds (alkylnitrosamines, alkylnitrosoureas, alkylnitro-nitroso-guanidines) have to a large extent been identified (Lawley 1976; Pegg 1977; O'Connor et al. 1979; Singer 1979; Rajewsky 1980). Adequate methods for the demonstration and quantification of specific reaction products of chemicals in DNA are a prerequisite for the analysis of molecular mechanisms underlying processes such as DNA repair, mutagenesis, or malignant transforma-tion. High sensitivity analytical methodology is, however, also needed for the dosimetry of the exposure of cells, tissues, or individuals, to agents affecting the integrity of the genome.

Until recently, conventional radiochromatography has been the method of choice for the analysis of chemically-modified DNA (Baird 1979; Beranek et al. 1980). The sensitivity of radiochromatographic techniques is, however, limited by the specific radioactivity of the agents in question and by the rela-tively large amounts of DNA thus required for analysis. With the exception of recently developed "post-labeling" techniques (Randerath et al. 1981; Haseltine, this volume), the application of radiochromatography is necessarily restricted to experiments using synthetic radiolabeled compounds. Therefore, analyses of small samples of cells (ideally of individual cells) exposed to low concentrations of nonradioactive (e.g., environmental or chemotherapeutic) agents cannot be performed.

As an alternative, immunoanalytical methods deserve particular attention (Müller and Rajewsky 1981; Poirier 1981; Müller et al. 1982). Immunoglobulins are characterized by an exceptional capability to recognize specifically subtle alterations of molecular structure; the sensitivity and specificity of immuno-analysis by highly purified or monoclonal antibodies of high affinity are out-standing; and the reaction products of chemicals in DNA (i.e., the haptens against which antibodies are produced) need not be radioactively labeled for analysis. Indeed, specific antisera have been raised against both naturally oc-curring modified nucleosides (for review, see Munns and Liszewski 1980) and DNA components structurally altered by carcinogens and mutagens (for review, see Müller et al. 1982), and immunological assays, in part highly sensitive, have been established for their detection and quantification (Poirier et al. 1977, 1979; Briscoe et al. 1978; Leng et al. 1978; Müller and Rajewsky 1978, 1980; Hsu et al. 1980, 1981; Rajewsky et al. 1980; Groopman et al. 1982; Saffhill et al. 1982; Van der Laken et al. 1982). We have recently developed, and are further expanding, a panel of high-affinity monoclonal antibodies specifically directed against DNA alkylation products with possible relevance in relation to the malignant transformation of mammalian cells (Rajewsky et al. 1980; J. Adamkiewicz and M.F. Rajewsky, in prep.). The characteristics of some of the monoclonal antibodies specific for O^6-ethyl-2'-deoxyguanosine (O^6-EtdGuo), O^6-butyl-2'-deoxyguanosine (O^6-BudGuo), O^6-isopropyl-2'-deoxy-guanosine (O^6-iProdGuo), and O^4-ethyl-2'-deoxythymidine (O^4-EtdThd), are described in the following.

EXPERIMENTAL PROCEDURES

Synthesis and Purification of Alkylated Nucleic Acid Constituents

1. Alkylated Ribonucleosides (Haptens):
   O^6-ethylguanosine (O^6-EtGuo), and O^4-ethylthymine riboside (O^4-EtrThd), were synthesized and purified by thin-layer chromatography on silica gel and Sephadex G-10 chromatography according to Müller and Rajewsky (1978, 1980) and Rajewsky et al. (1980). O^6-n-butylguanosine (O^6-BuGuo) was synthesized by Dr. R. Saffhill (Paterson Laboratories, Manchester, U.K.), and O^6-isopropylguanosine (O^6-iProGuo) was prepared from 6-chloroguano-sine (Pharma Waldhof, Mannheim, Germany) as described by Gerchman et al. (1972) for the synthesis of O^6-methylguanosine (O^6-MeGuo), and purified by recrystallization from acetone followed by Sephadex G-10 chromatography.

2. Other Alkylation Products

O^6-EtdGuo, O^6-methyl-2'-deoxyguanosine (O^6-MedGuo), O^6-MeGuo, O^6-BudGuo, O^4-EtdThd, O^6-ethylguanine (O^6-EtGua), O^6-ethyl-2'-dGMP (O^6-EtdGMP), 7-ethylguanine (7-EtGua), and 7-ethyl-2'-deoxyguanosine (7-EtdGuo), were synthesized and purified as described (Müller and Rajewsky 1978, 1980; Rajewsky et al. 1980). O^6-iProdGuo was prepared by reacting 2'-deoxyguanosine (dGuo) with 2-iodopropane (Merck, Darmstadt, Germany) following the procedure of Singer (1972) for the synthesis of various methylated and ethylated guanosine derivatives.

3. [^{3}H]-labeled Tracers for Use in Competitive Radioimmunoassay:

High specific [^{3}H]-activity tracers were synthesized according to Müller and Rajewsky (1978, 1980) and Rajewsky et al. (1980). The tracer O^6-iPro-[8,5'-^{3}H]dGuo was prepared analogous to the unlabeled deoxynucleoside, starting from [8,5'-^{3}H]dGuo (obtained by hydrolysis of [8,5'-^{3}H]dGTP, 33.0 Ci/mmol; NEN, Dreieich, Germany).

Hapten-Protein Conjugates

Alkylribonucleosides were coupled to keyhole limpet hemocyanin (KLH; Calbiochem; Marburg, Germany; molecular weight ~800,000, Stollar and Borel 1976) as a carrier protein (~100 hapten molecules bound per molecule KLH) (Erlanger and Beiser 1964; Müller and Rajewsky 1980).

Immunization Procedure

Adult female rats of the inbred BDIX strain (Druckrey 1971) were immunized by intracutaneous injection into about 20 sites with an emulsion of 50 μg of conjugate in 50 μl of phosphate-buffered saline (PBS), 50 μl of aluminium hydroxide (Alugel S; Serva, Heidelberg, Germany), and 100 μl of complete Freund's adjuvant. The animals were boosted by the same procedure 5 weeks later. A second intraperitoneal (i.p.) booster injection was administered after 8 weeks, and spleen cells were collected 3-4 days thereafter.

Cell Fusion, Culture, and Cloning of Monoclonal Antibody-Secreting Hybridoma Cells

Cells of the HAT (hypoxanthine/aminopterin/thymidine)-sensitive rat myeloma line Y3-Ag.1.2.3 (Galfré et al. 1979) were fused with spleen cells from immunized BDIX rats, using polyethylene glycol (PEG 4000; Roth, Karlsruhe, Germany) as a fusion reagent. Hybrid cells were cultured in multi-well tissue culture plates (Costar No. 3524; 1-2 × 10^6 cells/well) containing Dulbecco's modified Eagle's-HAT medium, with 1 mM sodium pyruvate, 5 × 10^{-5} M mercaptoethanol, and 20% fetal bovine serum (Lemke et al. 1978). Culture supernatants were tested for the presence of specific antibodies by an enzyme-linked immunosorbent assay (ELISA; see below). Cells from positive cultures were cloned, and once more recloned, without aminopterin in the presence of X-irradiated (40 Gy) BDIX rat "feeder" spleen cells. Positive clones were maintained in cell culture, and aliquots injected i.p. into Pristan-pretreated, X-irradiated (4 Gy) BDIX rats for growth and antibody production in the ascites. Isotype analysis of monoclonal antibodies was carried out by the Ouchterlony technique (immunoprecipitation in agar), using anti-rat-isotype antisera (Miles GmbH, Frankfurt am Main, Germany).

Antibody Concentrations and Affinity Constants

Antibody concentrations in cell culture or ascites fluid, respectively, and antibody affinity constants for the respective alkylated deoxyribonucleosides, were calculated from the data obtained by competitive radioimmunoassay (RIA) (Müller 1980). Antibodies were isolated with the aid of specific hapten-immunosorbents (haptens coupled to epoxy-activated Sepharose 6B (Pharmacia, Uppsala, Sweden) at acid or alkaline pH (Müller and Rajewsky 1980).

Enzymatic Hydrolysis of DNA

DNA isolated from rat tissues by a modified Kirby method (Goth and Rajewsky 1974) or by a hydroxylapatite adsorption technique (Müller and Rajewsky 1980), was enzymatically hydrolyzed to nucleosides with DNase I (EC 3.1.4.5; Boehringer Mannheim, Mannheim, Germany), snake venom phosphodiesterase (EC 3.1.4.1; Boehringer) and grade I alkaline phosphatase (EC 3.1.3.1; Boehringer), as described in Müller and Rajewsky (1980). In some cases, adenosine deaminase (EC 3.5.4.4; Boehringer; 0.3 units/ml) was used to convert deoxyadenosine (dAdo) to deoxyinosine (dino) in the DNA hydrolysates prior to analysis. This reaction is complete after 5 min (20°C) and does not lead to measurable dealkylation of O^6-EtdGuo or O^6-BudGuo (in contrast to O^6-MedGuo). 2'-deoxyguanosine (dGuo) and 2'-deoxythymidine (dThd) concentrations in DNA hydrolysates were determined by peak integration after separation of DNA hydrolysates by high pressure liquid chromatography (HPLC). HPLC was also used for the separation of different alkylation products from the same DNA sample, prior to analysis by competitive RIA (see Fig. 3).

Enzyme-Linked Immunosorbent Assay

The competitive ELISA was performed in 96-well microtiter plates (type M 129 A; Greiner, Nürtingen, Germany) coated with bovine serum albumin conjugates of the respective haptens (Müller and Rajewsky 1980). Cell culture supernatants to be assayed for monoclonal antibody activity, were added to duplicate wells (without or with the respective alkyldeoxynucleoside in a final concentration of 0.15 mM) and incubated for 90 min at 37°C. After washing with PBS containing 0.05% of Triton X-100, alkaline phosphatase-conjugated goat-anti-rat IgG was added. Following a 90 minute-incubation period at 37°C, the incubation mixture was removed, the wells were extensively washed, and a 10 mM solution of p-nitrophenyl phosphate (alkaline phosphatase substrate; Sigma-Chemie, Munich, Germany) was added. After incubation for 1 hour at 37°C, the plates were screened for positive wells (wells without added alkyl-deoxynucleoside, yellow; wells containing the alkyldeoxynucleoside colorless), or the degree of binding of antibody-alkaline phosphatase to the solid phase was determined spectrophotometrically (absorbance at 405 nm).

Competitive Radioimmunoassay

The competitive RIA, a modified Farr assay (Farr 1958), was carried out as described by Müller and Rajewsky (1978, 1980). In a total volume of 100 μl of Tris-buffered saline (with 1% bovine serum albumin [w/v] and 0.1% bovine IgG [w/v], each sample contained ~2.5 $\times$ 10^3 dpm of tracer, an antibody solution diluted to give 50% binding of tracer in the absence of inhibitor, plus varying amounts of hydrolyzed alkylated DNA, or of other natural or modified DNA constituents to be analyzed for cross-reactivity (inhibitor). After incubation for 2 hours at room temperature (equilibrium), 100 μl of a saturated ammonium sulphate solution (pH 7.0) were added. After 10 minutes the samples were centrifuged for 3 minutes at 10.000 x g. The [^{3}H]-activity in 150 μl of supernatant was measured by liquid scintillation spectrometry. The degree of inhibition of tracer-antibody binding (ITAB) was calculated as described (Müller and Rajewsky 1980; see Fig. 1 and 2).

RESULTS AND DISCUSSION

As shown in Figure 1, the detection limit of a competitive RIA is primarily dependent on the affinity of the antibody for the respective alkyldeoxynucleo-side. The specificity of the antibody (i.e., the degree of its cross-reactivity with other DNA constituents; see Fig. 2), comes into play whenever hydrolysates of alkylated DNA containing natural and modified deoxynucleosides are to be analyzed. Unlike antisera raised in animals, monoclonal antibodies by definition do not suffer from contamination with nonspecific antibodies causing reduced specificity. Nonetheless, a high degree of antibody purity can sometimes also be obtained by extensive affinity purification of specific antibodies from antisera (Müller and Rajewsky 1980). Antibody-affinity constants greater than 10^{10}-10^{11} 1/mol will not significantly lower the limit of detection by competitive RIA in a DNA hydrolysate, unless a tracer with a considerably higher specific radio-activity is used (which is practically not possible), or the level of cross-reactants is substantially reduced. The latter can, for example, be achieved by separating the alkylation products in question from a complete DNA hydrolysate with the

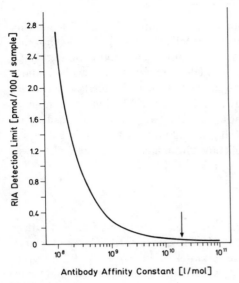

Figure 1
Theoretical detection limit for a given DNA alkylation product of a competitive RIA at 50% inhibition of tracer-antibody binding as a function of the affinity constant of the antibody. Assumed concentration of [^{3}H]-labeled tracer: 3×10^{-10} M. ($\rightarrow$) maximum affinity constants of currently available monoclonal antibodies directed against alkyldeoxynucleosides (see text).

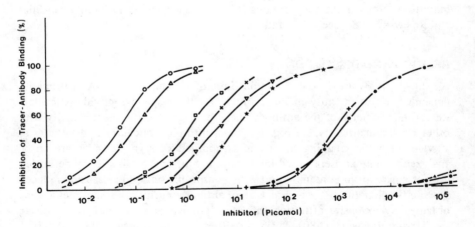

Figure 2
Competitive RIA O^6-EtdGuo, using rat $\times$ rat hybridoma-secreted monoclonal antibody ER-6 (immunogen: O^6-EtGuo-KLH; affinity constant for O^6-EtdGuo, 2×10^{10} liter/mol. Inhibition of tracer-antibody binding by different natural and alkylated nucleic acid constituents. Tracer: O^6-Et-[8,5'-^{3}H]-2'-dGuo. (O) O^{6}EtdGuo; ($\triangle$) O^6-EtdGMP; ($\square$) O^6-EtGuo; (x) O^6-EtGua; ($\triangledown$) O^6-BudGuo; ($\star$) O^6-MedGuo; (+) 7-EtdGuo; (*) O^4-EtdThd; ($\blacktriangle$) DNA hydrolysate; ($\bullet$) dGuo; ($\blacksquare$) dAdo.

270

aid of HPLC prior to the RIA (Fig. 3). However, in order to give reliable results, this procedure not only requires quantitative recovery of the respective fractions, but—in view of the exceedingly low amounts of the products to be measured—also meticulously clean (i.e., contamination-free) working conditions.

The best of our currently available monoclonal antibodies have affinity constants of $1\text{-}2 \times 10^{10}$ 1/mol (see Table 1 and Fig. 2). These antibodies also show a very low degree of (cross-)reactivity with normal deoxynucleosides, i.e., > 7 orders of magnitude lower (at 50% ITAB) as compared with the appropriate alkyldeoxynucleosides. However, not all monoclonal antibodies in our collection exhibit this extreme specificity; it is, therefore, recommendable to produce and characterize a sufficient number of monoclonal antibodies if maximum specificity and affinity is desired.

At 50% ITAB in the competitive RIA, the monoclonal antibodies listed in Table 1 detect 40-60 pmol of the respective alkyldeoxynucleosides in a 100 μl RIA sample (Table 1 and Fig. 2). When a probability grid is used to transform the ITAB-curves into straight lines (Müller and Rajewsky 1980), accurate readings are usually possible at ITAB values $< 50\%$. For example, reading at 20% ITAB reduces the detection limit for O^6-EtdGuo to $\sim$10 fmol/100 μl sample in the case of monoclonal antibody ER-6 (Table 1). When used in the competitive RIA monoclonal antibody ER-6 detects O^6-EtdGuo at an O^6-EtdGuo/dGuo molar ratio as low as $\sim$3 $\times$ 10^{-7} (i.e., the equivalent of $\sim$700 O^6-EtdGuo molecules per diploid genome) in a hydrolysate of 100 μg of DNA (equivalent to the DNA content of $\sim$1.6 $\times$ 10^7 diploid cells). This level of detection can be further reduced by separating the alkyldeoxynucleosides in question from the DNA hydrolysate prior to the RIA (see above).

One of the most important future applications, particularly of monoclonal antibodies, will be their use to detect and monitor the level of structural modifications in the DNA of individual cells by immunostaining procedures. This approach could be used to demonstrate specific DNA adducts (induced by, e.g., carcinogens or chemotherapeutic agents) in very small samples of (e.g., human) cells or tissue biopsies. Furthermore, comparative measurements would become possible, regarding the capacity for metabolic activation of DNA-binding chemicals or the elimination (and repair) of specific adducts from DNA, of different cell types at different stages of differentiation and/or development. Likewise, the detection of individuals with hereditary defects in enzymatic DNA repair mechanisms would be facilitated, as might be the definition of risk groups in terms of carcinogen exposure or the prediction of individual response to cancer chemotherapy. Applications of monoclonal antibodies for immuno-fluorescence measurements of alkyldeoxyguanosines in individual cells are still in their infancy. However, promising results have already been obtained at low levels of DNA modification, particularly with the aid of electronic image intensi-fication (J. Adamkiewicz et al., in prep.). Finally, first attempts have been successful to localize, with the aid of monoclonal antibodies and transmission electron microscopy, alkyldeoxynucleosides in doublestranded DNA molecules (P. Nehls et al., in prep.).

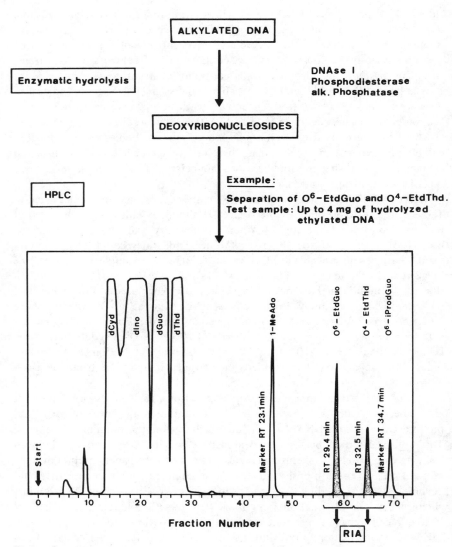

Figure 3

Simultaneous determination by competitive RIA of different alkylation products in the same DNA sample. Pre-separation of alkyldeoxynucleosides from hydrolyzed DNA by HPLC and selective sampling, followed by RIAs with monoclonal antibodies specific for the respective alkylation products. *Example*: Simultaneous determination of O^6-EtdGuo and O^4-EtdThd in DNA. Conditions of HPLC separation: Column, 250 × 4.8 mm (Lichrosorb RP-18. Particle size, 10 μm; Knauer, Berlin, Germany). Eluent, MeOH/H$_2$O. *Elution steps*: (i) 1-10% MeOH (linear), 10 min; (ii) 10-50% MeOH (linear), 30 min. Flow rate: 1.5 ml/min. *Fraction volume*: 750 μl. Absorbance measured at 254 nm. Liquid chromatograph, model 850 Dupont Instruments (Bad Nauheim, Germany). Due to the absence of cross-reactants in the final test samples, the limit of the RIA is in this case considerably lower than in the analysis of complete DNA hydrolysates.

Table 1
Monoclonal Antibodies Directed against Alkylated Deoxynucleosides

Alkyldeoxynucleoside	Immunogen	Designation of antibody (isotype)	Antibody affinity constant l/mol	Radioactive tracer (specific [³H]-activity)	Detection limit of RIA (pmol/100 µl-sample)[a]	Cross-reactivity with hydrolysate of unmodified DNA[b]
O^6-EtdGuo	O^6-EtGuo-KLH	ER-6 (IgG2b)	2.0×10^{10}	O^6-Et-[8,5'-³H]dGuo (~27 Ci/mmol)	0.04	2.5×10^6 [e]
O^6-BudGuo	O^6-EtGuo-KLH	ER-11 (n.d.)[c]	9.1×10^9	O^6-n-Bu-[8,5'-³H]dGuo (~22 Ci/mmol)	0.06	2.5×10^6 [e]
O^6-iProdGuo	O^6-iProGuo-KLH	ER-05 (IgG2a)	1.2×10^{10}	O^6-iPro-[8,5'-³H]dGuo (~17 Ci/mmol)	0.05	2.4×10^6 [e]
O^4-EtdThd	O^4-EtThd-KLH	ER-01 (IgG2a)	1.3×10^9	O^4-Et-[6-³H]dThd (~17 Ci/mmol)	0.24	1.3×10^6 [d]

[a]Determined at 50% ITAB.
[b]Value for ITAB by an adenosine deaminase-treated DNA-hydrolysate, or a mixture of dGuo, dIno, dCyd, and dThd, at the molar ratios for rat DNA; value given as the multiple of the 50% ITAB-value for the respective modified DNA component.
[c]n.d.: not determined
[d,e]Value given for < 10% ITAB (d) or 10-15% ITAB (e), respectively, because value for 50% ITAB not reached by the ITAB-curve within the concentration range of the DNA hydrolysate added to the RIA sample.

ACKNOWLEDGMENTS

This work was supported by the Deutsche Forschungsgemeinschaft (SFB 102, A9), by the Fritz Thyssen-Stiftung (1980/2/41), and by the Commission of the European Communities (ENV-544-D B). The authors gratefully acknowledge the valuable contributions of Dr. R. Müller (presently at the Tumor Virology Laboratory, The Salk Institute, San Diego, California), and the skillful technical assistance of Miss Ch. Knorr and Miss I. Spratte.

REFERENCES

Baird, W.M. 1979. The use of radioactive carcinogens to detect DNA modifications. In *Chemical carcinogens and DNA* (ed. P.L. Grover), p. 59. CRC Press, Boca Raton, Florida.

Beranek, D.T., C.C. Weis, and D.H. Swenson. 1980. A comprehensive quantitative analysis of methylated and ethylated DNA using high pressure liquid chromatography. *Carcinogenesis* 1:595.

Briscoe, W.T., J. Spizizen, and E.M. Tan. 1978. Immunological detection of O^6-methylguanine in alkylated DNA. *Biochemistry* 17:1896.

Druckrey, H. 1971. Genotypes and phenotypes of ten inbred strains of BD-rats. *Arzneim.-Forsch.* 21:1274.

Erlanger, B.F. and S.M. Beiser. 1964. Antibodies specific for ribonucleosides and ribonucleotides and their reaction with DNA. *Proc. Natl. Acad. Sci. U.S.A.* 52:68.

Farr, R.S. 1958. A quantitative immunochemical measure of the primary interaction between I BSA and antibody. *J. Infect. Dis.* 103:239.

Galfré, G., C. Milstein, and B. Wright. 1979. Rat X rat hybrid myelomas and a monoclonal anti-Fd portion of mouse IgG. *Nature* 277:131.

Gerchman, L.L., J. Dombrowski, and D.B. Ludlum. 1972. Synthesis and polymerization of O^6-methylguanosine 5'-diphosphate. *Biochim. Biophys. Acta* 272:672.

Goth, R. and M.F. Rajewsky. 1974. Molecular and cellular mechanisms associated with pulse-carcinogenesis in the rat nervous system by ethylnitrosourea: Ethylation of nucleic acids and elimination rates of ethylated bases from the DNA of different tissues. *Z. Krebsforsch.* 82:37.

Groopman, J.D., Å. Haugen, G.R. Goodrich, G.N. Wogan, and C.C. Harris. 1982. Quantitation of aflatoxin B_1-modified DNA using monoclonal antibodies. *Cancer Res.* 42:3120.

Grover, P.L., ed. 1979. *Chemical carcinogens and DNA*. CRC Press, Boca Raton, Florida.

Grunberger, D. and I.B. Weinstein. 1979. Conformational changes in nucleic acids modified by chemical carcinogens. In *Chemical carcinogens and DNA* (ed. P.L. Grover), p. 59. CRC Press, Boca Raton, Florida.

Hsu, I.C., M.C. Poirier, S.H. Yuspa, R.H. Yolken, and C.C. Harris. 1980. Ultrasensitive enzymatic radioimmunoassay (USERIA) detects femtomoles of acetylaminofluorene-DNA adducts. *Carcinogenesis* 1:455.

Hsu, I.C., M.C. Poirier, S.H. Yuspa, D. Grunberger, I.B. Weinstein, R.H. Yolken, and C.C. Harris. 1981. Measurement of benzo(a)pyrene-DNA adducts by enzyme immunoassays and radioimmunoassay. *Cancer Res.* 41:1091.

Lawley, P.D. 1976. Carcinogenesis by alkylating agents. *ACS Monogr.* **173**:83.

Lemke, H., G.J. Hämmerling, C. Höhmann, and K. Rajewsky. 1978. Hybrid cell lines secreting monoclonal antibody specific for major histocompatibility antigens of the mouse. *Nature* **271**:249.

Leng, M., E. Sage, R.P.P. Fuchs, and M.P. Daune. 1978. Antibodies to DNA modified by the carcinogen N-acetoxy-N-2-acetylaminofluorene. *FEBS Lett.* **92**:207.

Miller, J.A. and E.C. Miller. 1979. Perspectives on the metabolism of chemical carcinogens. In *Environmental carcinogenesis* (eds., P. Emmelot and E. Kriek), p. 25. Elsevier/North Holland, Amsterdam, Holland.

Müller, R. 1980. Calculation of average antibody affinity in anti-hapten sera from data obtained by competitive radioimmunoassay. *J. Immunol. Methods* **34**:345.

Müller, R. and M.F. Rajewsky. 1978. Sensitive radioimmunoassay for detection of O^6-ethyldeoxyguanosine in DNA exposed to the carcinogen ethylnitrosourea *in vivo* or *in vitro*. *Z. Naturforsch.* **33c**:897.

———. 1980. Immunological quantification by high-affinity antibodies of O^6-ethyldeoxyguanosine in DNA exposed to N-ethyl-N-nitrosourea. *Cancer Res.* **40**:887.

———. 1981. Antibodies specific for DNA components structurally modified by chemical carcinogens. *J. Cancer Res. Clin. Oncol.* **102**:99.

Müller, R., J. Adamkiewicz, and M.F. Rajewsky. 1982. Immunological detection and quantification of carcinogen-modified DNA components. *IARC Sci. Publ.* **39**:443.

Munns, T.W. and M.K. Liszewski. 1980. Antibodies specific for modified nucleosides: An immunological approach for the isolation and characterization of nucleic acids. *Prog. Nucleic Acid. Res. Mol. Biol.* **24**:109.

O'Connor, P.J., R. Saffhill, and G.P. Margison. 1979. N-nitroso-compounds: Biochemical mechanisms of action. In *Environmental carcinogenesis* (eds., P. Emmelot and E. Kriek), p. 73. Elsevier/North Holland, Amsterdam, Holland.

Pegg, A.E. 1977. Formation and metabolism of alkylated nucleosides: Possible role in carcinogenesis by nitroso compounds and alkylating agents. *Adv. Cancer Res.* **25**:195.

Poirier, M.C. 1981. Antibodies to carcinogen-DNA adducts. *J. Natl. Cancer Inst.* **67**:515.

Poirier, M.C., M.A. Dubin, and S.H. Yuspa. 1979. Formation and removal of specific acetylaminofluorene-DNA adducts in mouse and human cells measured by radioimmunoassay. *Cancer Res.* **39**:1377.

Poirier, M.C., S.H. Yuspa, I.B. Weinstein, and S. Blobstein. 1977. Detection of carcinogen-DNA adducts by radioimmunoassay. *Nature* **270**:186.

Pullman, B., P.O.P. Ts'o, and H. Gelboin, eds. 1980. *Carcinogenesis: Fundamental mechanisms and environmental effects*. Reidel, Dordrecht, Holland.

Rajewsky, M.F. 1980. Specificity of DNA damage in chemical carcinogenesis. *IARC Sci. Publ.* **27**:41.

Rajewsky, M.F., R. Müller, J. Adamkiewicz, and W. Drosdziok. 1980. Immunological detection and quantification of DNA components structurally modified by alkylating carcinogens (ethylnitrosourea). In *Carcinogenesis:*

Fundamental mechanisms and environmental effects (eds. B. Pullman et al.), p. 207. Reidel, Dordrecht, Holland.

Randerath, K., M. Reddy, and R.C. Gupta. 1981. ^{32}P-labeling test for DNA damage. *Proc. Natl. Acad. Sci. U.S.A.* **78**:6126.

Saffhill, R., P.T. Strickland, and J. Boyle. 1982. Sensitive radioimmunoassays for O^6-N-butyldeoxyguanosine, O^2-N-butylthymidine and O^4-N-butylthymidine. *Carcinogenesis* **3**:547.

Singer, B. 1972. Reaction of guanosine with ethylating agents. *Biochemistry* **11**:3939.

_____. 1979. N-nitroso alkylating agents: Formation and persistence of alkyl derivatives in mammalian nucleic acids as contributing factors in carcinogenesis. *J. Natl. Cancer Inst.* **62**:1329.

Stollar, B.D. and Y. Borel. 1976. Nucleoside specificity in the carrier IgG-dependent induction of tolerance. *J. Immunol.* **117**:1308.

Van der Laken, C.J., A.M. Hagenaars, G. Hermsen, E. Kriek, A.J. Kuipers, J. Nagel, E. Scherer, and M. Welling. 1982. Measurement of O^6-ethyldeoxyguanosine and N-(deoxyguanosine-8-μl)-N-acetyl-2-aminofluorene in DNA by high-sensitive enzyme immunoassays. *Carcinogenesis* **3**:569.

SESSION V: CYTOGENETICS AND SISTER CHROMATID EXCHANGE

The Use of Rat and Mouse Lymphocytes to Study Cytogenetic Damage After In Vivo Exposure to Gentoxoic Agents

ANDREW D. KLIGERMAN, JAMES L. WILMER, AND
GREGORY L. EREXSON
Department of Genetic Toxicology
Chemical Industry Institute of Toxicology
Research Triangle Park, North Carolina 27709

Cytogenetic methodologies offer some of the most sensitive approaches for detecting genotoxic effects of mutagenic carcinogens (Perry and Evans 1975; Latt et al. 1979). The analysis of both sister chromatid exchange (SCE) and chromosome breakage in an in vivo system should yield few false positive or false negative results and provide valuable information on the genotoxic potential of a compound.

Lymphocytes are excellent sources of potentially dividing cells for cytogenetic studies. They not only provide large numbers of metaphases for examination, but also permit the nonlethal sampling of blood for cytogenetic analyses of animals exposed in vivo. Thus, a subject can act as its own control before exposure to genotoxicants, and later be sampled for investigations of the persistence or repair of DNA damage. Furthermore, the vast majority of the peripheral lymphocytes are normally nondividing (G_o cells), long-lived cells that may accumulate chromosome damage over time and may serve as indicators of exposure to genotoxicants (Carrano et al. 1980).

The study of cytogenetic damage in lymphocytes does present some disadvantages. Lymphocytes are a heterogeneous cell population. The cell types may vary in number in individual animals and may show different sensitivities to genotoxic agents leading to variation in responses. Also, lymphocytes do not possess substantial capabilities to metabolize promutagenic xenobiotics. Therefore, they may not be damaged by such compounds unless the promutagens are metabolized in other tissues to relatively stable active intermediates, or the lymphocytes, themselves, come in close contact with metabolically proficient cells at the time of bioactivation.

Even with these disadvantages, lymphocytes are valuable for cytogenetic investigations because they are the only practical means to investigate chromosome damage in humans (see review by Lambert et al. 1982). Although information on the cytogenetic effects of genotoxicants can be gleaned from studies of patients exposed to chemotherapeutic agents, tobacco smokers, and occupationally exposed workers, such studies are difficult to conduct under carefully controlled conditions, and the construction of dose-response curves is all but impossible. Good animal models are needed. The laboratory rat and mouse are obvious choices for study because of their use as models for toxicological investigations, as well as the availability of purebred strains of known genetic composition. The small size, modest housing requirements, and low cost are additional criteria that favor their use.

Some reports have appeared in the literature during the last two decades describing methods for culture of lymphocytes from either rats (Metcalf 1965; Beltz and Benz 1980; Granberg-Öhman et al. 1980; Tates et al. 1980) or mice (Buckton and Nettesheim 1968; Bryan and Hybertson 1972; Farrow et al. 1974; Triman et al. 1975; de Boer et al. 1977; Barren et al. 1982). It has been difficult to assess the yield of mitotic figures or the reliability of some of these methods because of the scarcity of data and (or) methods presented in these papers. The paucity of papers on cytogenetic studies in rodent lymphocytes indicates the existence of methodological problems. This presentation describes the development and use of both rat and mouse lymphocyte culture systems for examining cytogenetic damage after in vivo exposures to genotoxic agents.

CULTURE METHODOLOGIES

The rat and mouse lymphocyte culture methods are designed for short-term culture of small quantities of blood so that repeated samples can be taken from the same animal if warranted by the constructs of the experiment. To prevent carry-over of xenobiotics from treated animals into the culture medium (DuFrain et al. 1980), each methodology incorporates the washing of the blood before inoculation.

Rat Lymphocyte Culture

Figure 1 depicts schematically the methodology for culturing lymphocytes in rat whole blood (Kligerman et al. 1981, 1982). Male Fischer-344 rats (CDF® [F-344] CrlBR, Charles River Breeding Laboratories, Kingston, NY) are anesthetized with Metofane® (Pitman-Moore), and the blood is removed using a 23-gauge, 1" needle fitted to a heparinized syringe. The amount of blood removed is determined by the needs of the experiment and the size of the animals. If the animal is to be sampled repeatedly over time, then as little as 0.2 ml blood can be removed. If the animal is to be killed, about 10-12 ml blood may be obtained by exsanguination. The blood is washed three times in phosphate-buffered saline (pH 7.4) containing 2% heat-inactivated fetal bovine serum to reduce hemolysis. This is a major factor in successful mitogenesis since the washing apparently removes an inhibitory factor or factors present in the blood plasma. Once the cells are washed and resuspended to their initial volume in medium, the number of leukocytes is determined using an automatic cell

Rat Lymphocyte Culture Methodology

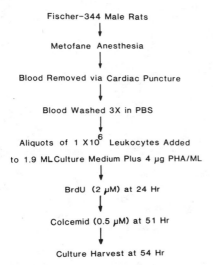

Figure 1
Schematic representation of the rat lymphocyte culture system.

counter (Coulter Counter Model Z_f), and whole blood containing from 1.0-1.5 million leukocytes is inoculated into 6 ml Falcon tubes (No. 2003; Scientific Products) containing 1.8 ml medium consisting of RPMI 1640 plus 25 mM HEPES buffer with L-glutamine (Gibco), 10% heat-inactivated fetal calf serum (Gibco), 100 units penicillin/ml, 100 μg streptomycin/ml (Gibco), an additional 292 μg L-glutamine (Gibco)/ml, 10 units of preservative-free sodium heparin (Upjohn) and either 4 μg phytohemagglutinin (PHA) (Burroughs-Wellcome HA-16) or 30 μg concanavalin A (Sigma; Type IV)/ml (Wilmer et al. 1982). The tubes are then loosely capped and incubated for 6 hours (5% CO_2, 37°C, 98% relative humidity) at which time the cells are resuspended gently. After 24 hours of incubation, 20 μl of a 200 μM stock solution of 5-bromo-2'-deoxyuridine (BrdU) (Sigma) are added to each culture to achieve a final concentration of 2 μM. Cultures are returned to the incubator and shielded from light until harvest. Three hours prior to harvest, Colcemid® (Gibco, 0.5 μg/ml) is added to arrest cells at metaphase, and the cultures are harvested after a total incubation period of 54 hours.

Mouse Lymphocyte Culture

The methodology for the culture of mouse lymphocytes from whole blood is shown in Figure 2 (G. Erexson et al. unpubl. results). The mice (C57BL/6NCrlBr,

MOUSE LYMPHOCYTE CULTURE METHODOLOGY

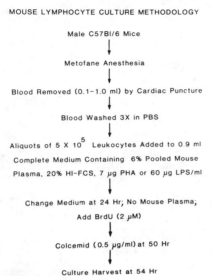

Figure 2
Schematic representation of the mouse lymphocyte culture system.

Charles River Breeding Laboratories, Kingston, NY) are anesthetized with Metofane®, and blood (0.2-1.2 ml) is removed by cardiac puncture using a 25-gauge, 5/8″ needle fitted on a heparinized syringe. The blood is washed 3 times as described above to prevent carry-over of any genotoxicants into the culture. Contrary to what was found with rat lymphocytes, washing the mouse blood actually inhibits mitogenesis. This indicates that a factor or factors present in mouse plasma can aid in stimulating the lymphocytes to divide. After the blood is washed and resuspended to its initial volume in medium, the number of leukocytes is counted. White blood containing 500,000 leukocytes is added to 0.9 ml medium in 6 ml Falcon culture tubes. The medium is similar to that used for rat lymphocyte culture except that 6% pooled mouse plasma (from control mice) is added to enhance cell stimulation, and the fetal calf serum concentration is increased to 20%. Either 7 µg PHA/ml or 60 µg lipopolysaccharide (LPS)(Sigma)/ml are added for T- or B-cell mitogenesis, respectively. The cultures are loosely capped and incubated for 24 hours (5% CO_2, 37°C, 98% relative humidity) without resuspension. Then, the medium is replaced with fresh medium (without 6% mouse plasma) containing 2 µM BrdU. The tubes are wrapped in aluminum foil, and the cultures are returned to the incubator for an additional 30 hours of culture. Four hours before harvest 0.5 µg Colcemid®/ml are added to each culture. The total incubation time is 54 hours.

For both rat and mouse lymphocyte cultures, the number of leukocytes in the inoculum and the inoculum volume are critical factors (Kligerman et al.

1982; Erexson et al., unpubl. results). For the rat the number of leukocytes added to the cultures affects both the SCE frequency and the mitotic activity (Fig. 3). As the number of cells increases, the SCE frequency and the mitotic activity decrease. The use of higher concentrations of BrdU (25, 33, 50, 100 μM) does not prevent variation in baseline SCE frequency caused by changes in the number of cells in the inoculum. Furthermore, a significant increase in chromosome breakage and evidence of cytotoxicity becomes apparent if BrdU levels are increased over 25 μM (Kligerman et al. 1982). For mouse lymphocyte cultures, the number of leukocytes in the inoculum has a pronounced effect on mitotic activity. A peak in mitotic activity occurs when 1 ml cultures are inoculated with whole blood containing 500,000 leucocytes. With greater or lesser numbers of leukocytes, the mitotic activity is reduced sharply.

Although inoculum volume has no significant effect on SCE frequency, it can have a significant effect on mitotic activity. If the leukocyte counts of the animals are low, then the volume which will contain sufficient numbers of leukocytes will be high. As the inoculum volume increases, the mitotic activity of the culture decreases (Kligerman et al. 1982). Thus, if possible, healthy animals with leukocyte counts $\geqslant$ 4 million/ml should be used, and the number of leukocytes added to cultures should be controlled carefully.

Similarly, the concentration of BrdU in the culture medium has a significant effect on SCE frequency and mitotic activity for both rat and mouse lymphocytes (Kligerman et al. 1981; Erexson et al. unpubl. results). When the initial number of leukocytes added to the cultures is held constant at 5 X

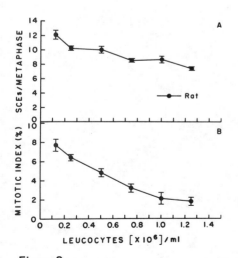

Figure 3
Effect of the number of leukocytes in the inoculum on the SCE frequency (A) and the mitotic index (B) of rat lymphocytes in culture. [mean ± S.E.M.]

10^5/ml, the BrdU dose-response curves are almost linear for both rat and mouse lymphocytes (Fig. 4A). However, varying the number of cells in the inoculum changes the shape of the dose-response curve (Fig. 5). With both lymphocyte culture systems, increases in BrdU concentration cause significant reductions in mitotic activity (Fig. 4B). Therefore, the lowest BrdU concentrations that give consistently good sister chromatid differentiation should be used.

Cell Harvest, Slide Preparation, and Staining

Cultures are harvested by centrifugation at 700 × g for 5 minutes, and the cell pellet is resuspended in 5 ml of hypotonic solution (10 mM sodium citrate, 50 mM potassium chloride; 37°C) for 15 minutes (rat) (Kligerman et al. 1981) or 10 minutes (mouse) (G. Erexson et al., unpubl. results). The cells are pelleted at 200 × g and fixed in three changes of 3:1 methanol:acetic acid. The fixed cultures are then stored overnight at 4°C. Cells are resuspended in 0.6 ml of fixative and dropped onto chilled wet slides. After the slides are aged at room temperature for 2 days, they are stained for SCE analysis by a modification of the procedure of Goto et al. (1978) as described by Kligerman et al. (1981). This involves a 20-minute staining under coverslips with 200 μg 33258 Hoechst (American Hoechst)/ml followed by a 30-35 minute exposure of the slides mounted

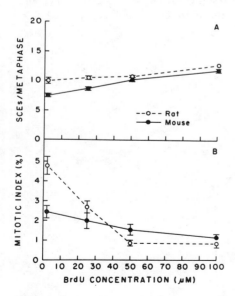

Figure 4
Effect of BrdU concentrations of the SCE frequencies (A) and mitotic indices (B) of rat and mouse lymphocytes when the initial whole blood inoculum contains 500,000 leukocytes/ml of culture. [mean ± S.E.M.]

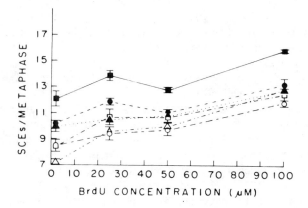

Figure 5
Dose-response curves for BrdU in 2 ml rat lymphocyte cultures when the initial number of leukocytes in the whole blood inoculum is varied. [mean ± S.E.M.] [(■) 125,000; (●) 250,000; (▲) 500,000; (□) 750,000; (○) 1,000,000; (△) 1,250,000 leukocytes/ml.]

under coverslips with 0.1 ml of McIlvaine's buffer (pH 8.0) to intense fluorescent light. The slides are then stained in 3% Giemsa dissolved in 20% McIlvaine's buffer (pH 7.0).

DISCUSSION

Lymphocyte cultures are an easy and reliable way to analyze the chromosomes of humans for cytogenetic damage; however, the meaning of results obtained from human studies is at times obscure because carefully controlled experimental manipulations of variables cannot be done with human subjects. Before large-scale testing of an exposed human population is undertaken, the value of lymphocyte culture systems for assessing in vivo exposure to genotoxicants must be determined. Thus, development and thorough characterization of a model mammalian lymphocyte culture assay was deemed necessary. Rodents were chosen for study because these animals serve as important models for predicting toxicological effects of xenobiotic exposures.

Cytogenetic responses of rat, mouse, and human lymphocytes to similar culture conditions are valuable for interspecific comparisons. When the red blood cells (RBCs) are removed from the blood on a Ficoll-Hypaque density gradient before culture, the baseline SCE frequency of rat lymphocytes more than doubles. When human blood is treated in a similar fashion, there is only a marginal increase in the baseline frequency (J. Wilmer et al., unpubl. results). Mouse lymphocytes show an intermediate response. Because of these differences, we recommend the use of whole blood instead of separated leukocytes for analysis of the effects of in vivo exposure to genotoxicants.

Subpopulations of lymphocytes show variation in SCE frequency and cell cycle kinetics. PHA, a mouse T-cell mitogen (Greaves and Janossy 1972; Stobo 1972) or LPS, a mouse B-cell mitogen (Peavy et al. 1970; Greaves and Janossy 1972; Janossy and Greaves 1973), can cause selective stimulation of these subpopulations of lymphocytes. When the SCE frequencies were analyzed in mouse B-and T-lymphocytes grown in the presence of 2 μM BrdU, the baseline SCE frequency was almost twice as high in T-cells (G. Erexson et al., unpubl. results). Santesson et al. (1979) and Lindblad and Lambert (1981) obtained similar results with human lymphocytes. They used a T-cell mitogen (PHA) to stimulate nonrosetting B-lymphocytes. These results suggest that chemicals which alter the ratio of B- to T-cells in the blood may influence the frequency of SCE (in mixed lymphocyte culture) without affecting directly the genetic material. In contrast to SCE frequencies, the baseline and γ-ray induced chromosome aberrations frequencies of T- and B-lymphocytes from humans are not significantly different (Schwartz and Gaulden 1980).

A primary concern in the use of lymphocytes is their sensitivity for detecting the effects of genotoxicants administered in vivo. Promutagens are of special interest because although lymphocytes can activate some polycyclic aromatic hydrocarbons (Rüdiger et al. 1976; Takehisa and Wolff 1978), they do not show other substantial bioactivation capabilities. In studies on human patients exposed to the promutagen cyclophosphamide, elevated SCE frequencies were found in peripheral lymphocytes after culture (Raposa 1978; Musilová et al. 1979). Similar results were obtained by Stetka and Wolff (1976) after injecting rabbits with cyclophosphamide. This is not surprising because the metabolites of cyclophosphamide are relatively long-lived (Wagner et al. 1977). The promutagens 2-aminofluorene (2-AF) and 2-acetylaminofluorene (2-AAF) induced elevated SCE frequencies in the peripheral lymphocytes of rabbits; however, with 2-AAF the response was variable from rabbit to rabbit (Takehisa and Wolff 1978). In preliminary studies with the rat lymphocyte culture system, we have shown that the promutagens dinitrotoluene (technical grade), aflatoxin B_1, and 7,12-dimethylbenz[a]anthracene produced elevated SCE frequencies when administered to rats by gavage (Table 1). Although some promutagens can elicit genotoxic effects in lymphocytes, many more compounds need to be examined in vivo using rodent lymphocyte cultures before a reliable estimate of the sensitivity of these assays can be made.

Table 1

Response of Cultured Rat Lymphocytes to In Vivo Exposures to Known and Suspected Genotoxicants

Compound	Dose[a]	Exposure method	SCE response	Cell cycle inhibition	Mitotic depression	Comments
EMS	30-300 mg/kg	ip	+++[d]	+[b]	+	blood cultured either 1, 3, 5, 7, 9, or 28 days after exposure
EO	288-7992 ppm × hrs	inhalation	++[c]	-	-	50, 140, or 450 ppm for 1 or 3 days
NB	60-7500 ppm × hrs	inhalation	-	+	+	10, 35, or 125 ppm for 1, 3, or 10 days
TCDD	20 and 200 μg/kg	gavage	-	-	-	blood cultured either 2, 12, or 48 hrs after exposure
DNT	100 mg/kg	gavage	+	-	-	blood cultured 48 hrs after exposure
MNNG	200 mg/kg	gavage	-	+	+	blood cultured 2 hrs after exposure
MMS	20 mg/kg	gavage	++	-	-	blood cultured 2 hrs after exposure
DMBA	200 mg/kg	gavage	+++	+	+	blood cultured 2 hrs after exposure
AFB$_1$	2 mg/kg	gavage	+	+	+	blood cultured 2 hrs after exposure

[a] Abbreviations: EMS, ethyl methanesulfonate; EO, ethylene oxide; NB, nitrobenzene; TCDD, 2, 3, 7, 8-tetrachlorodibenzo-p-dioxin; DNT, dinitrotoluene (technical grade); MNNG, N-methyl-N'-nitro-N-nitrosoguanidine; MMS, methyl methanesulfonate; DMBA, 7,12-dimethylbenz[a]-anthracene; AFB$_1$, aflatoxin B$_1$

[b] + < 50% increase over control SCE frequency

[c] ++ > 50% increase over control SCE frequency

[d] +++ > 100% increase over control SCE frequency

From some of our preliminary studies (Table 1) and from the magnitude of the responses that have been reported in the literature with laboratory animals (Stetka and Wolff 1976; Stetka et al. 1978; Takehisa and Wolff 1978), one suspects that cytogenetic damage seen in lymphocytes after in vivo exposures is less than would be expected if one could study cytogenetic damage simultaneously in either the bone marrow or liver. With promutagenic agents, the ultimate electrophile may be short-lived and not reach the lymphocytes because of reactions with plasma proteins, RBCs, and granulocytes. With direct-acting agents or long-lived active metabolites, some of the damage in the cells could be repaired between the time of exposure and the time the cells are replicated in the presence of BrdU. Evidence supporting repair of damage comes from liquid-holding studies with human lymphocytes. Obe et al. (1982) found significant decreases in SCE frequencies in human lymphocytes treated in vitro with Trenimon® and diepoxybutane when the cells were held in culture medium for 1-3 days prior to mitogenic stimulation. This is not unexpected because human lymphocytes can perform excision repair without going through blast transformation (Evans and Norman 1968; Darzynkiewicz 1971; Clarkson and Evans 1972). However, Evans and Vijayalaxmi (1980) failed to see a similar reduction in damage in human lymphocytes treated with mitomycin C. The persistence of lesions that lead to chromosome breakage is especially relevant when one is working with lymphocytes. Some human peripheral lymphocytes can remain viable in vivo for over 30 years as demonstrated by the occurrence of symmetrical chromosome exchanges (mainly reciprocal translocations) in the lymphocytes of persons exposed to the Hiroshima and Nagasaki atomic bomb blasts in 1945 (Awa 1975). Thus, knowledge of the lifetime or turnover time of the circulating lymphocytes in the species under study is important.

The situation is complicated further, especially after acute exposures, by the sequestering of damaged lymphocytes in the lymph nodes and the continual addition of newly differentiated lymphocytes into the peripheral circulation which may cause a "dilution" of the response. This is demonstrated in studies with patients on cytostatic therapy. An initial rise in SCE frequencies is followed by a sharp reduction in the level of damage with time after the cessation of treatment. In some patients the rise is transitory (Lambert et al. 1978; Nevstad 1978; Raposa 1978; Düker 1981) while in other patients the frequencies remain elevated (Lambert et al. 1978; Musilová et al. 1979; Lambert et al. 1979; Ohtsuru et al. 1980) over baseline. Studies with animal models again give contradictory results. Kligerman et al. (1981) found that lymphocytes of rats injected with ethyl methanesulfonate display highly elevated SCE frequencies that decline rapidly with time but remain 50% above baseline 28 days after exposure. Similarly, Stetka et al. (1978) saw SCEs produced from long-lived lesions in rabbit peripheral lymphocytes after administering multiple injections of mitomycin C to the animals. However, Stetka and Wolff (1976) and Takehisa and Wolff (1978) found short-lived (< 2 weeks) elevations in SCE frequencies in peripheral lymphocytes of rabbits injected with cyclophosphamide, ethyl methanesulfonate, methyl methanesulfonate, 2-AF, and 2-AAF. These differences in response are most likely due to differences in the chemicals examined, species investigated, and treatment protocols followed.

The contradictory nature of many of these findings points to the need for carefully controlled studies using well characterized lymphocyte culture systems. With such assays measurements of cytogenetic damage at various times after in vivo exposure can be made for different classes of genotoxicants administered over a range of doses. Such factors as route of exposure, persistence of damage, and efficacy of acute versus chronic exposure can be examined. By using small animals such as rats or mice, these studies could be accomplished with a minimal amount of time, space, or expense. Then, meaningful comparisons could be made between results obtained from humans and animal models exposed to the same genotoxicants. Studies of this nature should reveal the value of using lymphocyte cytogenetic studies as an indicator of genotoxic exposures.

ACKNOWLEDGMENTS

The authors would like to thank Drs. B. Butterworth, D. Couch, H. Garcia, and J. Mirsalis for blood samples from their treated and control animals, and Drs. R. Irons and J. Dean for critical review of the manuscript. The aid of Ms. Linda Smith in preparing the manuscript is appreciated.

REFERENCES

Awa, A.A. 1975. Review of thirty years study of Hiroshima and Nagasaki atomic bomb survivors. 2. Biological effects. B. Genetic effects. 2. Cytogenetic effects. *J. Radiat. Res.* **16**(*Suppl*):75.

Barren, P.R., S.M. Morris, H. Schol, and J.B. Bishop. 1982. A method for preparing chromosomes from peripheral blood of the mouse. *Mutat. Res.* **104**:159.

Beltz, P.A. and R.D. Benz. 1980. Nondestructive cytogenetic toxicologic testing with rats. *Environ. Mutagen.* **2**:297a.

Bryan, J.H.D. and R.L. Hybertson. 1972. The in vitro stimulation of lymphocytes from peripheral blood and lymph nodes of the laboratory mouse. *Cytogenetics* **11**:25.

Buckton, K.E. and P. Nettesheim. 1968. *In vitro* and *in vivo* culture of mouse peripheral blood for chromosome preparations. *Proc. Soc. Exp. Biol. Med.* **128**:1106.

Carrano, A.V., J.L. Minkler, D.G. Stetka, and D.H. Moore II. 1980. Variation in the baseline sister chromatid exchange frequency in human lymphocytes. *Environ. Mutagen.* **2**:325.

Clarkson, J.M. and H.J. Evans. 1972. Unscheduled DNA synthesis in human leucocytes after exposure to UV light, γ-rays and chemical mutagens. *Mutat. Res.* **14**:413.

Darzynkiewicz, Z. 1971. Radiation-induced DNA synthesis in normal and stimulated human lymphocytes. *Exp. Cell Res.* **69**:356.

deBoer, P., P.P.W. van Buul, R. van Beek, F.A. van der Hoeven, and A.T. Natarajan. 1977. Chromosomal radiosensitivity and karyotype in mice using cultured peripheral blood lymphocytes, and comparison with this system in man. *Mutat. Res.* 42:379.

DuFrain, R.J., L.G. Littlefield, and J.L. Wilmer. 1980. The effect of washing lymphocytes after in vivo treatment with streptonigrin on the yield of chromosome and chromatid aberrations in blood cultures. *Mutat. Res.* 69:101.

Düker, D. 1981. Investigations into sister chromatid exchange in patients under cytostatic therapy. *Hum. Genet.* 58:198.

Evans, H.J. and Vijayalaxmi. 1980. Storage enhances chromosome damage after exposure of human lymphocytes to mitomycin C. *Nature* 284:370.

Evans, R.G. and A. Norman. 1968. Radiation stimulated incorporation of thymidine into the DNA of human lymphocytes. *Nature* 217:455.

Farrow, M.G., A.M. Hawk, L.T. Wetzel, and G.C. Boxill. 1974. A microculture technique utilizing whole blood for mouse chromosome analysis. *Mutat. Res.* 26:401.

Goto, K., S. Maeda, Y. Kano, and T. Sugiyama. 1978. Factors involved in differential Giemsa-staining of sister chromatids. *Chromosoma (Berlin)* 66:351.

Granberg-Öhman, I., S. Johansson, and A. Hjerpe. 1980. Sister-chromatid exchanges and chromosomal aberrations in rats treated with phenacetin, phenazone and caffeine. *Mutat. Res.* 79:13.

Greaves, M. and G. Janossy. 1972. Elicitation of selective T and B lymphocyte responses by cell surface binding ligands. *Transplant. Rev.* 11:87.

Janossy, G. and M. Greaves. 1973. Functional analysis of murine and human B lymphocyte subsets. *Transplant. Rev.* 24:177.

Kligerman, A.D., J.L. Wilmer, and G.L. Erexson. 1981. Characterization of a rat lymphocyte culture system for assessing sister chromatid exchange after in vivo exposure to genotoxic agents. *Environ. Mutagen.* 3:531.

Kligerman, A.D., J.L. Wilmer, and G.L. Erexson. 1982. Characterization of a rat lymphocyte culture system for assessing sister chromatid exchange. II. Effects of 5-bromodeoxyuridine concentration, number of white blood cells in the inoculum, and inoculum volume. *Environ. Mutagen.* (in press).

Lambert, B., A. Lindblad, K. Holmberg, and D. Francesconi. 1982. The use of sister chromatid exchange to monitor human populations for exposure to toxicologically harmful agents. In *Sister chromatid exchange*, p. 149. John Wiley & Sons, Inc., New York.

Lambert, B., U. Ringborg, E. Harper, and A. Lindblad. 1978. Sister chromatid exchanges in lymphocyte cultures of patients receiving chemotherapy for malignant disorders. *Cancer Treat. Reports* 62:1413.

Lambert, B., U. Ringborg, and A. Lindblad. 1979. Prolonged increase in sister-chromatid exchanges in lymphocytes of melanoma patients after CCNU treatment. *Mutat. Res.* 59:295.

Latt, S., R. Schreck, K. Loveday, and C. Shuler. 1979. *In vitro* and *in vivo* analysis of sister chromatid exchange. *Pharmacol. Rev.* 30:501.

Lindblad, A. and B. Lambert. 1981. Relation between sister chromatid exchange, cell proliferation and proportion of B- and T-cells in human lymphocyte cultures. *Hum. Genet.* **57**:31.

Metcalf, W.K. 1965. Some experiments on the phytohaemagglutinin culture of leucocytes from rats and other mammals. *Exp. Cell Res.* **40**:490.

Musilová, J., K. Michalová, and J. Urban. 1979. Sister-chromatid exchanges and chromosal (sic) breakage in patients treated with cytostatics. *Mutat. Res.* **67**:289.

Nevstad, N.P. 1978. Sister chromatid exchanges and chromosomal aberrations induced in human lymphocytes by the cytostatic drug adriamycin *in vivo* and *in vitro. Mutat. Res.* **57**:253.

Obe, G., S. Kalweit, C. Nowak, and F. Ali-Osman. 1982. Liquid holding experiments with human peripheral lymphocytes. I. Effects of liquid holding on sister chromatid exchanges induced by Trenimon, diepoxybutane, bleomycin and x-rays. *Biol. Zbl.* **101**:97.

Ohtsuru, M., Y. Ishii, S. Takai, H. Higashi, and G. Kosaki. 1980. Sister chromatid exchanges in lymphocytes of cancer patients receiving mitomycin C treatment. *Cancer Res.* **40**:477.

Peavy, D.L., W.H. Adler, and R.T. Smith. 1970. The mitogenic effects of endotoxin and staphlococcal enterotoxin B on mouse spleen cells and human peripheral lymphocytes. *J. Immunol.* **105**:1453.

Perry, P. and H.J. Evans. 1975. Cytological detection of mutagen-carcinogen exposure by sister chromatid exchange. *Nature* **285**:121.

Raposa, T. 1978. Sister chromatid exchange studies for monitoring DNA damage and repair capacity after cytostatics *in vitro* and in lymphocytes of leukaemic patients under cytostatic therapy. *Mutat. Res.* **57**:241.

Rüdiger, H.W., F. Kohl, W. Mangels, P. von Wichert, C.R. Bartram, W. Wohler, and E. Passarge. 1976. Benzpyrene induces sister chromatid exchanges in cultured human lymphocytes. *Nature* **262**:290.

Santesson, B., K. Lindahl-Kiessling, and A. Mattsson. 1979. SCE in B and T lymphocytes. Possible implications for Bloom's syndrome. *Clin. Genet.* **16**:133.

Schwartz, J.L. and M.E. Gaulden. 1980. The relative contributions of B and T lymphocytes in the human peripheral blood mutagen test system as determined by cell survival, mitogenic stimulation, and induction of chromosome aberrations by radiation. *Environ. Mutagen.* **2**:473.

Stetka, D.G., J. Minkler, and A.V. Carrano. 1978. Induction of long-lived chromosome damage, as manifested by sister-chromatid exchange, in lymphocytes of animals exposed to mitomycin-C. *Mutat. Res.* **51**:383.

Stetka, D.G. and S. Wolff. 1976. Sister chromatid exchange as an assay for genetic damage induced by mutagen-carcinogens. I. *In vivo* test for compounds requiring metabolic activation. *Mutat. Res.* **41**:333.

Stobo, J.D. 1972. Phytohemagglutinin and concanavalin A: Probes for murine 'T' cell activation and differentiation. *Transplant. Rev.* **11**:60.

Takehisa, S. and S. Wolff. 1978. Sister-chromatid exchanges induced in rabbit lymphocytes by 2-aminofluorene and 2-acetylaminofluorene after *in vitro* and *in vivo* metabolic activation. *Mutat. Res.* **58**:321.

Tates, A.D., N. de Vogel, and I. Neuteboom. 1980. Cytogenetic effects in hepatocytes, bone marrow cells and blood lymphocytes of rats exposed to ethanol in the drinking water. *Mutat. Res.* **79**:285.

Triman, K.L., M.T. Davisson, and T.H. Roderick. 1975. A method for preparing chromosomes from peripheral blood in the mouse. *Cytogenet. Cell Genet.* **15**:166.

Wagner, T., G. Peter, G. Voelcker, and H.-J. Hohorst. 1977. Characterization and quantitative estimation of activated cyclophosphamide in blood and urine. *Cancer Res.* **37**:2592.

COMMENTS

ALLEN: Were your T-cell SCE levels higher than those in B-cells?

KLIGERMAN: The T-cells had an SCE frequency of 7.3, and the B-cells 4.6 SCEs/metaphase.

SORSA: The use of human whole blood cultures, the routine method used to look for SCEs and also for chromosome aberrations, may have some unexpected effects on the result. We have data on the effects of styrene, which is an indirect mutagen. However, it gives a very nice dose response for SCEs in human whole blood lymphocyte cultures. But when you take the red blood cells out, there is no response, which certainly points out the importance of red blood cells in the metabolic processing of styrene to styrene oxide.

KLIGERMAN: A report Ray and Altenburg (1978) indicates that sodium selenite can induce SCEs in vitro human blood cultures only when red blood cells are present.

CARRANO: When you remove the red cells, you put the blood sample on a gradient, I presume, so that you also remove the plasma. Perhaps something is carried in the plasma, as DuFrain et al. (1979) showed.

KLIGERMAN: We always wash the blood 3 times to remove the plasma.

SORSA: We have chromatographically measured that styrene is decreasing in the culture and styrene oxide is increasing, so there is metabolic activation by the red blood cells.

MOHRENWEISER: Did you do a dose-response curve with purified leukocytes and BrdU? Is it possible that purified leukocytes are simply more susceptible to BrdU-induced SCEs, and you just shifted the curve to the left?

KLIGERMAN: We thought of that, and we did radiotracer studies to determine if the red blood cells would sequester BrdU when various number of red blood cells were added to the cultures. We found a slight binding, but not nearly enough to account for the increase in SCE frequency seen in purified cultures.

References

DuFrain, R.J., L.G. Littlefield, and J.L. Wilmer. 1979. Cyclophosphamide induced SCEs in rabbit lymphocytes. *Environ. Mutagen.* **1**:283.

Ray, J.H. and L.C. Altenburg. 1978. Sister-chromatid exchange induction by sodium selenite: Dependence on the presence of red blood cells or red blood cell lysate. *Mutat. Res.* **54**:343.

SCE and Gene Mutation Studies with Ethyl Carbamate, Ethyl *N*-Hydroxycarbamate, and Vinyl Carbamate: Potencies and Species, Strain, Tissue Specificities

JAMES W. ALLEN, ROBERT LANGENBACH, SHARON LEAVITT
Genetic Toxicology Division
Health Effects Research Laboratory
United States Environmental Protection Agency
Research Triangle Park, North Carolina 27711

and

YOUSUF SHARIEF, JAMES CAMPBELL, KAREN BROCK
Northrop Services, Incorporated
Research Triangle Park, North Carolina 27709

It is well established that most carcinogens undergo metabolic activation, and that they may exhibit species-, strain-, and tissue-specific carcinogenic effects (Mirvish 1968; Langenbach et al. 1978; Miller and Miller 1979; Magee 1979). In general, the different effects among animals and tissues might be due to such factors as variable metabolic efficiencies, pharmacodynamics, or DNA replication and repair processes. Diverse genetic findings bear upon these specificities. For example, dissimilar levels of DNA adduct formation in different tissues following in vivo exposure to mutagenic and (or) carcinogenic substances have been found (Langenbach et al. 1982). Exogenous activation systems used with in vitro mutagenesis and DNA-binding assays have revealed species, strain, and tissue differences in effectiveness (Leavitt et al. 1979; Muller et al. 1980; Langenbach et al. 1981). Chromosome (Goetz et al. 1975; Allen et al. 1981), DNA damage (Petzold and Swenberg 1978), and gene mutation (Dean and Hudson-Walker 1979) endpoints have reflected similar animal- and organ-specific activities.

The following is an account of pertinent sister chromatid exchange (SCE) induction and gene mutagenesis (for 6-thioguanine [6-TG] and ouabain resistance) findings as they relate to in vivo and (or) in vitro exposures to ethyl, ethyl *N*-hydroxy, and vinyl carbamates. Ethyl carbamate (EC, urethane) is a classical carcinogen in experimental animals (Mirvish 1968), and one to which human exposure has historically been widespread due, in large part, to its industrial and medical uses (IARC 1974). Ethyl *N*-hydroxycarbamate is relatively weaker and vinyl carbamate (VC) much stronger, for carcinogenic activity (Dahl et al. 1978, 1980). The former is regarded as an interconvertible metabolite of EC (Mirvish 1968): the latter as a suspect metabolic intermediate (after EC exposure) with the potential for conversion to VC epoxide—possibly the ultimate carcinogen (Dahl et al. 1978, 1980). The collection of data presented here is concerned with the relative genotoxicities of these chemicals as indicated from broad assay system approaches.

METHODS

In vivo SCE analyses conducted in rodents relied upon DNA labeling by the bromodeoxyuridine (BrdU) tablet implantation procedure (Allen et al. 1978). Somatic or germ tissues were harvested (after in vivo colchicine treatment) at times suitable for the provision of differentially labeled chromosomes in second post-BrdU metaphase division cells. After cell hypotonic (KCl) and fixative (methanol:acetic acid) treatments, slides were prepared and chromosomes analyzed after differential fluorescent or fluorescent-plus-Giemsa staining techniques (Latt et al. 1981).

In vitro SCE analyses, carried out in Chinese hamster V-79 cells, were part of a multiple-endpoint assay system (Allen et al. 1982b). Eighteen hours after cells were seeded in Williams medium E, they were treated for 4 hours with the test chemical in either the presence or absence of a liver S9 reaction mixture. Positive and negative controls were included. Cells were then reseeded into separate dishes for growth under conditions appropriate for analyses of SCE induction, gene mutagenesis for 6-TG or ouabain resistance, and cytotoxicity. For SCE studies, $10\mu M$ BrdU was added to the medium; cells were harvested (after colchicine treatment) at 22-24 hours and prepared for analysis as noted above. Mutant frequencies were assessed after 2 days (ouabain) or 6 days (6-TG) mutation expression time, and 10 days in the presence of the selecting agent. Cytotoxicity was determined from cloning efficiencies after 6-7 days of growth.

RESULTS AND DISCUSSION

Although ethyl carbamate reportedly induces SCEs in cultured human lymphocytes (Csukas et al. 1979, 1981), an observation we have been unable to confirm (Allen et al. 1983), this chemical clearly does not cause substantial increases of SCE in cultured V-79 cells, with or without an added S9 mix (Table 1) (Allen et al. 1982a). In the absence of exogenous activation, increasing doses of EC—even to very high concentrations—fail to significantly affect SCE frequencies. Negative results (as compared with S9 controls) are also obtained when an S9 mix from aroclor or EC pretreated hamsters is included. Similarly, S9 mixes from either high-lung tumor susceptible (A) or low-lung tumor susceptible (C57BL/6) mouse strains have little apparent ability to activate ethyl carbamate.

In contrast, mice exposed to EC by a single intraperitoneal (i.p.) injection reveal extensive, tissue-specific increments in SCE frequencies (Roberts and Allen 1980; Cheng et al. 1981a). In Figure 1, relative SCE induction values are

Table 1
In Vitro SCE Analyses in Chinese Hamster V79 Cells Exposed to EC With and Without Various S9 Mixtures

Treatment	No S9	Golden Hamster S9 (Aroclor-induced)	Golden Hamster S9 (EC-induced)	Mouse S9 (Aroclor-induced)	
				Strain C57BL/6	Strain A
	mean SCE/cell[a]				
EC 20 mg/ml	12 ± 4.1	13 ± 5.2	–	17 ± 4.6	20 ± 6.2
10	12 ± 3.8	11 ± 3.8	18 ± 7.0	13 ± 3.8	16 ± 5.2
5	11 ± 3.6	12 ± 4.2	18 ± 5.9	16 ± 5.7	15 ± 4.9
1	8 ± 2.5	11 ± 3.6	–	16 ± 5.2	15 ± 6.5
–	9 ± 3.1	–	–	–	–
Rm	–	11 ± 3.8	13 ± 5.0	14 ± 5.2	15 ± 6.0
DMBA 0.015	–	42 ± 12.0	38 ± 16.2	33 ± 12.9	28 ± 9.1
MNNG 0.0001	52 ± 10.4	–	–	–	–

[a] mean and std. dev. of 30-40 cells/treatment
(Allen et al., 1982a)

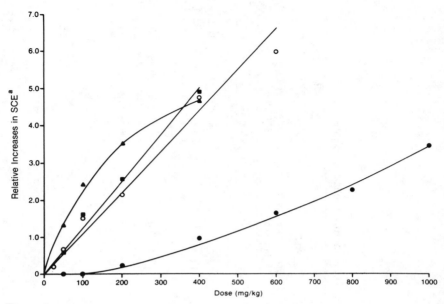

Figure 1
Induction of SCE by ethyl carbamate in hepatocytes from partially hepatectomized mice
(▲), bone marrow from partially hepatectomized mice (■), bone marrow from intact mice
(○), and spermatogonial cells from intact mice (●). [a]Relative increases in SCE over respective
control level (Roberts and Allen 1980).

illustrated when tissue discrepancies for baseline SCE levels are taken into
account. Spermatogonial cells show a much lower dose-responsiveness than do
normal marrow cells, marrow cells from partially hepatectomized mice, and regenerating liver cells. Liver cells generally showed the highest levels of SCE induction, and also cytotoxicity (at high doses) as indicated by lower mitotic
indices and enhanced proportions of cells with first metaphase division chromosome staining patterns. Other investigators have also confirmed the somewhat
higher SCE induction levels in liver, and have reported similar excesses in
alveolar macrophages (Cheng et al. 1981a). This is notable in view of the
especially high tumor susceptibility of liver and lung (Mirvish 1968), and the reportedly high levels of binding interactions with nucleic acids (Boyland and
Williams 1969; Chavan and Bhide 1972) which occur in those tissues after in
vivo exposure to EC.

The relatively more potent carcinogen, VC (Dahl et al. 1978, 1980), is
also a much stronger inducer of SCEs than is EC (Allen et al. 1982a) (Fig. 2).
For example, an i.p. dose of 25mg/kg VC causes approximately an eightfold
increase in marrow cell SCEs, while a similar dose of EC does not significantly
affect marrow cell SCE frequencies. Both of these chemicals, as well as ethyl *N*-

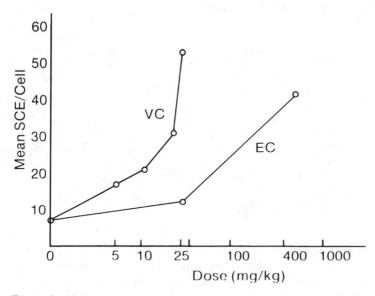

Figure 2
SCE induction in mouse bone marrow cells after in vivo i.p. exposure to EC or VC.

hydroxycarbamate which is relatively less effective an inducer of SCEs in vivo (see also, Cheng et al. 1981b), follow similar patterns of species specificity in their cytogenetic effects. That is to say, marrow cells from mice are more SCE-responsive than are marrow cells from Chinese hamsters, Golden hamsters, or rats. As notable in Table 2, 400mg/kg EC causes a fivefold increase in the SCE frequency of mouse cells, as compared with a threefold increase occurring in the other species (Y. Sharief et al., unpubl. results). Similarly, doses of 50 and 25-mg/kg VC are clearly harsher for cytotoxic and SCE effects in mice than they are in hamsters and rats. Although EC is known to be carcinogenic for all of these species (IARC 1974), and VC for mice and rats (Dahl et al. 1978, 1980), it is notable that the species (mouse) and tissues (i.e., liver) which are notoriously high in tumor formation are also high in expression of SCE induction. Yet, it is also interesting that high- (A) and low- (C57BL/6) lung tumor susceptible mouse strains exhibit comparable levels of SCE increment measurable in marrow cells (Table 2) (Y. Sharief et al., unpubl. results). As indicated below, the interpretation of these results as evidence for a lack of strain differences in metabolically processing these chemicals may be misleading.

An in vitro multiple endpoint approach was used to address several questions pertaining to the ethyl, ethyl N-hydroxy, and VCs: 1) their relative potencies for genetic activity in V-79 cells, 2) the comparative sensitivi-

Table 2
In Vivo SCE Induction by EC, VC, and ENHC in Various Rodent Species

		EC	VC		ENHC
Species	Control	400 mg/kg	50 mg/kg	25 mg/kg	400 mg/kg
		mean SCE/cell[a]			
Mouse					
Strain A	8 ± 3.9	41 ± 6.6	Toxic	47 ± 10.8	21 ± 7.4
C57BL	7 ± 3.6	42 ± 9.2	Toxic	47 ± 9.8	28 ± 8.7
Chinese hamster	6 ± 2.4	21 ± 6.1	54 ± 15.4	27 ± 8.2	18 ± 5.3
Golden hamster	5 ± 2.4	15 ± 5.8	54 ± 13.2	–	–
Rat	4 ± 2.0	11 ± 4.7	35 ± 11.4	27 ± 8.4	–

[a]mean and std. dev. values represent 20-40 cells/animal; minimum of 3 animals/ treatment. (Y. Sharief et al. unpubl. results)

ties of two endpoints: SCE induction and gene mutagenesis, and 3) the relative efficiencies of A strain vs. C57BL/6 strain mouse S9 mix used in the system for metabolically activating vinyl carbamate. Some results are discussed below.

Table 3 summarizes SCE induction and gene mutagenesis results for each of the test chemicals. EC produced no genetic effects in the V-79 cell system, regardless of the inclusion of aroclor induced hamster S9 mix. Ethyl N-hydroxycarbamate was generally ineffective in the presence of S9 mix; a mild response for both endpoints was evident when the activation mix was omitted. This trend towards lowered activity in the presence of liver activation fractions (possibly a matter of enzymatic detoxification) has also been noted in Salmonella (Dahl et al. 1980), and in human lymphocyte (Csukas et al. 1979) studies with this chemical. VC exhibited the same high genotoxic activity in vitro as it did in vivo. In the presence of hamster S9 mix, extensive SCE induction and gene mutagenesis occurred with clear dose-dependence. When the activation mix was left out, genotoxicity diminished; although, high doses which were toxic in the presence of S9 caused SCE induction to levels of 6 times baseline—without any accompanying gene mutagenesis. Human lymphocytes have also been reported to be much more sensitive to SCE induction by VC when S9 mix was included (Csukas et al. 1981). Mutagenic effects from this chemical in the presence, but not absence of, S9 mix has been reported in Salmonella systems (Dahl et al. 1978) as well. Different activities for gene mutation and SCE endpoints have been noted for other chemicals (Carrano et al. 1978; Bradley et al. 1979). In general, the observation here is suggestive of dissimilarities in the lesions and (or) their pathways of action leading to chromosomal and gene level effects. It is not clear whether the same ultimate chemical moieties and molecular interactions responsible for SCE induction in the presence of S9 are the same as those producing this effect in the absence of S9.

Table 3

SCE and Gene Mutation Induced by EC, VC, and ENHC in Chinese Hamster V79 Cells

Chemical	Dose (mg/ml)	Mean SCE/cell[a]		Mutagenesis (6-TG) mutants/10⁶ cells	
		+S9[b]	−S9	+S9[b]	−S9
EC	20.0	13 ± 4.4	9 ± 2.5	<10	<10
	10.0	12 ± 3.5	9 ± 2.7	<10	<10
	5.0	11 ± 4.0	9 ± 2.6	<10	<10
VC	1.0	Toxic	82 ± 12.4	Toxic	<10
	0.2	Toxic	30 ± 8.4	591	<10
	0.01	67 ± 8.5	15 ± 5.0	236	<10
	0.001	25 ± 5.9	14 ± 4.6	69	<10
ENHC	7.5	22 ± 9.0	Toxic	<10	62
	5.0	18 ± 7.5	39 ± 14.2	<10	55
	1.0	15 ± 3.6	15 ± 3.6	<10	51
	0.2	20 ± 8.4	18 ± 9.1	<10	72
	0.01	14 ± 4.6	14 ± 5.0	<10	20
Control	−	13 ± 4.3	11 ± 2.8	<10	<10
DMBA	0.015	46 ± 12.2	−	222	−
MNNG	0.0001	−	52 ± 14.3	−	82

[a] mean and std. dev. of 30-40 cells/treatment.
[b] Aroclor-induced hamster S9.
(Allen et al. 1982a)

It can be seen in Figures 3 and 4 that the use of mouse A strain S9 results in greater levels of VC induced genotoxicity than when C57BL/6 strain S9 (matched for protein content) is used (Y. Sharief et al., unpubl. results). The A strain enzyme mix is apparently more proficient for metabolically processing VC to result in higher levels of induced SCE, gene mutagenesis for both 6-TG and ouabain resistance, and cytotoxicity. S9 mixes prepared from two different animal group of each strain have consistently shown strain-specific activities. Thus, strain-specific influences upon genotoxic effects from this chemical are evident under these in vitro circumstances, despite a lack of such observations with in vivo studies. In animals, perhaps a dissipation (i.e., short-lived metabolite effect) occurs to result in an approximately equalized metabolite distribution to the marrow. Current studies are underway to determine if SCE induction in regenerating liver cells of A and C57BL/6 mice exposed to VC resembles the strain-similar SCE patterns obtained with marrow cells, or the strain S9-dissimilar patterns obtained with cultured V79 cells.

Several lines of evidence are now consistent with the possibility that vinyl carbamate may account for the carcinogenic activity of ethyl (and ethyl *N*-

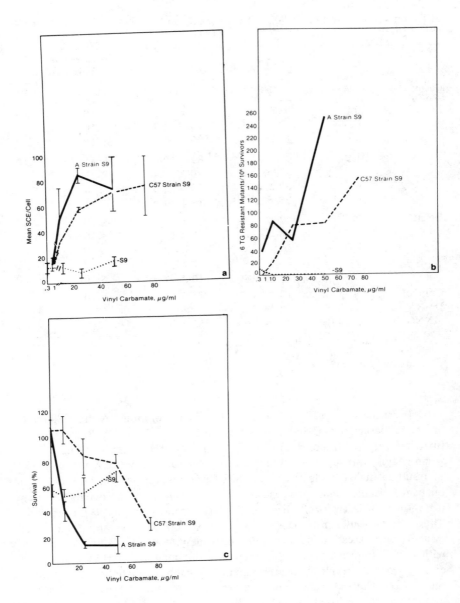

Figure 3
Mouse strain S9 differences for activating VC to induce SCE (a), gene mutation for 6-TG resistance (b), and cytotoxicity (c) in V79 cells. (——) dose-response values utilizing strain A S9; (- - -) utilizing strain C57BL/6 S9, and (. . .) without any S9 (Y. Sharief et al. unpubl. results). [Doses at which VC is very active for SCE induction in the absence of S9 mix are not included here].

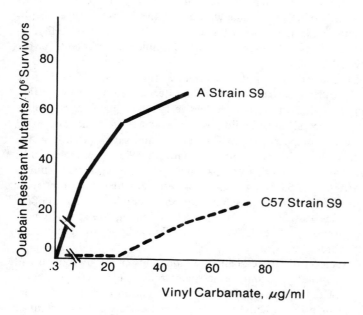

Figure 4
Mouse strain S9 differences for activating VC to induce gene mutation for ouabain resistance in V79 cells. (——) dose-response values utilizing strain A S9; (- - -) utilizing strain C57BL/6 S9.

hydroxy) carbamate. Relative to EC, VC is much more carcinogenic in rodents for similar types of tumors (Dahl et al. 1978, 1980); it is highly genotoxic (Dahl et al. 1978, 1980; Csukas et al. 1981; Allen et al. 1982a, and present study). Further, an enzyme fraction from the most tumor-susceptible (to EC) mouse strain appears to activate VC to induce genetic effects better than enzymes from a tumor resistant strain (present study; Y. Sharief et al., unpubl. results). Yet, the possibility that VC may relate to EC through parallel metabolic pathways and products, rather than as a direct metabolite, remains. Should a metabolic conversion product of VC (i.e., VC epoxide [Dahl et al. 1978]), in fact, prove the ultimate carcinogen after EC exposure, then the strain-specific S9 mediated genetic effects reported here are not supportive of the view (Mirvish 1968) that aspects of metabolism are unlikely to be important to the strain differences in tumor susceptibility.

SUMMARY

To summarize the in vivo findings, EC effectively induces SCEs in various rodent species; although, at equivalent doses mice are significantly more SCE responsive than are rats and hamsters. There is no difference for this effect between the marrow cells of mouse strains with alternative high- or low- susceptibility to lung adenoma formation from this chemical. Relative to EC, ethyl N-hydroxycarbamate is somewhat less potent and VC is much more potent, for the induction of SCEs.

With in vitro studies, EC is not very effective for inducing SCEs or gene mutations, with or without the inclusion of hamster or mouse S9 mix. Ethyl N-hydroxycarbamate is similarly rather ineffective for inducing these endpoints in the presence of S9 mix; although, mild positive responses are evident when S9 mix is omitted. VC is very potent for SCE induction and gene mutation in the presence of S9 mix. At high-dose exposures, without S9 mix, extensive SCE induction occurs without any accompanying gene mutagenesis. Tumor-susceptible strain A mouse S9 mix appears more efficient than tumor-resistant strain C57BL/6 mouse S9 mix for metabolically processing VC to result in SCE induction, gene mutagenesis, and cytotoxicity. These findings are generally consistent with the suggestion (Dahl et al. 1978) that VC may be a critical metabolic intermediate in EC carcinogenesis.

The above studies with the carbamate compounds point to some broad principles concerning genetic toxicology assessment systems in general. For some chemicals, in vivo SCE induction trials can provide clear evidence of genotoxicity when little or no such activity is seen in cell culture systems—with or without exogeneous forms of activation. Species and tissue differences in SCE responsiveness to chemical carcinogen exposures are demonstrable. With in vitro studies of genotoxic chemicals, the particular species and strain from which S9 mix is prepared can significantly influence induction levels of both SCE and gene mutation. The extent that gene and chromosomal endpoints correlate in dose-response relationships may vary with different experimental conditions (i.e., the optional inclusion of S9 mix) as well as with the particular chemical being investigated.

ACKNOWLEDGMENTS

Vinyl carbamate used in initial experiments was generously provided by Dr. James Miller, McArdle Laboratory for Cancer Research, University of Wisconsin, through the courtesy of Dr. Stephen Nesnow, Health Effects Research Laboratory, United States Environmental Protection Agency. We thank Ronnie Dunn, Karen Sasseville, and Bob Easterling for their able technical assistance.

REFERENCES

Allen, J.W., C.F. Shuler, and S.A. Latt. 1978. Bromodeoxyuridine tablet methodology for in vivo studies of DNA synthesis. *Somatic Cell Genet.* 4:393.

Allen, J.W., Y. Sharief, and R.J. Langenbach. 1982a. An overview of ethyl carbamate (urethane) and its genotoxic activity. In *Genotoxic effects of airborne agents* (eds. R.R. Tice et al.), p. 443. Plenum Press, N.Y.

Allen, J.W., K. Brock, J. Campbell, and Y. Sharief. 1983. SCE analysis in lymphocytes. In *Single cell mutation monitoring systems: Methodologies and applications* (eds. A.A. Ansari and F.J. de Serres). Plenum Press, N.Y. (In press).

Allen, J.W., R. Langenbach, S. Nesnow, K. Sasseville, S. Leavitt, J. Campbell, K. Brock, and Y. Sharief. 1982b. Comparative genotoxicity studies of ethyl carbamate and related chemicals: Further support for vinyl carbamate as a proximate carcinogenic metabolite. *Carcinogenesis* (in press).

Allen, J.W., E. El-Nahass, M.K. Sanyal, R.L. Dunn, B. Gladen, and R.L. Dixon. 1981. Sister chromatid exchange analyses in rodent maternal, embryonic and extra-embryonic tissues. Transplacental and direct mutagen exposures. *Mutat. Res.* **80**:297.

Boyland, E. and K. Williams. 1969. Reaction of urethane with nucleic acids in vivo. *Biochem. J.* **111**:121.

Bradley, M.O., I.C. Hsu and C.C. Harris. 1979. Relationships between sister chromatid exchange and mutagenicity, toxicity, and DNA damage. *Nature* **282**:318.

Carrano, A.V., L.H. Thompson, P.A. Lindl, and J.L. Minkler. 1978. Sister chromatid exchange as an indicator of mutagenesis. *Nature* **271**:551.

Chavan, B.G. and S.V. Bhide. 1972. Interaction of urethane with macromolecules in male and female newborn, adult, and tumor-bearing mice. *J. Natl. Cancer Inst.* **49**:1019.

Cheng, M., M.K. Conner, and Y. Alarie. 1981a. Multicellular in vivo sister-chromatid exchanges induced by urethane. *Mutat. Res.* **88**:223.

———. 1981b. Potency of some carbamates as multiple tissue sister chromatid exchange inducers and comparison with known carcinogenic activities. *Cancer Res.* **41**:4489.

Csukas, I., E. Gungl, F. Antoni, G. Vida, and F. Solymosy. 1981. Role of metabolic activation in the sister chromatid exchange-inducing activity of ethyl carbamate (urethane) and vinyl carbamate. *Mutat. Res.* **89**:75.

Csukas, I., E. Gungl, I. Fedorcsak, G. Vida, F. Antoni, I. Turtoczky, and F. Solymosy. 1979. Urethane and hydroxyurethane induced sister-chromatid exchanges in cultured human lymphocytes. *Mutat. Res.* **67**:315.

Dahl, G.A., J.A. Miller, and E.C. Miller. 1978. Vinyl carbamate as a promutagen and a more carcinogenic analog of ethyl carbamate. *Cancer Res.* **38**:3793.

———. 1980. Comparative carcinogenicities and mutagenicities of vinyl carbamate, and ethyl N-hydroxy carbamate. *Cancer Res.* **40**:1194.

Dean, B.J., and G. Hudson-Walker. 1979. Organ-specific mutations in Chinese hamsters induced by chemical carcinogens. *Mutat. Res.* **64**:407.

Goetz, P., J. Sram, and J. Dohnalova. 1975. Relationship between experimental results in mammals and man. Cytogenetic analysis of bone marrow injury induced by a single dose of cyclophosphamide. *Mutat. Res.* **31**:247.

IARC. 1974. Some antithyroid and related substances, nitrofurans, and industrial chemicals. In *IARC Monogr. Eval. Carcinog. Risk Chem. Hum.* **7**:111.

Langenbach, R., S. Nesnow, and J. Rice. 1982. *Symposium proceedings: Organ and species specificity in chemical carcinogenesis.* Plenum Press, N.Y. (In press).

Langenbach, R., H.J. Freed, D. Raveh, and E. Huberman. 1978. Cell specificity in metabolic activation of aflatoxin B_1 and benzo(a)pyrene to mutagens for mammalian cells. *Nature* 276:277.

Langenbach, R., S. Nesnow, A. Tompa, R. Gingell, and C. Kuszynski. 1981. Lung and liver cell-mediated mutagenesis systems: Specificities in the activation of chemical carcinogens. *Carcinogenesis* 2:851.

Latt, S.A., J. Allen, S. Bloom, A. Carrano, E. Falke, D. Kram, E. Schneider, R. Schreck, R. Tice, B. Whitfield, and S. Wolff. 1981. Sister chromatid exchanges: A report of the gene-tox program. *Mutat. Res.* 87:17.

Leavitt, R.C., O. Pelkonen, A.B. Okey, and D.W. Nebert. 1979. Genetic differences in metabolism of polycyclic aromatic carcinogens and aromatic amines by mouse liver microsomes. Detection by DNA binding of metabolites and by mutagenicity in histidine-dependent Salmonella typhimurium in vitro. *J. Natl. Cancer Inst.* 62:947.

Magee, P.N. 1979. Organ specificity of chemical carcinogens. In *Carcinogenesis,* vol. 1, p. 213. Pergamon Press, N.Y.

Miller, J.A. and E.C. Miller. 1979. Metabolic activation of chemicals to reactive electrophiles: An overview. In *Advances in pharmacology and therapeutics,* vol. 9, p. 3. Pergamon Press, N.Y.

Mirvish, S.S. 1968. The carcinogenic action and metabolism of urethane and N-hydroxyurethane. *Adv. Cancer Res.* 11:1.

Muller, D., J. Nelles, E. Deparade, and P. Arni. 1980. The activity of S9-liver fractions from seven species in the salmonella/mammalian-microsome mutagenicity test. *Mutat. Res.* 70:279.

Petzgold, G.L. and J.A. Swenberg. 1978. Detection of DNA damage induced in vivo following exposure of rats to carcinogens. *Cancer Res.* 38:1589.

Roberts, G.T. and J.W. Allen. 1980. Tissue-specific induction of sister chromatid exchanges by ethyl carbamate in mice. *Environ. Mutagen.* 2:17.

COMMENTS

MIRSALIS: Have you looked at SCE levels in liver of partially hepatectomized mice in both strains?

ALLEN: That is what we have ongoing now. I want to find out if this tissue agrees with the marrow or if it agrees with the S9 results. We don't have an answer yet.

HSIE: Chinese hamster cells have been used by the geneticists for cytogenetic studies, and rat and mouse have been used for carcinogenesis studies. In this case your initial interest was from the carcinogenic point of view, but you found that the mouse showed a higher level of SCE induction than the hamster.

ALLEN: Certainly the Chinese hamster is very useful for these kinds of studies although, again, with the understanding that cytogenetic responses to some chemicals may be lower in magnitude relative to those occurring in the mouse. I know of other studies which have indicated a similar lower level of micronucleus induction and chromosome aberrations in the Chinese hamster as compared with the mouse after exposure to cyclophosphamide. I don't know what enzymic or other possible explanation there might be for this.

CALLEMAN: Have you tried to identify some reaction products of VC and EC with proteins or DNA? I ask this since I think VC may very well introduce the same group which is being introduced by vinyl chloride.

ALLEN: No, I haven't.

Sister Chromatid Exchange as an Indicator of Human Exposure

ANTHONY V. CARRANO
Biomedical Sciences Division
Lawrence Livermore National Laboratory
Livermore, California 94550

There is currently a considerable debate among scientists, government agencies, and industry concerning the utility and application of short-term assays to the evaluation of human exposure to potential environmental mutagens. Part of this controversy arises from the uncertainty of the biological significance of the measured endpoint, i.e., whether the measured events have clinical significance for the individual at risk (Dabney 1981). The cytogenetic endpoints, chromosomal aberrations and sister chromatid exchange (SCE), are no exception to this debate. I believe it is important to state at the outset that, given the present state of our knowledge, we should not expect any single human bioassay to provide all the answers. Thus, even though an endpoint cannot predict clinical outcomes, this should not preclude its use as an indicator of exposure to potentially harmful agents. Any indication that an individual or group of individuals has received an exposure is imperative for the proper management of occupational hygiene and, in a broader sense, environmental regulations at all levels of government. A better criteria for application of a short-term endpoint to humans might then be whether it is a measure of exposure. Simultaneous studies at the basic research level might ultimately provide information relevant to clinical consequences. At the same time, follow-up epidemiological studies on individuals or populations that were analyzed previously for exposure using the short-term endpoints could be scientifically meaningful. This manuscript will focus on only one of the cytogenetic endpoints, SCE, as it relates to the quantification of human exposure. In human population studies, however, it would be optimal to utilize a battery of assays.

The ability of an assay to quantify exposure will be dependent upon the type of lesion induced and the capability of the cell to repair that lesion or convert it to a measurable alteration. SCEs can be induced by a wide variety of lesions both in vitro and in vivo (see reviews by Wolff 1977; Gebhart 1981; Latt 1981). The most efficient inducers of SCE appear to be those agents that can form adducts or otherwise distort the DNA backbone. Agents that directly break the DNA-forming strand breaks are less efficient inducers of SCE.

Even though the molecular mechanism of SCE has not been established, this should not preclude its use in quantifying exposure. In general, for most substances tested, the induction of SCEs is linearly dependent upon dose. Moreover, linear relations between induced SCE and single-gene mutation have been demonstrated in vitro for a series of physical and chemical agents (Carrano et al. 1978; Carrano et al. 1979; Sirianni and Huang 1980; Okinaka et al. 1981; Carrano and Thompson 1982). The slope of the SCE-mutation relation, however, differs for each agent, suggesting that the conversion of chromosome damage to an SCE or mutation is lesion dependent. Recent experiments by Jostes (1981) imply that, at least for ethylnitrosourea, the mutagenic lesions may be a subset of the lesions that produce SCE.

In order for a chromosome lesion to be converted to an SCE, the lesion must be present during DNA synthesis when bromodeoxyuridine (BrdU) is also present (Kato 1974; Wolff et al. 1974). This presents a potential problem for the measurement of SCEs induced in peripheral lymphocytes in vivo. Since the lymphocyte must be stimulated to undergo DNA synthesis in vitro after it is removed from the circulating blood, the SCE-inducing lesions must persist from the time of exposure in vivo until the time of DNA synthesis in vitro. During this time, the lesions could be removed through efficient repair or could be diluted in the circulating blood due to normal lymphocyte turnover. In reality both effects may be operative with the result being a low or nonmeasurable increase in SCEs. This will be particularly true if SCEs are induced by low-level chronic exposures or if blood is sampled a long time after an acute exposure.

Early experiments in rabbits demonstrated that it was possible to measure an increase in SCEs in peripheral blood lymphocytes up to a few days following acute in vivo exposure (Stetka and Wolff 1976). This has also been observed in humans following exposure to a variety of chemotherapeutic agents (Lambert et al. 1978a; Nevstad 1978; Raposa 1978; Musilova et al. 1979). In order to determine whether SCEs would persist and be measureable in lymphocytes following repeated in vivo exposure to small doses of chemicals, we injected rabbits weekly for up to 8 weeks with mitomycin C (Stetka et al. 1978), with methylcholanthrene, or with benzo[a]pyrene (A. Carrano, unpubl. results). The results demonstrated that the SCE lesions could be removed within a few days after each exposure but that several exposures produced a small but consistent increase in SCEs that was evident even 6 months after termination of the exposures. More significantly, this increase appeared to be due to the presence of a small subpopulation of the lymphocytes that possessed very high SCE frequencies. This suggested that an important parameter to quantify for chronic exposures might be the proportion of lymphocytes with a high SCE frequency, i.e., high frequency cells (HFCs). These HFCs might represent either lymphocytes that are deficient in repair of lesions, lymphocytes that are long-lived and have accumulated damage, or a combination of both. Their presence has also been observed in humans (Lezana et al. 1977; Lambert et al. 1978b; Raposa 1978; Kowalczyk 1980).

The conclusion to be drawn from the above discussion is that large increases in lymphocyte SCE frequencies will generally not be observed with chronic exposures. This, coupled with the known variability in baseline SCE frequency among humans (Carrano et al. 1980), suggests at least two approaches for further study: standardization of cell culture procedures in order to minimize variation and development of more sensitive methods of data analysis. Data relevant to each of these are presented below.

METHODS

Whole blood collected in acid:citrate:dextrose is cultured in Minimal Essential Medium with 100μM BrdU for a period of 72 hours according to protocols described previously (Carrano et al. 1980). These conditions were chosen to minimize the variation in SCE frequency among individuals as well as between replicate cultures of a single individual. Care is taken to add the same aliquot of blood to each culture vessel in order to maintain the number of lymphocytes per 5 ml culture at approximately 3×10^6. This choice of cell number and BrdU concentration was chosen since the dose response curve for the induction of SCEs by BrdU is relatively flat at this BrdU/base pair ratio. This means that, if the cell counts vary by a factor of 2-3 or the fraction of lymphocytes stimulated by PHA varies by the same amount, the SCE frequency will not be greatly altered. At low doses of BrdU, or high lymphocyte concentrations, the SCE frequency is most variable since the amount of incorporated BrdU fluctuates considerably. These culture conditions consistently yield approximately 40% first-, 50% second-, and 10% third-division cells at 72 hours.

Data analysis is performed according to procedures recently outlined by Carrano and Moore (1982). Briefly, both the mean SCE frequency and number of HFCs are calculated for the 80 cells scored per individual. The number of HFCs is determined by first establishing a 95% tolerance level for the SCE frequencies in the historical and (or) concurrent controls. The tolerance level is defined as the SCE frequency in the control population below which 95% of the cells fall. The number of cells having SCE frequencies beyond this tolerance level is then determined. The 95% tolerance level for our historical controls is 0.349 SCE/chromosome or 16 SCE/cell. An individual who has 8 or more cells (out of the 80 cells scored) above this value has a significantly elevated SCE frequency based on HFCs alone.

Because large amounts of data are accumulated in these studies, we developed a microprocessor-based data acquisition and analysis system. The scorer enters data into the microprocessor using a numerical keypad adjacent to the microscope. The number of SCEs for each chromosome in the cell is stored on a floppy disk and a simultaneous printout of the number of SCEs, number of chromosomes, and SCE/chromosome is produced for each cell. The data on the floppy disk can then be compared with control data on the microprocessor to test for differences in the mean SCE frequency and the number of HFCs. Questionnaire information for each individual is stored and encoded on a PDP 11/34 for later sorting and retrieval.

RESULTS AND DISCUSSION

Our initial studies on a randomly selected cohort of 42 nonsmoking individuals failed to demonstrate any dependence of SCE frequency on sex or age (at least from ages 21-63). The pooled mean SCE frequency for this cohort of non-smokers is 0.200 SCE/chromosome or 9.2 SCE/cell. There was also no significant difference in the number of HFCs per individual as a function of age. The lack of an age effect in lymphocytes of healthy individuals has also been reported by other investigators (Galloway and Evans 1975; Morgan and Crossen 1977; Hollander et al. 1978; Cheng et al. 1979; Lambert and Lindblad 1980) but Waksvik et al. (1981) noted an increase in SCEs in old compared to young twin pairs.

The stability of an individual's SCE frequency over a long period of time is an important consideration prior to the initiation of any longitudinal study. Table 1 lists the variation associated with the mean SCE frequency and number of HFCs for four nonsmokers accumulated over a period of about 3 years. The coefficients of variation (standard deviation divided by the mean) varies from 2.2% for individual 46 to 15.7% for individual 50 over the 3-year period. This variation in the mean SCE frequency with time for any individual is generally greater than the cell-to-cell variation for the same individual but less than the variation observed among different individuals. The range of means observed for each individual can be quite large (as much as 60% difference for individual 19) and emphasizes the need for large sample sizes in order to detect small differences among cohorts (Carrano and Moore 1982). Table 1 also illustrates the variation in HFCs with time for the same individuals. The coefficients of variation for each individual is greater than for the means. This is primarily attributed to the fact that only 80 cells were scored and the HFCs are only a small percentage of that sample. Scoring a greater number of cells would reduce the sampling variation but may not be worthwhile in a practical sense if economy of time is important. For the 3-year period, individual 19 has a higher mean SCE frequency and a significantly elevated HFC frequency. This individual takes medication to control a hypertensive condition. All other individuals are medication free. A more extensive evaluation of those factors that contribute to variation was reported (Carrano et al. 1980).

If SCEs are an indicator of in vivo exposure, ideally they should exhibit a dose-related increase in the lymphocytes of exposed individuals. In some cases of chronic exposure a dose dependence may not be observed due to the complex kinetics of lesion removal and cellular turnover. Figure 1 illustrates that for cigarette smokers the mean SCE frequency increases with the number of cigarettes smoked per day. The increase is about 12% for 1 pack per day, 20% for 2 packs per day, and 35% for 3-pack-per-day smokers. The range of

Table 1
Variation in SCE Frequencies and the Number of HFCs in a Longitudinal Study

Person #	Duration of study	SCE/chromosome		HFC/80 cells	
		Range	Mean	Range	Mean
19	4/78-12/81	0.181-0.292 (19)[a]	0.240 (.031)[b]	2-24 (15)[a]	11 (6.6)[b]
46	8/78-2/82	0.160-0.245 (22)	0.198 (.019)	0-10 (18)	2.8 (3.1)
43	8/78-8/81	0.148-0.206 (15)	0.187 (.019)	0-5 (12)	1.8 (1.6)
50	3/78-11/80	0.157-0.239 (8)	0.188 (.031)	1-11 (4)	4 (4.6)
48	3/78-8/81	0.178-0.220 (10)	0.193 (.016)	2-8 (3)	5 (1.7)

[a]Number of samples
[b]Standard deviation

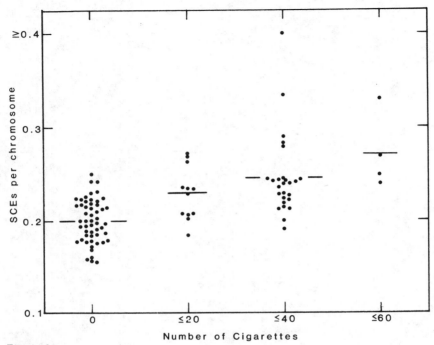

Figure 1
The frequency of SCE in lymphocytes of cigarette smokers as a function of the number of cigarettes smoked per day. Each point represents an individual. Eighty cells were scored per individual and the bars indicate the mean for each group.

individual means within any group is large and, for the 1 and 2 pack-per-day group, not all individuals have an increased mean. A regression line fitted to these data yields the relation:

$$\text{SCE/chromosome} = 9.2 + 0.054\ x, \tag{1}$$

where x is the number of cigarettes smoked per day. The slope is very similar to the value of 0.076 reported by Husum et al. (1982) and is about 20-fold less than the slope for in vitro exposure of lymphocytes to low-tar cigarette smoke condensate as calculated from the data of Hopkin and Evans (1980). Other recent studies of cigarette smokers have shown an increased SCE frequency (Lambert et al. 1978b; Ardito et al. 1980; Hopkin and Evans 1980; Husgafvel-Pursiainen et al. 1980) but normal SCE frequencies were also reported (Hollander et al. 1978; Crossen and Morgan 1980).

In Figure 2, the cigarette smokers are pooled for comparison to nonsmokers and a small group of pipe and cigar smokers. It is evident that the mean of the cigarette smoker cohort is significantly elevated. The pipe and cigar

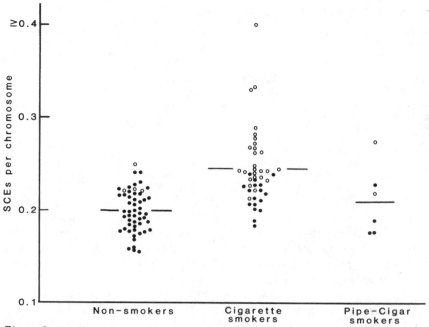

Figure 2
The mean SCE frequency and individuals with HFCs in pooled groups of nonsmokers, cigarette smokers, and pipe and cigar smokers. (·) an individual; (—) mean SCE frequency for each group; (○) those individuals with a significantly elevated frequency of HFCs.

smokers have a lower mean but with this small sample a definitive statement would be premature. This figure also illustrates those individuals that possess a significant number of HFCs. In this case the tolerance level was calculated from the nonsmoker data. HFCs are significantly elevated in only 4 of the 53 non-smokers (7.5%) but in 26 of the 43 smokers (60.5%). The data also demonstrate that an individual with an increased mean is likely to have a significant number of HFCs (e.g., the high means within the smoker cohort) but that an individual with a normal mean can also possess an elevated HFC frequency.

It would be of interest to determine whether the lesions that lead to the formation of SCEs persist in the lymphocytes of cigarette smokers as we observed in rabbits treated with known mutagens. We addressed this question by initiating a longitudinal study on individuals who stopped smoking as part of a group program. This study is still in progress but preliminary data from three individuals is shown in Table 2. There appears to be no dramatic decrease in the mean SCE frequency for any of these individuals up to 6 or 10 months after they quit smoking. There may be a tendency for the number of HFCs to de-

Table 2
Persistence of SCEs in Individuals Who Stopped Smoking

Months since last smoked	Individual #47 SCE/chromosome	Individual #158 SCE/chromosome	Individual #170 SCE/chromosome
0	0.297	0.201	0.239
0.5	0.243	0.226	0.188
1.0	–	0.176	0.204
1.5	0.297	0.193	0.230
2.5	0.272	0.174	0.244
4.5	–	0.198	–
5.5	0.269	–	0.226
10	–	–	0.308

crease but further evaluation of these results is necessary before any conclusions can be made. The individuals continue to be sampled 1 year after the study was initiated. It is interesting to point out that Anderson et al. (1981) could not demonstrate a persistent increase in SCE frequencies of vinyl chloride workers 18 months subsequent to reduction in exposure.

There are several reports in the literature that SCEs can indicate exposure in an occupational setting (see reviews by Dabney 1981; Carrano and Moore 1982). For estimating acute or high-level chronic exposures to many chemicals, the use of the SCE assay in human lymphocytes appears to be a sensitive and feasible approach. A considerable amount of further research to understand normal variation as well as lesion persistence in the circulating lymphocyte is necessary before definitive conclusions can be made concerning the utility of this assay for low-level chronic exposures. The need for additional studies, however, should not rule out the application of SCE methods to cases of low-level chronic exposures. An increased SCE frequency in such investigations would indicate exposure. On the other hand, failure to observe an increase does not rule out exposure but may help resolve the limitations of this method.

ACKNOWLEDGMENTS

This work was supported by the U.S. Department of Energy under contract number W-7405-Eng-48 to Lawrence Livermore National Laboratory. The excellent technical assistance of Ms. Linda Ashworth is appreciated.

REFERENCES

Anderson, D., C.R. Richardson, I.F.H. Purchase, H.J. Evans, and M.L. O'Riordan. 1981. Chromosomal analysis in vinyl chloride exposed workers: Comparison of the standard technique with the sister-chromatid exchange technique. *Mutat. Res.* 83:137.

Ardito, G., L. Lamberti, E. Ansaldi, and P. Ponzetto. 1980. Sister chromatid exchanges in cigarette-smoking human females and their newborns. *Mutat. Res.* 78:209.

Carrano, A.V. and D.H. Moore II. 1982. The rationale and methodology for quantifying sister chromatid exchange in humans. In *Mutagenicity: New horizons in genetic toxicology* (ed. J.A. Heddle). Academic Press, New York (In press).

Carrano, A.V. and L.H. Thompson. 1982. Sister chromatid exchange and single-gene mutation. In *Sister Chromatid Exchange* (ed. S. Wolff), p. 59. John Wiley & Sons, New York.

Carrano, A.V., L.H. Thompson, P.A. Lindl, and J.L. Minkler. 1978. Sister chromatid exchanges as an indicator of mutagenesis. *Nature* 271:551.

Carrano, A.V., J.L. Minkler, D.G. Stetka, and D.H. Moore II. 1980. Variation in the baseline sister chromatid exchange frequency in human lymphocytes. *Environ. Mutagen.* 2:325.

Carrano, A.V., L.H. Thompson, D.G. Stetka, J.L. Minkler, J.R. Mazrimas, and S. Fong. 1979. DNA crosslinking, sister chromatid exchange and specific locus mutations. *Mutat. Res.* 63:175.

Cheng, W., J.J. Mulvihill, M.H. Greene, L.W. Pickle, S. Tsai, and J. Whang-Peng. 1979. Sister chromatid exchanges and chromosomes in chronic myelogenous leukemia and cancer families. *Int. J. Cancer* 23:8.

Crossen, P.E. and W.F. Morgan. 1980. Sister chromatid exchange in cigarette smokers. *Hum. Genet.* 53:425.

Dabney, B.J. 1981. The role of human genetic monitoring in the workplace. *J. Occup. Med.* 23:626.

Galloway, S. and H.J. Evans. 1975. Sister chromatid exchange in human chromosomes from normal individuals and patients with ataxia telangiectasia. *Cytogenet. Cell Genet.* 15:17.

Gebhart, E. 1981. Sister chromatid exchange (SCE) and structural chromosomal aberration in mutagenicity testing. *Hum. Genet.* 58:235.

Hollander, D.H., M.S. Tockman, Y.W. Liang, D.S. Borgaonkar, and J.K. Frost. 1978. Sister chromatid exchanges in the peripheral blood of cigarette smokers and in lung cancer patients and the effect of chemotherapy. *Hum. Genet.* 44:165.

Hopkin, J.M. and H.J. Evans. 1980. Cigarette smoke-induced DNA damage and lung cancer risks. *Nature* 283:338.

Husgafvel-Pursiainen, K., J. Maki-Paakkanen, H. Norppa, and M. Sorsa. 1980. Smoking and sister chromatid exchange. *Hereditas* 92:247.

Husum, B., H.C. Wulf, and E. Niebuhr. 1982. Increased sister chromatid exchange frequency in lymphocytes in healthy cigarette smokers. *Hereditas* 96:85.

Jostes, R.F., Jr. 1981. Sister-chromatid exchanges but not mutations decrease with time in arrested Chinese hamster ovary cells after treatment with ethylnitrosourea. *Mutat. Res.* 91:371.

Kato, H. 1974. Possible role of DNA synthesis in formation of sister chromatid exchanges. *Nature* 252:739.

Kowalczyk, J. 1980. Sister-chromatid exchanges in children treated with nalidixic acid. *Mutat. Res.* 77:371.

Lambert, B. and A. Lindblad. 1980. Sister chromatid exchange and chromosome aberrations in lymphocytes of laboratory personnel. *J. Toxicol. Environ. Health* **6**:1237.

Lambert, B., U. Ringborg, E. Harper, and A. Lindblad. 1978a. Sister chromatid exchanges in lymphocyte cultures of patients receiving chemotherapy for malignant disorders. *Cancer Treat. Rep.* **62**:1413.

Lambert, B., A. Lindblad, M. Nordenskjold, and B. Werelius. 1978b. Increased frequency of sister chromatid exchanges in cigarette smokers. *Hereditas* **88**:147.

Latt, S. 1981. Sister chromatid exchange formation. *Ann. Rev. Genet.* **15**:11.

Lezana, E.A., N.O. Bianchi, M.S. Bianchi, and J.E. Zabala-Suarez. 1977. Sister chromatid exchanges in Down syndromes and normal human beings. *Mutat. Res.* **45**:85.

Morgan, W.F. and P.E. Crossen. 1977. The incidence of sister chromatid exchanges in cultured human lymphocytes. *Mutat. Res.* **42**:305.

Musilova, J., K. Michalova, and J. Urban. 1979. Sister-chromatid exchanges and chromosomal breakage in patients treated with cytostatics. *Mutat. Res.* **67**:289.

Nevstad, N.P. 1978. Sister chromatid exchanges and chromosomal aberrations induced in human lymphocytes by the cytostatic drug adriamycin in vivo and in vitro. *Mutat. Res.* **57**:253.

Okinaka, R.T., B.J. Barnhart, and D.J. Chen. 1981. Comparison between sister chromatid exchange and mutagenicity following exogenous metabolic activation of promutagens. *Mutat. Res.* **91**:57.

Raposa, T. 1978. Sister chromatid exchange studies for monitoring DNA damage and repair capacity after cytostatics in vitro and in lymphocytes of leukemic patients undergoing cytostatic therapy. *Mutat. Res.* **57**:241.

Sirianni, S.R. and C.C. Huang. 1980. Comparison of induction of sister chromatid exchange, 8-azaguanine- and ouabain-resistant mutants by cyclophosphamide, ifosfamide and 1-(pyridyl-3)-3,3-dimethyltriazene in Chinese hamster cells cultured in diffusion chambers in mice. *Carcinogenesis* **1**:353.

Stetka, D.G. and S. Wolff. 1976. Sister chromatid exchange as an assay for genetic damage induced by mutagen-carcinogens. Part I. In vivo test for compounds requiring metabolic activation. *Mutat. Res.* **41**:333.

Stetka, D.G., J. Minkler, and A.V. Carrano. 1978. Induction of long-lived chromosome damage, as manifested by sister chromatid exchange, in lymphocytes of animals exposed to mitomycin-C. *Mutat. Res.* **51**:383.

Waksvik, H., P. Magnus, and K. Berg. 1981. Effects of age, sex and genes on sister chromatid exchange. *Clin. Genet.* **20**:449.

Wolff, S. 1977. Sister chromatid exchange. *Ann. Rev. Genet.* **11**:183.

Wolff, S., J. Bodycote, and R.B. Painter. 1974. Sister chromatid exchanges induced in Chinese hamster cells by UV irradiation of different stages of the cell cycle: The necessity for cells to pass through S. *Mutat. Res.* **25**:73.

COMMENTS

KLIGERMAN: Do you have any information on the life cycle of the human lymphocyte? There are stories that aberrations will last 35 years. Is that rare?

CARRANO: Karen Buckton and John Evans have accumulated information from people irradiated for ankylosing spondylitis. In some cases the lymphocytes were found to last as long as 20-30 years.

KLIGERMAN: Can you induce lesions in stem cells which have the potential of being transformed later into SCEs?

CARRANO: No, there is no information on this point. I would doubt that for all the divisions a cell would have to go through from the progenitor cell to circulating lymphocyte, that a lesion would remain around that long. Most of the lesions—there may be exceptions—are removed after a few cell divisions, at least in terms of our ability to detect them as inducing SCEs. But there may be exceptional compounds.

MOHRENWEISER: What is the initial chemical reaction or the initial lesion that may result in SCE?

CARRANO: There have been a few studies done in terms of some specific agents, particularly the simple ethylating agents. The data are conflicting and suggest that many types of DNA adducts can induce SCEs. There does not appear to be a common lesion that induces an SCE. Rather, there are a variety of lesions.

HEDDLE: It wasn't clear to me that your experiments on multiple exposure ruled out the possibility that if you had given a single exposure and waited 6 months you might not have seen an elevated frequency in SCEs.

CARRANO: In the rabbits a single acute exposure of mitomycin C did not seem to induce a persistent effect.

RAJEWSKY: Do you believe that the lesion that produces the SCE persists, or that the initial mechanical event that is the dormant SCE, so to speak, is the thing that persists, and is only brought to expression by cell division later?

CARRANO: Until we know the exact mechanism of the SCE, I can't really answer that question. As a first response to this, I would think that the lesion would have to persist. If it is a configurational change in the DNA or a modification of the DNA structure, I am not certain as to how that might persist for many, many months.

RAJEWSKY: Is there any correlation between DNA breaks, strand breaks, and SCE frequency?

CARRANO: John Evans will refer to that later.

PARODI: The compound is a methylating agent. The number of breaks is, diacetic diacetamide, we found 500 times more single-strand breaks than SCEs.

CARRANO: That doesn't surprise me. With radiation, you will observe a very similar effect. In general, agents that directly break the backbone of DNA are going to be less efficient at producing SCEs than agents which will form an adduct to the DNA, or somehow distort the backbone. If the substance breaks DNA directly, it is probably going to be a very inefficient inducer of SCEs.

PARODI: The compound is a methylating agent. The number of breaks is, again, much less than the amount of methylation. So if the breaks are 500 times more frequent than SCEs, then the methylation will probably be more than 10,000 times more frequent.

CARRANO: The question is, what lesions are present when a cell goes through S phase? Are there breaks present? Or are there methylations present?

PARODI: Probably what are present are apurinic sites, because the majority of breaks are apurinic sites.

HEDDLE: I think it is probably worthwhile just making a note that it is possible to manipulate exposure to 8-methyoxypsoralen plus light in such a way that all the effects that you get are attributable to cross-links induced in the DNA. Although there is more than one kind of cross-link, under these conditions, one gets an enormous production of both SCEs and chromosomal aberrations, and, in the mouse lymphoma cells we looked at anyway, no increase in mutation frequency. But under those circumstances, we know, to a relatively specific degree, that a particular lesion is leading to production of an SCE and a chromosome aberration. It must be a cross-link between pyrimidines in the DNA. We don't know which pyrimidines or whether indeed that matters.

Controlled Human Exposure Studies:
Cytogenetic Effects of Ozone Inhalation

WENDELL H. MC KENZIE
Department of Genetics
North Carolina State University
Raleigh, North Carolina 27650

Public concern has been expressed regarding human exposure to elevated levels of ozone (O_3), a key component of oxidant air pollution. O_3 is created by the interaction of nitrogen oxides, oxygen, and solar radiation.

Respiratory (Kagawa and Toyama 1975), circulatory (Buckley, et al. 1975), and visual (Lagerwerff 1963) influences have been associated with O_3 inhalation. O_3 reportedly alters the absorption spectra of nucleic acids (Christiansen and Giese 1954), produces specific mutants in *Escherichia coli* (Hamelin and Chung 1975), causes chromosome aberrations in embryonic chick fibroblasts (Sachsenmaier et al. 1965) and human lymphocytes in vitro (Fetner 1962). Zelac et al. (1971) reported an increase in chromosome aberrations in Chinese hamster lymphocytes exposed to O_3 in vivo. However, Tice et al. (1978) in repeating the experiment of Zelac et al. observed no increase in any chromosome aberration type. Similarly, Gooch et al. (1976) found no cytogenetic effect of O_3 inhalation in Chinese hamster marrow cells.

Cytogenetic studies of college students living in dense smog (i.e., high O_3) areas have yielded conflicting results (Scott and Burkart 1978; Magie et al. 1980). Merz et al. (1975) in a small-scale controlled human O_3 exposure study reported increased chromatid-type aberrations in lymphocytes. In contrast, however, Guerrero et al. (1979) found no sister chromatid exchange (SCE) effect in circulating lymphocytes from O_3-exposed humans.

The present series of investigations was designed to obtain additional cytogenetic data from larger-scale studies directly from human subjects following controlled in vivo exposure to varied O_3 concentrations for varied durations, single or multiple exposures and with potential interactions (i.e., exposure situations which would be similar to those experienced by individuals residing or working in many air polluted environments).

METHODS

Subjects

Healthy, nonsmoking young adult males served as the subjects in these investigations. The experimental design as well as possible short-term and long-term effects were explained in detail to each subject prior to his participation. After obtaining informed consent, a physical examination was made and a medical history taken.

Exposure Facility and Conditions

The plexiglass exposure facility used in these studies is located at the Environmental Protection Agency, Health Effects Research Laboratory (Chapel Hill, NC). O_3 was generated from oxygen using an OREC O_3 generator, mixed with ambient air and dispersed into the chamber. O_3 content was monitored throughout each exposure as were the levels of sulfur dioxide, nitrogen oxides, carbon dioxide and total suspended particulates. Temperature and relative humidity were maintained at 21.1-24.4°C and 40-60%, respectively.

Subjects remained seated during the exposure period except for two 15-minute periods of moderate exercise on a bicyle ergometer.

Experimental Approach

Four independent O_3 exposure studies were conducted. The experimental design was as follows:

1. 10-30 Nonsmoking human male subjects/study
2. Exposure:
   Study 1: 0.4 ppm ozone for 4 hours (1 day)
   Study 2: 0.6 ppm ozone for 2 hours (1 day)
   Study 3: 0.4 ppm ozone for 4 hours (4 days)
   Study 4: 0.6 ppm ozone + histamine for 2 hours (1 day)
3. Blood samples collected: prior to exposure, immediate post-exposure, at 3 days, at week 2, and week 4 post-exposure
4. Lymphocyte cultures set, harvested, slides prepared
5. 100 Metaphases/blood sample/subject analyzed for Chromosome aberrations
6. 50 Metaphases/blood sample/subject analyzed for sister chromatid exchanges (SCEs)

Cytogenetic and Statistical Methods

Culturing, harvesting, staining and scoring methods were those used routinely and reported previously (McKenzie et al. 1977). The data from these experiments were analyzed using an analysis of variance for repeated measurements on both the raw observations and appropriately arcsin transformed observations.

RESULTS

Tables 1, 2, 3 and 4 present the summary data from Studies 1, 2, 3, and 4, respectively. Additional, more detailed, data from these studies have been reported elsewhere (McKenzie and Hall 1977; McKenzie et al. 1977; McKenzie et al. 1979).

No statistically significant differences in frequencies of numerical aberrations, structural aberrations, or SCEs were found when comparing ozone pre-exposure and post-exposure values. These nonsignificant differences were found consistently across all studies at all concentrations and durations tested, single and multiple exposures, and post-exposure sampling times. Further, no effect of histamine exposure or a histamine-O_3 interaction was noted.

Table 1

Study 1: Mean Chromosome Aberrations in Lymphocytes of Subjects Prior and Subsequent to 0.4 ppm Ozone—4 hr Single Exposure

	Pre-exposure	Immediate post-exposure	3 days post-exposure	2 weeks post-exposure	4 weeks post-exposure
Numerical aberrations (% cells)	7.0	6.4	8.3	5.5	5.8
Structural aberrations (% cells)	2.3	1.8	2.4	1.7	1.6

Table 2

Study 2: Mean Chromosome Aberrations and SCEs in Lymphocytes of Subjects Prior and Subsequent to 0.6 ppm Ozone—2 hr Single Exposure

	Pre-exposure	Immediate post-exposure	3 days post-exposure	2 weeks post-exposure	4 weeks post-exposure
Numerical aberrations (% cells)	5.6	6.2	6.1	5.0	5.8
Structural aberrations (% cells)	3.3	3.8	3.0	2.6	3.2
SCEs (per cell)	6.2	6.0	6.2	6.2	7.1

Table 3
Study 3: Mean Chromosome Aberrations and SCEs in Lymphocytes of Subjects Prior and Subsequent to 0.4 ppm Ozone—4 hr/day—4 Consecutive Day Multiple Exposure

	Pre-exposure	Day 1 post initial exposure	Day 4 post initial exposure	7-Day post initial exposure	31-Day post initial exposure
Numerical aberrations (% cells)	6.3	7.6	7.5	6.6	5.1
Structural aberrations (% cells)	3.7	4.7	3.4	4.7	3.0
SCEs (per cell)	8.4	9.2	8.6	9.5	8.4

Table 4
Study 4: Mean Chromosome Aberrations and SCEs in Lymphocytes of Subjects Prior and Subsequent to Histamine, Histamine-0.6 ppm Ozone—2 hr Single Exposure

	Pre-exposure	Post-histamine exposure	Post-histamine and ozone exposure
Numerical aberrations (% cells)	5.6	5.2	3.9
Structural aberrations (% cells)	3.3	3.8	3.7
SCEs (per cell)	10.9	10.6	9.0

DISCUSSION

Although these studies are far from exhaustive, they point to no chromosome damaging effect which can be attributed to ozone exposure in humans. It is possible, however, that given another set of experimental conditions, or more sensitive mutagenic test systems such an effect could be detected if present. It is not clear to what extent our findings influenced EPA to relax by 50% the maximum permissible ozone exposure level from 0.08 ppm to 0.12 ppm/hour.

Controlled human exposure studies, such as these, offer recognizable experimental and interpretive advantages and disadvantages. Some advantages:

1. Control. Subjects serve as their own controls: pre- vs. post-exposure.
2. Direct Results: Human Exposure → Human Effect. Avoid extrapolation from other species and in vitro exposures.
3. Coordinate Studies. Opportunities to correlate genetic damage with other biological, physiological and health parameters.

Some disadvantages:

1. Inflexible experimental design (ethical/practical considerations)
   —low, short-term exposures, limited follow-up
   —limited tissues for analysis
   —limited subject availability (no ♀♀, embryos, elderly, ill, etc.)
   —somatic, but not genetic effects assessed
2. Expense
   —exposure facility: construction, operational costs
   —cytogenetic laboratory input/data output = large
3. Over-weighting
   Danger of taking results too seriously, (i.e., without adequately recognizing shortcomings and without simultaneously considering results obtained from other experimental systems and approaches).

ACKNOWLEDGMENTS

The author is indebted to Susan H. Hall, Lynn G. Bumgarner and Carolyn L. Doerr for their excellent technical assistance. This study was supported in part by EPA Contract 68-02-1283 and EPA Grant R-805739. The use of trade names in this publication does not imply endorsement by the North Carolina Agricultural Research Service of the products named nor criticism of similar ones not mentioned.

This is paper no. 8373 of the Journal Series of The North Carolina Agricultural Research Service, Raleigh, NC 27650.

REFERENCES

Buckley, R.D., J.D. Hackney, K. Clark, and C. Posin. 1975. Ozone and human blood. *Arch. Environ. Health* **30**:40.

Christensen, E. and A.C. Giese. 1954. Changes in absorption spectra of nucleic acids and their derivatives following exposure to ozone and ultraviolet radiation. *Arch. Biochem. Biophys.* **51**:208.

Fetner, R.H. 1962. Ozone-induced chromosome breakage in human cell cultures. *Nature* **194**:793.

Gooch, P.C., D.A. Creasia, and J.G. Brewen. 1976. The cytogenetic effects of ozone: Inhalation and *in vitro* exposures. *Environ. Res.* **12**:188.

Guerrero, R.R., D.E. Rounds, R.S. Olson, and J.D. Hackney. 1979. Mutagenic effects of ozone in human cells exposed *in vivo* and *in vitro* based on sister chromatid exchange analysis. *Environ. Res.* **18**:336.

Hamelin, C. and Y.S. Chung. 1975. The effect of low concentrations of ozone on *Escherichia coli* chromosome. *Mutat. Res.* 28:131.

Kagawa, J. and T. Toyama. 1975. Effects of ozone and brief exercise on specific airway conductance in man. *Arch. Environ. Health* 30:36.

Lagerwerff, J.M. 1963. Prolonged ozone inhalation and its effects on visual parameters. *Aerosp. Med.* 34:479.

Magie, A.R., D.E. Abbey, and W.R. Centerwall. 1980. Chromosome aberrations in peripheral lymphocytes of college students as a response to photochemical air pollution. *EPA Report* 600/1-81-007:1.

McKenzie, W.H. and S.H. Hall. 1977. Conventional aberration and sister chromatid exchange analyses of human lymphocytes exposed to ozone *in vivo*. *Am. J. Hum. Genet.* 29:73A.

McKenzie, W.H., S.H. Hall, and C.L. Doerr. 1979. Ozone and humans: A summary of *in vivo* cytogenetic studies. *Am. J. Hum. Genet.* 31:104A.

McKenzie, W.H., J.H. Knelson, N.J. Rummo, and D.E. House. 1977. Cytogenetic effects of inhaled ozone in man. *Mutat. Res.* 48:95.

Merz, T., M.A. Bender, H.D. Kerr, and T.J. Kulle. 1975. Observations of aberrations in chromosomes of lymphocytes from human subjects exposed to ozone at a concentration of 0.5 ppm for 6 and 10 h. *Mutat. Res.* 31:299.

Sachsenmaier, W., W. Siebs, and Tjong-an-tan. 1965. Effects of ozone upon mouse ascites tumor cells and upon chick fibroblasts in tissue culture. *Z. Krebsforsch.* 67:113.

Scott, C.D. and J.A. Burkart. 1978. Chromosomal aberrations in peripheral lymphocytes of students exposed to air pollutants. *EPA Report* 600/1-78-054:1.

Tice, R.R., M.A. Bender, J.L. Ivett, and R.T. Drew. 1978. Cytogenetic effects of inhaled ozone. *Mutat. Res.* 58:293.

Zelac, R.E., H.L. Cromroy, W.E. Bolch, B.G. Dunavant, and H.A. Bevis. 1971. Inhaled ozone as a mutagen. I. Chromosome aberrations induced in Chinese hamster lymphocytes. *Environ. Res.* 4:262.

Cytogenetic Studies on Industrial Populations Exposed to Mutagens

H. JOHN EVANS
Medical Research Council
Clinical and Population Cytogenetics Unit
Western General Hospital
Edinburgh, EH4 2XU Scotland

All other papers in this section deal with sister chromatid exchange (SCE) induction, but since we are concerned with indicators of genotoxic exposure and because the most unambiguous cytogenetic manifestation of genotoxicity is the induction of chromosome aberrations, I think it important that I spend some time discussing chromosome aberration induction in industrially exposed populations and then comment, by way of discussing some examples, on SCE induction in these and other relevant populations.

INDUSTRIAL POPULATIONS EXPOSED TO IONIZING RADIATIONS

Not surprisingly, the industrial mutagens that have been the subject of most interest and of most study are the ionizing radiations; since not only are they the first man-made agents that were shown to be mutagenic and carcinogenic, but they also form a part of our natural environmental background and are in widespread use in medicine and industry as well as being potentially powerful, destructive and incapacitating agents in warfare.

Following the development in the late 50s and early 60s of techniques which enabled human blood lymphocytes, which are in a G_0 interphase stage while in the peripheral circulation, to be stimulated to enter into a mitotic cycle and proceed to mitosis in short-term culture, it was quickly shown that the chromosome damaging effects of X-rays could be readily demonstrated in blood cells from patients who had undergone radiation therapy (Buckton et al. 1962). Since that time there have been a large number of studies on the response of human lymphocyte chromosomes to a variety of mutagens following exposure in vitro or in vivo (Evans and Lloyd 1978; Buckton and Evans 1982). The in vivo studies with radiations have largely been concerned with individuals or populations exposed to relatively high doses of X- or γ-rays delivered to the whole or to part of the body. Over the last decade or so,

however, we have had the opportunity to study a population of nuclear dockyard workers occupationally exposed to low levels of γ-irradiation within the
maximum permissible level of 5 rads per annum. I refer to certain aspects of our
results from this and certain other studies to illustrate some points which are
important when we consider the merits, or indeed the relevance, of the endpoints of aberrations of chromosome structure as opposed to SCE as indicators
of genotoxic exposure in the occupationally exposed.

In the nuclear dockyard project some 200 men were studied at 6-month
intervals over a 10-year period so that some of the men had accumulated doses
of up to 30 or 40 rads over that time. The results of our findings on chromosome aberrations in the blood cells of these men (Evans et al. 1979) are summarized in Figure 1. In brief, they show: (i) that, if we count sufficient cells,
we can readily demonstrate that exposure to γ-rays within the allowable

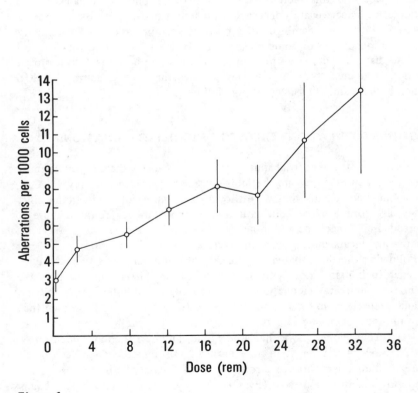

Figure 1
Total unstable chromosome aberrations in peripheral blood cells of nuclear dockyard workers occupationally exposed to cumulative doses of gamma radiation over a 10-year period.

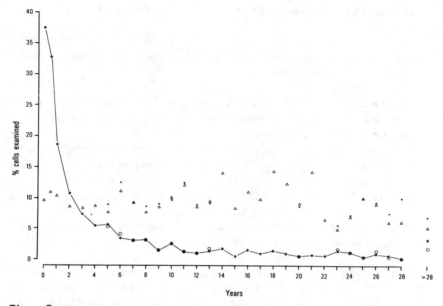

Figure 2

Frequencies of cells with unstable (C_u) and stable (C_s) abberations, at various times following X-irradiation, in peripheral blood samples from patients treated with X-ray therapy for ankylosing spondylitis. (*) C_u cells: conventional staining; (O) C_u cells; all staining; (Δ) C_s cells: conventional staining; (·) C_s cells: all staining.

occupational limits results in a significant increase in chromosome aberrations in blood cells of exposed individuals, and (ii) that the aberration frequency increases linearly with increasing exposure. The rate of aberration increase is 3.0 ± 0.6/cell $\times 10^{-4}$/rem of total dose; however, if we analyze the data in terms of the dose accumulated in the 12 months prior to sampling then the rate of increase is 4.4 ± 1.4/cell $\times 10^{-4}$/rem. The reason for this difference is that the aberration frequency in blood cells of an individual decreases with increasing time after sampling. There is little decrease in the first few months following exposure, but thereafter a steady decline as a result of cell turnover is evident as shown in our data from patients exposed to radiation therapy through X-rays as illustrated in Figure 2. These data show that the frequency of dicentric aberrations, but not the stable translocation rearrangements, decreases by around 42% per year for the first 4 years and then reduces at a lower rate (14% per year) over longer time periods. However, it is still possible to detect an elevated aberration yield in people exposed to ionizing radiations many years after their exposure as indeed is well illustrated by the studies of the survivors of the atomic bombings at Hiroshima 25 years or so after their exposure (Sasaki and Miyata 1968).

From these, and other studies on the response of man's own chromosomes to ionizing radiations we can draw three important conclusions which we must bear in mind when considering the data on SCEs.

1. That exposure to ionizing radiation at occupationally accepted levels results in increasing the incidence of true genetic damage exemplified by the formation of chromosome aberrations.
2. That the aberrations are induced in all cells, including "resting" G_o cells, and that they are formed rapidly after exposure and do not require the cells to proceed through a DNA replicating phase for their development.
3. That because of their "immediate" formation, and in spite of the reduction in the number of damaged cells with time after exposure, a proportion of the lymphocytes are long-lived cells so that aberrations can be observed months or years after an acute exposure to radiation.

These three conclusions are in stark contrast to the general findings when we come to consider cytogenetic aspects of man's exposure to most chemical mutagens. But first we should ask whether we can detect the effects of low level exposure to ionizing radiations by studying the incidence of SCEs in irradiated individuals? The answer is no. X-ray-induced DNA damage does not appear to lead to the development of SCEs so that cells exposed to relatively high doses of radiation in vivo or in vitro show virtually no increases in SCE (Galloway 1977; Littlefield et al. 1979; Cavaglia 1981); therefore, these exchanges cannot be used as end-points to monitor or detect radiation exposure. Similarly, those chemical mutagens that produce most of their genetic damage like X-rays, by inducing DNA strand scissions as opposed to DNA adducts, e.g., bleomycin, are also very inefficient at inducing SCEs relative to aberrations (Perry and Evans 1975; Cavaglia 1981). However, most chemical mutagens are efficient in inducing SCEs and induce such events more frequently than structural aberrations and there are now a number of studies in which both end-points have been analyzed in populations exposed to chemical mutagens.

INDUSTRIAL POPULATIONS EXPOSED TO CHEMICAL MUTAGENS

Few studies on populations occupationally exposed to chemical mutagens have been as extensive (or as exhaustive) as the studies on irradiated populations. There have, nevertheless, been quite a number of studies, some of which, like our own on populations exposed to lead oxides (O'Riordan and Evans 1974)

Table 1

Some Occupationally Exposed Populations Showing Clearly
Increased Aberration Frequencies, Relative to Controls, in
Peripheral Blood Lymphocytes

Agent	Population size	Selected references
X or γ-rays	>1000	Evans et al. (1979) Lloyd et al. (1980)
Arsenic	33	Nordenson et al. (1978)
Benzene	>190	Tough et al. (1970) Forni et al. (1971)
Chloromethylether	12	Zudova and Landa (1977)
Chloroprene	>50	Zhurkov et al. (1977)
Epichlorhydrin	>100	Kucerova et al. (1977) Picciano et al. (1977)
Organophosphates	>180	Van Bao et al. (1974) Kiraly et al. (1977)
Pesticide/Herbicide	>40	Yoder et al. (1973)
Styrene	>50	Meretoja et al. (1977) Andersson et al. (1980)
Vinyl chloride	>500	Hansteen et al. (1978) Anderson et al. (1981)
Ziram	9	Pilinskaya (1970)

or to cadmium (O'Riordan et al. 1978) have yielded negative findings; others
which have provided suggestive, but not very convincing, positive findings such
as in groups exposed to phenolformaldehyde (Suskov and Sazonova 1982);
and others which are clearly positive. Some of the studies recording clearly significant increases in aberration frequencies in cultured blood lymphocytes
from exposed workers, as compared with relevant matched controls, are listed
in Table 1.

In many of the studies referred to in Table 1 no attempt was made to
determine the frequencies of SCE in the populations, but in others, such as those
on populations exposed to vinyl chloride or to styrene, elevated incidences of
SCE have been reported. In one instance, that involving the alkylating agent
epichlorhydrin, the study was prospective in nature, workers being sampled

before being subjected to exposure to epichlorhydrin and then followed up after one year and then two years occupational exposure. In this study (Kucerova et al. 1977) an increasing aberration frequency with increasing time of exposure was clearly demonstrated. With the exception of the radiation studies, however, all the other surveys listed are retrospective and although the results are positive there is usually little or no evidence of a clear dose response. The most studied populations have been those exposed to vinyl chloride (VCM). Below I present some of the problems associated with data on aberration and SCE incidence in workers exposed to chemical mutagens by reference to some of the results obtained from workers in polyvinylchloride (PVC) plants.

In 1975 various groups showed that the incidence of chromosome aberrations in peripheral blood lymphocytes of workers exposed to VCM, a substance considered to be a causal factor in the development of angiosarcomas in certain workers in the plastics industry, was considerably higher than in workers who were not so exposed (Ducatman et al. 1975; Funes-Cravioto et al. 1975). Later, original studies on a population of VCM workers in England (Purchase et al. 1978) showed that the aberration frequency in autoclave workers who were exposed to the highest dose levels were almost an order of magnitude greater than in matched controls. Following these studies, plant exposure levels were reduced to below 5 ppm and the exposed and control groups studied again 18 months and 42 months later (Anderson et al. 1980). In the second 18-month sample SCEs were also scored in addition to aberrations (Anderson et al. 1981) and these results are summarized in Table 2. In the samples taken almost 4 years after exposure the aberration frequencies in exposed individuals were still somewhat higher than those in controls, but the difference was not significant. However, in samples taken 18 months after exposure the aberration frequencies were still high and clearly significantly above background levels, but when SCEs were determined in these samples their frequencies were seen to be not significantly different from control levels. In vitro studies (Anderson et al. 1981) readily demonstrate that VCM, or rather a metabolite, is an important inducer of SCEs in human lymphocyte chromosomes exposed in vitro and indeed Kucerova et al. (1979) have also demonstrated that the SCE frequency in vivo is also enhanced in workers exposed to high levels of VCM. The data in Table 2 tell us that although the frequency of chromosomal aberrations and of SCEs will decrease with time after exposure, the rate of decrease in SCE frequency is far more rapid than the rate of decrease in aberration frequency. Similar findings of normal SCE frequencies have also been reported in other populations originally exposed to fairly high levels of VCM and monitored over succeeding years (Hansteen et al. 1978; Natarajan et al. 1978) so that this pattern of results seems well established.

Table 2
Frequencies of Chromosome Aberrations and SCE (Per Cell) in Workers Occupationally Exposed to Vinyl Chloride Monomer VCM

Group	Number of people	Number of cells	Chromosome Aberrations								SCE	
			Chromatid-type				Chromosome-type				Number of cells	SCE/cell
			Gaps	Breaks	Inter-changes	Total structure	Dicentric + rings	Frag-ments	Symmetric translation	Total structure		
VCM Autoclave Workers	10	1000	9.1	3.2	0.4	3.6	0.2	2.1	1.0	3.3	265	8.2
VCM or PVC Workers	8	800	6.4	1.8	0.2	2.0	0.2	1.6	0.9	2.8	215	8.0
Former PVC Workers	3	300	6.3	1.0	0	1.0	0	1.7	0.3	2.0	90	7.4
Controls	6	600	3.7	0.7	0	0.7	0	0.5	0	0.5	170	6.7

Data from Anderson et al. 1981.

These findings on workers occupationally exposed to VCM are entirely in line with expectation from studies on patients treated with cytotoxic drugs and from experiments on animals where SCE levels are high in the few days or weeks following treatment, but usually, but not invariably (Stetka et al. 1978; Lambert et al. 1982), then decline down to control levels. The principal reason for this decline is simply a consequence of two facts: SCEs are only produced at the time of replication (Wolff et al. 1974); and that at least some of the lesions that may give rise to SCE have a high probability of being repaired if the cells are not engaged in a replicative cycle (Jostes 1981), although there is evidence for the persistence and modification of some SCE-inducing lesions in unstimulated human lymphocytes exposed to an alkylating agent in vitro (Evans and Vijayalaxmi 1980).

In practical terms, these findings tell us that the actions of a genotoxic agent that will produce SCEs in vivo will only be detected if the work force is being chronically exposed at a sufficiently high level up until the time of sampling, or if we sample blood cells shortly following an acute exposure and not many weeks or months later. A similar qualification will apply to those so-called 'S-dependent' genotoxic chemicals that will largely induce chromosome aberrations in exposed G_o lymphocytes as a consequence of misreplication of unrepaired lesions induced in the non-DNA-synthetic phases of the cell cycle. This delayed development of SCE and aberrations in the case of chemical agents that do not induce double-strand breaks in DNA, is a major factor which, in practice, makes the cytogenetic detection of genotoxic damage with chemical agents more difficult than with X- or γ-rays.

SUMMARY

Many of the more recent studies in which SCE incidence has been determined in occupationally exposed populations, and in which proper regard was paid to undertaking a blind study with carefully matched controls, have frequently shown a rather more elevated incidence of SCEs in cigarette smokers relative to nonsmokers (e.g., Andersson et al. 1980; Husgafvel-Pursiainen et al. 1980; Lambert et al. 1982). The difference is often of the order of ~30% and cigarette smoking is but one of a number of compounding factors in any population study. Although it is entirely possible that this increased SCE frequency in cigarette smokers reflects smoke-induced DNA damage (Evans 1981), we should note that there is a lot of evidence that background SCE rates are strongly influenced by cell proliferation rates, with for example slowly cycling human lymphocytes showing up to twice as many SCEs as more rapidly developing cells (Santesson et al. 1979; Snope and Rary 1979; Lindblad and Lambert 1981). Any nonmutagenic factor that alters the proportions of lymphocyte subsets in the peripheral blood with regard to their stimulation and proliferation rate in vitro, will alter the overall SCE frequency. Indeed, we can readily demonstrate this in a variety of individuals not exposed to mutagenic drugs, but with abnormal lymphocyte profiles where we observe up to ~30% increase in background SCE frequency relative to normal controls. We are careful to emphasize that the occurrence of a normal SCE frequency is most certainly not indicative of the absence of exposure to a mutagen, but we need to be equally careful in noting that the presence of a small increase in SCE frequency may also not be indicative of the occurrence of such an exposure.

REFERENCES

Anderson, D., C. R. Richardson, I. F. H. Purchase, H. J. Evans, and M. L. O'Riordan. 1981. Chromosomal analysis in vinyl chloride exposed workers: Comparison of the standard technique with the sister-chromatid exchange technique. *Mutat. Res.* 83:137.

Anderson, D., C. R. Richardson, T. M. Weight, I. F. H. Purchase, and W. G. F. Adams. 1980. Chromosomal analyses in vinyl chloride exposed workers. Results from analysis 18 and 42 months after an initial sampling. *Mutat. Res.* 79:151.

Andersson, H. C., E. A. Tranberg, A. H. Uggla, and G. Zetterberg. 1980, Chromosomal aberrations and sister-chromatid exchanges in lymphocytes of men occupationally exposed to styrene in a plastic-boat factory. *Mutat. Res.* 73:387.

Buckton, K. E. and H. J. Evans. 1982. Human peripheral blood lymphocyte cultures: An in vitro assay for the cytogenetic effects of environmental mutagens. In *Cytogenetic assays of environmental mutagens* (ed. T. C. Hsu), p. 183. Allanheld, Osmun, Totowa, New Jersey.

Buckton, K. E., P. A. Jacobs, W. M. Court Brown, and R. Doll. 1962. A study of the chromosome damage persisting after X-ray therapy for ankylosing spondylitis. *Lancet* 2:676.

Cavaglia, A. M. V. 1981. *Sister chromatid exchange induction in mammalian cells.* Ph.D. Thesis, University of Edinburgh, Scotland.

Ducatman, A., K. Hirschhorn, and I. J. Selikoff. 1975. Vinyl chloride exposure and human chromosome aberrations. *Mutat. Res.* 31:163.

Evans, H. J. 1981. Cigarette smoke induced DNA damage in man. In *Progress in mutation research* (ed. A. Kappas), vol. 2, p. 111. Elsevier/North-Holland, Amsterdam.

Evans, H. J., K. E. Buckton, G. E. Hamilton, and A. Carothers. 1979. Radiation-induced chromosome aberrations in nuclear-dockyard workers. *Nature* 277:531.

Evans, H. J. and D. C. Lloyd (eds.) 1978. *Mutagen-induced chromosome damage in man.* University Press, Edinburgh, Scotland.

Evans, H. J., and Vijayalaxmi. 1980. Storage enhances chromosome damage after exposure of human leukocytes to mitomycin C. *Nature* 284:370.

Forni, A., E. Pacifico, and A. Limonta. 1971. Chromosome studies in workers exposed to benzene or toluene or both. *Arch. Environ. Health* 22:373.

Funes-Cravioto, F., B. Lambert, J. Lindsten, L. Ehrenberg, A. T. Natarajan, and S. Osterman-Golkar. 1975. Chromosome aberrations in workers exposed to vinyl chloride. *Lancet* 1:459.

Galloway, S. M. 1977. Ataxia telangiectasia: The effects of chemical mutagens and X-rays on sister chromatid exchanges in blood lymphocytes. *Mutat. Res.* 45:343.

Hansteen, I.-L., L. Hissestad, E. Thiis-Evensen, and S. S. Heldaas. 1978. Effects of vinyl chloride in man. A cytogenetic follow-up study. *Mutat. Res.* 51:271.

Husgafvel-Pursiainen, K., J. Mäki-Paakkanen, H. Norppa, and M. Sorsa. 1980. Smoking and sister chromatid exchange. *Hereditas* 92:247.

Jostes, R. F. 1981. Sister-chromatid exchanges but not mutations decrease with time in arrested Chinese hamster ovary cells after treatment with ethylnitrosourea. *Mutat. Res.* 91:371.

Kiraly, J., A. Czeizel, and I. Szentesi. 1977. Genetic study on workers producing organophosphate insecticides. *Mutat. Res.* 46:224.

Kucerova, M., Z. Polivkova, and J. Batora. 1979. Comparative evaluation of the frequency of chromosomal aberrations and the sister chromatid exchange numbers in peripheral lymphocytes of workers occupationally exposed to vinyl chloride monomer. *Mutat. Res.* 67:97.

Kucerova, M., V. S. Zhurkov, Z. Polivkova, and J. E. Ivanova. 1977. Mutagenic effect of epichlorohydrin. II. Analysis of chromosomal aberrations in lymphocytes of persons occupationally exposed to epichlorohydrin. *Mutat. Res.* 48:355.

Lambert, B., A. Lindblad, K. Holmberg, and D. Francesconi. 1982. The use of sister chromatid exchange to monitor human populations for exposure to toxicologically harmful agents. In *Sister chromatid exchange* (ed. S. Wolff), p. 149. Wiley, New York.

Lindblad, A. and B. Lambert. 1981. Relation between sister chromatid exchange, cell proliferation and proportion of B and T cells in human lymphocyte cultures. *Hum. Genet.* 57:31.

Littlefield, L. G., S. P. Colyer, E. E. Joiner, and R. J. Dufrain. 1979. Sister chromatid exchanges in human lymphocytes exposed to ionizing radiation during G_0. *Radiat. Res.* 78:514.

Lloyd, D. C., R. J. Purrott, and E. J. Reeder. 1980. The incidence of unstable chromosome aberrations in peripheral blood lymphocytes from unirradiated and occupationally exposed people. *Mutat. Res.* 72:523.

Meretoja, T., H. Vainio, M. Sorsa, and H. Härkönen. 1977. Occupational styrene exposure and chromosomal aberrations. *Mutat. Res.* 56:193.

Natarajan, A. T., P. P. W. Van Buul, and T. Raposa. 1978. An evaluation of the use of peripheral blood lymphocyte systems for assessing cytological effects induced in vivo by chemical mutagens. In *Mutagen-induced chromosome damage in man* (eds. H. J. Evans and D. C. Lloyd), p. 268. University Press, Edinburgh, Scotland.

Nordenson, I., G. Beckman, L. Beckman, and S. Nordström. 1978. Occupational and environmental risks in and around a smelter in northern Sweden. I. Chromosomal aberrations in workers exposed to arsenic. *Hereditas* **88**: 47.

O'Riordan, M. L. and H. J. Evans. 1974. Absence of significant chromosome damage in males occupationally exposed to lead. *Nature* **247**:50.

O'Riordan, M. L., E. G. Hughes, and H. J. Evans. 1978. Chromosome studies on blood lymphocytes of men occupationally exposed to cadmium. *Mutat. Res.* **58**:305.

Perry, P. and H. J. Evans. 1975. Cytological detection of mutagen-carcinogen exposure by siter chromatid exchange. *Nature* **258**:121.

Picciano, D. J., R. E. Flake, P. C. Gay, and D. J. Kilian. 1977. Vinyl chloride cytogenetics. *J. Occup. Med.* **19**:527.

Pilinskaya, M. 1970. Chromosome aberrations in the persons contacted with Ziram. *Genetika* **6**:157.

Purchase, I. F. H., C. R. Richardson, D. Anderson, G. M. Paddle, and W. G. F. Adams. 1978. Chromosomal analyses in vinyl chloride exposed workers. *Mutat. Res.* **57**:325.

Santesson, B., K. Lindahl-Kiessling, and A. Mattsson. 1979. SCE in B and T lymphocytes. Possible implications for Bloom's syndrome. *Clin. Genet.* **16**:133.

Sasaki, M. S. and H. Miyata. 1968. Biological dosimetry in atomic bomb survivors. *Nature* **220**:1189.

Snope, A. J. and J. M. Rary. 1979. Cell-cycle duration and sister-chromatid exchange frequency in cultured human lymphocytes. *Mutat. Res.* **63**: 345.

Stetka, D. G., J. Minkler, and A. V. Carrano. 1978. Induction of long-lived chromosome damage, as manifested by sister-chromatid exchange, in lymphocytes of animals exposed to mitomycin-C. *Mutat. Res.* **51**:383.

Suskov, I. I. and L. A. Sazonova. 1982. Cytogenetic effects of epoxy, phenol-formaldehyde and polyvinylchloride resins in man. *Mutat. Res.* **104**: 137.

Tough, I. M., P. G. Smith, W. M. Court Brown, and D. G. Harnden. 1970. Chromosome studies on workers exposed to atmospheric benzene. The possible influence of age. *Eur. J. Cancer* **6**:49.

Van Bao, T., I. Szabo, P. Ruzicska, and A. Czeizel. 1974. Chromosome aberrations in patients suffering acute organic phosphate insecticide intoxication. *Humangenetik* **24**:33.

Wolff, S., J. Bodycote, and R. B. Painter. 1974. Sister chromatid exchanges induced in Chinese hamster cells by UV irradiation of different stages of the cell cycle: The necessity for cells to pass through S. *Mutat. Res.* **25**:73.

Yoder, J., M. Watson, and W. W. Benson. 1973. Lymphocyte chromosome analysis of agricultural workers during extensive occupational exposure to pesticides. *Mutat. Res.* **21**:335.

Zhurkov, V. S., B. S. Fichidzhyan, G. G. Batikyan, R. M. Arutyunyan, and V. N. Zil'fyan. 1977. Cytogenetic examination of persons in contact with chloroprene under industrial conditions. *Tsitol. Genet.* **11**:13.

Zudova, Z. and K. Landa. 1977. Genetic risk of occupational exposures to haloethers. *Mutat. Res.* **46**:242.

COMMENTS

KLIGERMAN: In our ethylene oxide study, though it was a short exposure, we didn't pick up any breakage after 3 days, but we did pick up SCEs.

EVANS: Well, SCEs are certainly a more sensitive indicator, but the trouble is they are short-lived. There are not very many long-lived lesions that give rise to SCE. There are some, but not very many so that to detect exposures that result in lesions that lead to SCEs, one will need to sample cells within days rather than months after exposure.

KLIGERMAN: That is why we probably should do both.

EVANS: Tony [Carrano], your presentation showed some long-lived effects that resulted in SCEs so perhaps you'd like to comment.

CARRANO: I didn't point this out, but we are doing a study of smokers that have quit smoking. We have got them for only 1 year, but we have only analyzed a few of them up to about 6 months since giving up smoking. In those individuals we only see about a 10% decrement in SCE frequency from 6 months from the time they stopped smoking. That is probably not significant but we don't have enough data yet to really make that certain.

I would like some clarification. For example, ethylene oxide workers have been shown to have an increase in SCE frequency. Now, at what level do you say that it is not useful for industry?

EVANS: I am not saying it isn't useful, but it is very dependent upon the timing of your sampling. Thus far there are very few clear studies where people have sampled cells to look at SCEs and aberrations in a population being exposed, and then looked again at various times after removal of the agent to which they were exposed. I think the lesson we get from our VCM study is fairly clear. If you sample 3 years after the removal of VCM from the atmosphere there is no elevated SCE frequency or alteration frequency. If you go back 3 years after a radiation exposure, which would initially give you the same aberration frequency as VCM, you can still see a significant excess of aberrations.

What I am saying is that, provided you look shortly after exposure or during exposure, yes, an analysis of SCE levels is clearly a good approach.

CARRANO: When you looked at those people initially, did you have high SCE frequencies?

EVANS: Yes.

CARRANO: Now you remove the insult, and if you did an epidemiological study on current workers or workers that have worked in the industry for a long time versus workers that have left the industry, are they more at risk or less at risk?

EVANS: I can't answer regarding the risk. All I can say is that those who have left the industry will have less SCEs.

WATERS: Wasn't there a report in the literature by Hansteen about an accidental exposure to vinyl chloride, about 1,500 ppm? She did a follow-up study in about 4 days and found no increase. Is that corredt?

EVANS: No. She found an increase in aberration frequency, but she found that this increase disappeared when the workers were sampled 2-2½ years later after the VCM levels were reduced to a minimum. SCEs were not studied in the initial sampling, but they showed the same frequencies in workers and controls in the later samples. The same kind of result has been obtained in Holland. Again, 1 or 2 years after exposure the aberration level is back down to baseline level.

WATERS: But there was an initial increase?

EVANS: Yes, there was certainly an initial increase.

EISENSTADT: What happens to the SCE frequency in *Xeroderma pigmentosum* individuals? Is it up?

EVANS: Not usually, but it depends on what they are exposed to.

EISENSTADT: Take an agent that induces SCEs normally, is the frequency raised among XPs?

EVANS: If you expose XP cells to UV light, you will get a large SCE frequency.

EISENSTADT: Does it take a lower dose to give you the same level?

EVANS: Yes, it certainly does because there is less DNA repair.

EISENSTADT: Have lymphocytes from living patients been looked at?

EVANS: Yes.

EISENSTADT: And do they have higher SCE frequencies, as if they were a sentinel population, sensitively expressing the effects of environmental mutagens?

EVANS: Not, not in general.

PRESTON: It is varied, isn't it? Some have higher SCE frequencies and some have the same frequencies as normal individuals. It depends on the report. There is very little overall increase above normal spontaneous levels.

SORSA: My data on styrene SCE induction was an in vitro study, an unpublished report showing data on workers and styrene exposure. I am convinced that there is no SCE increase, if you exclude totally the smoking effect. But there certainly is a clastogenic effect on chromosome aberrations.

EVANS: A very important point in all these studies on occupationally exposed people is that if you divide the groups into cigarette smokers and nonsmokers, usually the more significant effect is smoking, not occupational exposure. That is true, I think, in most of the studies. But I am very worried about smoking in terms of the interpretation of this up to 30% increase of SCE frequency in smokers against nonsmokers. I am concerned because we can detect up to a 30% increase in SCE frequency in people who don't smoke, but who show a different lymphocyte profile from normal. We see this level of increase in people who have disease states that change the proportion of T–cells and B–cells in the peripheral blood, e.g., patients with muscular dystrophy or Huntington's chorea. I think there is no question that there are subpopulations of T-lymphocytes that are not homogeneous in their response to exposure to BrdU. We obviously need to know a great deal more about the structure of the lymphocyte populations if we have to interpret very small changes in SCE frequency.

ALLEN: Is it clear yet whether the T-cells or the B-cells are showing the higher frequencies in human blood? I have seen a lot of different reports with different conclusions.

EVANS: Well, it isn't the right question to ask. The question to ask is, do T 'helper cells' and T-'suppressor cells:, and other T-cell subclasses differ in their response? It isn't just T cells versus B; they are all different. So I think we have to categorize those cells and then ask, how do the sensitivities and patterns differ?

GARCIA: Are the populations of lymphocytes different for smokers and non-smokers?

EVANS: The inference is that they might well be, but we need to have firm independent evidence before drawing that conclusion.

Induction of Sister Chromatid Exchanges
Among Nurses Handling Cytostatic Drugs

MARJA SORSA, HANNU NORPPA, AND HARRI VAINIO
Institute of Occupational Health
Haartmaninkatu 1, SF-00290
Helsinki 29, Finland

Recently attention has been paid to the possible occupational hazards of cyto-static drugs not only at oncology units, but also at other hospital wards and health care centers. Both the number of people handling cytostatic drugs and also the frequency with which they are handled have increased with more wide-spread use of cancer chemotherapy. Consequently, the biological monitoring of possible occupational exposure is important, especially because the carcino-genicity and the mutagenicity of many cytostatic drugs have been recently con-firmed (IARC 1981).

The practical methods available for biological monitoring depend on the character of the exposing agent (Vainio et al. 1981; Sorsa et al. 1982). Only seldom is there previous experimental information to suggest cytogenetic methodologies to be used for the biological monitoring of exposure. Exposure to antineoplastic agents offers such a rare possibility.

Both in experimental systems and in patients, many studies have shown the in vivo induction of sister chromatid exchanges (SCEs) by the alkylating cytostatic drugs and by some of the antibiotic type chemotherapeutics (Gebhart et al. 1980b; Latt et al. 1981). Studies on patients have also demonstrated that some other types of antineoplastic drugs (e.g. antimetabolites, antimitotic agents, and drugs affecting the cellular nucleotide pool) do not induce a response detectable as chromosome damage in the peripheral blood lymphocytes (Lambert et al. 1982).

The SCE method has been used for the biological monitoring of exposure to cancer chemotherapeutic drugs among hospital personnel. The SCE method was chosen because many of the drugs handled daily (e.g., cyclophosphamide, adriamycin, *cis*-platinum (*cis*-platinum (II) diamine dichloride), and CCNU(1-[2-chloroethyl-3-(4-methylcyclohexyl)]-1 nitrosourea) are known to induce SCEs both in patients and in experimental systems (Nevstad 1978; Raposa 1978; Lambert et al. 1978, 1979a,b; Banerjee and Benedict 1979; Musilova et al. 1979; Turnbull et al. 1979; Wiencke et al. 1982).

SUBJECTS AND METHODS

Blood samples were taken from 20 nurses working at three oncological units. The nurses' duties frequently included the preparation of patients' chemotherapy solutions and infusion syringes. A personal interview was carried out so that information could be obtained about the frequency of handling and about the types of cytostatics handled as well as about the health status and personal habits, including smoking, the consumption of alcohol, and the use of drugs.

Three referent groups were included in the study: five patients under chemotherapy (cyclophosphamide as the main drug); 10 nurses who worked at other units of the same hospital and were not involved with chemotherapy; and 10 office workers whose main tasks were paperwork. All of the subjects were nonsmokers, except for one of the oncology nurses and one office worker control, both of whom smoked 10-15 cigarettes daily. According to blood cell counts and the personal interviews, all the subjects except the cancer patients were healthy.

The whole blood microculture method was used. The culture time at +37°C was 68 hours, and bromodeoxyuridine (BrdU, Calbiochem, 5.0 μg/ml) was present from the beginning of the culture. Harvesting and staining according to the FPG-technique (Perry and Wolff 1974) was performed as described earlier (Mäki-Paakkanen et al. 1980). The blood samples, which were cultured in two groups, were all analyzed together by one person, on coded slides, 30-second division metaphases per subject.

RESULTS

The SCE frequencies among the patients under treatment (four of the five cultures were successful) were 4-5 times higher than the frequencies among the healthy subjects. Among the healthy subjects, the highest individual SCE frequencies were found among the nurses handling cytostatic drugs at the oncology unit where most of the patients are treated (unit C, see Table 1). Each of the two smokers (one of the control subjects and a nurse working at the oncology unit) had SCE values among the highest for the particular group (Fig. 1).

Even though the SCEs of a few control persons (hospital nurses and office workers) were also above the mean for all the healthy subjects, the distribution of the individual SCE frequencies among the oncology nurses clearly leaned towards a mean SCE value higher than for the other hospital nurses or the office workers (see Fig. 1 and Fig. 2).

The increase of the mean SCE among the oncology nurses (9.4 ± 0.3) deviated statistically significantly ($P < 0.01$, one-tailed t-test) from the mean SCE of the office personnel (8.1 ± 0.3) and all referents (8.4 ± 0.2). However, the difference was not statistically significant ($P < 0.1$) when compared with the mean SCE of the other hospital nurses (8.7 ± 0.2).

Table 1

SCEs in Peripheral Lymphocytes of Control Persons, Oncology Nurses, and Patients on Chemotherapy[a]

Group Person Number	Age (years)	Sex	Contact with cytostatics Duration (years)	Last contact (days before sampling)	Mean SCEs/cell ± S.E.	Range
Office controls						
1	41	f	—	—	7.2 ± 0.7	1-16
2	24	f	—	—	7.3 ± 0.5	2-14
3	35	f	—	—	7.4 ± 0.6	2-17
4	35	m	—	—	7.6 ± 0.5	4-13
5	27	m	—	—	7.8 ± 0.5	2-15
6	24	f	—	—	7.8 ± 0.5	3-15
7	45	f	—	—	8.0 ± 0.6	4-21
8	29	f	—	—	8.2 ± 0.6	1-16
9	35	f	—	—	9.3 ± 0.6	4-21
10[b]	34	f	—	—	9.9 ± 0.9	2-20
Mean (1-10)	33				8.1 ± 0.3	
Hospital controls						
11	28	m	—	—	7.4 ± 0.5	3-15
12	38	f	—	—	8.0 ± 0.7	2-16
13	39	f	—	—	8.0 ± 0.7	2-17
14	26	f	—	—	8.1 ± 0.5	3-15
15	32	f	—	—	8.7 ± 0.7	2-16
16	39	f	—	—	8.9 ± 0.7	3-16
17	37	f	—	—	9.1 ± 0.7	4-20
18	29	f	—	—	9.2 ± 0.7	4-20
19	41	f	—	—	9.6 ± 0.7	4-18

Table 1 *(Continued)*

Group Person Number	Age (years)	Sex	Contact with cytostatics Duration (years)	Contact with cytostatics Last contact (days before sampling)	Mean SCEs/cell ± S.E.	Range
20	40	f	—	—	9.7 ± 0.8	2-21
Mean (11-20)	35				8.7 ± 0.2	
Mean (1-20)	34				8.4 ± 0.2	
Oncology nurses						
Unit A						
21	41	f	19.0	1	8.2 ± 0.6	3-16
22	32	f	4.0	0	8.3 ± 0.7	2-17
23	37	f	12.0	0	9.4 ± 0.7	4-19
24	33	f	1.0	0	9.6 ± 0.7	5-17
25	42	f	6.0	0	9.6 ± 0.6	5-20
26	34	f	3.5	3	9.7 ± 0.8	2-20
Mean (21-26)	37		7.6		9.1^c ± 0.3	
Unit B						
27	32	f	9.0	0	7.9 ± 0.6	3-17
28	38	f	7.0	7	8.3 ± 0.6	3-17
29	42	f	12.0	0	8.4 ± 0.7	2-17
30	30	f	4.0	0	8.5 ± 0.6	3-18
31	35	f	4.5	0	9.3 ± 0.8	2-23
32	38	f	16.0	0	10.0 ± 0.9	2-22

					SCE frequency (mean ± S.E.)	Range
33	26	f	0.3	0	10.6 ± 0.7	5-21
34	23	f	0.8	0	10.7 ± 0.8	3-21
Mean (27-34)	33		6.7		9.2 ± 0.4[c]	
Unit C						
35	27	f	0.3	0	7.1 ± 0.6[d]	2-15
36	22	f	0.3	0	9.1 ± 0.8	1-19
37	33	f	2.7	42	9.5 ± 0.7	5-18
38	34	f	9.0	0	10.4 ± 1.0	3-26
39[b]	34	m	1.3	0	10.9 ± 1.0	4-22
40	38	f	17.3	0	11.8 ± 0.8	4-27
Mean (35-40)	31		5.2		9.8 ± 0.7[e]	
Mean (21-40)	34		6.5		9.4 ± 0.3[f]	
Patients						
41	75	m	3 days	1	35.8 ± 1.7	24-67
42	69	m	2 days	1	36.0 ± 2.6[g]	24-55
43	57	f	2 days	1	36.8 ± 2.2	14-66
44	68	f	3 days	1	38.4 ± 1.6	21-50
45	42	m	5 days	1	no result	
Mean (41-44)	67				36.8 ± 0.6[h]	—

[a]30 cells were analyzed per person
[b]Smokes 10-15 cigarettes per day
[c]$P < 0.05$ in one-tailed t-test as compared to office controls and all controls
[d]A second sample 3 months later showed the SCE frequency of 10.6 ± 0.8 (mean ± S.E.)
[e]$P < 0.01$ in one-tailed t-test as compared to office controls, $P < 0.05$ as compared to all controls
[f]$P < 0.01$ in one-tailed t-test as compared to office controls and all controls
[g]Only 11 cells could be analyzed
[h]$P < 0.001$ in one-tailed t-test as compared to all other groups

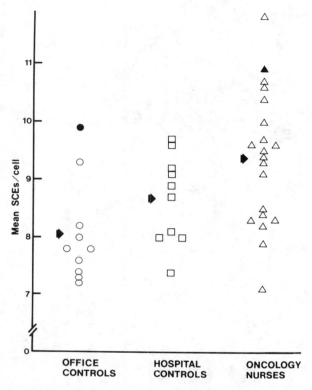

Figure 1
Individual mean number of SCEs/cell among office-worker controls, hospital controls and oncology nurses. (→) group mean. (▲) (●) subjects who smoked.

DISCUSSION

Several recent studies have confirmed that hospital personnel preparing and administering cytostatic drugs may be exposed to these drugs. This possibility for exposure was first pointed out in studies which measured the mutagenicity of the urine of oncology nurses (Falck et al. 1979). Detected by bacterial tester strains (*Salmonella typhimurium* TA 98, *Escherichia coli* WP 2 uvr A), the mutagenicity of the urine samples was higher among the oncology nurses than among the office personnel. Subsequent improvements in the work conditions at this oncology unit (instructions for handling, protective clothing, laminar flow hoods) minimized the exposure, as shown by the decreased urinary mutagenic activity (Vainio et al. 1982). Even though the method of urinary mutagenicity analysis can be considered a sensitive indicator of recent exposure, especially when the bacterial fluctuation assay is used, positive findings of increased urinary

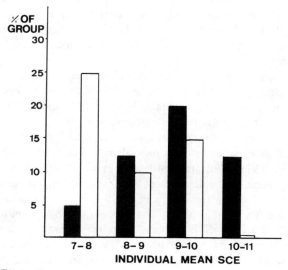

Figure 2
Distribution of individual SCE frequencies (mean of 30 cells) among nurses handling cyto-static drugs and among the two referent groups (other hospital nurses and office personnel). (■) oncology nurses; (□) controls.

mutagenicity have been observed only when cytostatics were handled frequent-ly, with poor protective measures (K. Falck, et al., unpubl. results).

Also cytogenetic methods have been used to monitor possible occupation-al exposure to cytostatics. Some of the group results reported earlier (Norppa et al. 1980; Sorsa et al. 1981) have indicated that the SCE frequencies among the oncology nurses were higher than among the referent group. Similar findings—not only an increased frequency of SCEs but also an increased fre-quency of structural chromosome aberrations (including gaps)—have been re-ported by Waksvik et al. (1981). Furthermore, the unpublished results of E. Nikula (University of Oulu, Finland) indicated a significant increase of structural chromosome aberrations among nurses who had worked for several years (mean of estimated handling time with cytostatics 2800 hours) at an oncology unit preparing parenteral antineoplastic agents.

The above findings from the monitoring of urinary mutagenicity and from cytogenetic studies thus confirm that occupational exposures to antineoplastic drugs may occur at oncology units. Mutagenicity in the urine can be used as a device to monitor exposure, whereas effect monitoring can be done by cyto-genetic methods such as SCEs or chromosomal aberrations.

The limitations of cytogenetic surveillance possibilities should be pointed out. It is obvious that both structural chromosome aberrations and SCEs are rather insensitive measures of in vivo exposure which are further complicated by partly uncontrollable technical and interindividual variability (Sorsa et al. 1982; Vainio and Sorsa 1982). Furthermore, the qualitative and quantitative definition of the true exposing agent responsible for the possibly observed positive result is rarely achievable.

Our study could not decisively single out any particular causative agent for the observed SCE increase among the oncology nurses. All of the nurses had been handling cyclophosphamide, which is quantitatively the antineoplastic drug most frequently used at these oncology units, and cyclophosphamide is known to induce SCEs in vivo (Raposa 1978; Musilova et al. 1979; Düker 1981). Positive findings of SCE induction in patient studies have also been obtained with adriamycin (Musilova et al. 1979), busulfan (Musilova et al. 1979), CCNU (Gebhart et al. 1980a), melphalan (Lambert et al. 1979b), and cis-platinum (Wiencke et al. 1982), which are less frequently handled cytostatics. Consequently, this group of cytostatics most probably contains the potential candidates for the causative agents of the SCE increase. Negative data about the capacity to induce SCEs in experimental or in patient studies is available for the following chemotherapeutics handled by the nurses: vincristine and vinblastine of the Vinca alkaloids (Lambert et al. 1978; Morgan and Crossen 1980); bleomycin (Lambert et al. 1978) and actinomycin D (Lambert et al. 1979b) of the antibiotics; and methotrexate (Lambert et al. 1978; Düker 1981) and 5-fluorouracil (Musilova et al. 1979) of the antimetabolites.

At the level of present knowledge the results of cytogenetic surveillance methods, either structural chromosome aberrations or SCEs, should be evaluated on the basis of the exposed group rather than on an individual basis (Sorsa et al. 1982; Vainio and Sorsa 1982). Some studies of patients have reported individual variability in the response to chromosome damage induced by chemotherapeutic agents (Gebhart et al. 1980a,b; Wiencke et al. 1982). Individual susceptibility may also affect the interindividual variation in the SCE responses caused by occupational exposures to cytostatics. The increase of SCEs among the oncology nurses was similar to that found for smokers studied at our laboratory (Husgafvel-Pursiainen et al. 1980). A further study should be carried out among the personnel of the oncology units in order to detect the expected decrease back to a normal SCE range once the new hygienic improvements have been put into practice at these units.

ACKNOWLEDGMENTS

Thanks are due to our colleagues Dr. Kai Falck and Ms. Hilkka Järventaus for help in arrangements to get the blood samples and Ms. Eeva Nikula, M.Sc., University of Oulu, for permission to use her unpublished data. Ms. Sheryl Hinkkanen has checked the English language and Ms. Leila Turunen typed the manuscript for which we express our gratitude.

REFERENCES

Bannerjee, A. and W.F. Benedict. 1979. Production of sister chromatid exchanges by various cancer chemotherapeutic agents. *Cancer Res.* **39**:797.

Düker, D. 1981. Investigations into sister chromatid exchange in patients under cytostatic therapy. *Hum. Genet.* **58**:198.

Falck, K., P. Gröhn, M. Sorsa, H. Vainio, E. Heinonen, and L.R. Holsti. 1979. Mutagenicity in urine of nurses handling cytostatic drugs. *Lancet* i:1250.

Gebhart, L., J. Lösing, and F. Wopfner. 1980a. Chromosome studies on lymphocytes of patients under cytostatic therapy. I. Conventional chromosome studies in cytostatic interval therapy. *Hum. Genet.* **56**:53.

Gebhart, E., B. Windolph, and F. Wopfner. 1980b. Chromosome studies under cytostatic therapy. II. Studies using the BUDR-labelling technique in cytostatic interval therapy. *Hum. Genet.* **56**:157.

Husfagvel-Pursiainen, K., J. Mäki-Paakkanen, H. Norppa, and M. Sorsa. 1980. Smoking and sister chromatid exchange. *Hereditas* **92**:247.

International Agency for Research on Cancer. 1981. Some antineoplastic and immunosuppressive agents. *IARC Monogr. Eval. Carcinog. Risk Chem. Hum.* **26**:411.

Lambert, B., V. Ringborg, E. Harper, and A. Lindblad. 1978. Sister chromatid exchanges in lymphocyte cultures of patients receiving chemotherapy for malignant disorders. *Cancer Treat. Rep.* **62**:1413.

Lambert, B., U. Ringborg, A. Lindblad, and M. Sten. 1979a. Prolonged increase of sister chromatid exchanges in lymphocytes of melanoma patients after CCNU treatment. *Mutat. Res.* **59**:295.

––––––. 1979b. The effects of DTIC, melphalan, actinomycin D and CCNU on the frequency of sister chromatid exchanges in peripheral blood of melanoma patients. In *Adjuvant therapy of cancer II* (eds. S.E. Jones and W.E. Salmon), p. 55. Grune & Stratton, New York.

Lambert, B., A. Lindblad, K. Holmberg, and D. Francesconi. 1982. Use of sister chromatid exchange to monitor human populations for exposure to toxicologically harmful agents. In *Sister chromatid exchange* (ed. S. Wolff), J. Wiley & Sons, New York. (In press).

Latt, S.A., J. Allen, S.E. Bloom, A. Carrano, E. Falck, D. Kram, E. Schneider, R. Schreck, R. Tice, B. Whitfield, and S. Wolff. 1981. Sister-chromatid exchanges: A report of the Gene-Tox Program. *Mutat. Res.* **87**:17.

Mäki-Paakkanen, J., K. Husgafvel-Pursiainen, P.-L. Kalliomäki, J. Tuominen, and M. Sorsa. 1980. Toluene-exposed workers and chromosome aberrations. *J. Toxicol. Environ. Health* **6**:775.

Morgan, W.F. and P.E. Crossen. 1980. Mitotic spindle inhibitors and sister-chromatid exchange in human chromosomes. *Mutat. Res.* **77**:283.

Musilova, J., K. Michalova, and J. Urban. 1979. Sister-chromatid exchanges and chromosomal breakage in patients treated with cytostatics. *Mutat. Res.* **67**:289.

Nevstad, N.P. 1978. Sister chromatid exchanges and chromosomal aberrations induced in human lymphocytes by the cytostatic drug adriamycin in vivo and in vitro. *Mutat. Res.* **57**:253.

Norppa, H., M. Sorsa, H. Vainio, P. Gröhn, E. Heinonen, L. Holsti, and E. Nordman. 1980. Increased sister chromatid exchange frequencies in lymphocytes of nurses handling cytostatic drugs. *Scand. J. Work Environ. Health* **6**:299.

Perry, P. and S. Wolff. 1974. New Giemsa method for the differential staining of sister chromatids. *Nature* **251**:156.

Raposa, T. 1978. Sister chromatid exchange studies for monitoring DNA damage and repair capacity after cytostatics in vitro and in lymphocytes of leukaemic patients under cytostatic therapy. *Mutat. Res.* **57**:241.

Sorsa, M., K. Falck, H. Norppa, and H. Vainio. 1981. Monitoring genotoxicity in the occupational environment. *Scand. J. Work Environ. Health* **7**(Suppl. 4):61.

Sorsa, M., K. Hemminki, and H. Vainio. 1982. Biological monitoring of exposure to chemical mutagens in the occupational environment. *Teratog. Carcinog. Mutagen.* (in press).

Turnbull, D., N.C. Popescu, J.A. DiPaolo, and B.C. Myhr. 1979. *cis*-Platinum-(II)diamine dichloride causes mutation, transformation, and sister-chromatid exchanges in cultured mammalian cells. *Mutat. Res.* **66**:267.

Vainio, H. and M. Sorsa. 1982. Application of cytogenetic methods for biological monitoring. *Ann. Rev. Public Health* (in press).

Vainio, H., K. Falck, and M. Sorsa. 1982. Mutagenicity in urine of workers occupationally exposed to mutagens and carcinogens. In *Biological monitoring of workers exposed to chemicals* (eds. A. Aitio et al.). Hemisphere Publ. Co., Washington DC (in press).

Vainio, H., M. Sorsa, J. Rantanen, K. Hemminki, and A. Aitio. 1981. Biological monitoring in the identification of the cancer risk of individuals exposed to chemical carcinogens. *Scand. J. Work Environ. Health* **7**:241.

Waksvik, H., A. Brøgger, and P. Klepp. 1981. Chromosome analyses of nurses handling cytostatic drugs. *Cancer Treat. Rep.* **65**:607.

Wiencke, J.K., J. Cervenka, B.J. Kennedy, J. Prlina, and R. Gorlin. 1982. Sister chromatid exchange induction by *cis*-platinum/adriamycin cancer chemotherapy. *Mutat. Res.* **104**:131.

COMMENTS

KLIGERMAN: Do you have any evidence in the cancer registries that the incidence of cancers for hospital personnel of any type are significantly elevated?

SORSA: No, as far as I know. I know that there are at least two studies which are being done among the nurses handling cytostatics, one in the United States, and we also have a small cohort in Finland. But the problem is the latency time. These are often the problems of epidemiological studies.

EVANS: These include a relatively high background incidence, relatively low exposure levels and the long latency period and problem of follow-up.

MOHRENWEISER: You chose your words very carefully, but you left the impression that time of exposure really related to an increased risk.

SORSA: No. I think I really chose my words carefully by not trying to say this, because I think this is something which we don't know. It just happened to be so that the person with the highest SCE frequency happened to have been working the longest time. But the numbers are so small that I think one should not draw conclusions from this about the dose or exposure time and the response. I can't discuss the risk because we cannot quantify it.

MOHRENWEISER: Did you happen to look at regression of time of exposure versus decrease of SCE frequency?

SORSA: No, there are no data, unfortunately.

EVANS: But the differences are quite small. They are statistically significant, but they are really very small.

SORSA: Yes, the differences are always very small.

EVANS: I think we have to be careful here, although, of course, in general one accepts that length of exposure is related to level of exposure in an integrated sense. Nevertheless, an apparent enhanced response to a mutagen seen in workers who have been employed for long periods need not necessarily reflect the increased cumulative exposure over staff who have worked for shorter periods. Length of service may sometimes be associated with increased laxity in handling toxic agents so that levels of recent exposure may be higher in longer service workers—and, in the case of SCE's, recent exposures may be more relevant than earlier ones.

SORSA: Unfortunately there was no possibility to have another sample from the nurses. I would expect that in one year's time, because of the improvements in the hygienic conditions, the SCEs will decrease in frequency. But there was no possibility to prove this.

CARRANO: You are using your data in the context of industrial hygiene. What do you tell the individuals?

SORSA: In this case—and this is a special case, because they are persons with training in a medical field—we informed each individual of the results, and we also told them that the highest SCE frequencies are similar to those we observed in cigarette smokers. But we also confessed that at present we do not know if there is any individual health impairment for such findings.

HEDDLE: Has anyone given consideration to finding some artificial way of enhancing the response, such as adding inhibitors of DNA repair, in an attempt to magnify the sensitivity of the assay?

SORSA: The SCE assay?

HEDDLE: Any assay.

SORSA: I don't think so.

EVANS: That is an interesting possibility, but what kind of inhibitors would you suggest should be used?

HEDDLE: Well, you might for example try heat shock or higher BrdU concentrations.

SORSA: Would it make us happier?

HEDDLE: Well, that is another question. As an index, it might, just as an index. The absolute basic number of SCEs doesn't mean anything to you.

PRESTON: I tried to do such a repair-inhibition experiment with ionizing radiation, which was not necessarily the best choice. But waiting three days after irradiating and then exposing the lymphocytes to cytosine arabinoside didn't give any increase in aberrations, probably because all the repair had taken place in a G_0 cell within that three-day period.

ALBERTINI: What was the heterogeneity and range of the SCE response in the treated cancer patients?

SORSA: Not terribly great, because all of these patients had had the last cyclophosphamide injection the previous day.

ALBERTINI: So there are none of them that are in the normal range following cyclophosphamide.

SORSA: Oh, no. They were all extremely high.
There are two other studies which I know of, which also have shown increases in chromosome aberrations among the nurses. One unpublished study comes from Finland. The other from Norway, from Waksvik and Brøgger, who found increases in SCEs and also structural chromosome aberrations.
If we are very critical about the occupational groups in which there is incontrovertible evidence about either increased structural chromosome aberrations or SCEs, the number is really small. John Evans showed us a list, but I think if one is severely critical, there are only five agents where there is clear confirmatory evidence from independently repeated studies. They would be studies on benzene, ethylene oxide, styrene, vinyl chloride, and epichlorohydrin.

EVANS: I would not disagree, for much of the published data is equivocal and some downright contradictory. However, the question often arises as to whether positive results can be really ascribed to the substance that is considered to be the occupational mutagen.

SORSA: Well, that is another problem of the occupational exposures, because people are usually exposed to many substances.

THILLY: Is it possible that a chemical mutagen will be specific with regard to the distribution pattern of induced chromosome breakage or sister chromatid exchange? If so, why not look for the "fingerprint" for diagnostic evidence?

SORSA: It is possible, and some people have been interested in this clustering possibility but the information on patterns is by no means clear-cut.

PETRAKIS: Nurses aren't really necessarily representative of the general population, and many people go into medicine and nursing because they have strong family histories of cancer and that kind of disease. If you were to go in and ask the average nurse, "Do you have a first-degree relative with cancer?" you would probably find more than you would in the office group. Would that make a difference here?

SORSA: You mean susceptibility?

PETRAKIS: Yes.

SORSA: I think our chairman can answer that.

EVANS: There are clear differences in susceptibility between different individuals, and certain genetic carriers for cancer proneness may show higher responses, but this is another topic that we have no time to discuss now as we need to move on to our next speaker.

Measuring Prenatal Genotoxic Effects in Mice and Men

ROBIN J. COLE AND LEIGH HENDERSON
Developmental Genetics Laboratory
School of Biological Sciences
University of Sussex
Falmer
Brighton
Sussex BN1 9QG
England

Some 50 chemicals have been shown to initiate malignancy in cells of prenatal mammals when administered via the mother during pregnancy (Bailar 1979, Rice 1981, Kleihues 1982). In man, cancer has become the major nonaccidental cause of death in childhood in the economically developed world (c 1 death/1.5 cases/1000 births, in the United Kingdom [Draper et al. 1982]) and although there is direct evidence implicating only two carcinogens, diethylstilbestrol (Herbst and Cole 1978) and ionizing radiation (Kneale and Stewart 1976), in the prenatal initiation of human childhood tumors, it is clear that levels of human fetal exposure to potentially genotoxic agents should be a matter of concern. The incidence of some childhood malignancies (e.g., acute lymphoid leukemia) has risen in the United Kingdom in recent years (Birch et al. 1981) but it is, of course, unknown if this is a continuing trend, or if the frequency will 'settle' at the higher, but apparently more stable levels found in Scandinavia and the United States. While substantial advances in the therapy of childhood tumors have occurred in recent years, current treatments carry risks of severely adverse effects on physical, social, and intellectual development, and may themselves increase the frequency of tumors in later life, as well as heritable germ-cell anomalies (Van Eys and Sullivan 1980).

Thus far, no chemical studied has been shown to be a transplacental carcinogen without being genotoxic in adult tissues, but in some cases (e.g., ethylnitrosourea [ENU]) high frequencies of prenatally induced tumors arise from doses < 1% of those needed to demonstrate tumorigenicity in adults (Rice 1981). Tests designed to minimize human exposure to carcinogens, but based exclusively on adult animals might therefore seriously underestimate risks from prenatal exposure to genotoxic agents.

The biological factors which distinguish fetal from adult sensitivity to carcinogenesis are complex, and often the outcome of counter-balancing phenomena. These factors also determine the developmental stage and organ specificity for effective transplacental carcinogenesis, which may be very precise, so that only a narrow 'window' is presented for experimental studies. In general, maximum sensitivity to carcinogens occurs in the period following organo-genesis, i.e., after the period of maximum sensitivity to teratogenic agents. While some activating enzymes, such as arylhydrocarbon hydroxylase, are present from early embryonic stages (Galloway et al. 1980) other pathways needed to activate procarcinogens; for example, the cytochrome P-450 based 'mixed function oxidases' develop relatively late in gestation, and often have not reached maximal levels by birth. On the other hand, enzymes linked to these pathways may detoxify potential carcinogens, and so, in the absence of excretory mechanisms, late development of some enzyme systems may enhance fetal sensitivity.

In the case of direct-acting agents, balance between maternal and fetal effects will largely be determined by transport and relative binding of the agent. Where metabolic activation is required, the fetus is at risk if the products are relatively stable, even without endogenous activation potential. If the ulti-mate genotoxic form is very unstable, then fetal cells are only vulnerable if they, or their near neighbors, possess activation potential. In some cases maternal-fetal relativities can be modified by genetically determined effects on activation, e.g., the *Ah* locus, which regulates metabolism of polycyclic aromatic hydro-carbons. In this situation, a "responsive fetus" is at greatest risk in a "non-responsive" mother, who is herself relatively protected (Shum et al. 1979). Patterns of cell migration may present hazards to particular cell types in the fetus which are avoided in postnatal life. Adult bone marrow is unable to activate nitrosamines efficiently, and, since the metabolic products of these carcinogens are very short-lived, primitive blood-forming cells in the bone marrow are protected from their effects. However, significant ability to activate nitrosamines appears in the mouse fetal liver while this organ still contains substantial numbers of long-lived primitive blood cells, which then colonize bone marrow and provide a pool of 'stem-cells' for life, so that nitrosamine-induced leukemogenesis could more likely result from transplacental exposure. The clastogenic activity of diethylnitrosamine (NDEA) is easily demonstrated transplacentally in mouse fetal liver (Cole et al. 1982a).

The high frequency of dividing cells, and the short cell-cycle times charac-teristic of fetal organs also enhance risks of primary DNA lesions and the probability of their 'fixation'. In adults, the kinetic organization of self-renewing cell populations into 'stem-cell', 'progenitor cell' and 'mature' (often post-mitotic) compartments, and, at least in epithelial tissues, their physical organ-ization as well (Potten et al. 1979) limit risks from malignant transformation (Cairns 1975), but in the fetus these systems are still being organized and variant cell clones have a much greater chance of becoming established. Because only a small proportion of fetal cells may be vulnerable to initial genotoxic effects at particular developmental stages the significance of correlations between DNA repair capacity and tumorigenic potential have been difficult to assess (Kleihues 1982). The extraordinary sensitivity of the fetal rat brain to carcinogenesis by

ENU has been associated with the low rate of excision of O^6-ethylguanine, relative to that found in 'resistant' organs, such as the liver (Goth and Rajewsky 1974) but definitive studies on this relationship require observations at the single-cell level (see Rajewsky, this volume).

The most frequent sites of prenatally induced tumors in rodents are lung, brain, and spinal cord. In man, childhood tumors of the nervous system are only slightly less frequent than the most common cases, which are found in the lymphoid and blood-forming systems (but lung tumors are very rare). Short-term cellular tests for potentially genotoxic effects of chemicals, currently applicable in vivo to the fetal rodent tissues of particular interest (i.e., brain, lung, and fetal liver, which is the major site of prenatal blood formation), are analyses of chromosome aberrations (or, in the case of fetal liver, the micronucleus test) and sister-chromatid exchange (SCE). These tests can also be applied to human cord lymphocytes, obtained at birth, and present an opportunity to monitor long-lasting effects of prenatal exposure in man. It is also possible to measure gene-mutational endpoints in these cells. Together, these techniques provide a useful package to complement epidemiological analyses of the role of prenatal exposure to genotoxic agents, in the initiation of human disease.

METHODS

We have used random-bred Swiss-albino mice ('Porton strain') from day 16 of gestation. The performance of the transplacental micronucleus test, (which examines polychromatic erythrocytes in the fetal liver or in the circulation of fetal or neonatal mice) and the rationale behind choice of time courses etc., are described by Cole et al. (1981, 1982a). To determine SCE levels in tissues of fetal mice after transplacental exposure, cells were explanted into tissue culture for two cell cycles in the minimum concentrations of bromodeoxyuridine (BrdU), consistent with clear sister-chromatid differentiation. Erythropoietin was used to maintain cell proliferation in erythroblasts, and colony stimulating activity (prepared from pregnant mouse uterus) to maintain proliferation of granulocyte-macrophage progenitor cells (Cole et al. 1982b). Fetal calf serum provided the growth factors necessary for fetal lung and fetal brain cells. Only fetal lung required trypsinization to provide cell suspensions; other tissues were dissociated by gentle pipetting. Cultures were harvested for analysis of second metaphases at the time of their maximum representation in vitro (fetal erythroblasts, c 16 hours; granulocytes, c 28 hours; fetal brain and lung, c 72 hours); optimal in vivo exposure times were determined for each agent and cell type (Cole et al., unpubl. results). Human lymphocytes were obtained from the clamped umbilical vessels shortly after birth. Frequencies of micronucleated cells were determined after PHA-stimulation of Ficol-hypaque density-gradient enriched lymphocytes. Cells were exposed to very mild hypotonic treatment, so that cytoplasmic integrity was maintained and micronucleus frequencies determined from intact cells only. To determine frequencies of 6-thioguanine-resistant (TG^r) lymphocytes, PHA-stimulated whole blood cultures were exposed to 6-thioguanine in the presence of BrdU. After 72 hours (including 12 hours of metaphase arrest induced by colchicine) cultures were harvested. The numbers of 1st, 2nd, 3rd, etc., metaphases, relative to those in parallel control cultures were determined and the frequency of TG^r lymphocytes in the original blood sample was calculated. SCE frequencies in PHA-stimulated cord lymphocytes were determined by standard techniques.

RESULTS AND DISCUSSION

Animal Studies

The micronucleus test is a particularly convenient method to compare maternal and fetal effects of clastogenic agents (Fig. 1). The responses to ionizing radiation in fetal liver and maternal bone marrow are similar but the responses to the six chemicals tested comprehensively thus far (as part of the collaborative environmental program supported by the Commission of the European Communities) show considerable differences. NDEA, which is not effective in maternal bone marrow, is clastogenic in fetal liver, on and after day 16 of gestation, and this is consistent with observations on the development of fetal

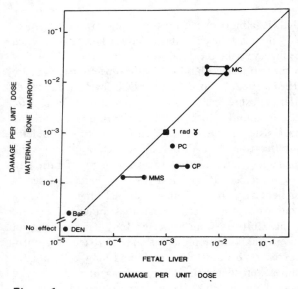

Figure 1
Relationships between induction of micronuclei in mouse maternal bone marrow and fetal liver erythroid cells by γ radiation (11 rad/min from a CS^{137} source) MC, PC, CP, MMS, B[a]P, and NDEA. Damage/unit dose is the frequency of induction of damage leading to a micronucleated polychromatic erythrocyte/μmole/kg. A range of values indicates that the dose response was nonlinear.

ability to metabolize this carcinogen (Cole et al. 1982). Of the other agents, cyclophosphamide (CP) shows the greatest difference between maternal and fetal response, being about 10 times more effective in the latter, whereas mitomycin C (MC), and benzo[a]pyrene (B[a]P), are both more effective in maternal bone marrow.

The ability to use differential cell explantation techniques, specific growth stimulators, and to manipulate experimental protocols in the light of cell-kinetic data, allow a more confident identification of specific cell types in the in vivo in vitro SCE technique we have developed (Cole et al., unpubl. results) than is possible in the wholly in vivo technique utilized by other authors in transplacental studies (Kram et al. 1980; Allen et al. 1981). This has enabled us to test the degree of correlation between induction of micronucleated cells, and induction of SCEs in transplacentally exposed prenatal erythroblasts. Spontaneous SCEs (which must include some induced by the BrdU used to visualize them), at a frequency of c 8/fetal erythroblast, occur some 2,000 times more frequently than the 'spontaneous' chromosomal events which cause micronuclei (c 0.4% of fetal erythroblasts 'spontaneously' give rise to micronucleated descendants). For five of the six tested compounds, there is a fairly high correlation, at a 100-fold difference, in their ability (on a μmole/kg basis) to induce SCEs and micronuclei, but procarbazine (PC) is significantly less effective as an inducer of SCEs than as a clastogen (see Fig. 2). The metabolic products of PC include hydrogen peroxide, formaldehyde, formylhydrazine, and N-hydroxy methyl derivatives, so this resemblance to the genotoxicity pattern of

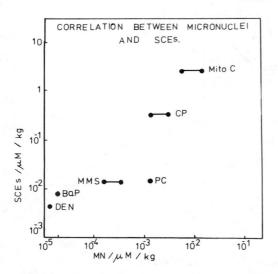

Figure 2
Relationships between induction of micronuclei and induction of SCEs in mouse fetal liver erythroblasts.

ionizing radiation is of interest. As a hydrazine derivative it represents a class of chemicals of considerable environmental importance, especially as its geno-toxicity is difficult to demonstrate in in vitro systems dependent on exogenous metabolic activation.

All the agents studied here are proven carcinogens and their relative 'oncogenic potencies' have been established in adult animals, and, in some cases, after transplacental exposure, so the predictive value of these short-term assays for potential carcinogens can be critically evaluated. We have used the 'oncogenic potency' data prepared by Clive et al. (1979) (to test carcinogenicity in adult mice and rats, with mutagenicity in vitro), as the basis for our compari-son (Fig. 3). The transplacental oncogenic potency of B[a]P (induction of lung adenomas after maternal i.p. injection) is c 1/50 that in adults. The transpla-cental oncogenic potency of NDEA during day 16 of gestation (5×10^{-4} tumor-bearing (lung adenoma and leukemia) offspring/μmole/kg,) is within the range observed for adult rodents. The transplacental oncogenic potency of methyl methane sulphonate (MMS) (3×10^{-4} tumor-bearing offspring/μmole/kg,) and PC (1×10^{-3}/μmole/kg), are also near the mean of the ranges ob-served in adult rodents. (Data from Tomatis 1979). With MC, MMS, and CP, micronucleated cells and SCEs in the fetal liver correlate closely with 'oncogenic potency'. In the case of PC there is a good correlation with induction of micro-nucleated cells, but less good for SCE induction. Access of B[a]P to the fetus

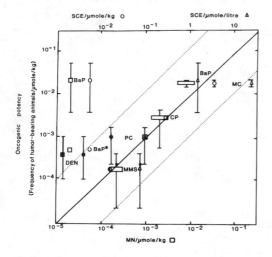

Figure 3

Relationships between oncogenic potency and (□) induction of micronuclei and (○) SCEs in erythroblasts in mouse fetal liver in vivo. (*) oncogenic potency in transplacental experi-ments (Tomatis 1979), otherwise the values presented by Clive et al. (1979), from observa-tions on adult mice and rats are used; (△) induction of SCEs by MMS, MC, and B[a]P by direct exposure in organ cultures of fetal liver.

after maternal exposure is probably limited, so induction of SCEs and micro-nucleated cells in fetal liver in vivo and its transplacental carcinogenicity are low. Comparisons of the ability of B[a]P, MC, and MMS to induce SCEs in directly-exposed fetal liver explants show that in these conditions the SCE-inducing capacity of B[a]P correlates more closely with its higher 'oncogenic potency' in adults, while the relative positions of MC and MMS are only slightly changed.

We have also used the in vivo in vitro SCE technique to compare induction of SCEs by PC and CP in maternal and fetal granulocyte progenitors, and fetal erythroblasts, lung, and brain cells (Cole et al., unpubl. results) (Fig. 4). There are cell-type specific differences in the response to both chemicals. CP induced fewest SCEs in fetal brain, and most in fetal erythroblasts, while PC was least effective in fetal erythroblasts and most effective in granulocyte progenitors.

Human Studies

The use of peripheral lymphocytes for analyses of exposure to genotoxic chem-icals in humans is well established (Natarajan and Obe 1980) and is particularly appropriate for monitoring prenatal exposure, since lymphocytes formed during fetal life are easily obtained from the placenta at birth and, when activated in vitro, will express lesions initiated prenatally. As long-lived circulating cells they will be exposed to unstable genotoxic metabolites produced in fetal organs and the placenta, as well as those present in the blood, and themselves have con-siderable potential for metabolic activation. We have examined frequency of micronucleated cells, frequency of TG^r lymphocytes, and SCE frequencies, in samples of cord blood obtained at term. The optimum time to harvest PHA-

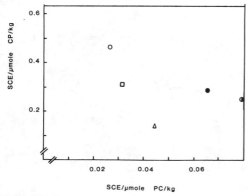

Figure 4
Induction of SCEs by PC and CP in fetal and maternal blood cells, and fetal brain and lung cells. (○) fetal erythroblast; (△) fetal brain; (●) adult GM cell; (□) fetal lung; (◐) fetal GM cell.

stimulated cord lymphocytes to measure micronucleated cells was fixed by following cultures set up from γ-irradiated whole blood (200 rads). The most reliable results are obtained after 96 hours in vitro. Using this technique the dose required to double the spontaneous frequency is c 8 rads γ radiation, and 1.5 μg/ml bleomycin. The frequency distribution for micronucleated lymphocytes from 24 newborns is shown in Figure 5. The mean frequency is 0.34 ± 0.06%. Whether or not the outlying values have any medical significance has still to be determined.

We have measured the frequency of TG^r lymphocytes in cord blood by a technique which relies on PHA-stimulated 'variant' cells progressing to metaphase. This technique, if successful, is more likely than the autoradiographic technique of Albertini (1979) to be applicable to other selective systems, although the latter has been substantially validated (e.g., Evans and Vijayalaxmi 1981). Use of selective agents such as methotrexate, ouabain, cytosine arabinoside, high cAMP, etc., could, in principle, detect a range of mutagenic lesions. Automated microscopy techniques for detection of metaphase preparations are already available, and even without automation, metaphases are easily measured. A toxicity curve for TG, determined by this 'metaphase' method is shown in Figure 6. The frequency of variants resistant to TG $\geqslant$ 1 × 10^{-4}M is c 4.5 × 10^{-5}. Although the number of infants studied is so far very limited it is of interest that this is the birth-value to which the age-dependent effect noted by Evans and Vijayalaxmi (1981) extrapolates. We have investigated the hypoxanthine guanine phosphoribosyl transferase (HGPRT) status of TG^r variant

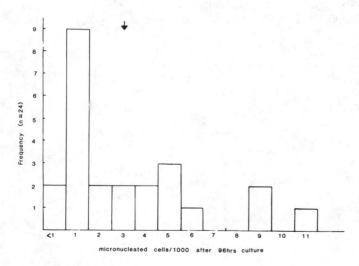

Figure 5
The frequency of micronucleated lymphocytes after PHA-stimulation of human cord blood. Data from 24 individual infants are shown (mean 0.34 ± 0.06%).

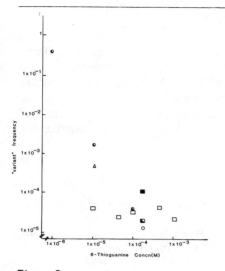

Figure 6

TG toxicity curve based on the ability of PHA-stimulated human cord lymphocytes to proceed through one or more metaphases. Each symbol represents an individual infant.

lymphocytes, by assessing autoradiographically, at metaphase, previous incorporation of ^{3}H-hypoxanthine. Some metaphases selected by TG showed no grains, others substantially less heavy labeling than parallel controls, thus providing preliminary evidence that the cells proceeding to metaphase in selective conditions are indeed HGPRT variants.

The SCE technique has been widely applied to human lymphocytes and is generally accepted as providing a sensitive indicator of interactions between genotoxic agents and chromatin. In a study of 50 births we have found the mean SCE/cell/individual to be 4.17 ± 0.13 (range 2.9-6.6; 20 cells/individual). Cells with > 10 SCEs were rare (c 1.0% of all cells studied) and only one individual has been found with a lymphocyte with > 15 SCEs.

Since these studies were developed sequentially we have as yet little information on correlations between the different endpoints in individual humans. This study is now in progress, together with measurements on maternal lymphocytes, and will be linked to epidemiological assessments of individual pregnancies and their outcome. Cellular defense mechanisms impose severe limitations on the numbers of genetically damaged cells which can proceed through overt malignant transformation to form tumors. Short-term in vivo genetic tests, which in effect detect changes which predispose cells to malignant transformation are therefore inherently more responsive to genotoxic agents than are assays based on whole organisms, as well as offering savings in time and cost. However, in studies of transplacental effects where cell specificity is high, the nature of the cells harvested for evaluation is obviously important.

ACKNOWLEDGMENTS

Data in this report were provided by Zeinab Aghamohammadi, Jane Cole, Terry Regan, Natalie Taylor, and John Whittaker. Our work is supported by the Medical Research Council and the Commission of the European Communities Environmental Programme. Erythropoietin was provided through the Blood Resources Programme, U.S.D.H.E.W. We are grateful for the collaboration of staffs of the Royal Sussex County Hospital and Royal Alexandra Hospital for Sick Children, Brighton.

REFERENCES

Albertini, R.J. 1979. Direct mutagenicity testing with peripheral blood lymphocytes. *Ban. Rep.* **2**:359.

Allen, J.W., E. El Nahass, M.K. Sanyal, R.L. Dunn, B. Gladen, and R.L. Dixon. 1981. Sister chromatid exchange analysis in rodent maternal, embryonic and extra embryonic tissues; transplacental and direct mutagen exposures. *Mutat. Res.* **80**:297.

Bailar, J.C., Ed. 1979. Perinatal Carcinogenesis. *Natl. Cancer Inst. Monogr.* **51**.

Birch, J.M., R. Swindell, M.B. Marsden, and P.H. Morris Jones. 1981. Childhood leukemia in N.W. England 1954-1977. Epidemiology, incidence and survival. *Br. J. Cancer* **43**:324.

Cairns, J. 1975. Mutation, selection and the natural history of cancer. *Nature* **255**:197.

Clive, D., K.O. Johnson, J.F. Spector, A.G. Batson, and M.J. Brown. 1979. Validation and characterisation of the L5178Y/TK^{+}/$^{-}$mouse lymphoma mutagen assay system. *Mutat. Res.* **59**:61.

Cole, R.J., N.A. Taylor, J. Cole, and C.F. Arlett. 1981. Short-term tests for transplacentally active carcinogens. Micronucleus formation in fetal and maternal mouse erythroblasts. *Mutat. Res.* **80**:141.

Cole, R.J., N.A. Taylor, J. Cole, L. Henderson, and C.F. Arlett. 1982a. Short-term tests for transplacentally active carcinogens. Sensitivity of the transplacental micronucleus test to diethylnitrosamine. *Mutat. Res.* **104**:165.

Cole, R.J., J. Cole, L. Henderson, N.A. Taylor, C.F. Arlett, and T. Regan. 1982b. Short-term tests for transplacentally active carcinogens. A comparison of sister chromatid exchange and the micronucleus test in mouse fetal liver erythroblasts. *Mutat. Res.* (in press).

Draper, G.J., J.M. Birch, J.F. Bithel, L.M. Kinnier-Wilson, I. Leck, H.B. Marsden, P.H. Morris-Jones, C.A. Stiller, and R. Swindell. 1982. *Childhood Cancer in Britain. Incidence Survival and Mortality.* Her Majesty's Stationery Office, London.

Evans, H.J., and Vijayalaxmi. 1981. Induction of 8-azaguanine resistance and sister chromatid exchange in human lymphocytes exposed to mitomycin C and x-rays *in vitro. Nature* **292**:601.

Galloway, S.M., P.E. Perry, J. Meneses, D. Nebert, and R. Pedersen. 1980. Cultured mouse embryos metabolize benzo[*a*]pyrene during early gestation. Genetic differences detectable by sister-chromatid exchange. *Proc. Natl. Acad. Sci. U.S.A.* **77**:3524.

Goth, R. and M.F. Rajewsky. 1974. Resistance of O^6 ethyl guanine in rat brain DNA. Correlation with nervous system specific carcinogenesis by ethyl-nitrosourea. *Proc. Natl. Acad. Sci. U.S.A.* **71**:639.

Herbst, A.L. and P. Cole. 1978. Epidemiologic and clinical aspects of clear cell adenocarcinoma in young women. In *Intrauterine exposure to diethyl-stilbestrol in the human* (ed. L. Herbst), p. 2. American College of Obstetrics and Gynecology, Chicago, Illinois.

Kleihues, P. 1982. Developmental carcinogenicity. In *Developmental toxicology* (ed. K. Snell), p. 211. Croom Helm, London, England.

Kneale, G.W. and A.M. Stewart. 1976. Mantel-Haenszel analyses of Oxford data I. Independent effects of several birth factors including fetal irradiation. *J. Natl. Cancer Inst.* **56**:879.

Kram, D., G.D. Bynum, G.C. Senula, C.B. Bickings, and E.L. Schneider. 1980. In utero analysis of sister chromatid exchange. Alterations in susceptibility to mutagenic damage as a function of fetal cell type and gestational age. *Proc. Natl. Acad. Sci. U.S.A.* **77**:4784.

Natarajan, A.T. and G. Obe. 1980. Screening of human populations for mutations induced by environmental pollutants. Use of human lymphocyte system. *Ecotoxicol. Environ. Safety* **4**:468.

Potten, C.S., R. Schofield, and L.G. Lajtha. 1979. A comparison of cell replacement in bone marrow, testis and three regions of surface epithelium. *Biochim. Biophys. Acta* **560**:281.

Rice, J.M. 1981. Effects of prenatal exposure to chemical carcinogens and methods for their detection. In *Developmental Toxicology* (ed., C.A. Kimmel and J. Buelkesam), p. 191. Raven Press, New York.

Shum, S., N.M. Jensen, and D.W. Nebert. 1979. The murine Ah locus. In utero toxicity and teratogenesis associated with genetic differences in benzo[a]-pyrene metabolism. *Teratology* **20**:365.

Tomatis, L. 1979. Prenatal exposure to chemical carcinogens and its effect on subsequent generations. *Natl. Cancer Inst. Monogr.* **51**:159.

Van Eys, J. and M.R. Sullivan. 1980. (eds.) *Status of the curability of childhood cancers*, Raven Press, New York.

COMMENTS

KLIGERMAN: At what stage in the fetal liver is the erythropoietin effective in causing stimulation?

HENDERSON: Up to about day 16, and after that you still get a few cells responding, but not too many, whereas before that the response is extremely high.

HEDDLE: First, a comment. I know you are using someone else's carcinogenicity data, but I think it is asking a bit much of an assay in one tissue and given by one route to correlate very well with carcinogenic potency measured, often, by other routes and in other tissues. But when I looked at the data there, I wondered whether you would say that the SCE correlated significantly better than the micronuclei did as far as carcinogenic potency was concerned.

HENDERSON: I was a bit concerned to say that. But the answer is yes in all cases except for PC, where obviously the micronuclei data did correlate better than the SCE. However, I am aware of the limitations of that sort of correlation, so I have no wish to emphasize it.

SORSA: Did you check for the spontaneous level of SCEs in the cord blood as compared to the mothers?

HENDERSON: No. That is what we are doing now but I don't have any data.

SORSA: I would expect that the SCE frequencies of young children, at least, are a lot lower than those of adults, which may depend on the difference in the proportion of B- and T-cells in the lymphocyte populations.

HENDERSON: Yes, that is right. There is a paper published which compares the newborns from smoking and nonsmoking mothers, where there is no difference between the babies, but there is a difference between the mother and the baby populations (Ardito et al. 1980).

CARRANO: There is also a paper that just came out by Norma Hatcher and Ernie Hook which compared SCE frequencies in the cord blood versus neonatal blood, and showed that cord blood had higher levels.

References

Ardito, G., L. Lamberti, E. Ansaldi, and P. Ponzetto. Sister-chromatid exchanges in cigarette-smoking human females and their newborns. *Mutat. Res.* 78:209.

Micronuclei and Related Nuclear Anomalies as a Short-term Assay for Colon Carcinogens

JOHN A. HEDDLE, DAVID H. BLAKEY, ALESSANDRA M. V. DUNCAN,
MARK T. GOLDBERG, H. NEWMARK, MICHAEL J. WARGOVICH,
AND W. ROBERT BRUCE
Ludwig Institute for Cancer Research
Toronto, Ontario M4Y 1M4
Canada

In Canada, as in other countries of the Western world, the colon is a major site of cancer, whereas in Japan and many developing countries the age-specific rates are much lower (Doll and Peto 1981). Studies of migrant populations by Haenszel and Kurihara (1968) show that the reason for this difference is not the genetic susceptibility of the populations but rather is environmental (i.e., anything other than the genetic susceptibility). Unfortunately, the exact nature of this environmental difference is unknown. Of the various factors that have been studied, dietary factors are those most strongly associated with colon cancer (Armstrong and Doll 1975; Higginson and Muir 1977; Wynder and Gori 1977). These studies show that the mortality from colon cancer is positively correlated with the consumption of some foods and nutrients, such as meat, and negatively associated with others, such as cereals. Since the presence of apparently harmful components of the diet is inversely correlated with the presence of apparently protective components, these correlations could, in principle, be explained in two ways: First, there could be some factor associated with the consumption of cereal that is protective against a ubiquitous carcinogen; second, there could be some factor associated with the consumption of meat that is carcinogenic. Regardless of whether the first or second explanation is correct, or a combination of the two, it should be possible to identify one or more carcinogens, presumably in the diet, and, possibly, one or more anticarcinogens as well.

The strong correlation between the carcinogenic and mutagenic properties of chemicals (Ames et al. 1973) suggests that short-term assays for genotoxicity could be used in such studies. Numerous attempts are now being made to use such assays to classify environmental contaminants, to detect occupational hazards, and to screen drugs and other man-made chemicals to which people are exposed (c.f. Heddle 1982). Unfortunately, none of the existing in vivo nor in vitro short-term assays for carcinogenesis is entirely satisfactory for our purpose. In vitro assays do not reflect important aspects of the uptake, distribution, metabolism, and excretion of chemicals; and in vivo assays, such as the bone marrow micronucleus and sperm abnormality assays, each involve the use of only one tissue, and so often fail to detect tissue-specific carcinogens that are active in other tissues (e.g., Tates et al. 1980). In addition, all assays suffer from our lack of knowledge concerning which genetic endpoints are relevant to carcinogenesis or promotion (Cairns 1981). We report here preliminary results with a new assay which, we believe, bypasses some of these defects with respect to colon cancer.

Since cancers of the large bowel arise primarily in the epithelial cells that line the colon, we decided to develop a short-term assay in this cell population and thus to avoid the problem of tissue specificity, or even to turn it to our advantage. Of the many possible genetic events that could be measured, micronuclei seemed to be the easiest. It has been accepted for some time that micronuclei are a rapid index of chromosomal damage in the bone marrow and in many other dividing cell populations. In these populations, micronuclei arise from acentric chromosomal fragments that, lacking a spindle attachment site, are often outside the reforming nuclei of the daughter cells (Heddle and Carrano 1977). In addition, a few chemicals are known that induce micronuclei as a result of the disruption of cell division. In such instances the micronuclei arise from whole chromosomes or groups of chromosomes and thus are larger than those arising from chromosomal fragments (Schmid 1975; Yamamoto and Kikuchi 1980). Although micronuclei are ordinarily cell-lethal events, they arise from damage to DNA that also leads to viable chromosomal aberrations, gene mutations, and possibly to altered gene expression. Thus, micronuclei should be an index of the carcinogenic exposure.

MATERIALS AND METHODS

All of the results below were obtained on C57BL mice, although we have also made some observations on other strains of mice and on human colonic cells. The mice were supplied with food and water ad libitum and were housed under a 12-hour light and dark cycle with constant temperature and humidity. Three methods of preparing colonic epithelium have been used. In the first method, standard histological sections of formalin-fixed, paraffin-embedded colons have been used. For most of the work a standard Feulgen stain with a fast-green counterstain has been found to be suitable. This technique enables one to determine where the damage occurred within the crypts of Lieberkuhn. These crypts, which are the structural and functional units within the epithelium, contain both proliferating and nonproliferating cell populations. Micronuclei would be expected to arise only in the proliferating part of the crypt but to appear later in the nonproliferating cell population which is proximal to the lumen of the colon, as these cells are replenished from the proliferating pool.

The second method of sample preparation has been the isolation of the crypts as a suspension which is essentially free of other tissue, followed by conversion to a single cell suspension by passage through a needle, and then preparation of slides by standard cytogenetic techniques. These slides can also be stained satisfactorily by the Feulgen reaction and counterstained with fast-green. Cells in such preparations can be scored much more rapidly than they can be scored in sectioned material, but the anatomical location of the damage within the crypt cannot be determined.

The third technique which we have used is fixation and staining of small pieces of the colon, again by the Feulgen reaction. After staining these are cut into small pieces and squashed out between a coverslip and the microscope slide. This preparation can then be made permanent by the method of Conger and Fairchild (1953). This method preserves some of the morphological information if the squashing is not carried too far, but its main advantage is that it is the quickest of the methods.

RESULTS AND DISCUSSION

Characteristics of the Assay

These recent results must be regarded as preliminary since we have been determining the characteristics of the assay and refining the techniques simultaneously. Consequently, a number of interesting results have not been replicated under identical conditions. Nevertheless, the general characteristics of the assay are now evident, together with some unresolved questions concerning it. As expected, micronuclei are observed following whole-body irradiation with Cs^{137} gamma rays. The overall time-course is more or less as expected: Micronuclei increase for 1-2 cell cycles (20-30 hours) and then return to control levels as cells containing micronuclei are lost from the population. There were, however, three factors which came as a surprise:

1. Although the incidence of micronuclei as a function of time seemed to rise and then fall, as expected (see Fig. 1), experiments with more frequent sampling suggest that there may be more rapid changes than were expected;
2. the frequency of micronuclei declined almost to control levels by 72 hours post-irradiation although the lifetime of cryptal cells is about 4 days (Chang and LeBlond 1971);
3. many cells contained structures which resembled micronuclei, but which were atypical. These atypical structures were micronuclei that appeared to be in vacuoles and clusters of micronuclei, some of which occurred in cells which did not contain a normal nucleus.

Together these observations have led us to inquire what fraction of the events that we have been scoring have arisen by the "classical" mechanism of micronucleus formation, i.e., from acentric fragments or whole chromosomes that have not been distributed to the sites of the reforming nuclei during cell division. Indeed, two other explanations for these structures in the colonic

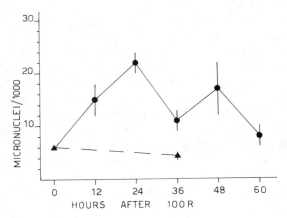

Figure 1
Incidence of typical micronuclei (per 1000 cells) as a function of time after 100R whole-body irradiation, as observed in preparations made from suspensions of colonic epithelium. 1000 cells/animal were scored; 5 animals/point.

epithelium have been proposed by others who have observed them previously. One of these, karyorrhexis, is nuclear disintegration involving the subdivision of the nucleus of the cell into a number of smaller bodies (Maskens 1979). The other mechanism that has been suggested, apoptosis, involves phagocytosis of one cell by another accompanied by digestion of its nucleus (Kerr et al. 1972). Both of these mechanisms differ from "classical" micronucleus formation by the fact that micronuclei can arise without cell division. Maskens (1979) has, however, emphasized the association between mitotic activity and the frequency of karyorrhexis. This association is also evident in our observations which show that both typical and atypical micronuclei are most frequent in the proliferating cells at the bottom of the crypt. This association with proliferation is, of course, just what would be expected for micronuclei arising by fragment loss during mitosis, so that our results are not necessarily incompatible with the classical mechanism. The unexpectedly rapid increase in the frequency of micronuclei in the colonic epithelium, for example, could arise in several different ways. The most obvious is that there is more extensive chromosomal damage in the cells closest to mitosis than in other cells, as is the case in several cell populations. A second possibility is that there is a subpopulation of cells in the epithelium that divides very rapidly and thus experiences fragment loss early after treatment. A third possibility is that the fraction of chromosomal fragments that are lost, which is thought to be 20% or so in other systems (Carrano and Heddle 1973), is higher in the colonic epithelium. This would lead to a more rapid rise and fall in micronucleus frequency. Experiments designed to discriminate amongst these possibilities are now in progress. Regardless of the mechanisms involved, however, we think the assay is measuring consequences of DNA damage in the colonic epithelium. Accordingly we now think that scoring all nuclear anomalies (i.e. both typical and atypical micronuclei) is justified without distinguishing between them.

The results obtained in irradiated mice are remarkably similar to the results obtained in mice treated with the colon carcinogen 1,2-dimethylhydrazine, in both cases at doses well below the LD_{50}. While this finding does not help to explain the origin of the clusters of micronuclei, it does support our intended use of the assay as a predictor of colon carcinogenesis. Also, the fact that both the clusters of micronuclei and the micronuclei in vacuoles occur after irradiation strongly suggests that they arise from DNA damage and, hence, are an index of it. Our results suggest that the index is a relatively reliable one when the frequency of micronuclei is not changing rapidly, such as at 24 hours after treatment; at other times, such as at 12 hours, the results are more variable (Fig. 2). Although the variation is greater than that expected from sampling alone, as shown by the deviations from the solid line, at 24 hours most of the variation seems to be from sampling rather than from animal-to-animal variation. Hence, while continuing to investigate the nature and origin of the cytological events seen in treated colonic epithelium, we have begun experiments designed to test whether or not the assay will be useful as an early index of carcinogenesis.

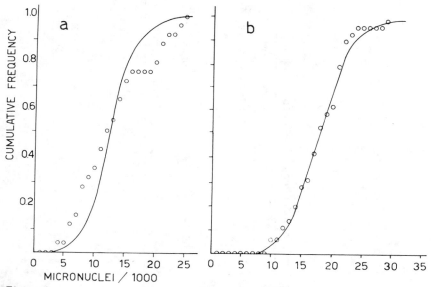

Figure 2
Cumulative frequency of animals with various numbers of typical micronuclei; (0) observed and expected for a Poisson Distribution with the same mean (–) as measured by the cell suspension technique (1000 cells/animal). (Panel A) 12 hr after 100R; (Panel B) 24 hr after 100R.

Table 1
Compounds Tested for Induction of Aberrations (Results)

Noncarcinogen	Noncolon carcinogen	Colon carcinogen
I Hydrazine (–)	1,1-Dimethylhydrazine (–)	1,2-Dimethylhydrazine (+)
II Nitrosoproline (–)	N-nitrosodimethylamine (–)	N-nitrosobis (2-oxopropyl)amine (+)
III 2-Aminobiphenyl (–)	4-Aminobiphenyl (–)	3,2'-Dimethyl-4-aminobiphenyl (+)
IV Benzo[e]pyrene (–)	Benzo[a]pyrene (+)	3-Methylcholanthrene (+)

Verification of the Assay

As outlined earlier, our rationale for the development of this assay is to investigate the human diet in hopes of discovering and then controlling the factors responsible for the country-to-country correlation between diet and colon cancer. There are two prototypical explanations for this correlation: variation in the level of dietary carcinogens from one population to another, variation in the level of anticarcinogens, or both. Hence an ideal assay must respond specifically to colon carcinogens and the magnitude of the response must be modified by protective agents. To test the specificity of the response of the colonic micronucleus assay to colon carcinogens, we have tested four structurally diverse colon-specific carcinogens and two structural analogues of each. In each case, one of the analogues is believed to be noncarcinogenic and the other analogue is carcinogenic, but not for the colon. The results that have been obtained are summarized in Table 1. As can be seen, the four colon carcinogens are positive (i.e., they showed a dose-related increase in nuclear abnormalities), the four noncarcinogens are negative, but one of the four carcinogens thought to be inactive in the colon was positive. This result for benzo[a] pyrene, while not in perfect accord with the data on carcinogenicity, is not surprising if the colon micronucleus assay is more sensitive than the cancer bioassay. The magnitude of the response after small doses of ionizing radiation or of 1,2-dimethylhydrazine suggests that this is the case. Thus it may be that the carcinogen thought to be inactive in the colon is merely weaker in the colon than at other sites and is weaker than the active colon carcinogen or has a different ratio of carcinogenic events to micronuclei.

A second test of the colon micronucleus assay is based on the work of Wattenberg (1975) who found that a 0.5% dietary supplement of disulfiram prevents formation of tumors produced by 1,2 dimethylhydrazine. We have obtained a similar inhibition of 1,2 dimethylhydrazine-induced nuclear anomalies (Fig. 3) produced by a 1% dietary supplement. This result, together with the dose-response curves we have obtained, indicates that the assay responds quantitatively to the carcinogenic damage induced by a particular colon carcinogen.

CONCLUSION

In a series of experiments we have investigated the usefulness of measuring cytological events in the colonic epithelium as an index of carcinogenesis. Although our results are preliminary, to date they indicate that the frequency of micronuclei and related nuclear anomalies is correlated qualitatively and quantitatively with carcinogenesis in the same cell population.

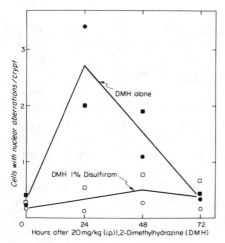

Figure 3
Frequency of nuclear anomalies measured on histological sections of colons of mice on diets (○, □) with or (●, ■) without a supplement of 1% disulfiram beginning 1 week prior to DMH treatment. Ten crypts were scored from each of 5 mice at each point. Different shapes of symbols represent different experiments.

REFERENCES

Ames, B.N., W.E. Durston, E. Yamasaki, and F.D. Lee. 1973. Carcinogens are mutagens: A simple test system combining liver homogenates for activation and bacteria for detection. *Proc. Natl. Acad. Sci. U.S.A.* **70**:2281.

Armstong, B. and R. Doll. 1975. Environmental factors and cancer incidence and mortality in different countries with special reference to dietary practices. *Int. J. Cancer* **15**:617.

Cairns, J. 1981. The origin of human cancers. *Nature* **289**:353.

Carrano, A.V. and J.A. Heddle. 1973. The fate of chromosome aberrations. *J. Theor. Biol.* **38**:289.

Chang, W.W.L. and C.P. LeBlond. 1971. Renewal of the epithelium in the descending colon of the mouse. I. Presence of three cell populations: Vacuolated columnar, mucous and argentaffin. *Am. J. Anat.* **131**:73.

Conger, A.D. and L.M. Fairchild. 1953. A quick freeze method for making smear slides permanent. *Stain Technol.* **28**:281.

Doll, R. and R. Peto. 1981. The causes of cancer: Quantitative estimates of avoidable risks of cancer in the United States today. *J. Natl. Cancer Inst.* **66**:1196.

Haenszel, W. and M. Kurihara. 1968. Studies of Japanese migrants. I. Mortality from cancer and other diseases among Japanese in the United States. *J. Natl. Cancer Inst.* **40**:43.

Heddle, J.A., Ed. 1982. *Mutagenicity: New horizons in genetic toxicology.* Academic Press, New York.

Heddle, J.A. and A.V. Carrano. 1977. The DNA content of micronuclei induced in mouse bone marrow by X-irradiation: Evidence that micronuclei arise from acentric chromosomal fragments. *Mutat. Res.* **44**:63.

Higginson, J. and C.S. Muir. 1977. Détermination de l'importance des factors environmentaux dans le cancer humain: Rôle de l'épidémiologie. *Bull. Cancer* **64**:365.

Kerr, J.F.R., A.H. Wyllie, and A.R. Currie. 1972. Apoptosis: A basic biological phenomenon with wide-ranging implications in tissue kinetics. *Cancer* **26**:239.

Maskens, A.P. 1979. Significance of the karyorrhectic index in 1,2-dimethylhydrazine carcinogenesis. *Cancer Lett.* **8**:77.

Schmid, W. 1975. The micronucleus test. *Mutat. Res.* **31**:9.

Tates, A.D., I. Neuteboom, M. Hofker, and L. den Engelse. 1980. A micronucleus technique for detecting clastogenic effects of mutagens/carcinogens (DEN, DMN) in hepatocytes of rat liver in vivo. *Mutat. Res.* **74**:11.

Wattenberg, L.W. 1975. Inhibition of dimethylhydrazine-induced neoplasia of the large intestine by disulfiram. *J. Natl. Cancer Inst.* **54**:1005.

Wynder, E.L. and G.B. Gori. 1977. Contribution of the environment to cancer incidence: An epidemiologic exercise. *J. Natl. Cancer Inst.* **58**:825.

Yamamoto, K.I. and Y. Kikuchi. 1980. A comparison of diameters of micronuclei induced by clastogens and by spindle poisons. *Mutat. Res.* **71**:127.

COMMENTS

FURIHATA: How do you give the dimethylhydrazine?

HEDDLE: It was given intraperitoneally in the background investigations and per os in the verification study, as were the other agents.

COMBES: Is the technique for studying colonic cells applicable to human tissue?

HEDDLE: We have only taken a very preliminary look at human tissue but, as far as we can see, the methods work perfectly well. One can often use histological sections made for other purposes or stain new sections cut from paraffin blocks that have been retained. We have also taken a biopsy from an already excised colon and been able to make cell suspensions and score micronuclei on them. At the moment, however, our experience is minimal; we have merely convinced ourselves that the techniques will work.

WOGAN: I missed how you collected the crypts.

HEDDLE: I didn't tell you. It is rather simple, actually. One treats the everted colon or biopsy with 20 mmole EDTA in 0.075 M KCl for 20 minutes, then takes a 1 ml syringe, pulls on the plunger until one catches the colon in the middle, sucks the colon until it pops in, and then squirts it out again. This is repeated five times after which the colon is removed and what remains is a suspension of crypts. A cell suspension can be made from this by passing the crypts through a 20G needle.

MOHRENWEISER: Would you say the micronuclei have a normal life span, given that you didn't get the expected number?

HEDDLE: No, we think the cells are dying and sloughing off, are being digested by other cells, are losing their micronuclei as mucin is excreted, or perhaps are digesting their own micronuclei. In addition one can find actual holes in the crypt after treatment. Instead of a continuous epithelial layer some cells are missing. Whether they are the ones that had the micronuclei or not, I couldn't say.

BRIDGES: I noticed the yields of micronuclei from ionizing radiation in your study were much lower per rad then in the Henderson study. It is a different system, of course. You used 100 rads, but your micronucleus counts were not very high.

HEDDLE: You must remember that the colonic cell population that we are looking at here is a mixture of dividing and nondividing cells. Therefore one gets an obvious dilution. Furthermore the micronuclei in bone marrow are very well defined structures. With any experience at all and with any sort of reasonable criteria, one cannot doubt their reality. In contrast, there is a large subjective factor in scoring some of the events that we are seeing in the colon. The absolute frequencies would vary depending on the technique used. We haven't yet sorted out the scorer-to-scorer variation, nor have we been able to get the standard settled for the laboratory with respect to exactly what defines a typical versus an atypical micronucleus. That is one of the reasons these results are preliminary. Most of the data were for "typical" micronuclei. If we put in all the "atypical" micronuclei as well and we counted each one, rather than just as a cluster, we would push that up by at least a factor of 4.

EVANS: Where do your atypical micronuclei turn up in relation to cell position in the crypt? Each of these crypts has quite a small number of cells at the bottom which are turning over.

HEDDLE: There is a small number at the bottom, but cells are dividing at least half way up the sides, too.

EVANS: With decreasing frequency.

HEDDLE: Yes.

EVANS: Now, where do these atypical micronuclei appear in relation to the architecture of the crypt?

HEDDLE: They have the same distribution as micronuclei: They are primarily in the lower third of the crypt early on, and they don't move up the crypt very far.

EVANS: In other words they occur in association with mitosis?

HEDDLE: That is true, but that is not to say they arise that way.

COMBES: Could your clustering be due to differential metabolism in the cells? Some cells metabolize at different rates.

HEDDLE: It is unlikely, because we see the same clustering with radiation.

EVANS: The only other perhaps relevant information I have on mouse crypts is the work of Bruce Ponder, which is to be published shortly and which traces the origin of cells in the crypt using chimeras. The two genetically different cells in the chimeras can be distinguished by immunohistochemical techniques, by virtue of their differing cell surface antigens.

SESSION VI:
MUTAGENESIS

Relevance of the Mouse Spot Test as a Genotoxicity Indicator

LIANE BRAUCH RUSSELL
Biology Division
Oak Ridge National Laboratory
Oak Ridge, Tennessee 37830

Indicators of genotoxicity do not all carry the same weight in the assessment of potential health hazard posed by environmental agents. Obviously, the various organisms that have been used for mutagenicity tests resemble human beings to very different degrees, both with respect to organization of the genetic material, and with respect to the cellular and organismic environment within which the genetic targets are contained. Furthermore, certain tests detect actual genome alterations (e.g., gene mutations or gross chromosomal damages), while others detect effects thought to be related to genome alterations (e.g., unscheduled DNA synthesis [UDS], sister chromatid exchange [SCE], transformation, carcinogenesis).

This paper will examine the mammalian spot test (MST) as an indicator of genotoxicity and evaluate the weight that may be given to results from this test with respect to their predictiveness for human health hazards. The evaluation will be based on theoretical considerations, and on the performance of the test to date.

The spot test was developed 25 years ago in a radiation experiment (Russell and Major 1957). Over the past few years, it has been applied to chemical mutagenesis studies, and these results have recently been reviewed (Russell et al. 1981a). Protocols for conducting the assay have been established (Russell et al. 1981a; Braun et al. 1981; L.B. Russell 1982). Mouse embryos heterozygous for a number of coat-color recessives are exposed to the test agent in midgestation, when they possess roughly 200 pigment precursor cells. Certain genetic alterations result in an "uncovering" of the recessive in one of these cells, which multiplies to form a clone of melanocytes that produce hair pigment distinguishable from that in the remainder of the fur. The spots of altered color, designated RS (recessive spots), can be detected at about 12 days after birth of the animals that were exposed in utero. Other features of these animals, including other types of spots, can be used as indicators of cell killing, embryotoxicity, and teratogenicity.

THEORETICAL CONSIDERATIONS

Genotoxicity can present health hazards both to the exposed individual himself (herself) and to his (her) progeny. In the former case, the altered genetic material is somatically transmitted to form clones of mutant cells which might become cancerous or might have some other deleterious impact on the whole organism. In the latter case, the genetic alteration must be induced in a reproductive cell which survives to participate in fertilization.

The MST has the following features that place it relatively high on the scale of parallelisms to both of these human pathways (Table 1).

1. The genome exposed is a mammalian one, the organization of the chromatin being presumably similar to that in man.
2. Exposure occurs in vivo, the chemical thus being subject to native activation systems and other metabolic influences. Since the target cells of the MST are in the embryo, the agent or its metabolites must cross the placenta, but there is no evidence that this presents a barrier to anything other than unusually large molecules.
3. The endpoints scored are alterations of the genome itself, rather than other effects that are thought merely to have some relation to genome changes.

As regards parallism with the somatic-mutation pathway in human beings, the MST thus requires only the extrapolation from mouse to man. As regards genetic hazards, an additional extrapolation must be made, namely, that from somatic cells to germ cells. Germ cells can present special conditions, as indicated by the different responses of different germ cell stages and by the evidence for efficient repair systems (Russell et al. 1982). The germ-cell response can be different for gene mutations and for gross chromosomal lesions. The MST, while not employing a germ-cell target, does, however, have the capability of detecting, both, gene mutations and certain chromosomal lesions.

The MST can detect a considerable array of events that affect the marker loci (Russell and Major 1957; Russell 1983). Among these are

1. point mutations either to an amorph or to certain hypomorphs;
2. deficiencies of the wild-type allele that normally "covers" the marker;
3. multilocus deficiencies of varied lengths that include the wild-type allele and adjacent regions;
4. losses (as a result either of breakage or nondisjunction) of any of the chromosomes that bear the wild-type allele of any of the markers;
5. mutations anywhere in the genome that can produce a dominant pigment phenotype;
6. mitotic recombination (if this occurs in mammals) with crossing-over between the centromere and a marker.

These various types of events are inferred on theoretical grounds on the basis of the phenotypes observed. The mutant clone that is responsible for a RS is very

Table 1
Weighting of Indicators of Genotoxicity for the Assessment of Health Hazards, and Role of the MST

Proposed weighting of indicators of genotoxicity	MST performance
For heritable or somatic damage:	
Eukaryote > prokaryote	+
Mammal > other eukaryote	+
In vivo > in vitro	+
Genome alterations > possibly related endpoints	+
Detects several types of genotoxicity > detects	
single type of genotoxicity	+
For heritable damage specifically:	
germ cell > somatic cell	–
For somatic damage specifically:	
many target tissues > single target tissue	–
Ancillary results provided by test > genotoxicity	+ (cytotoxicity
result only	teratogenicity)

unlikely to have contributed to the animal's gonad as well, and genetic testing of the nature of the mutations cannot, therefore, be carried out.

In addition to providing indicators of genotoxicity, the MST also yields information on toxic effects whose mechanistic bases are not precisely known. Since the test requires exposure of embryos at mid-gestation, any reduction in the average litter size or in the number of litters born indicates embryotoxicity or toxic effects on the mother. Day 10¼ postconception, the optimum treatment time for the scoring of genotoxic effects (RS), happens to coincide with the critical period for limb and tail malformations (Russell 1950). Therefore, simple external examination of the newborns in an MST can yield ancillary information on teratogenicity of an agent. Finally, indicators of cytotoxicity are provided by the induction of white midventral spots (WMVS), which result from melanocyte insufficiency (Russell and Major 1957). Such cytotoxicity could be the result of massive damage to the genetic material or to other constituents of cells (e.g., membranes).

PERFORMANCE OF THE MOUSE SPOT TEST

In the absence of epidemiological information on the action of all but a handful of environmental agents, experimental indicators of genotoxicity cannot be compared directly with human endpoints. The standard for comparison must therefore be provided by experimental mammals.

Genotoxic Indicators of Heritable Damage

To determine the degree of success with which the MST can predict the induction of heritable damage, the mouse specific-locus test (SLT) was used as a standard. The SLT is the only direct test for heritable gene mutations induced in a mammal for which data from several chemicals are available. Moreover, quantitative comparisons can be made between SLT and MST results, since the MST is essentially a type of specific-locus test carried out in in vivo somatic, rather than reproductive, cells. Both tests detect mutational changes if they involve a locus opposite one at which there is a recessive marker. The SLT and MST crosses have several markers in common.

Of the approximately 40 and 30 agents for which MST and SLT results, respectively, have been reported, 16 were tested in both assays (Table 2). In the following comparison, Gene-Tox criteria (Russell et al. 1981a,b) have been used in classifying results. For 11 agents, positive results were obtained in both the MST and the SLT (positive either in stem cells, in postspermatogonial cells, or both); and for one, negative results in both MST and SLT. For one, the MST

Table 2

Comparison of MST Results with Results from Assays that Directly Measure Heritable Damage in Mammals[a]

Agent	MST[b]	SLT[c]	HTT[c]
X-rays	+	+	+
TEM	+	+	+
MC	+	+	+
ENU	+	+	
ENUth	+	+	
Procarbazine	+	+	+
INH	+	+	
CP	+	+	+
MMS	+	+	+
MNU	+	+	
EMS	+	+	+
MNNG	+	inc	
B[a]P	+	−	−
NDEA	−	−	
HC	inc	−	
Caffeine	inc	−	−

[a]Other abbreviations not mentioned in text as follows: HTT = heritable translocation test, TEM = triethylenemelamine, ENU = ethylnitrosourea, ENUth = ethylnitrosourethane, INH = isoniazid, CP = cyclophosphamide, MMS = methylmethanesulfonate, EMS = ethylmethanesulfonate, NDEA = diethylnitrosamine, HC = hycanthone methanesulfonate.
[b]Results listed only for agents that have also been tested in the SLT or HTT.
[c]Result listed as + if either spermatogonial stemcells, or post stemcells, or both, yielded a positive result, regardless of results in other stages.

was positive and the SLT inconclusive; and in two cases, the MST result was inconclusive while the SLT was negative. Only one (benzo[a] pyrene) gave a clearly divergent result: negative in the SLT, positive in the MST. Nine of the agents studied in the MST have also been tested in the heritable-translocation test (Generoso et al. 1980); correspondence is found for 7 of these (Table 2).

The comparison between the findings of the MST and SLT can be taken beyond the simple plus-minus to a more quantitative level by calculating a "unit" mutation rate in each case. This figure is the induced rate (experimental minus appropriate control) per locus and per mole (or per R, in the case of X-rays) multiplied by 10^4 (or, 10^7, in the case of X-rays). Where multiple results for the MST are available, those chosen for the comparison come from MST Cross 1 (which most closely corresponds to the SLT cross as to marker loci) and from day-10¼ exposures (for which evidence is available concerning the number of cells at risk). The SLT "unit" rates have been calculated separately for results from treated stemcells and those from treated post-stemcell stages, since these may differ from each other qualitatively (i.e., as to nature of lesions), as well as quantitatively. Data were taken from the Gene-Tox reviews (Russell et al. 1981a,b) and from papers that have appeared since that time (Ehling 1980; Russell and Montgomery 1982; W.L. Russell 1982; Russell and Hunsicker 1981, 1982); all are documented in more detail by Russell (1983).

The control rates used for subtraction from experimental rates are historical controls as compiled in the Gene-Tox reviews (Russell et al. 1981a,b). The number of loci is 7 for the SLT (a, b, c^{ch}, p, d, se, s), and 4 for MST Cross 1 (b, c^{ch}, p, d). The latter 4 loci are marked in each of about 200 melanoblasts at risk in a 10¼-day embryo (Russell and Major 1957; Fahrig 1978); and thus, each animal observed for spots in the MST scores 800 loci (i.e., 4 × 200). Calculation of mutation rates on a per-mole basis implies linearity of dose-response curves. (While this simplifying approximation had to be made in order to permit comparisons, it should be noted that at least one *non*linear dose-response relation has been clearly demonstrated [Russell et al. 1982].) Additional details of the calculations, and discussion of the assumptions involved, are given elsewhere (Russell 1983).

"Unit" mutation rates are listed in Table 3 for SLT-stemcell and SLT-poststemcell results, as well as for MST results. In 26 of 27 comparisons, the "unit" rate for the MST is higher than that for the SLT. This is not surprising in view of the ability of the MST to detect, in addition to gene mutations, certain types of genetic damage—mostly of a gross chromosomal nature—that would not be detected in the SLT. The MST increment should—all other things being equal—be greatest in the case of agents that are good inducers of chromosomal damage. Another obvious difference between the two tests lies in the type of target cells: in the MST, rapidly dividing melanoblasts; and, in the SLT, slow-cycling spermatogonial stemcells, or nondividing cells for most post-stemcell stages.

Table 3
Comparison of "Unit" Mutation Rates for the SLT and MST

Agent[a]	"Unit" mutation rates[b]			Comparisons for:					
	SLT spermatogonial stemcells	SLT post stemcells	MST	Spermatogonial stemcells			Post-stemcell stages		
				Finding[c]		MST/SLT[d]	Finding[c]		MST/SLT[d]
				SLT	MST		SLT	MST	
X-rays	2.4[e]		7.0[e]	+	+	low	+	+	low[f]
TEM	28.8	106.9	120.5	+	+	low	+	+	low
MMC	24.4	1.5[g]	54.2	+	+	low	inc	+	high
ENU	4.9		5.5	+	+	low			
ENUTh	0.25		0.3	+	+	low			
Procarbazine	0.19	0.6	1.9	+	+	low		+	low
Isoniazid	0.10		1.2	+	+	low			
CP		0.8	3.9				+	+	low
MMS	0.08[g]	1.9	0.6	inc	+	low	+	+	low
MNNG	0[h]	0	0.8	inc	+	low	inc	+	low

384

MNU	0.07[g]	2.9	7.4	inc	+	high	+	+	low
B[a]P	0.02[g]	0	0.8	−	+	high	inc	+	low
EMS	0	0.3	0.9	−	+	high	+	+	low
HC	0	0	0.8[g]	−	inc		inc	inc[i]	
NDEA	0.00	0	0.3[g]	−	inc[i]		inc	inc[i]	
Caffeine	0	0	0.3[g]	−	inc				

[a] For abbreviations, see footnote a to Table 2.

[b] See text for calculation.

[c] Classified by Gene-Tox criteria (Russell et al. 1981a,b); (+) = positive, (−) = negative; inc = inconclusive.

[d] Ratio of "unit mutation rates." low, < 12; high, > 35. Where an SLT "unit" mutation rate was 0, and the MST/SLT ratio thus ∞, the low vs. high classification was made on the basis of an alternate calculation, namely by assuming 1, or 2 mutations (possible numbers within the 90% confidence limits) in the SLT sample.

[e] Calculated per R (rather than per mole) and multiplied by 10^7 (rather than 10^4). See Russell and Major (1957) for method of calculation, which differs somewhat from method used for chemicals.

[f] Based on unpublished data of W. L. Russell for X-irradiated post-stemcell stages in the SLT. This rate is about twice that of the stemcell rate.

[g] Not significantly different from zero.

[h] Values smaller than 0 are listed as 0.

[i] NDEA is listed as negative in Table 2. However, for purposes of the quantitative comparisons in Table 3, only the 10¼-day results were used, and these were inconclusive.

The MST/SLT ratios were calculated separately for stemcell and post-stem-cell SLT results, and these values are presented in a semi-quantitative manner in Table 3. Because of the uncertainties of some of the parameters used in calculating "unit" mutation rates, little importance should be attached to small differences among the MST/SLT ratios.) Wherever both the SLT and MST are positive ("+ +" result), the MST/SLT ratio is of the order of 1-10 (a range referred to as "low"); in fact, in 11 of 14 comparisons of "+ +" results, the ratio ranges from 1 to 5. In the two cases where the SLT was clearly negative and the MST positive ("- +" results), the MST/SLT ratio is high (> 35). For these chemicals (B[a]P and EMS in spermatogonial stemcells), it may be concluded that special conditions in (or affecting) the stemcells prevent induction or transmission of heritable mutations, even though mutagenicity can be demonstrated in melanoblasts; EMS is also mutagenic in post-stemcell stages tested in the SLT.

The pattern of a relatively low MST/SLT ratio in the case of "+ +" results and a high MST/SLT ratio in the case of "- +" results may be tentatively applied in making predictions about future SLT findings for chemicals that have to date given inconclusive results. If the pattern is a valid indicator, N'-methyl-N'-nitrosoguanidine (MNNG) and methylmethanesulfonate (MMS) will turn out to be positive in spermatogonial stem cells (low MST/SLT ratio) and MNNG and B[a]P will be positive in post-stemcell stages; however, methyl nitrosourea (MNU) will be negative in spermatogonial stemcells (high MST/SLT ratio), and Mitomycin C (MC) (which is positive in stemcells) will be negative in post-stemcell stages.

In its performance to date, the MST has given no false negatives with respect to agents that induce heritable gene mutations or translocations; further, MST results can apparently be used to make broad quantitative predictions about upper limits for SLT frequencies. Because of special conditions (e.g., repair systems) in germ cells, especially spermatogonial stemcells, the SLT can be negative when the MST is positive. In looking at the practical question of what role the MST can play in prescreening for agents that might be most potent in inducing heritable damage, one may, on the basis of the available evidence, conclude that the priority for initiating SLTs should be low when the MST is clearly negative, and should be high where the MST "unit" mutation rate is high.

Genotoxic and Other Indicators of Somatic Damage

Genotoxicity in somatic cells is thought to lead to cancer, but the evidence is mainly circumstantial at this time. Somatic genotoxicity could presumably produce other types of ill effects as well, but these have not been clearly identified or systematically investigated. Finding a somatic standard against which to judge the performance of the MST as an indicator of genotoxicity is therefore much more difficult than was the finding of a standard for heritable damages.

When MST results are compared with those for carcinogenicity obtained in long-term mammalian studies (Tomatis et al. 1978; Griesemer and Cueto 1980), it becomes apparent that no clear noncarcinogens have as yet been tested in the MST. Therefore, specificity cannot be calculated. Of 16 known carcinogens tested in the MST, 13 yielded positive results; the sensitivity is thus 81.3%. The three carcinogens for which genotoxicity results in the MST were negative are vinyl chloride, 2,4,6-trichlorophenol, and diethylnitrosamine (which is negative in other whole-mammal mutagenicity tests as well).

The MST, as discussed above, provides a small window for detecting teratogenicity directly. An absence of obvious morphological malformations in the MST clearly cannot be taken as evidence for nonteratogenicity of an agent. Conversely, however, the finding of malformations is valid evidence that the agent has teratogenic properties. These can subsequently be investigated in more detail in experiments designed expressly for this purpose. Among publications of MST studies, only some state whether or not malformations were looked for. It would be well to add this simple observation routinely in the future to exploit the test's ancillary capabilities.

While an increase in WMVS can hardly be regarded as serious developmental damage, it is, however, probably indicative of cytotoxicity. Cytotoxicity, if it occurs in other, more critical, tissues, might result in some as yet unknown feature of ill health. It will be important to keep a catalog of WMVS results to be able to compare this array with any standard of somatic ill health that may emerge in the future.

SUMMARY

The MST is an in vivo mammalian assay that can detect a considerable array of genetic events affecting a set of marker loci. It thus provides indicators of genotoxicity germane to predictions of human health hazards that arise from either heritable or somatic mutations.

To determine the degree of success with which the MST predicts the induction of heritable damage, quantitative comparisons were made between the MST and the SLT by calculating "unit" mutation rates (rates per locus per mole of exposure). In 26 of 27 comparisons, the "unit" rate for the MST (a broad-spectrum test) was higher than that for the SLT. When both tests had yielded clearly positive results, the MST/SLT ratio was of the order of 1-10 (usually 1-5). Future MST results can thus be used with some confidence to make broad quantitative predictions about upper limits for SLT frequencies. Priority for initiating SLTs should be low when the MST is clearly negative, and should be high when the MST "unit" mutation rate is high.

The MST has ancillary capabilities for directly detecting embryotoxicity, teratogenicity, and cytotoxicity. In limited comparisons between MST results and those from carcinogenicity bioassays, the MST sensitivity was 81.3%, but no negative carcinogens have as yet been tested in the MST.

ACKNOWLEDGMENTS

I am grateful to Drs. W. L. Russell and W. M. Generoso for a critical reading of the manuscript.

REFERENCES

Braun, R., L.B. Russell, and J. Schöneich. 1981. Meeting report: Workshop on the practical applications of the mammalian spot test in routine mutagenicity testing of drugs and other chemicals. *Mutat. Res.* **97**:155.

Ehling, U.H. 1980. Induction of gene mutations in germ cells of the mouse. *Arch. Toxicol.* **46**:123.

Fahrig, R. 1978. The mammalian spot test: A sensitive in vivo method for the detection of genetic alterations in somatic cells of mice. *Chem. Mutagens* **5**:151.

Generoso, W.M., J.B. Bishop, D.G. Gosslee, G.W. Newell, C.-J. Sheu, and E. von Halle. 1980. Heritable translocation test in mice. *Mutat. Res.* **76**:191.

Griesemer, R.A. and C. Cueto. 1980. Toward a classification scheme for degrees of experimental evidence for the carcinogenicity of chemicals for animals. In *Molecular and Cellular Aspects of Carcinogen Screening Tests* (ed. R. Montesano et al.), p. 259. International Agency for Research on Cancer, Lyon, France.

Russell, L.B. 1950. X-ray induced developmental abnormalities in the mouse and their use in the analysis of embryological patterns. I. External and gross visceral changes. *J. Exp. Zool.* **114**:545.

_____. 1982. The mouse spot test: Procedures and evaluations of results. In *Handbook of mutagenicity test procedures* (ed. B. Kilbey), (In press.) Elsevier/North Holland.

_____. 1983. The mouse spot test as a predictor of heritable genetic damage and other endpoints. *Chem. Mutagens* (in press).

Russell, L.B. and M.H. Major. 1957. Radiation-induced presumed somatic mutations in the house mouse. *Genetics* **42**:161.

Russell, L.B. and C.S. Montgomery. 1982. Supermutagenicity of ethylnitrosourea in the mouse spot test; comparisons with methylnitrosourea and ethylnitrosourethane. *Mutat. Res.* **92**:193.

Russell, L.B., P.B. Selby, E. von Halle, W. Sheridan, and L. Valcovic. 1981a. Use of the mouse spot test in chemical mutagenesis: interpretation of past data and recommendations for future work. *Mutat. Res.* **86**:355.

_____. 1981b. The mouse specific-locus test with agents other than radiations; interpretation of data and recommendations for future work. *Mutat. Res.* **86**:329.

Russell, W.L. 1982. Factors affecting mutagenicity of ethylnitrosourea in the mouse specific-locus test and their bearing on risk estimation. In *Proceedings of the 3rd International Conference on Environmental Mutagens* (T. Sugimura et al.), 59. University of Tokyo Press.

Russell, W.L. and P.R. Hunsicker. 1981. Comparative mutagenicity of methyl- and ethylnitrosourea in the mouse specific-locus test. *Environ. Mutagen.* **3**:372.

_____. 1982. Effect of dose fractionation on specific-locus mutation induction by ethylnitrosourea in mouse spermatogonia. *Environ. Mutagen.* (in press).

Russell, W.L., P.R. Hunsicker, G.D. Raymer, M.H. Steele, K.F. Stelzner, and H.M. Thompson. 1982. Dose-response curve for specific-locus mutations induced by ethylnitrosourea in mouse spermatogonia. *Proc. Natl. Acad. Sci. U.S.A.* **79**:3589.

Tomatis, L., C. Agthe, H. Bartsch, J. Huff, R. Montesano, R. Saracci, E. Walker, and J. Wilbourn. 1978. Evaluation of the carcinogenicity of chemicals: A review of the monograph program of the international agency for research on cancer (1971 to 1977). *Cancer Res.* **38**:877.

COMMENTS

EVANS: You said that, for a certain set of loci, the ratio of mutation rate in the MST to that in the SLT is about 10 for some agents. Was that the figure you quoted?

RUSSELL: In 11 out of 14 comparisons that can be made, the MST/SLT ratio was in the range from 1 to 5. For a couple of chemicals, it was about 10; and in one case, it was less than 1.

EVANS: What I wanted to point out was: If you looked at the data in man for chromosomal changes and asked "what is the spontaneous mutation rate for structural rearrangements in offspring?" this comes out to about 0.4×10^{-4} per gamete per generation. If you were to ask "what is the frequency of that kind of change in somatic cells in blood?" it is about ten times higher.

Do you have any idea about the frequency of chromosomal rearrangements observed cytologically in mouse lymphocytes and in mouse embryos?

RUSSELL: These types of data are being obtained by I.-D. Adler and others and are as yet unpublished. I would have to look up the actual figures. In the case of the spot test, we don't really know what proportion of the observed spots is the result of gene mutations, and what proportion is due to gross chromosomal changes that affect the marked loci. Spots can't be analyzed by breeding tests because the chances are very small that the gonad also will be involved in the clone of mutant cells that form the visible spot. Where two of the markers are on the same chromosome, one may be able to exclude certain spot causes on the basis of the spot color or hair histology; e.g., one might rule out loss of the whole chromosome, or somatic recombination. Where there is only one marker on the chromosome, phenotype alone can give little or no information regarding the type of genetic lesion that caused the spot.

HEDDLE: There are two things that strike me about this. The first item is just a comment. All tests, when they start out, have quite high success rates, because you start looking at the most potent carcinogens. The more you do, even if your test gets better—I am not saying you should make your test better—the success rate doesn't go up because you are dealing with more and more difficult chemicals. That is one thing.

The second item is that I don't really understand why the ratios come out the way they do. I understand there are the problems of distribution and metabolism but when one looks at the agents that are potent chromosome breakers which ought, one would think, to give high ratios, that is not always true. Some of them are high and some of them are not. X-rays, for example, the ratio you characterized as being low, and yet, relatively, it is a potent chromosome breaker. I just wondered if you had any explanations for why.

RUSSELL: Well, as to your first comment, I agree with you that when the percentage of positives is very high in two sets of data being compared, one has little discriminatory power in determining whether there are correlations. That is why I was interested in going beyond the plus-minus line-up to a quantitative comparison.

As regards your second comment, I think that I may have left some confusion as to the meanings of "high" and "low." "High" ratios (> 35) were *not* found for cases in which both tests gave positive results. Within the "low" category, MST/SLT ratios ranged from 0.3 to about 12. In accordance with the expectation you mention, ethylnitrosourea, which seems to produce predominantly intragenic mutations in spermatogonia, gave one of the lowest ratios (about 1.1), while the ratios for X-rays and triethylenemelamine, good inducers of gross chromosome damage, were 3 and 4 times higher. However, because of the several uncertainties that enter into the calculations of "unit" mutation rates, not too much importance should be attached to relatively minor differences in MST/SLT ratios.

Studies with T-Lymphocytes:

An Approach to Human Mutagenicity Monitoring

RICHARD J. ALBERTINI
Department of Medicine
College of Medicine
University of Vermont
Burlington, Vermont 05405

There is a great deal of current interest in developing mutagenicity monitoring tests that will be capable of defining the somatic and germinal genetic effects of environments, foods, drugs, or other medical treatment on human populations. Suitable monitoring tests should detect genetic damage—or the potential for such damage—that occurs in vivo in man. Thus, these tests should use, as indicators of genotoxicity, cells or other materials obtained directly from the body.

Human mutagenicity monitoring can provide information concerning genetic hazards that cannot be obtained in other ways. I have detailed elsewhere the several potential advantages of human monitoring (Albertini and Allen 1981), two of which are repeated here: First, monitoring may define human population heterogeneity as regards susceptibility to specific mutagens. It is entirely possible, given probable human exposures, that unusual sensitivities to particular environmental genotoxins, rather than doses, may determine who actually suffers somatic or transmissible genetic disorders resulting from exposures. Susceptible individuals, detected by monitoring, may require special means to protect them from untoward genetic risks. Second, it is now feasible to correlate mutagenicity monitoring test results for specific individuals at risk from genotoxin exposure with the health outcome of these same individuals. Such correlations will allow the validation of monitoring test systems in terms of their relevance as predictors of human genetic disease. The public health goal to be realized is the substitution of a laboratory test result for an ill individual— the currently required epidemiological indicator of a genetically hazardous environment.

There are several tests available or under development for human mutagenicity monitoring. Standard cytogenetic tests (Evans and O'Riordan 1977; Brogger 1979), tests of sister chromatid exchange (Perry and Evans 1975; Stetka and Wolff 1976; Latt et al. 1977; Wolff 1977), tests of DNA damage and (or) repair (Pero and Mitelman 1979), the detection of sperm morphological abnormalities (Wyrobek and Bruce 1978), or double Y bodies (Kapp et al. 1979), and measurements of protein alkylation in vivo (Osterman-Golkar 1981) will all be useful for this purpose. As is apparent, the several tests measure several end points. To be included in this list are tests that detect specific locus somatic cell mutants arising in vivo in humans. Two such test systems have been or are being developed. One is the mutant hemoglobin test (Papayannopoulou et al. 1976, 1977; Stammatoyannopoulos 1980). I will discuss here the detection of mutant T-lymphocytes that arise in vivo in man. These cells may be obtained from peripheral blood or from specific tissues and organs.

THE HUMAN T-LYMPHOCYTE SYSTEM

At another Banbury Conference I reviewed the 6-thioguanine resistant (TG^r) peripheral blood lymphocyte (PBL) system as it had been developed to that time (Albertini 1979). Here I briefly restate the rationale for that system and summarize the results of early studies. What was only dimly suspected then was the existence of "phenocopies." The "phenocopies" are essentially cells which artifactually mimic the mutant TG^r PBLs being sought. Our current concept of a major source of these interfering cells and also our method for eliminating their effect is presented below. Additionally, I summarize our current method for the autoradiographic detection of variant TG^r PBLs.

With the development of T-cell growth factors, it has become possible to clone human T-lymphocytes directly in vitro. Furthermore, the resultant cultured T-cells (CTCs) can be propagated in vitro for prolonged periods. Thus, as a further development of the lymphocyte system, I present a brief account of our cloning assays for mutant T-lymphocytes arising in vivo and summarize some studies with the resultant mutant cells.

Rationale

The Lesch-Nyhan (LN) mutation, a naturally occurring human mutation of the chromosomal gene controlling the enzyme hypoxanthine-guanine phosphoribosyltransferase (HPRT), produces a striking clinical disorder (Lesch and Nyhan 1964; Seegmiller et al. 1967). Its fundamental biochemical defect is decreased to absent HPRT activity in all cells of affected males (Balis 1968; Kelley 1968; Strauss et al. 1981).

HPRT is a constitutive but dispensable enzyme that converts the normal substrates hypoxanthine and guanine to inosine monophosphate (Kelley 1968; Caskey 1979). The enzyme is also necessary for the phosphorylation of purine analogs such as 8-azaguanine (AG) and TG to render them cytotoxic (Elion and Hitchings 1965; Elion 1967).

The LN mutation produces a convenient natural prototype mutation for human mutagenicity studies (DeMars 1971; Albertini and DeMars 1973). It produces a clear cellular phenotype characterized by: purine analog resistance; an inability to utilize exogenous hypoxanthine as a purine source; and deficiency of HPRT activity.

Several years ago, we exploited this mutation in developing the human diploid fibroblast system for in vitro mutagenicity screening (Albertini and DeMars 1973). We showed at that time that LN peripheral blood lymphocytes (PBLs) also express the mutation, manifesting the same mutant cellular phenotype as do fibroblasts (Albertini and DeMars 1974). Furthermore, we showed that LN heterozygous females showed a decided minority population of LN PBLs, as opposed to the 50% expected on the basis of single X-chromosome inactivation (Lyon 1961; Russell 1963). This apparently results from selection in vivo against LN lymphocytes (Dancis et al. 1968; Nyhan et al. 1970; McDonald and Kelley 1972).

We proposed then that purine analog resistant PBLs, arising in vivo in non-LN individuals, be quantitated as indicators of somatic cell mutation occurring in vivo—i.e., as a specific locus mutagenicity monitoring test. We considered the selection in vivo against mutant lymphocytes to be an advantage. Changes in purine analog resistant "LN-like" PBL frequencies in non-LN individuals resulting from environmentally induced somatic cell mutation should be temporary, and related in time to the influences producing them.

Early Studies

We initially defined AG^r or TG^r PBLs based on their ability to incorporate tritiated thymidine (^{3}H-TdR) following lectin (phytohemagglutinin = PHA) stimulation in vitro in the presence of cytotoxic concentrations of purine analogs. For the scoring of rare resistant cells, scintillation spectrometry proved far too insensitive when used for mass lymphocyte cultures (Albertini and DeMars 1974). However, as will be noted later, it may be useful for the scoring of cloning assays. By contrast, autoradiography does allow the enumeration of rare AG^r- or TG^r-resistant PBLs present in mass PHA-stimulated lymphocyte cultures (Strauss and Albertini 1979; Albertini 1980).

We reported a median TG^r PBL variant frequency (V_f) as determined autoradiographically of 1.1×10^{-4} for normal, nonmutagen-exposed individuals (Strauss and Albertini 1979; Albertini 1979, 1980) (see Table 1, Albertini and Sylwester 1982). By contrast, TG^r PBL V_fs were elevated over this when determined for cancer patients and others receiving known mutagenic therapies (Table 1) (Strauss and Albertini 1979; Strauss et al. 1979; Albertini 1979, 1980). The distribution of V_fs for treated patients was such that most values were higher than the highest values seen in normal controls. This was clearly according to expectations if the TG^r PBL V_f elevations in these patients resulted from somatic cell mutation in vivo.

There were, however, unexpected results in these early studies, some of which are shown also in Table 1 (Strauss et al. 1979; Albertini 1980). Many cancer and other patients had elevated V_fs even before they received mutagenic therapies. Furthermore, some treated patients showed TG^r PBL V_f values of the order of 10^{-2}—clearly too high to be attributed to somatic cell mutation at a single genetic locus. On repeat testing, normal individuals showed wide day-to-

Table 1
TG^r PBL V$_f$ Values Determined in 2 × 10^{-4}M TG

Fresh PBLs	Number of Tests	TG^r PBL V$_f$(×10^{-6}) Median	Range
Normal 1[a]	98[b]	110[c]	25, 380[d]
Normal 2	46[e]	160[d]	30, 1150[d]
Psoriasis$_1$ (conventional treatment)	16	1400	120, 2900
Psoriasis (PUVA)[a]	18	610	100, 4200
Cancer (pretreatment)[a]	14	500	58, 5000
Cancer (treated)[a]	12	850	40, 9900
Cryopreserved PBLs			
Normal (adult)	11[f]	4.5[c]	0, 26.4[d]
Cancer (treated)	11	14.0	0, 50.5
Multiple skin cancer	8	5.8	2.9, 27.1

[a]Possible underestimate of V$_f$ due to error in electronic counting
[b]98 tests in 63 individuals
[c]Median value for number of individuals
[d]Median value of range of values for number of tests
[e]46 tests in 11 individuals
[f]11 tests in 10 individuals

day variation in their V$_f$s. Placental cord blood samples could not be tested because they showed inexplicably high TG^r PBL V$_f$s. Finally, we learned that TG^r PBL V$_f$ values were markedly lower when cryopreserved as opposed to fresh lymphocytes from the same individual were tested (Albertini et al. 1981). Since cryopreservation had no unusual cytotoxic effects on prototype LN mutant PBLs (Albertini 1980), it did not appear that this treatment selectively removed mutant PBLs from test populations. All of this suggested the presence of "phenocopies" in the system—i.e., nonmutant cells that mimic, under our conditions of assay, the "LN-like" TG^r mutant PBLs.

Phenocopies

On a forced review of our early results, it appeared as though unexpectedly high TG^r PBL V$_f$s were frequently found in clinical settings where a large fraction of PBLs might be in "cell cycle" in vivo. Under basal conditions, lymphocytes in the peripheral blood are in an arrested G$_0$ stage of the cell cycle, and enter into cycle only following a stimulus. This stimulus is provided in vitro by lectin or antigen. However, an immunological stimulus in vivo will also induce PBLs into cycle. Cancer and other disorders might have provided this in vivo immunological stimulus in the patient groups studied initially.

When put into culture, the great majority of PBLs that are in G$_0$ are induced by lectin to undergo a G$_0$ to G$_1$ transformation, with the acquisition of T-cell growth factor receptors on the transformed cells (Maizel et al. 1981). This is associated with profound metabolic shifts in these cells (Allison et al. 1977; Hovi et al. 1977). We have postulated that it is this transformation that is inhibited by TG in normal resting PBLs in short-term PHA cultures when the TG-sensitive cells fail to incorporate ^{3}H-TdR during their first DNA synthesis in vitro (Albertini et al. 1982a). However, PBLs that are already in cycle in the peripheral blood have no need to undergo this transformation in vitro. Thus, TG is operating on quite different cell functions in this minority population of cells. These cells too are inhibited by TG. However, since they do not require this profound metabolic alteration prior to DNA synthesis in vitro, they may accomplish this function, and even undergo cell division in TG prior to arrest. Thus, a fraction of these cells, while not mutant, may become labeled in the short-term cultures used to assay for TG^r PBLs. These cells are then scored as mutants, thereby producing the phenocopies.

We performed experiments in which fresh and cryopreserved PBLs from the same individuals were tested in parallel for TG^r PBL V_fs (Albertini et al. 1981). In every case, the V_f value determined with cryopreserved cells was significantly lower than the corresponding value determined with fresh cells. This difference was often one or two orders of magnitude. We postulated that cryopreservation was exerting its effect on the spontaneously cycling PBLs.

In order to quantify the effect of cryopreservation on the ability of cycling, nonmutant PBLs to incorporate ^{3}H-TdR in TG under our conditions of assay, we tested in parallel fresh and cryopreserved PBLs from five normal young individuals. PHA was not added to lymphocyte cultures, so that the frequency of spontaneously cycling cells could be determined from the frequency of cells labeling with ^{3}H-TdR added at different times of culture. Also, parallel cultures without PHA but containing TG were established. This allowed the determination of the frequency of spontaneously cycling PBLs that would label with ^{3}H-TDR added at different times in culture in the presence of TG. These we believe to be "phenocopies." The ratio of spontaneously labeling cells in TG to spontaneously labeling cells without TG is the frequency of these interfering cells in the spontaneously cycling subpopulation of PBLs.

Table 2 gives a summary of the results of this experiment, which has been reported in different forms elsewhere (Albertini et al. 1981; Albertini 1982). For the five individuals studied, PBLs spontaneously labeling in TG were found at all culture intervals when fresh PBLs were tested. These are "phenocopies." When fresh cells were incubated for 30 hours in 2×10^{-4} M TG followed by a 12-hour label interval (the conditions used for the standard autoradiographic assay—see below), the spontaneously labeling cells ("phenocopies") ranged in frequency from 4.6×10^{-6} (subject 3) to 2.1×10^{-5} (subject 2). The frequencies of these labeling cells among all of the spontaneously cycling cells of fresh PBLs varied from 0.2%-4%.

Cryopreserved PBLs showed quite a different pattern of response. In four of the five cases, there were no spontaneously labeling PBLs in the 30-hour TG-

Table 2

Autoradiographically Labeled Nuclei Appearing in Non-pha Stimulated Lymphocyte Cultures in the Presence of 2×10^{-4}M TG Fresh and Cryopreserved PBLS

| Subject | Time in culture[a] (4 hr) | | LI in 2×10^{-4}M TG[b] | | Apparent TG^r PBL V_f^c among cycling cells | |
	Incubation	Label	Fresh	Cryopreserved	Fresh	Cryopreserved
1	0	12	2.2×10^{-4}	2.4×10^{-4}	0.090	0.300
	20	12	9.3×10^{-6}	3.3×10^{-6}	0.006	0.004
	30	12	1.0×10^{-5}	0	0.004	0
2	0	12	4.9×10^{-4}	7.0×10^{-4}	0.310	0.580
	20	12	3.0×10^{-5}	6.5×10^{-6}	0.150	0.008
	30	12	2.1×10^{-5}	0	0.008	0
3	0	12	1.0×10^{-3}	1.4×10^{-3}	0.280	0.500
	20	12	8.7×10^{-6}	3.4×10^{-6}	0.004	0.001
	30	12	4.6×10^{-6}	0	0.002	0
4	0	12	1.6×10^{-4}	—	0.150	—
	20	12	1.8×10^{-5}	9×10^{-6}	0.020	0.010
	30	12	1.5×10^{-5}	0	0.040	0
5	0	12	4×10^{-4}	3.6×10^{-4}	0.170	0.900
	20	12	4×10^{-4}	7.4×10^{-6}	0.005	0.006
	30	12	1.3×10^{-5}	3.4×10^{-6}	0.002	0.009

[a] In vitro cultures did not contain PHA; labeling was with 5μC ^{3}H-TdR.
[b] "Phenocopies" among all recovered PBL nuclei.
[c] "Phenocopies" among cycling PBLs (See text).

containing cultures as assayed by a 12-hour subsequent labeling interval. In one 30-hour culture, labeled cells were present at a frequency of 3.4×10^{-6}. Here, they comprised 0.9% of the spontaneously cycling subpopulation of cryopreserved PBLs. Importantly, in all four instances where it could be determined, the incidences of spontaneously cycling PBLs that labeled in TG was much higher in cryopreserved than in fresh cells when assayed early in culture—i.e., following only 12 hours of label immediately after cultures were established.

We interpret these results to indicate that cryopreservation moves those cycling cells that are capable of labeling in TG through their S phase (DNA synthesis) earlier in the culture interval than is the case with fresh PBLs. These cells then are not in S phase during labeling when the usual assay procedure is followed. Thus, the frequency of labeled cells determined from cryopreserved samples more nearly approximates the frequency of G_0 cells that are resistant to TG inhibition of their transformation to G_1. We feel that this value more closely approximates the "true" TG^r PBL V_f.

Current Autoradiographic Method

The autoradiographic method for detecting rare TG^r PBLs has been described in detail (Strauss and Albertini 1979; Albertini 1980; Albertini et al. 1982a) and is outlined here. The mononuclear cell (MNC) fraction of whole heparinized blood is obtained by the Ficoll-Hypaque method (Boyum 1968). MNCs are suspended in 7.5% dimethylsulfoxide (DMSO) and cryopreserved in the vapor phase of liquid nitrogen prior to test. For test, the MNCs are cultured with PHA in replicate flasks, with or without $2 \times 10^{-4}M$ TG, for 30 hours under standard conditions of 37°C and 5% humidified CO_2 atmosphere. Cultures are then labeled with ^3H-TdR, and cultured for an additional 12 hours. The total culture interval of 42 hours does not allow significant cell division to occur in vitro (Strauss and Albertini 1979).

Cultures are terminated by preparing free nuclei. The nuclei, suspended in fixative, are counted with a Coulter Counter and added in measured volumes to coverslips affixed to microscope slides. The slides are then stained and autoradiographed.

Autoradiographed coverslips are scored by light microscopy. A labeling index is determined for the PHA-stimulated lymphocytes from cultures not containing TG by scoring 5000 nuclei. The labeling index for control cultures (LI_c) is:

$$LI_c = \frac{\text{Number of labeled nuclei in 5000 scored}}{5000} \qquad (1)$$

Similarly, a labeling index is determined for the PHA-stimulated lymphocytes from the parallel TG-containing cultures. This labeling index for test cultures (LI_t) is determined by scanning all coverslips made with all nuclei from all test cultures, and counting the labeled nuclei. The total number of nuclei on coverslips is known from the Coulter counts. Thus:

$$LI_t = \frac{\text{Total number of labeled nuclei on all test coverslips}}{\text{Total number of nuclei on all test coverslips}}$$

The TG^r PBL V_f is the ratio of these two LIs:

$$V_f = \frac{LI_t}{LI_c}$$

Elsewhere we show how confidence intervals (CIs) for true variant frequencies estimated by V_fs obtained as above, are determined by assuming that the number of labeled nuclei from all test cultures is a Poisson variable and that LI_c is a binomial fraction (Albertini et al. 1982a).

Results with Cryopreserved Cells

We now routinely use cryopreserved cells for autoradiographic TG^r PBL V_f determinations. A summary of results, reported elsewhere (Albertini 1982; Albertini et al. 1982c,d), for 29 adults is given in Table 1 and a listing of the individual test values with their 95% confidence intervals is provided in Table 3. The range of V_f values for the normal adults in this group was from 1.1×10^{-6}-26.4×10^{-6}. The single high value in this group (26.4×10^{-6}) was determined for an individual who is exposed to organic chemicals. Untreated psoriatic

Table 3
TG^r PBL V_f Values Determined in 2×10^{-4}M TG Cryopreserved Cells ($\times 10^{-6}$)

Normal controls		Treated cancer patients		Multiple skin cancer patients	
V_f	95% CI[a]	V_f	95% CI[a]	V_f	95% CI[a]
2.9	(1.1, 6.4)	14.0	(1.4, 76.7)	27.1	(2.1, 97.3)
<1.1[b]	(0, 4.3)	13.3	(4.9, 29.0)	2.9	(1.1, 6.3)
26.4	(15.9, 43.4)	21.6	(12.0, 39.1)	5.4	(3.2, 9.4)
1.8	(0.2, 6.5)	7.0	(0.7, 25.2)	5.5	(1.1, 16.1)
2.5	(1.0, 5.2)	13.0	(2.5, 36.4)	10.8	(4.6, 21.3)
8.5	(5.7, 12.7)	<2.0	(0, 7.6)	6.1	(1.5, 15.0)
5.4	(1.1, 15.6)	13.8	(8.8, 21.8)	11.7	(3.7, 27.3)
1.1	(0.4, 2.7)	53.5	(37.7, 76.0)	16.3	(7.2, 31.0)
3.8	(1.6, 7.4)	53.5	(37.1, 76.1)		
6.1[c]	(0.6, 34.1)	51.2	(30.8, 85.2)		
12.1[c]	(1.2, 43.6)	57.5	(21.2, 126.0)		

[a]CI = Confidence intervals
[b]Replicate studies on the same sample
[c]Individuals with psoriasis

patients did not have values markedly out of the normal range when cryopreserved samples were tested. By contrast, cancer patients receiving mutagenic therapies still showed elevated values when compared to normals, but these elevations were less when determined with cryopreserved cells. The range of TG^r PBL V_f values for the treated cancer patients was from 2×10^{-6} to 57.5×10^{-6}, with 9 of the 11 values being higher than 10 of the 11 values for normals. Elevations in this range are much more consistent with a mutational origin for the variant cells. Also, we have tested patients with multiple skin cancers, who were not receiving chemotherapy, in an attempt to find individuals with elevated variant frequencies occurring in vivo. An increase in mutation may underlie their multiple skin cancers. The data in Table 3 suggest that such individuals may exist, but the 95% confidence intervals of the tests performed thus far do not allow definite conclusions.

Thus, the use of cryopreserved PBLs allows testing with the autoradiographic method. We are now performing larger experiments in flask cultures to achieve greater precision in the assay. Monitoring studies of placental cord blood samples are in progress, and the TG^r PBL V_f values for this group are in the range of values found for normal adults.

Cloning of TG^r T-PBLs in vitro

In order to demonstrate by the usual criteria (Chu and Powell 1976) the mutant nature of the human TG^r PBLs that arise in vivo, as well as to establish an outside reference for the "true" TG^r PBL V_f, we have developed cloning assays for these unusual cells. This work, reported in detail elsewhere (Albertini and Borcherding 1982; Albertini et al. 1982b), is outlined briefly below.

T-lymphocytes are easily grown in vitro using crude T-cell growth factors (Paul et al. 1981). Mass populations of cultured T-cells can be developed from the peripheral blood, or from other tissues and organs. Furthermore, T-PBLs can be cloned in vitro directly from the peripheral blood. Cloning in vitro can be accomplished in the presence or in the absence of selective agents requiring only, for the former, that large numbers of cells be plated. This is the basis of our cloning assays.

We obtain PBLs for cloning by Ficoll-Hypaque separation as described above. The cells are then primed in vitro with lectin for a sufficiently short interval so that significant cell division does not occur. The primed PBLs are inoculated in limiting dilutions into the wells of microtiter plates in the presence or absence of TG. An average of 1 cell/well is inoculated into nonselected wells; 10^5 cells per well are inoculated into TG-containing wells. We select with $5 \times 10^{-6}M$ TG—a concentration which is toxic to normal cultured T-lymphocytes.

Wells can be scored after approximately 2 weeks of culture. Wells receiving a single cell per well are scored by microscopy or scintillation spectrometry; wells receiving 10^5 cells/well in TG are scored by cell transfer and clonal expansion, or by scintillation spectrometry. Scintillation spectrometry is useful here when it was not for mass cultures, because, for the cloning assay, it is necessary only to discriminate wells containing growing cells from those not containing growing cells. Furthermore, the incorporation of ^{3}H-TdR into DNA gives unequivocal evidence of cell division in the cultures scored as positive.

The cloning efficiency of T-PBLs in vitro is determined from the plates receiving 1 cell/microtiter well by scoring that fraction of wells containing no growing cells. By assuming a Poisson distribution of cells in wells, the average number of clonable cells in the wells is derived from the P_o class of that distribution—i.e., the observed fraction of wells without growing cells. Similarly, the incidence of TG^r cells in wells receiving 10^5 primed T-PBLs in TG is determined from the P_o class of wells in the TG-containing plates. The frequency of TG^r T-PBLs is that incidence, divided by 10^5 and corrected by the cloning efficiency.

Illustrative results from a portion of a recent experiment are shown in Figures 1 and 2. The 96 wells of two microtiter plates were preloaded with 0.1 ml volumes each of "feeder-mix" containing T-cell growth factors, X-irradiated lymphoblastoid feeder cells and medium RPMI 1640—the whole supplemented with 15% fetal calf serum as described in detail elsewhere (Albertini et al. 1982b). The "feeder-mix" added to one plate (Fig. 2) contained TG. Primed PBLs from a normal individual were then added to the 96 wells of each plate so that each well received an average of 1 cell/well in that plate without TG, used to determine cloning efficiency (Fig. 1), or 10^5 cells/well in that

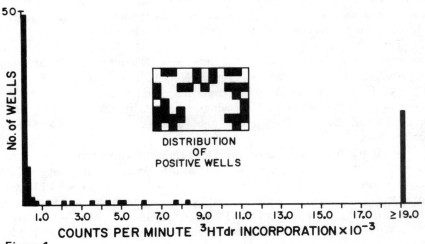

Figure 1

The distribution of CPM ^{3}H-TdR incorporation among the 96 wells of a microtiter plate receiving 1 cell/well in nonselective medium. Wells with $\geq$ 2000 CPM are considered to be positive for growing cells. The insert shows that distribution of the 33 wells with $\geq$ 2000 CPM. In addition, three wells had been "picked" prior to label, giving a total of 36 positive wells, for a calculated cloning efficacy of 0.47 (see text).

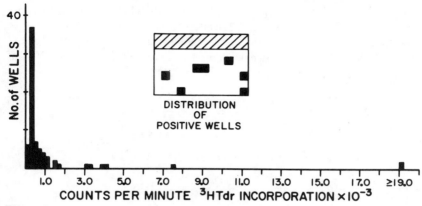

Figure 2

The distribution of CPM ^{3}H-TdR incorporation among the 72 "unpicked" wells of a micro-titer place receiving 10^5 cells/well in TG selective medium. Wells with ≥ 2000 CPM are considered to be positive for growing cells. The insert shows the distribution of CPM in the 7 wells with ≥ 2000 CPM. The incidence of TG^r cells/well was 0.102. The hatched area of the insert represents two rows of wells (24 wells) that were "picked" blindly prior to label. Three of these 24 wells were positive for growing cells, yielding an incidence of 0.134 TG^r cells/well for the subset of wells. TG^r mutant frequencies were calibrated at 2.2×10^{-6}, as determined from the 72-well subset, and 2.9×10^{-6}, as determined from the 24-well subset.

plate with TG, used to determine mutant frequency (Fig. 2). After 10 days, all wells were "fed" by replacement of 0.1 ml of the appropriate medium. Some wells from the cloning efficiency plate were "picked" for clonal expansion, as were the first two horizontal rows (Fig. 2) of the selection plate. Wells of both plates were labeled with ^{3}H-TdR (0.77 μC in 0.05 ml/well) on day 13, and counted on day 14 by scintillation spectrometry as described (Albertini et al. 1982b).

Figure 1 shows the distribution of CPM ^{3}H-TdR incorporation among the 96 wells of the cloning efficiency plate—i.e., the plate containing 1 cell/well in nonselective medium. The distribution of CPM among wells is clearly other than unimodel. Wells with less than 2000 CPM are considered to be negative; those with ≥ 2000 CPM are considered to be positive for growing cells. Thirty-three wells have counts ≥ 2000 CPM. Also, three wells with < 2000 CPM had been "picked" prior to label, giving a total of 36 positive wells among the 96 wells inoculated with 1 cell/well. The cloning efficiency can be calculated:

$$P_o = e^{-x} = \frac{96 - 36}{96} = \frac{60}{96} \tag{2}$$

$$X = 0.47$$

Where X, the average number of clonable cells/well, is the cloning efficiency.

Figure 2 shows the distribution of CPM ^{3}H-TdR incorporation among the 72 "unpicked" wells of the plate containing 10^5 primed PBLs/well in TG. The distribution of CPM among these 72 wells shows the majority (65) to have < 2000 CPM. These are considered to be negative for growing cells. However, 7 wells have CPM $\geqslant 2000$, and are considered to be positive. The incidence (X) of TG^r cells per well can be calculated as:

$$P_o \, e^{-x} = \frac{72 - 7}{72} = \frac{65}{72} \tag{3}$$

$X = 0.102$

From which the TG^r mutant frequency (M_f) can be derived:

$$M_f = \frac{0.102}{0.47 \times 10^5} = 2.2 \times 10^{-6} \tag{4}$$

The inserts in both Figures 1 and 2 show the distribution of positive wells in the respective microtiter plates.

The two horizontal rows of wells (24 wells) that were picked blindly from the selection microtiter plate (Fig. 2) were scored independently. Three clones that were capable of continuous growth in vitro in TG were obtained from these 24 wells. Thus, three of these 24 wells were positive for growing cells, yielding an incidence (X) of TG^r cells/well of:

$$P_o = e^{-x} \frac{24 - 3}{24} = \frac{21}{24} \tag{5}$$

$X = 0.134$

The mutant frequency (M_f), determined independently by the recovery of cells growing in vitro in TG, determined from this subset of wells, can be derived as:

$$M_f = \frac{0.134}{0.47 \times 10} = 2.9 \times 10^{-6} \tag{6}$$

Cloning has allowed us to demonstrate that the human TG^r T-lymphocytes that arise in vivo are mutant somatic cells. I reported on our temporary propagation in TG in vitro of PBLs obtained from an individual with an elevated TG^r PBL V_f, using crude T-cell growth factors (Albertini 1979). We since have expanded several TG^r T-lymphocyte clones recovered from cloning experiments, and have fully characterized these TG^r cells. These results, reported in detail elsewhere (Albertini et al. 1982d), allow us to state with confidence, by the usually accepted criteria (Chu and Powell 1976) that the TG^r PBLs that arise in vivo are mutant somatic cells. Thus, with this method, we are truly measuring TG^r T-PBL mutant frequencies (M_f).

The TG^r T-PBL M$_f$s of normal adults determined thus far by cloning are in the range of the values given above for autoradiographically determined TG^r PBL V$_f$s using cryopreserved cells. For testing, we will be using both methods in parallel so that large numbers of individual comparisons can be made. By using multiple automated sample harvesters (Hartzman et al. 1971) and scintillation spectrometry, cloning assays can be semi-automated.

Future Directions for Lymphocyte Studies

Clearly, the availability of T-cell growth factors extends greatly the utility of T-lymphocytes for human mutagenicity monitoring. Since these cells are easily propagated in mass culture, they provide individualized cell populations for human in vitro mutagenicity testing. The question of population heterogeneity as regards susceptibility to specific mutagens can be addressed with these cells and complimentary in vitro—in vivo mutagenicity testing schemes can be developed. Also, we have grown T-lymphocytes in vitro directly from human tissues other than blood, and have demonstrated by surface markers that the T-lymphocyte populations derived from these tissues may differ from those of the peripheral blood. Questions of differential mutagenicity in vivo in different organs may be approached with this material.

It is obvious that with cloning assays performed as described, mutagenicity studies in human T-lymphocytes need not be limited to the HPRT locus. We recently have recovered diphtheria toxin resistant T-lymphocytes from human PBL populations. Figure 3 shows that diphtheria toxin resistant clones maintain this phenotype in vitro. We will soon characterize similar cells as to the genetic basis of their diphtheria toxin resistance. Quantitative cloning assays using this and a host of other markers should greatly extend the range of human specific locus mutagenicity monitoring studies with lymphocytes.

A Proposal To Validate Human Mutagenicity Monitoring Tests

This proposal, made elsewhere, is appropriate also for consideration here. (Albertini and Allen 1981).

As mentioned, one advantage of human mutagenicity monitoring tests is that they provide data on individuals that can later be correlated with the health outcomes of these same individuals. Mutagenicity tests can be performed repeatedly and clinical observations can be made over time. This advantage should be exploited. Long-term, cooperative studies designed to validate mutagenicity monitoring tests in terms of their ability to predict human disease should be implemented. Once validated, mutagenicity monitoring tests will truly be useful for making direct, quantitative, individualized human health risk assessments.

There are now effective chemotherapies and X-irradiation therapies for many human cancers. Also, the selective use of adjunctive chemotherapy following apparent surgical removal of cancer has reduced recurrence rates for some

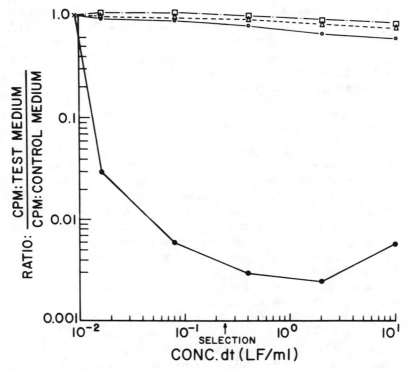

Figure 3
The resistance of human cultured T-lymphocytes, originally isolated in selective medium containing 0.25 LF diphtheria toxin/ml, to diphtheria toxin. For test, cells were inoculated at varying densities into replicate microtiter wells containing T-cell growth factors, X-irradiated lymphoblastoid feeder cells, medium RPMI 1640, 15% fetal calf serum and varying concentrations of diphtheria toxin. Wells were labeled with $1.0 \, \mu C \, {}^3H$-TdR after approximately 24 hours and counted by scintillation spectometry at approximately 42 hours. The abscissa ($\log_{10}$) shows diphtheria toxin concentration in wells; the ordinate ($\log_{10}$) shows the ratio of CPM in wells containing diphtheria toxin test medium to that in wells containing growth medium without diphtheria toxin. ($\square$———$\square$), ($\triangle$———$\triangle$), ($\circ$———$\circ$) select in 0.25 LF diphtheria toxin/ml; ($\bullet$———$\bullet$) isolated under selective conditions.

malignancies (Bonnadonna et al. 1976). However, with the benefits of therapy have come some apparent sequelae of the genotoxicity of agents used in treatment—many of which are known to be mutagens. The incidence of second malignancies has increased, sometimes dramatically, in some treated patient groups (Bender and Young 1978; Casciato and Scott 1979).

The benefits of treatment of these cancer patients clearly outweigh the attendant genetic risks. Nonetheless, these treated patients constitute a group of individuals who are knowingly and ethically exposed to potent mutagens. Many of these patients are young; many will have children following their successful therapies. Furthermore, many of these patients are treated in large, cooperative chemotherapy or radiotherapy trials in which there exists excellent clinical and epidemiological follow-up as well as infrastructures capable of sharing information and samples.

These treated cancer patients are a precious resource for study and should be studied repeatedly, and with as many mutagenicity monitoring tests as are available. A repository should be created to store at least cryopreserved blood cells and possibly sperm or other materials. A computerized data bank should be developed, which can be continually updated, to contain the results of the mutagenicity tests performed, as well as the clinical and reproductive outcomes of the individuals tested. Clearly, a cooperative study of this magnitude will not be done to confirm again the mutagenicity of agents used to treat cancer. Rather—and most importantly—it will be done to demonstrate unequivocally which mutagenicity monitoring tests have validity as predictors of human health and which are not valid predictors.

Epidemiology provides the traditional and presently most realistic endpoint for documenting human health hazards. Mutagenicity monitoring tests may allow the recognition of genetic damage occurring in vivo in humans. It must be demonstrated that quantitation of the latter allows predictions regarding the former. The ideal and goal of validation of mutagenicity monitoring tests is to replace the occurrence of disease with a test result as a useful indicator of the health or hazard associated with a human environment.

REFERENCES

Albertini, R.J. 1979. Direct mutagenicity testing with peripheral blood lymphocytes. *Ban. Rep.* 2:359.

———. 1980. Drug resistant lymphocytes in man as indicators of somatic cell mutation. *Teratog. Carcinog. Mutagen.* 1:25.

———. 1982. Human mutagenicity monitoring: Studies with 6-thioguanine resistant lymphocytes. In *Utilization of mammalian specific locus studies in hazard evaluation and estimation of genetic risk* (ed. F.J. deSerres), Plenum Press, New York. (In press).

Albertini, R.J. and E.F. Allen. 1981. Direct mutagenicity testing in man. In *Health risk analysis: Proceedings of the third life sciences symposium* (ed. C. Richmond et al.), p. 131. The Franklin Institute Press, Philadelphia.

Albertini, R.J. and W.R. Borcherding. 1982. Cloning *in vitro* of human 6-thioguanine resistant peripheral blood lymphocytes arising *in vivo*. *Environ. Mutagen.* (in press).

Albertini, R.J. and R. DeMars. 1973. Detection and quantification of X-ray induced mutation in cultured diploid human fibroblasts. *Mutat. Res.* 18:199.

———. 1974. Mosaicism of peripheral blood lymphocyte populations in females heterozygous for the Lesch-Nyhan mutation. *Biochem. Genet.* 11:397.

Albertini, R.J. and D.L. Sylwester. 1982. Thioguanine resistant lymphocytes as indicators of somatic cell mutation in man. *Environ. Sci. Res.* **25**:489.

Albertini, R.J., D.L. Sylwester, and E.F. Allen. 1982a. The 6-Thioguanine resistant peripheral blood lymphocyte assay for direct mutagenicity testing in man. In *Mutagenicity: New horizons in genetic toxicology* (ed. J.A. Heddle), p. 305. Academic Press, New York.

Albertini, R.J., K. Castle, and W.R. Borcherding. 1982b. T-cell cloning to detect the mutant 6-thioguanine resistant lymphocytes present in human peripheral blood. *Proc. Natl. Acad. Sci. U.S.A.* **79**:6617.

Albertini, R.J., E.F. Allen, A.S. Quinn, and M.R. Albertini. 1981. Human somatic cell mutation: *In vivo* variant lymphocyte frequencies as determined by 6-thioguanine resistance. In *Population and biological aspects of human mutation: Birth defects institute symposium XI* (ed. E.B. Hook and I.H. Porter), p. 235. Academic Press, New York.

Albertini, R.J., D.L. Sylwester, E.F. Allen, and B.D. Dannenberg. 1982c. Detection of somatic mutation in man. In *Carcinogens and mutagens in the environment* (ed. H.F. Stich), (In press). CRC Press, Boca Raton, Florida.

Albertini, R.J., D.L. Sylwester, B.D. Dannenberg, and E.F. Allen. 1982d. Mutation *in vivo* in human somatic cells: Studies using peripheral blood mononuclear cells. In *Genetic toxicology, an agricultural perspective* (ed. R. Fleck). Plenum Press, New York. (In press).

Allison, A.D., T. Hovi, R.W.E. Watts, and A.B.D. Webster. 1977. The role of *de novo* purine synthesis in lymphocyte transformation. In *Purine and pyrimidine metabolism* (eds. K. Elliot and D.W. Fitzsimmons), p. 207. Elsevier/North Holland/Excerpta Medica, Amsterdam.

Balis, M.E. 1968. Enzymology and biochemistry. B. Aspects of purine metabolism. *Fed. Proc.* **27**:1067.

Bender, R.A. and R.C. Young. 1978. Effects of cancer treatment on individual and generational genetics. *Semin. Oncol.* **5**:46.

Bonadonna, G., E. Brusamolino, P. Valagussa, A. Rossi, L. Brugmatelli, C. Brambilla, M. DeLena, G. Tacini, E. Bajetta, R. Musumeci, and V. Veronesi. 1976. Combination chemotherapy as an adjacent treatment in operable breast cancer. *N. Engl. J. Med.* **294**:404.

Boyum, A. 1968. Separation of leukocytes from blood and bone marrow. *Scand. J. Clin. Lab. Invest. Suppl.* 97 **21**:51.

Brogger, A. 1979. Chromosome damage in human mitotic cells after *in vivo* and *in vitro* exposure to mutagens. In *Genetic damage in man caused by environmental agents* (ed. K. Berg), p. 87. Academic Press, New York.

Caskey, C.T. and G.D. Kruh. 1979. The HPRT locus. *Cell* **16**:1.

Casciano, D.A. and J.L. Scott. 1979. Acute leukemia following prolonged cytotoxic agent therapy. *Medicine* **58**:32.

Chu, E.H.Y. and S.S. Powell. 1976. Selective systems in somatic cell genetics. *Adv. Hum. Genet.* **7**:189.

Dancis, J., P.H. Berman, V. Jansen, and M.E. Balis. 1968. Absence of mosaicism in the lymphocyte in X-linked congenital hyperuricemia. *Life Sci.* **7**:587.

DeMars, R. 1971. Genetic studies of HGPRT deficiency and the Lesch-Nyhan syndrome with cultured human cells. *Fed. Proc.* **30**:944.

Elion, G.B. 1967. Biochemistry and pharmacology of purine analogs. *Fed. Proc.* **26**:898.

Elion, G.B. and G.H. Hitchings. 1965. Metabolic basis for the actions of analogs of purines and pyrimidines. *Adv. Chemother.* 2:91.

Evans, H.J. and M.L. O'Riordan. 1977. Human peripheral blood lymphocytes for the analysis of chromosome aberrations in mutagen tests. In *Handbook of Mutagenicity Test Procedures* (eds. B. Kilbey et al.), p. 261. Elsevier/North Holland, Amsterdam and New York.

Hartzman, R.J., M. Segall, M.L. Bach, and F.H. Bach. 1971. Histocompatibility matching. VI. Miniaturization of the mixed leukocyte culture test, a preliminary report. *Transplantation* 11:268.

Hovi, T., A.C. Allison, K.O. Raivio, and A. Vaheri. 1977. Purine metabolism and control of cell proliferation. In *Purine and pyrimidine metabolism* (eds. K. Eliot and D.W. Fitzsimmons), p. 225. Elsevier/North Holland/Excerpta Medica, Amsterdam.

Kapp, R.W., Jr., D.J. Picciano, and C.B. Jacobson. 1979. Y-chromosomal nondisjunction in dibromochloropropane-exposed workmen. *Mutat. Res.* 64:47.

Kelley, W.N. 1968. Enzymology and biochemistry. A HGPRT deficiency in the Lesch-Nyhan syndrome and gout. *Fed. Proc.* 27:1047.

Latt, S., J.W. Allen, W.E. Rogers, and L.A. Juerglus. 1977. *In vitro* and *in vitro* analysis of sister chromatid exchange formation. In *Handbook of mutagenicity test procedures* (eds. B. Kilbey et al.), p. 275. Elsevier/North Holland, Amsterdam.

Lesch, M. and W.L. Nyhan. 1964. A familial disorder of uric acid metabolism and central nervous system function. *Am. J. Med.* 36:561.

Lyon, M.F. 1961. Gene action in the X-chromosome of the mouse (mus Musculus L.). *Nature* 190:372.

Maizel, A.L., S.R. Mehta, S. Hauft, D. Franzini, L.B. Lachman, and R.J. Ford. 1981. Human T-lymphocyte/monocyte interaction in response to lectin: Kinetics of entry into S-phase. *J. Immunol.* 127:1058.

McDonald, J.A. and W.N. Kelley. 1972. Lesch-Nyhan syndrome: Absence of the mutant enzyme in erythrocytes of heterozygote for both normal and mutant hypoxanthine-guanine phosphoribosyltransferase. *Biochem. Genet.* 6:21.

Nyhan, W.L., B. Bakay, J.D. Connor, J.F. Marks, and D.K. Kelle. 1970. Hemizygous expression of glucose-6-phosphate dehydrogenase in erythrocytes of heterzygotes for the Lesch-Nyhan syndrome. *Proc. Natl. Acad. Sci. U.S.A.* 65:214.

Osterman-Golkar, S. 1981. Dosimetry of electrophilic compounds by means of their reaction products with hemoglobin: A method directly applicable to man. In *Health risk analysis: Proceedings of the third life sciences symposium* (eds. C.R. Richmond, et al.), p. 147. The Franklin Institute Press, Philadelphia.

Papayannopoulou, T.H., G. Lim, T.C. McGuire, V. Ahern, P.E. Nute, and G. Stamatoyannopoulos. 1977. Use of specific fluorescent antibody for the identification of a hemoglobin C in erythrocytes. *Am. J. Hematol.* 2:95.

Papayannopoulou, T.H., T.C. McGuire, G. Lim, E. Garzel, P.E. Nute, and G. Stamatoyannopoulos. 1976. Identification of hemoglobin S in red cells and normoblasts using fluorescent anti-Hb antibodies. *Br. J. Haematol.* 34:25.

Paul, W.E., B. Sredni, and R.H. Schwartz. 1981. Long-term growth and cloning of non-transformed lymphocytes. *Nature* **294**:697.

Pero, R.W. and F. Mitelman. 1979. Another approach to *in vivo* estimation of genetic damage in humans. *Proc. Natl. Acad. Sci. U.S.A.* **76**:462.

Perry, P. and H.J. Evans. 1975. Cytological detection of mutagen-carcinogen exposure by sister chromatid exchange. *Nature* **258**:121.

Russell, L.B. 1963. Mammalian X-chromosome action: Inactivation limited in spread and in region of origin. *Science* **40**:986.

Seegmiller, J.E., F.M. Rosenbloom, and W.N. Kelley. 1967. Enzyme defect associated with a sex-linked human neurological disorder and excessive purine synthesis. *Science* **155**:1682.

Stamatoyannopoulos, G., P.E. Nute, T.H. Papayannopoulos, T. McGuire, G. Lim, H.F. Bunn, and D. Racknagel. 1980. Development of a somatic mutation screening system using Hb mutants. IV. Successful detection of red cells containing the human frameshift mutants HB Wayne and Hb Cranston using monospecific fluorescent antibodies. *Am. J. Hum. Genet.* **32**:484.

Stetka, D.G. and S. Wolff. 1976. Sister chromatid exchange as an assay for genetic damage induced by mutagen-carcinogens. I. *In vivo* test for compounds requiring metabolic activation. *Mutat. Res.* **41**:333.

Strauss, G.H. and R.J. Albertini. 1979. Enumeration of 6-thioguanine resistant peripheral blood lymphocytes in man as a potential test for somatic cell mutations arising *in vivo*. *Mutat. Res.* **61**:353.

Strauss, M., L. Lubbe, and E. Geissler. 1981. HG-PRT structural gene mutation in the Lesch-Nyhan syndrome as indicated by antigenic activity and reversion of the enzyme deficiency. *Hum. Genet.* **57**:185.

Strauss, G.H., R.J. Albertini, P. Krusinski, and R.D. Boughman. 1979. 6-thioguanine resistant peripheral blood lymphocytes in humans following psoralen long-range UV light therapy. *J. Invest. Dermatol.* **73**:211.

Wolff, S. 1977. Sister chromatid exchanges. *Ann. Rev. Genet.* **11**:183.

Wyrobek, A.J. and W.R. Bruce. 1978. The induction of sperm shape abnormalities in mice and humans. *Chem. Mutagens* **5**:257.

COMMENTS

ZETTERBERG: Through the freezing effect you have for variant frequency, would it be possible that just more thioguanine gets into the frozen cells?

ALBERTINI: Well, it is possible. However, with very short culture intervals, such as a 12-hour labeling interval alone, there are actually more of the spontaneously cycling cells in the cryopressed samples that label in TG. It is not that we lose these cells by freezing; it is just that they enter their S period earlier in culture. You see, the "pseudo" TG^r cells are still there, but they incorporate label early. But at every time point after the first 12 hours there are fewer of these "pseudo" resistant cells that label until, at 30 hours incubation and 12 hours of label in most cases, there are none.

MOHRENWEISER: Have you had a chance to determine if any of these are CRM-positive?

ALBERTINI: I haven't looked at them yet, no. One of the advantages of cloning is that you have the mutant cells for their studies as well as their DNA, to look at that level of molecular damage. Of course we have performed the enzyme assays to show that the TG^r mutants are HPRT-deficient.

EVANS: How many clones have you looked at in terms of classifying T4s or T8s? You showed us eight, and they were all T4s, weren't they? (Note: Surface markers determination data were shown for several cultured T-cell clones—both normal and mutant.)

ALBERTINI: Just those. Those were all T4s. But I have only looked at that particular group. We are going to continue to do that. It is possible that T8 lymphocytes will not clone as well in vitro.

EVANS: You are not going to place too much weight on that, are you?

ALBERTINI: No. Again, you have seen the data.

EVANS: Going back to the question we just had about freezing, what surprised me was to see that the variance of the variant frequency in the frozen populations is quite large. You went to 10^{-7}, I think, with two people. I wouldn't expect that.

ALBERTINI: This is the variance we place on the estimate. We calculate CIs for single estimates based on the fact that the numerator in the variant frequency ratio is a Poisson variable and the denominator is a binomial variable. So the precision of an estimate rests very much on the number of labeled nuclei counted. We must score at least 14 positive TG^r nuclei to be able to detect as significant with 95% confidence, a doubling. In some of the studies shown we did not count that many labeled nuclei from TG-containing cultures. So we now use flask cultures, where we culture 10 million cells. We are trying to base our variant frequency estimates on 20, 30, or 40 labeled nuclei. This will narrow CIs. At present we often have statistically large CIs because the numbers of positives are so small.

EVANS: Can I ask you about the cloning up to get 10^{12} lymphocytes, the growth factor or lymphoblastoid cell line feeder? Do you pick up more T4s in this kind of growth? Are they selective populations from that point of view or are they all helper cells?

ALBERTINI: When you look at the mass populations, both T4 and T8 cells are present. The mass population has both helpers and suppressors. It is only in the cloning. I have gotten some T8s out by cloning. But two of six of normals were T8, but one clearly was a doublet, because it had both T8 and T4. Actually one out of seven was T8—that is close to what they exist in the blood.

EVANS: And these are not immortal, are they?

ALBERTINI: Well, at 10^{13}, I think most people say they are immortal—the mouse people have kept them going for 2 years with mice. In the humans I think they have quit. I talked to the people who have done it. Clearly, if you stop adding growth factor, they die.

Use of Flow Cytometry to Concentrate 6-Thioguanine-Resistant Variants of Human Peripheral Blood Lymphocytes

GÖSTA ZETTERBERG AND HERMAN AMNÉUS
Department of Genetics
University of Uppsala
S-75007
Uppsala, Sweden

PER MATSSON
The Gustaf Werner Institute
University of Uppsala
S-75007 Uppsala
Sweden

Since mutation rates are low, very large numbers of individuals or cells have to be screened in all mutation test systems. Test systems based on selection procedures, e.g., selective growth media, often make possible the handling of enough cells to detect mutant frequencies as low as 10^{-8} or 10^{-7}. In addition, methods to determine such low frequencies always rely on a very efficient and simple scoring criteria for mutant clones, like colonies of bacteria on agar plates. Whenever the detection of mutants involves an actual inspection and classification of all the individual units in the test, be it single cells or whole animals, more work is required and the sensitivity of the test will usually be inversely proportional to the required investment of work, time, and money.

The method developed by R.J. Albertini and his group (Strauss and Albertini 1979) for scoring human lymphocyte variants resistant to 6-thioguanine (TG) involves as the endpoint the microscopic counting of rare nuclei labeled with ^{3}H-TdR. This is tedious work and would be even more so if the frequency of variant cells were much lower than has actually been demonstrated. The accuracy in the determination of an increase of the variant frequency is furthermore dependent on the total number of variants scored. In order to increase the potential resolution of this method we have developed a technique to concentrate the variant lymphocytes through use of flow cytometry.

The lymphocyte system makes use of a genetic marker, the X-chromosome-linked HGPRT locus, well known from experiments with established mammalian cell lines. In the presence of TG or AG cells deficient in HGPRT activity are resistant to the base analogs and can synthesize DNA and proceed through several cell cycles while the HGPRT proficient cells are sensitive and stop DNA synthesis early in S-phase. At the start of a lymphocyte culture the cells are non-dividing, i.e., in G_0 phase. A mitogen like phytohemagglutinin (PHA) stimulates a certain proportion of the cells to divide fairly synchronously. Thus, after a certain time of culturing in the presence of TG the resistant cells have a higher DNA content than the sensitive cells. Theoretically, an instrument that can effectively register the DNA content of single cells could be used to discriminate the resistant cells from the sensitive ones. A flow cytometer is such an instrument that can be used to precisely measure the DNA content of individual cells. It can handle a great number of cells in a short time, many thousands per second. In principle, this seemed to be an easy method to determine the frequency of TG-resistant (TG^r) variants of mononuclear blood cells. However, this approach also has some difficulties.

METHODS

In general we followed the procedures described by Strauss and Albertini (1979) for the culturing and processing of the lymphocytes to obtain suspensions of nuclei for flow cytometric analysis.

The cells were harvested after 48 hours of culturing, including radioactive labeling for the last 6 hours in medium with ^{3}H-TdR. After washing 150 μl propidium iodide (1 mg/ml H_2O) was added to the cell pellets which were re-suspended in 0.1% sodium citrate. Hypotonic treatment lasted for 20 minutes upon which an equal amount of 40% ethanol was added during simultaneous agitation on a Vortex mixer for 15 seconds. The suspension of nuclei was then analyzed by flow cytometry with a FACS-III apparatus (Becton-Dickinson Electronics Laboratory) equipped with a 5 W argon laser and dual fluorescence detection. The 488 nm laser line was used for excitation at a power of 160-300 mW. Cut-off filters at 520 nm and 620 nm were used for fluorescence detection. The set up allowed a 256-channel display. As sheath fluid we used a 0.1% sodium citrate in 20% ethanol. The flow rate used was 3000-5000 nuclei per second. The nuclei were outsorted directly onto slides which were dried at room temperature and fixed overnight at 4°C in methanol:glacial acetic acid (5:1; v/v). The nuclei were stained with acetoorcein. For autoradiography the slides were dipped in Ilford K5 emulsion and exposed for 6 days at room temperature. (A more detailed description of the methods is given in Amnéus et al. 1982.)

RESULTS AND DISCUSSION

In order to obtain a good discrimination between nuclei from cells in $G_0 + G_1$, presumably being sensitive to TG_1 or from nonstimulated cells, and nuclei from cells in $S + G_2$, the spectrum of the control culture without TG was first analyzed. A region from the minimum between the $G_0 + G_1$ peak and the G_2

peak and extending beyond the G_2 peak of the control (Fig. 1) was selected for the counting of nuclei from TG^r cells. The high proportion of cells in the selected region indicated that the suspension contained doublets of $G_0 + G_1$ nuclei. The proportion of nuclei registered in the selected region ranged between 1×10^{-4}-5×10^{-3} which was much higher than the expected frequency of TG^r variants. Thus, in order to discriminate $G_0 + G_1$ doublets from nuclei in $S + G_2$ it was necessary to include an autoradiographic analysis as described by Strauss and Albertini (1979). Unlike the earlier approach, however, the outsorting provided us with an enrichment of nuclei from TG^r cells. The enrichment factor was usually about 500-fold and was influenced largely by how well we succeeded in obtaining a suspension of nuclei with a low number of doublets.

The time required for scoring a given number of labeled nuclei is considerably reduced by the outsorting procedure. Furthermore, the scoring was easier because the outsorting eliminated most of the disturbing small particles that contaminated slides with nuclei from nonsorted suspensions.

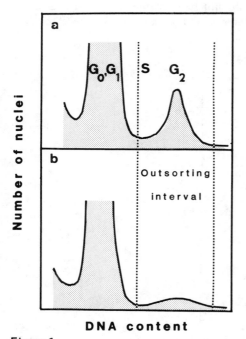

Figure 1

Distribution of nuclei according to DNA content as registered with the flow cytometer. (a) lymphocytes grown for 48 hr in the absence of TG (b) lymphocytes grown for 48 hr in the presence of TG. Nuclei in the indicated range are outsorted giving an enrichment of nuclei from TG^r cells.

The labeling index (LI) was determined in the control culture without TG and in the culture containing TG using nuclei from identical intervals of the spectrum. From the control culture about 500 nuclei were analyzed after autoradiography while from the TG culture all nuclei in the outsorted interval were analyzed. The variant frequency (Vf) was calculated according to:

$$Vf = \frac{\text{Labeling index in presence of TG}}{\text{Labeling index in absence of TG}} \tag{1}$$

The outsorting in the $S + G_2$ region resulted in a loss of about 40% of the labeled nuclei seen in nonsorted preparations from parallel cultures. The losses were identical both in the presence and absence of TG. For concentrations of TG between 1×10^{-6} M and 1×10^{-4} M it was found that the outsorting gave a lower Vf than in nonsorted preparations. The largest difference was noted for concentrations of TG around 1×10^{-5} M (Fig. 2). It seems likely that around this concentration of TG a fraction of sensitive cells can synthesize enough DNA so that their nuclei will appear labeled in a nonsorted preparation. It is also possible that some of the labeled cells in the nonsorted material are variants which are only partly resistant to TG and have a prolonged S-phase under selective conditions. Thus they would not have synthesized enough DNA to enter into the standard outsorting interval and would therefore be eliminated

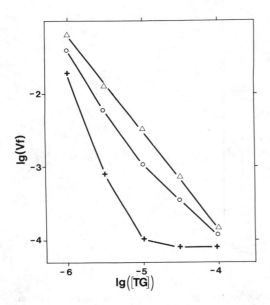

Figure 2
The frequency of resistant cells (with ^{3}H-TdR-labeled nuclei) as a function of the concentration of TG. The variant frequency is indicated for nonsorted cells $(G_0 + G_1 + S + G_2)$, [O]; for outsorted cells in $(G_0 + G_1 + S)$, [△]; and for outsorted cells in $(S + G_2)$, [X].

from the scoring. In support of this, quantitative flow cytometry of cells from cultures with TG concentrations in the range 10^{-6}-10^{-5} M showed an additional peak in early S-phase representing a fraction of sensitive cells with incompletely inhibited DNA synthesis at those TG concentrations.

In samples from two blood donors, serving as interval standards in all culturing experiments, we studied the inhibition profile of TG-sensitive cells as a function of the TG concentration. The results are shown in Fig. 3. Each Vf value is based on more than 1×10^7 analyzed cells. Qualitatively identical profiles were also reported by Strauss and Albertini (1979) and by Evans and Vijayalaxmi (1981) not using flow-cytometric enrichment of variants.

The variant frequency showed a variation of about 20% in six samples taken at about monthly intervals from a person serving as one of our internal standards (Table 1). This variation is about the same as the calculated methodo-logical scatter determined by analysis of parallel cultures from 76 blood donors. The scatter around the mean of the variant frequency was ± 20% for cultures started at the same time and ± 30% for cultures from the same persons but started on different occasions.

In a group of 33 healthy blood donors the variant frequency was de-termined at two concentrations of TG, 1×10^{-4} M and 2×10^{-4} M. There was a great inter-individual variation. The standard deviation was ± 0.83 × the

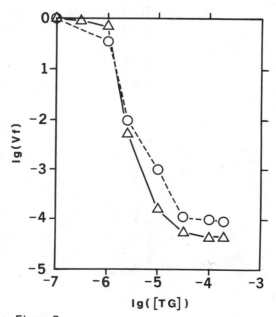

Figure 3
Vf as a function of the concentration of TG for two healthy blood donors.

Table 1
Data for Calculating Variant Frequency in Six Cultures[a]

No TG		2 × 10⁻⁴ M TG					
LI_{tot}[b]	LI_{sort}[b]	Total number of nuclei ($\times 10^7$)[c]	Number of outsorted nuclei ($\times 10^4$)[c]	Enrichment factor[d]	Number of labeled nuclei[c]	LI_{TG} ($\times 10^{-5}$)[e]	V_f ($\times 10^{-5}$)
0.28	0.18	1.09	5.86	× 186	122	1.12	6.26
0.25	0.16	0.96	1.68	× 573	108	1.13	7.08
0.26	0.16	1.10	3.45	× 319	139	1.27	7.95
0.26	0.16	0.86	1.77	× 486	98	1.15	7.20
0.24	0.15	0.88	2.30	× 382	73	0.84	5.63
0.27	0.17	1.30	4.55	× 286	139	1.07	6.31

The blood samples were taken at about monthly intervals for 6 mo.
[a] This is represented by △ in Figure 3.
[b] The labeling indices in nonsorted and outsorted nuclei from cultures without TG.
[c] These figures are the sums from triplicate cultures.
[d] The figures for Enrichment factor, LI_{TG}, and V_f are the mean values from triplicate cultures.
[e] The labeling indices in outsorted nuclei from cultures with 2 × 10⁻⁴ M TG.

calculated mean value at 1×10^{-4} M TG and 0.59 $\times$ the calculated mean value at 2×10^{-4} M TG. This is significantly higher than the standard deviation $\pm$ 0.25 $\times$ the calculated mean value due to methodological scatter.

The HGPRT enzyme phosphorylates the base analog TG and the product causes the death of the sensitive HGPRT-proficient cell. The exact mechanism of the toxic effect of TG is not known. It seems likely that the inhibiting effect of TG is proportional to the activity of the HGPRT enzyme at a given concentration of TG. Thus, cells with less HGPRT activity would be less sensitive than cells with a higher HGPRT activity. Increasing the concentration of TG causes more cells to be killed in a population of cells with different HGPRT activity. This is reflected in the shape of the selection curve showing the fraction of surviving (resistant) cells as a function of the TG concentration. We observed that the selection profiles vary from person to person. This might reflect differences in overall HGPRT activity. It follows that at a concentration of TG, high enough to inhibit also cells with a low HGPRT activity, only fully resistant variants will be scored. If the concentration of TG is further increased it should have no effect on those variants and the selection curve then reaches a plateau. Strauss and Albertini (1979) and Evans and Vijayalaxmi (1981) found that their selection curves reached a plateau at a concentration of about 2×10^{-4} M TG or AG, respectively. For many of the blood donors investigated by us a similar function was established but there was a considerable scatter between different persons.

The frequencies of spontaneous TG^r or AG^r mutants in cultures of established mammalian cell lines are usually lower (Caskey and Kruh 1979) than has been reported for human lymphocyte cultures (Strauss and Albertini 1979; Evans and Vijayalaxmi 1981; Amnéus et al. 1982). The majority of the mutants from mammalian cell lines have been identified as structural mutants due to point mutations (Caskey and Kruh 1979). Evans and Vijayalaxmi (1981) concluded that the AG^r lymphocyte variants they obtained after in vitro X-irradiation were due to loss of the HGPRT gene by chromosome breakage. It is possible that in mammalian cell lines HGPRT mutants due to chromosomal deficiencies are eliminated during the cloning. Albertini and Borcherding (1982) recently reported that they have succeeded in cloning TG^r human lymphocytes showing that these completely lack HGPRT activity. Thus, the difference in mutation frequency observed for the same locus in different systems might reflect differences in opportunities for selection to occur.

Albertini et al. (1981) have observed that in human peripheral blood there is a low frequency of lymphocytes dividing in vivo and that these are resistant to TG. By a freezing-thawing procedure this fraction of cycling lymphocytes could be eliminated from being scored as resistant variants. This decreased the observed variant frequency in controls by an order of magnitude. For persons exposed to mutagenic cytostatics the same handling of the lymphocyte cultures drastically reduced the induced variant frequency previously determined in blood samples from patients treated with cytostatics but not using the freezing procedure.

If by modifications of the method a considerably lower frequency of variants can be expected, the number of cells that have to be screened will have to be increased proportionally in order to determine accurately the variant frequency. The application of flow cytometry for this mutation system will therefore become important.

ACKNOWLEDGMENT

Work supported by grants 79/1069 and 80/1375 from the Swedish Council for Planning and Coordination of Research and by grant B-2468-101 from the Swedish Natural Science Research Council.

REFERENCES

Albertini, R.J., E.F. Allen, A.S. Quinn, and M.R. Albertini. 1981. Human somatic cell mutation: *In vivo* variant lymphocyte frequencies as determined by 6-thioguanine resistance. In *Population and biological aspects of human mutation* (ed. E.B. Hook and J.H. Porter), p. 235. Academic Press, New York.

Albertini, R.J. and W.R. Borcherding. 1982. Cloning *in vitro* of human 6-thioguanine resistant (TG^r) peripheral blood lymphocytes (PBLs) arising *in vivo*. Abstract 13th Annual Meeting Environ. Mutagen Society. *Environ. Mutagen.* (in press).

Amnéus, H., P. Matsson, and G. Zetterberg. 1982. Human lymphocytes resistant to 6-thioguanine: Restrictions in the use of a test for somatic mutations arising *in vivo* studied by flow-cytometric enrichment of resistant cell nuclei. *Mutat. Res.* (in press).

Caskey, C.T. and G.D. Kruh. 1979. The HGPRT locus. *Cell* 16:1.

Evans, H.J. and Vijayalaxmi. 1981. Induction of 8-azaguanine resistance and sister chromatid exchange in human lymphocytes exposed to mitomycin C and X rays *in vitro*. *Nature* 292:601.

Strauss, G.H. and R.J. Albertini. 1979. Enumeration of 6-thioguanine-resistant peripheral blood lymphocytes in man as a potential test for somatic cell mutations arising *in vivo*. *Mutat. Res.* 61:353.

COMMENTS

PARODI: You labeled the cells and tried to count the labeled cells. Have you abandoned the idea of counting the cells that have the double amount of weaker chelating dye stain. Do you label the cells and measure a kind of absorbing parameter?

ZETTERBERG: No. "Abandoned" may be too much to say. But I think the chances of having the original idea working—that is, to just let the flow cytometer count nuclei that have double amounts, or nearly double amounts, of DNA—is not going to work, just for practical purposes. The clumping of nuclei occurs and it looks like we can't get rid of that. Now we label the DNA with tritiated thymidine and run the nuclei through the flow cytometer. This gives an enrichment of nuclei, because we score now for TG^r variants in the region where they should be, among nuclei with an increase in DNA content. This gives a reduced number of nuclei to score on our slides, compared to taking a sample directly from an unsorted culture. With this method there are more nuclei to look for.

PARODI: Do you select the ones that get labeled?

ZETTERBERG: You can't select for radioactive labeling in a flow cytometer. I wish there were an instrument like that. A combination of scintillation and flow cytometry would be nice.

MALLING: There are monoclonal antibodies to BrdU for DNA. You could put a fluorescence marker in that antibody to see which cells have incorporated BrdU.

ZETTERBERG: Yes, maybe the signal would be strong enough for the flow cytometer.

EVANS: We found exactly the same problems you have with cell clumping so we gave up on flow cytometry for this work. We have a slide scanning system which does 1,000 cells/sec, which is as fast as a flow cytometer. This retains all the cells on the slide. The answer probably is a slide-based scanning system, not a flow system since there are too many complications with this.

Direct Mutagenicity Testing: The Development of a Clonal Assay to Detect and Quantitate Mutant Lymphocytes Arising In Vivo

GARY H. S. STRAUSS*
Medical Research Council
Cell Mutation Unit
Sussex University
Falmer, Brighton
United Kingdom

Owing to the technologically oriented nature of our society, we genetic toxicologists need to develop methods for making valid ascertainments of hazards for present populations and future generations. Our ultimate concern must be to provide useful data for sound risk-benefit predictions and decisions. We are now faced with the serious problem of developing relevant indicators of genotoxic exposures for direct estimations of human health risk. Robert DeMars has called attention to the need for direct testing methods (DeMars 1973). During the second Banbury conference he stated that we must develop methods for . . . "the quantification of mutant cells immediately upon the removal of endogenous somatic cells from the body of an animal after they have been subjected to mutagenic influences within their normal context in the animal (DeMars 1979)." The Strauss-Albertini test was developed as a direct method to detect and enumerate 6-thioguanine (TG) resistant peripheral blood lymphocytes (TG PBLs) arising in vivo in humans and animals (Strauss and Albertini 1979). Figure 1 provides a schematic outline of the method. This test is relatively simple, rapid, and inexpensive. The variants which incorporate $[^3H]$-TdR in vitro in response to phytohemagglutinin (PHA) despite the presence of TG can be counted by light microscopy after autoradiography. This test is well suited to longitudinal monitoring of individuals due, in part, to the brevity of both latency and patency periods, determining variant frequencies following exposures to mutagens. Recently it was discovered that "phenocopies," cells which are artifactually resistant to TG in the test, exist but effectively can be eliminated (Albertini, this volume) and the method has been modified accordingly. The variants are assumed to be HGPRT deficiency mutants by virtue of various characterizations and analogy with the naturally occurring Lesch-Nyhan (LN) mutation.

*Present Address: United States Environmental Protection Agency, Health Effects Research Laboratory, Genetic Toxicology Division, Research Triangle Park, North Carolina 27711

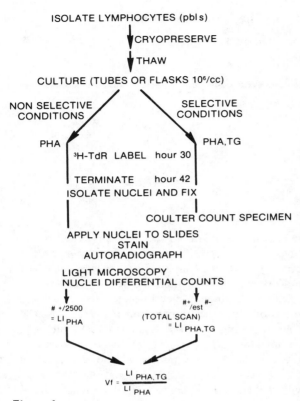

Figure 1
Strauss-Albertini Test (Autoradiographic Assay)

Confirmation of the mutant nature of the variants is now possible using an adaptation of popular immunological technique for cloning nontransformed T-lymphocytes from blood and other tissues including lymph nodes, spleen, bone marrow, and skin. The T-cell can be grown continuously under the influence of T-cell growth factor (TCGF) (Ruscetti et al. 1977; Gillis et al. 1978; Bonnard et al. 1980). Rare drug resistant precursor cells can be isolated and expanded to form visible colonies in appropriate selective media containing TCGF.

The purpose of this report is to present preliminary results from a system under development which was designed to clone and quantitate mutant lymphocytes taken directly from the body.

EXPERIMENTAL PROCEDURES

Human Studies

Two normal, healthy individuals (a female, age 24 and a male, age 30) volunteered to be studied and donated blood repeatedly as required. Dr. Gillian McCarthy kindly provided blood specimens from 4 LN hemizygous boys under her care at the Chailey Heritage Hospital, North Chailey, Sussex.

Preparation of Peripheral Blood Lymphocytes

Materials and methods employed in the preparation of PBLs for these studies have been presented in detail previously (Strauss 1982a) and will be described in brief now. Mononuclear cells (mainly PBLs) were extracted from venous blood specimens by means of density gradient centrifugation. Cells of the lymphocyte band were saline washed and resuspended in aliquots, some with cryopreservation medium and the remainder with various culture media. Some PBLs were frozen, then thawed with careful attention to efficient recovery of viable cells and were resuspended in appropriate culture media for parallel tests with non-cryopreserved aliquots. PBLs were prepared for use in several applications:

1. TCGF production;
2. establishment and characterization of continuous T-lymphocytes (CTLs);
3. TCGF quality and CTL growth assessment;
4. Strauss-Albertini test (fresh and frozen PBLs) and
5. clonal assay (fresh PBLs and CTLs).

TCGF Production

TCGF containing supernates were prepared for these studies by culturing PBLs from the two normal volunteers as follows: 5.0×10^7 cells were placed in flasks of 50cc RPMI-1640 medium, 24 mM Hepes supplemented with 1% penicillin/streptomycin (100 units/100μg/cc), 300μg/cc fresh L-glutamine, and 1% autologous serum. Optimal mitogenic doses of Difco Phytohemagglutinin-A, type M(PHA) of 1% v/v were added to PBL suspensions which were incubated at 37°C in an humidified 5% CO_2 atmosphere. 48-hour cultures were harvested by decanting through coarse filters then stored at 4°C, concentrated fivefold by Amicon filtration, .45μm Swinnex filter sterilized and again stored at 4°C. The crude lymphokine-containing conditioned media concentrates were tested for TCGF activity (described below) and used accordingly. For our purposes the simple term TCGF suffices to describe these reagents.

Cultured T-Lymphocytes

Continuous cultures of T-cells were initiated from PBL suspensions derived from the bloods of the two normal and four LN subjects. The cells were resuspended in T-cell growth medium (TCGM) at $.5 \times 10^6$/cc in various volumes ranging from 1–100 cc. TCGM is the combination of complete medium (CM) and TCGF. CM consists of RPMI-1640 medium containing the supplements already mentioned with 20% (v/v) fetal calf serum (FCS) or AB human plasma substituted as the serum source. A simple assay, described below, had been employed to determine that 20% TCGF (v/v) in CM represents an average optimal stimulatory dose (data not presented here) for continued growth of T-lymphocytes incubated at 37°C in humidified 5% CO_2 atmospheres. Cells were fed by dilution with TCGM after the first 4-5 days of incubation. Thereafter, cell densities routinely were maintained between .5 to 1.0×10^6/cc by viability counting and refeeding every 2-3 days.

TCGF Quality and CTL Growth Assessments

TCGF preparations were assessed for biological activity by measuring their capacities to cause concentration dependent stimulation of ^{3}HTdR incorporation by established CTLs. CTLs which had been cultivated at least a fortnight were washed of TCGM and were resuspended at $.5 \times 10^6$ cells/cc in tubes and reincubated. After 36 hours under these conditions serial dilutions of various TCGF samples in CM were added to the CTLs and the cultures plated in .2cc volumes and replicates of 6 among 96 flat bottom microtiter wells. Following a 24-hour incubation period, 1.0 μCi of ^{3}H-TdR was added to each well and the plates were returned to the incubator for an additional 6 hours. The contents of the wells were then harvested onto glass filter strips through a cell harvester. The filter-strips were dried and added to scintillation fluid in counting vials for liquid scintillation spectrophotometry. Based on reports from those who introduced the original version of this method (Bonnard et al. 1980; Lotze et al. 1980) and our own results, dilutions of TCGF inducing at least 25,000 cpm/10^5 CTLs can be expected successfully to support prolonged CTL growth. A variant of this assay was used to test the proliferative responses of CTLs in TCGM from two normal and two LN individuals in the presence of TG at various concentrations. The details of incubations, labeling, harvesting, and counting were identical to those described above. Growth assessments over time were also accomplished by viability counting. Trypan blue dye exclusion and sometimes fluorescein diacetate activation were used to assess cultures for cell viability and extent of proliferation. 2×10^{-5}M was deemed to be an optimal selective concentration of TG on the basis of kill curves produced as outlined above. TG used in these experiments was prepared from swinnex filtered stock solution of 2×10^{-2}M in 0.5% Na_2CO_3; the diluent alone was used as a control additive. In one set of growth experiments evaluated on the basis of viable cells per culture, TG^R CTLs from a normal individual gradually were grown up in TCGM containing 2×10^{-5}M TG, then placed in TCGM without TG and finally challenged to grow in selective medium.

STRAUSS-ALBERTINI TEST

The schematic (Fig. 1) presented with the introductory remarks outlines the means by which PBLs from two normal and two LN individuals were assessed for the presence of TG^r variants. A more complete description of this method is presented elsewhere (Strauss 1982a). The present studies compared frequencies of TG^r PBLs in both fresh and frozen specimens from each individual mentioned on four separate occasions using the Strauss-Albertini test. As is usual, the final concentration of TG in this test was 2×10^{-4}M; however, one should note that extra cells under selection were cultured and spread on slides for counting owing to the anticipated fall in variant frequencies among cells frozen to eliminate phenocopies. Concurrently TG^r PBL frequencies were also determined by means of our new clonal assay using both fresh PBLs and CTLs derived from these subjects.

CLONAL ASSAY

The assay is an adoption of a technique for immunological cloning from limiting dilutions without the use of soft agar or preincubations for priming. Sheep red blood cells (S-RBCs) were received in Alsever's solution, stored at 4°C until use (within 2 weeks of bleeding), when they were irradiated with 2000 rads (^{137}cesium source) and washed several times in HBSS. Cells, either fresh PBLs, frozen and thawed PBLs, or CTLs were diluted to appropriate numbers in TCGM containing .05% x-irradiated sheep red blood cells (x-SRBC) either with or without TG at a final concentration of 2×10^{-5}M. Cells under nonselective conditions were plated at an average of 1 cell/.2cc medium into the .2cc wells of a flat-bottom 96-well microtiter plate. The counterpart cells under selection were plated out at 10^5 cells/well in .2cc volumes over 96 wells. Paired plates thus initiated representing each test cell, and sometimes replicates thereof, were incubated at 37°C, in an humidified 5% CO_2 in air atmosphere. After 5 days and again after 9 days, .1cc of medium was removed from each well and replaced with fresh medium with or without TG as before, although on the latter occasion 1.0μCi ^{3}H-TdR/well was included. On day 10 (or sometimes a day or 2 later) plates were ready for sequential assessment by three comparable methods:

1. counting of macroscopic colonies by visual inspection. Although one might have thought the S-RBC feeders would tend to obscure colonies from enumeration, on the contrary, lysis by acids produced by proliferating lymphocytes created distinctive halos around colonies. Colonies generally contained 50-500 cells at this stage and were not difficult to score.
2. colourimetric indication of activity per well. Sufficient pH effects occur (and these can be further enhanced by addition of extra methyl red at the time of assay) to allow differentiation of wells on the basis of colour. This mode of

evaluation has potential for automation via state-of-the-art ELISA plate readers. But the final method provided the greatest confidence in accuracy.

3. scintillation counting of $[^3H]$-TdR incorporation after harvesting of cellular material from wells onto filters as already described.

Some colonies detected by methods 1 or 2 were picked and subcloned for use as expanded CTLs and for further characterizations including classifications of cell type on the basis of surface antigens. Results (not presented here) of scintillation counting correlate favourably with those of the other methods. Virtually no ambiguous results have been in evidence since scintillation counting results were absolutely bimodal in distribution and could be related directly to visible effects. By scoring wells positive or negative on the basis of these assays it was possible to calculate cloning efficiency (CE) assuming ideal Poisson distribution. For each 96-well plate the number of cells per well resulting in the percent positive wells was divided by the number of cells per well plated. In this way, for each pair of plates, CEs were determined for cells TG-exposed and cells not TG-exposed such that the ratio $CE_{TCGM,TG}/CE_{TCGM}$ approximates in vivo (or for CTLs, in vitro) TG^r PBL mutant frequencies.

CHARACTERIZATION OF CTL SURFACE MARKERS

Determinations of lymphocyte surface antigenic markers were performed generously by J. A. Thomas, Immunology Department, Royal Free Hospital, London. According to the standard labeling method, test cells were first treated with mouse monoclonal antibodies and exposed then to anti-mouse fluorescein isothiocyanate labeled goat antibodies. Cells were differentially scored by fluorescence microscopy. Briefly, all CTLs bore pan T3 or OK T4 markers and, at least, can be considered to be normal, mature thymus derived lymphocytes on this basis.

RESULTS AND DISCUSSION

TG^r of Normal and LN CTLs

CTL microcultures were established for scintillation spectrophotometry as described from 2 normal on 2 LN individuals. Replicate cultures were set up to contain molar concentrations 2×10^{-7}, 2×10^{-6}, 2×10^{-5}, 2×10^{-4}, and 2×10^{-3} TG. The results are shown in Figure 2 where growth of 10^4 CTLs is assessed in terms of cpm ^3H-TdR incorporation as indicated on the ^{10}log scale ordinate. All cultures not containing TG showed vigorous activity, whereas the presence of TG at $2 \times 10^{-5}M$ resulted in complete inhibition of growth in CTLs from normals. In contrast, CTLs from the LN boys were resistant to TG concentrations as high as $2 \times 10^{-4}M$. On the basis of these findings TG at $2 \times 10^{-5}M$ was used to select against TG sensitive (TG^s) cells in subsequent cloning assays. Growth curves for CTLs are plotted in Figure 3. The results are presented

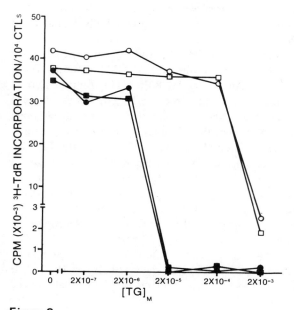

Figure 2
Results of scintillation spectrometry of CTLs derived from 2 normals (•, ■) and 2 LNs (○, □) cultivated in medium containing TCGF and various concentrations of TG.

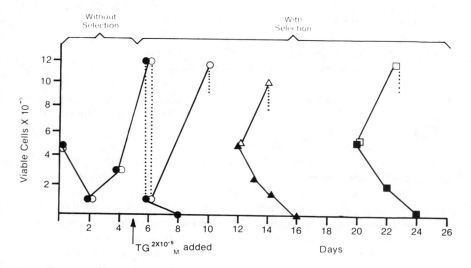

Figure 3
Representative kinetics of growth in TCGM for lymphocytes from various sources: (•) normal PBLs, and (○) LN PBLs; (▲) normal CTLs; (△) LN CTLs and (■) normal CTLs; and (□) normal TG^r CTLs (see text).

in terms of viable cells, on the ordinate, as surveyed over time in days of culture. The first plot compares the growth of PBLs from one normal individual and one LN cultured in TCGM. Viable cell counts fell from $.5 \times 10^6$ to below 10^5 during the first 2 days and by day 5 rose to 1.2×10^6 in each case. On day 5, cultures were split by dilution to densities of 10^5/culture with TCGM containing TG at selective levels (TCGM, TG). As one can see, by day 8 the CTLs from the normal subject were killed whereas those from the LN were unaffected. The second plot offers results of planting nonselected 12 day old CTLs from one normal and one LN in selective medium. The CTLs from the LN doubled within 2 days; however, those from the normal died off completely within 2 days. A third experiment in this series was intended to question the persistence of TG^r phenotype in CTLs developed from one normal individual in the absence of selection. One set of 10^8 CTLs was challenged with TCGM, TG, and survivors raised under these selective conditions until about 10 days later when levels of about 10^6 cells were reached (data not shown here). At this point, the TG^r CTLs were placed in nonselective medium and cultivated for another 10 days. At this stage, $.5 \times 10^6$ TG^r CTLs (□) thus derived were planted in selective medium $.5 \times 10^6$ and compared with nonselected CTL (■) counterparts, also at 20 days of culture. The results suggest that if selection against LN-like cells from normal individuals indeed occurs in vitro, it is far from complete.

Reconstruction Experiments—Clonal Assay

Cultures containing artificial mixtures of known numbers of TG^s and TG^r lymphocytes were established and analyzed to test the efficiency with which the clonal assay detects rare TG^r PBLs in majority populations of TG^s PBLs. In one set of experiments, serial dilutions of minority fresh PBLs from a LN boy were added to large numbers of fresh PBLs from a normal individual prior to plating, culture, and analysis as described. As depicted in Table 1, expected mutant frequencies (mfs) were determined as the sum of the observed mf for the PBLs from the normal and the predicted optimal recovery of LN PBLs in each addition based upon the known CE for these cells alone (in this example, .13). The high efficiencies of recovery (EoR) determined from these data give further evidence in favor of the conclusion that in vitro negative selection through metabolic cooperation, for example, does not occur to any significant degree in this culture system. Further, to assess the sensitivity of the clonal assay a second set of experiments was initiated. The intention was to avoid complications possible (but not borne out) in the first EoR test owing to histo-incompatibility between the immunocompetent allogenic cells. This experiment, presented in Table 2, was performed with mixtures of minority TG^r and majority TG^s CTLs developed from the pbls of one normal individual as shown. The CTLs employed here also contributed to the experiment described by the 3rd plot of Figure 3. EoR results again are consistent with the conclusion that in vitro selection against rare TG^r lymphocytes probably does not interfere with

Table 1
Efficiency of Recovery—Clonal Assay

Mixtures of fresh cells from individuals: LN/normal	TCGM plate 1 cell/well		TCGM, TG 2×10^{-5} M plate 10^5 cells/well		observed Mf	expected Mf	EoR = $\dfrac{\text{observed MF}}{\text{expected MF}}$
	#+ wells/96	%+[a]	#+ wells/96	%+[a]			
$0/10^6$	12	.13	5	$.54 \times 10^{-6}$	4.1×10^{-6}	—	—
$6.25/10^6$			7	$.76 \times 10^{-6}$	5.8×10^{-6}	5.1×10^{-6}	1.14
$12.50/10^6$			7	$.76 \times 10^{-6}$	5.8×10^{-6}	6.1×10^{-6}	.95
$25.0/10^6$			10	1.1×10^{-6}	8.5×10^{-6}	8.1×10^{-6}	1.05
$50.0/10^6$			12	1.3×10^{-6}	10.0×10^{-6}	12.1×10^{-6}	.83

[a]Poisson = (P_o)

Table 2
Efficiency of Recovery—Clonal Assay

Mixtures of CTLs from 1 normal individual	TCGM plate 1 cell/well		TCGM,TG 2×10^{-5} M plate 10^5 cells/well		observed Mf	expected Mf	EoR = $\dfrac{\text{observed Mf}}{\text{expected Mf}}$
	#+ wells/96	%+[a]	#+ wells/96	%+[a]			
CTL-TG^r/CTL-TG^s							
0/10^6	32	.41	13	1.9×10^{-6}	3.4×10^{-6}	—	—
2.4/10^6			19	2.2×10^{-6}	5.4×10^{-6}	4.4×10^{-6}	1.22
4.8/10^6			20	2.3×10^{-6}	5.6×10^{-6}	5.4×10^{-6}	1.04
9.6/10^6			26	3.2×10^{-6}	7.8×10^{-6}	7.4×10^{-6}	1.05
19.2/10^6			31	3.9×10^{-6}	8.5×10^{-6}	11.4×10^{-6}	.75

[a]Poisson = P_o

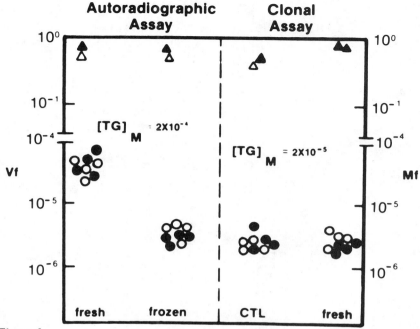

Figure 4
Comparative results from parallel experiments using lymphocytes from 2 LNs (▲, △) and 2 normals (●, ○)

the performance of this assay. It is also likely that cell-mediated cytolysis among alloreactive immunocytes was not problematic in the first EoR experiment. To express themselves destructively in this situation PBLs from the normal individual would have to be both primarily responsive or recruited in sufficient number and TG[r]. Given time restraints and relatively large spaces, the probability that these requirements can be satisfied to any observable effect is slight.

Parallel Experiments

The Strauss-Albertini test and the new clonal assay were performed in parallel (once for two LNs and on four separate occasions for two normal individuals) directly to compare in vivo and in vitro TG[r] lymphocyte frequencies, respectively, among fresh and cryopreserved pbls and CTLs. Table 3 provides tabulated Strauss-Albertini test results from one test occasion to show absolute values and demonstrate data handling. The variant frequencies (vfs) from normals calculated for three additional test occasions are also given. LNs, as expected, produced vfs near unity in every case. All results are plotted on the scattergram of Figure 4 wherein they are plotted against the ^{10}log scale abcissa in columns representing vfs from fresh or frozen PBLs. Though they are few, the data would

Table 3
Strauss-Albertini Test (autoradiographic assay)

Individuals normal	Test cells	PHA		PHA, TG$^2 \times 10^{-4}$M					Vf^e
		LI^a	Average LI^a	#+^{b}	#-^{c} ×10^{-6}	LI^a	Average LI^a	Vf^d	
GS	fresh	0.072	0.076	6	1.039	5.8×10^{-6}	5.7×10^{-6}	7.5×10^{-5}	8.4×10^{-5}, 6.6×10^{-5},
		0.080		5	0.975	5.1×10^{-6}			9.1×10^{-5},
	frozen	0.074	0.074	1	3.121	3.2×10^{-7}	3.3×10^{-6}	4.5×10^{-6}	3.1×10^{-6}, 4.2×10^{-6},
		0.073		1	2.931	3.4×10^{-7}			5.4×10^{-6},
PS	fresh	0.080	0.079	4	0.769	5.2×10^{-6}	4.5×10^{-6}	5.8×10^{-5}	7.2×10^{-5}, 6.3×10^{-5},
		0.077		3	0.802	3.7×10^{-6}			5.1×10^{-5},
	frozen	0.074	0.074	2	4.677	4.3×10^{-7}	4.0×10^{-7}	5.4×10^{-6}	5.5×10^{-6}, 5.8×10^{-6},
		0.074		1	2.786	3.6×10^{-7}			8.3×10^{-6},

Lesch-Nyhan		LI[a]	Average LI[a]	#+[b]	#−[c]	LI[a]	Average LI[a]	vf[d]
PD	fresh	0.080	0.080	180	2320	0.072	0.076	9.5×10^{-1}
		0.079		201	2299	0.080		
	frozen	0.070	0.072	165	2335	0.071	0.070	9.7×10^{-1}
		0.073		171	2329	0.068		
AC	fresh	0.062	0.062	140	2360	0.036	0.045	8.6×10^{-1}
		0.061		128	2372	0.051		
	frozen	0.054	0.053	126	2374	0.050	0.048	9.1×10^{-1}
		0.051		116	2384	0.046		

[a] % + nuclei
[b] determined by visual scan on count over entire slide
[c] determined by coulter counting sample nuclei in suspension
[d] $vf = \dfrac{LIPHA,TG}{LIPHA}$
[e] vfs from 3 additional tests on separate occasions

seem to support the possibility that phenocopies can be eliminated from assay by cryopreservation prior to testing (Albertini et al., this volume). As Figure 4 also shows, clonal assay mfs plotted on the ^{10}log scale abcissa and representing CTLs or fresh PBLs from LNs and normals compare favourably between themselves and with vfs from frozen pbls. Here again data used to calculate these representative results were prepared in tabular form and are shown in Table 4. The mfs determined for the same normal individuals on three separate occasions are given as well.

FUTURE PROSPECTS

It is lamentable that the reading of texts such as this rarely, if ever, can inspire much confidence in new, scarcely tried methods. Nevertheless, I can assure you that the clonal assay as performed here, and now in the laboratories of others, has good prospects as a system under development. TCGF is available from commercial suppliers and also can be produced in large quantities in the laboratory from the leukophoresed PBLs of normal donors (Bonnard et al. 1980; Lotze et al. 1980). Though X-SRBCs were used as feeders in the work reported here, x-irradiated cells of various lymphoblastoid lines can be used. These become more attractive when one considers that the lymphocyte cloning assay may be employed to assess, by comparison, damage from mutagens over a broad spectrum of markers where suitable selective agents can be used in conjunction with feeders developed for resistance or multiple resistancy (see Thilly, this volume).

One easily can envisage the adaptation of the clonal assay for evaluating genotoxic damage after in vitro exposures to various agents. Individual susceptibility to mutagens and relevancy in terms of assessible long-term and short-term health risks may be approached from this direction. Also in this regard, mutagenicity testing in the mouse, both in vivo and in vitro, by means of lymphocyte cloning should be entirely possible since the first immunological studies employed mice (Gillis et al. 1978). Various target lymphocytes from diverse tissues may be assayed in the mouse (and in man to a lesser degree) and of course the animal presents more approachable disease state endpoints. We need to develop the means to learn realistically to extrapolate results from rodent to man (and back). Figure 5, an idealized prospective picture, attempts to place our work in perspective as I wish to see it. As an example of a newly proposed approach to monitoring of short-term health effects in humans exposed to mutagens, I can offer our studies of patients undergoing photochemotherapy (Bridges and Strauss 1980; Strauss 1982b). It is my opinion that as we begin directly to monitor humans for mutagenic effects we should also learn to use this information to predict individual susceptibility in terms of disease state endpoints.

Table 4
Clonal Assay

Individuals Normal	test cells	TCGM plate 1 cell/well			TCGM, TG 2×10^{-5}M plate 10^5 cells/well				mf^c
		#+ wells/96	$(P_o)\%+^a$	Average $(P_o)\%+^a$	#+ wells/96	$(P_o)\%+^a$	Average $(P_o)\%+$	mf^b	
GS	CTL	34	0.44	0.41	14	1.6×10^{-6}	1.4×10^{-6}	3.4×10^{-6}	3.9×10^{-6}, 5.7×10^{-6}
		30	0.37		10	1.1×10^{-6}			6.0×10^{-6}
	fresh	13	0.15	0.13	6	6.5×10^{-7}	5.4×10^{-7}	4.1×10^{-6}	4.3×10^{-6}, 5.5×10^{-6}
		10	0.11		4	4.3×10^{-6}			5.4×10^{-6}
PS	CTL	23	0.27	0.31	8	8.7×10^{-7}	1.0×10^{-6}	3.3×10^{-6}	3.9×10^{-6}, 4.5×10^{-6}
		28	0.34		11	1.2×10^{-6}			4.7×10^{-6}
	fresh	15	0.17	0.15	9	9.8×10^{-7}	7.6×10^{-7}	5.0×10^{-6}	4.2×10^{-6}, 5.1×10^{-6}
		11	0.12		5	5.3×10^{-7}			5.7×10^{-6}
Lesch-Nyhan									
PD	CTL	35	0.45	0.40	26	0.31	0.34	8.5×10^{-1}	
		28	0.34		29	0.36			
	fresh	12	0.13	0.16	13	0.15	0.16	1.0	
		16	0.18		15	0.17			
AC	CTL	26	0.32	0.37	21	0.25	0.31	8.4×10^{-1}	
		32	0.41		30	0.37			
	fresh	7	0.08	0.11	10	0.11	0.11	1.0	
		12	0.13		9	0.10			

[a] $P_o\%+$
[b] $mf = \dfrac{CE_{TCGM,TG}}{CE_{TCGM}}$
[c] mfs from 3 additional tests on separate occasions

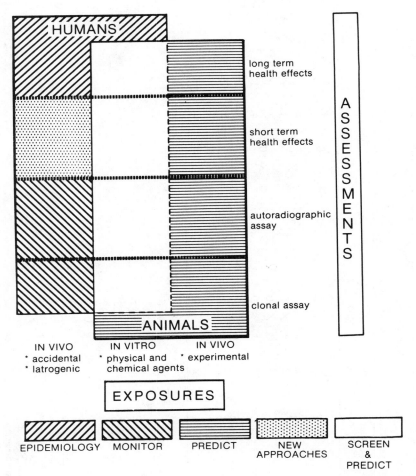

Figure 5
Anthropocentric evaluation of risks from environmental mutagens.

ACKNOWLEDGMENTS

I am pleased to acknowledge the support of the British Cancer Research Campaign. I am grateful for the expert assistance of Sandra Staton and Elaine Whisnant who helped prepare this manuscript. I am also pleased formally to thank my wife Pamela Alix who gave so much blood and encouragement too.

REFERENCES

Bonnard, G.D., K. Yasaka, and R.D. Maca. 1980. Continued growth of functional human T-lymphocytes: Production of human T-cell function by a T-all product. *Nat. New Biology.* **237**:15.

Bridges, B.A. and G.H. Strauss. 1980. Possible hazards of photochemotherapy. *Nature* **983**:523.

DeMars, R. 1979. The maturation of test systems. *Banbury Rep.* **2**:329.

_____. 1973. Mutation studies with human fibroblasts. *Environ. Health Perspect.* **6**:127.

Gillis, S., M.M. Ferm, W. Ou, and K.A. Smith. 1978. T-cell growth factor: Parameters of production and a quantitative microassay for activity. *J. Immunol.* **120**:2027.

Lotze, M.T., J.L. Strausser, and S.A. Rosenberg. 1980. In vitro use of cytotoxic human lymphocytes II. Use of T-cell growth factor (TCGF) to clone human T-cells. *J. Immunol.* **124**:2972.

Ruscetti, F.W., D.A. Morgan, and R.C. Gallo. 1977. Functional and morphologic characterization of human T-cells continuously grown in vitro. *J. Immunol.* **119**:131.

Strauss, G.H. and R.J. Albertini. 1979. Enumeration of 6-thioguanine-resistant peripheral blood lymphocytes in man as a potential test for somatic cell mutations arising in vivo. *Mutat. Res.* **61**:353.

Strauss, G.H.S. 1982a. The Strauss-Albertini test for enumerating drug resistant peripheral blood lymphocytes. In *The use of human cells for the assessment of risk from physical and chemical agents.* Plenum Press, New York.

_____. 1982b. PUVA Therapy—immunology and genotoxic approaches to risk estimation. In *The use of human cells for the assessment of risk from physical and chemical agents.* Plenum Press, New York.

COMMENTS

THILLY: I guess this is a question for all of us, but Dick and Gary in particular. I know one of the problems you always have when you try to get very precise, that is to say, reproducible, measurements of mutant fractions or variant fractions, is when you divide the plating efficiency by one means or another into plating efficiency under selective conditions. One of the things that has been bothering me, and I am sure you, is, how do we know that the actual efficiency of clone formation or tritiated thymidine uptake, or any end point marker, is the same under the low-density conditions of plating efficiency and the high-density conditions under selective conditions?

ALBERTINI: We really don't. It's a luxury though to be able now with cloning to ask this question for an in vivo system the way we did in the past for in vitro systems. I'm not sure we ever knew it in vitro.

THILLY: I think it has been overcome in vitro now, I think it can be extended to your approach. An example would be, very simply, to take a rare marker and include it in a mixture at an artificially high level and put it under both conditions.

STRAUSS: To the extent it is possible, I did that. It didn't register with you as data flashed by too quickly. I took this question to task within a series of EOR experiments. In one set I added minority Lesch-Nyhan cells to normal cells. In another set I added TG^r CTLs to TG^s CTLs, both types were developed from the PBLs of the same normal individual. When you see the data again, you will realize that I used separate cloning efficiencies, from each set of conditions, to calculate the mutant frequencies. I assumed the various cells would grow at different rates. In fact they did.

ALBERTINI: Even though they should be done, there might be a problem in using artificial mixtures of lymphocytes from two individuals—allogeneic mixtures—as reconstruction experiments for long-term lymphocyte cultures, unless they came out exactly as expected. I don't think that such experiments that show fewer times expected recovered mutants can be used to calculate correction factors, or to make judgements about relative plating efficiencies at high-and low-cell densities. This is because lymphocytes are also immunocompetent cells. In allogeneic combinations majority populations may kill rare histoincompatible cells.

THILLY: Following up on that very point: you mentioned that all the mutants fell into this category of T4. I don't even know what T4 means. Does that mean it is possible that some of the cells that were showing up under nonselective conditions are T4 plus other kinds? Your mutants may be arising from one cell population, but your clone-forming units could be arising from a mixture of populations, thus giving a different thing in the numerator and the denominator. Is that possible?

ALBERTINI: I think it is too early to tell but this is one of the things that we hope the cloning procedures will allow us to determine. This will allow us to make a general question concerning direct in vivo mutagenicity testing— to in vivo cell populations contain subregulations sensitive to and subpopulations resistant to mutagenesis.

STRAUSS: It would be exciting if it is possible, because subclasses of lymphocytes may be mutable at different levels. In fact, mutation and selection may well be required in the development and maintenance of normal immunocompetency. This is a question I intend to pursue.

ALBERTINI: The cutaneous T-cell lymphomas are T4s. These may arise from naturally occurring somatic cell mutations. But this is a tremendous leap.

Detection of Phenotypically Variant (Thioguanine-Resistant) Rat Spleen Lymphocytes

HECTOR D. GARCIA AND DAVID B. COUCH
Department of Genetic Toxicology
Chemical Industry Institute of Toxicology
Research Triangle Park, North Carolina 27709

Many investigators have attempted to develop means for detecting gene mutations that incorporate the pharmacokinetic and metabolic factors operative in vivo with the speed and sensitivity of in vitro detection of mutants. Host-mediated assays, in which target cells of various types are inoculated into host animals for exposure to mutagens, then recovered for in vitro selection for the mutant phenotype, represent one approach to this goal (Gabridge and Legator 1969; Legator and Malling 1971; Capizzi et al. 1974). Other procedures use primary cell cultures from treated animals to detect mutations. For example, Dean and Senner (1977) established primary lung cell cultures from Chinese hamsters treated with mutagens in vivo and scored for mutation to 8-azaguanine (AG) or ouabain resistance. In man efforts have been made to detect somatic cell variants of peripheral blood cells by immunologic or cytochemical means, but the genetic basis of many of these procedures has been questioned (Albertini 1980).

Albertini and co-workers (Strauss and Albertini 1977, 1979; Strauss et al. 1979; Albertini 1979, 1980) have developed a direct mutagenicity test capable of determining the fraction of 6-thioguanine (TG) resistant cells in a population of lymphocytes obtained from human blood. Lymphocytes are stimulated with mitogen in the presence of tritiated thymidine with and without TG, and the TG resistant fraction determined by autoradiography. The rationale for and the development of this procedure have been described earlier in this volume (Albertini, Strauss). This assay was developed for retrospective screening of humans for mutation induced in somatic cells as a result of accidental or therapeutic exposure to known or suspected mutagens. However, Albertini (1980) has shown that thioguanine-resistant (TG^r) peripheral blood lymphocytes (PBLs) can also be detected in rats, suggesting the feasibility of a direct mutagenicity assay for prospective testing. A preliminary report of efforts to detect thioguanine-resistant variants of mouse spleen lymphocytes by autoradiography has also recently appeared (Gocke et al. 1982). We report here on our efforts to develop a rat lymphocyte/hypoxanthine-guanine phosphoribosyl transferase (HGPRT) assay modeled after that of Albertini and coworkers.

METHODS

Animals

Male Fischer CDF®(F344)/CrlBR rats (Charles River, Kingston, NY) weighing 150-200 g were obtained and maintained as specific-pathogen-free (except as noted for preliminary experiments); that is, sera from these animals were tested and found to be free of the viruses in the standard virus screen (Microbiological Associates, Bethesda, MD). Animals were fed open formula NIH-07 lab diet (Zeigler Brothers, Gardner, PA) and water ad libitum. Animals were housed in 8m³ University of Rochester style inhalation chambers in plastic cages with hardwood chip bedding, one rat per cage.

Lymphocyte Purification and Culture

The following protocols are essentially as described by Albertini (1979), with slight modification:

1. Rats were anesthetized with i.p. injections of 60 mg sodium pentobarbital/ kg body weight and were exsanguinated by cardiac puncture using heparinized syringes.
2. Spleens were excised aseptically, trimmed of extraneous tissue and placed in 100 mm petri dishes containing 10 ml of cold HBSSc [Hanks' balanced salt solution containing 0.2% (v/v) sodium heparin (Uphohn), 1% (v/v) penicillin-streptomycin (100 units/ml and 100 μg/ml, respectively; GIBCO]. Lymphocytes were "massaged" from the spleen using sterile blunt forceps and the cell suspension was diluted to 20 ml with HBSSc.
3. Lymphocytes were isolated from blood or spleen cell suspensions as a layer of cells on Ficoll-hypaque (Accurate Chem. Co.) gradients (600xg, 30 min). Cells were then washed with HBSSc and suspended in complete medium consisting of RPMI-1640 with 25 mM HEPES buffer (GIBCO), 0.2% heparin, 1% penicillin-streptomycin (GIBCO), 20 μM 2-mercaptoethanol (Sigma), 0.2 mM L-glutamine (GIBCO), 2% fetal bovine serum (FBS, GIBCO) and 2 μg phytohemagglutinin (PHA; Burroughs-Wellcome HA-16)/ml. Replicate 1 ml cultures of 1 $\times$ 10⁶ lymphocytes/ml either with (+TG) or without (-TG) 0.2 mM TG were incubated in 24-well dishes (Costar) at 37°C, 5% CO_2 in a humidified incubator for 42 hours. In selective conditions (+TG) 9-21 replicate cultures/animal were established; 3 replicate cultures/animal were used to establish the labeling index in the absence of TG. ³H-Thymidine (1 μCi/ ml) was added for the final 3 hours of incubation.

Slide Preparation

Harvesting of cultures and slide preparation were as described by Albertini (1979), with minor modifications. At termination 1 ml of cold 0.1 M citric acid was added to each culture, and the cell suspensions were transferred to 15 ml tubes. Each well was rinsed with three additional 1 ml aliquots of citric acid, and the combined cell suspension centrifuged 10 min at 600xg. The cell pellets were suspended in 4 ml fresh fixative (5:1 methanol:glacial acetic acid), stored at 4°C overnight, centrifuged and then suspended in 100-200 μl of fixative. A 10 μl aliquot was diluted to 50 μl with H_2O and the cell concentration determined by hemocytometer. Aliquots (10 μl) of the cell suspension were placed onto a coverslip (18 mm square for -TG; 9 mm square for +TG) previously mounted on a microscope slide. The slides were allowed to dry between addition of successive aliquots, and, when the desired volume of cell suspension was placed on the coverslips, the slides were dried further overnight.

Autoradiography

Slides were dipped in NTB-2 (Eastman Kodak) emulsion prewarmed to 43°C, allowed to dry in a horizontal position, and exposed for 24-72 hours at room temperature in a dark box. Slides were developed in D-19 (Kodak) developer at 17°C for 3 minutes, rinsed 30 seconds in H_2O at 17°C, fixed (Kodak fixer) for 8 minutes at 17°C and washed for 20 minutes in H_2O. The wet slides were then stained with methyl green-pyronin Y (Schmid GMBH and Company).

Scoring and Calculation of the Variant Fraction

The labeling index (LI) in the absence of -TG is defined as the ratio of labeled cells to total cells on a slide from a mitogen stimulated culture. This ratio is determined for successive random microscope fields at 400-800x magnification until the percent change in the standard errors (determined with the aid of a computer) between the cumulative LIs for successive fields is minimized (usually about 30 fields encompassing 50-150 cells/field).

The LI in the presence of +TG is defined as the ratio of labeled cells (as determined by scanning the entire coverslip at 160x magnification) to the total number of cells on the slide as determined by hemocytometer count.

The variant frequency (Vf) is defined as the LI (+TG) divided by the LI (-TG) to correct the LI (+TG) for the fraction of cells which could have been labeled in the absence of TG.

RESULTS AND DISCUSSION

In initial experiments with rat peripheral blood lymphocytes (PBLs), it was found that, by including 20 μM 2-mercaptoethanol in the culture medium, -TG labeling indices (LIs) of approximately 0.3 could be obtained. One initial objective was that the assay should have sufficient sensitivity to measure accurately Vfs as low as 1×10^{-5}. In order to achieve 95% confidence that the assay would be able to detect one variant in 10^5 cells, statistical considerations require that $> 2.99 \times 10^5$ mitogen-responsive cells/animal be scored. The yields of lymphocytes from the peripheral blood of specific-pathogen-free rats were not always adequate to maintain this level of sensitivity. The problem of inadequate yields of PBLs can be overcome by pooling cells from individual animals. In one trial, the Vf in pooled PBLs from three animals was found to be 3.9×10^{-4}, which is in agreement with results obtained by Albertini (1980) in PBLs from individual rats. However, the use of pooled lymphocytes precludes examination of animal to animal variation in background and induced Vfs. As a means of increasing the yield of lymphocytes per rat, the isolation of lymphocytes from rat spleen was therefore investigated.

The use of rat spleen lymphocytes (RSLs) necessitated a change in the composition of the culture medium. For PBLs equivalent results were obtained by supplementing the RPMI-1640 medium with either FBS or autologous rat serum, but rat sera caused severe clumping and lysis of rat spleen cells. The use of FBS eliminated this problem. The average yield of spleen cells after Ficoll-hypaque isolation was $> 6 \times 10^7$ cells/rat, consisting of $> 98\%$ small lymphocytes. PHA stimulation of RSLs gave LIs of 0.2-0.5. The optimal PHA concentration for stimulating RSLs was 2.5 μg/ml.

After basic parameters for culturing RSLs were established, an attempt was made to determine if treatment of rats with a known mutagen could increase the Vf in spleen lymphocytes. In addition, an attempt was made to determine the optimal time for scoring for TG^r RSL following treatment. In vitro mutagenicity assays using purine analogue resistance as the genetic marker require up to several days for optimal expression of the mutant phenotype (van Zeeland and Simons 1976; O'Neill et al. 1977; Thilly et al. 1978). It was felt that lymphocytes might require shorter expression times than in assays using continuously growing cell lines, and that in vivo selection against TG^r lymphocytes (Albertini and DeMars 1974) might also be a factor in determining the optimal time for scoring for TG^r. Forty rats were divided into five treatment groups receiving 0, 30, 100, 300, or 1000 mg ethylmethanesulfonate (EMS)/kg; EMS was injected i.p. in a vehicle of phosphate buffered saline-dimethylsulfoxide (DMSO) (1:1; v/v). RSLs from one rat of each treatment group were prepared for measurement of the Vf at days 1, 3, 5, 7, 9, 11, 14, and 28 after treatment.

No increase in Vf over background levels could be reliably established (Table 1) even though the doses of EMS used were large (indeed, all animals treated with 1000 mg/kg EMS died one day after treatment) and did produce a genotoxic effect in rat peripheral lymphocytes (RPLs) (Kligerman et al. 1981). Peripheral blood was taken from these animals at sacrifice, and EMS treatment was found to induce up to four-fold increases in sister chromatid exchanges over controls. The difficulty in demonstrating an increase in Vf of RSLs from EMS treated rats is illustrated by the representative data in Table 1. When Vfs were determined 3 days after treatment, there appeared to be an increase in Vfs of treated animals, up to five times the control value. However, the control

Table 1

Variant Frequency (Vf) of Spleen Cells from Rats Treated with EMS

Days after treatment	Dose of EMS (mg/kg)	LI −TG	LI +TG ($\times 10^4$)	No. of cells labeled (+TG)	Total cells scored ($\times 10^{-6}$)	Vf
3	0	0.22	0.97	25	0.52	0.45×10^{-3}
3	30	0.20	2.8	151	0.53	1.4×10^{-3}
3	100	0.15	1.9	108	0.60	1.3×10^{-3}
3	300	0.11	2.6	27	0.25	2.3×10^{-3}
9	0	0.12	5.1	475	1.34	4.1×10^{-3}
9	30	0.11	4.4	495	1.28	4.0×10^{-3}
9	100	0.13	5.6	378	1.64	4.3×10^{-3}
9	300	0.18	5.6	1196	2.28	3.2×10^{-3}

Vf at day 9 was 20 times that at day 3 and was as high as that of RSL from treated animals. One explanation for this high background is that a subpopulation of spleen cells might be escaping the selection process, possibly because it consists of cycling, rather than resting cells. While, for in vitro assays, selection against TG-sensitive cells occur over several days, quantitation of mutants in the lymphocyte/HGPRT system requires that efficient selection be accomplished within the 42-hour culture time (before the stimulated lymphocytes divide). The LI of RSL in the absence of mitogen is approximately 4% (Table 2), indicating the presence of cycling cells, most of which do not incorporate tritiated thymidine in the presence of TG. The fact that selection against TG-sensitive lymphocytes occurs suggests that selection in this system may involve a biochemically different mechanism than that which is operating in the *in vitro* systems. We therefore considered several approaches to try to reduce the background TG^r frequency in the RSL system.

The first approach was simply to increase the time the cells were exposed to TG. This was done by preincubating the cells in medium +TG but without PHA for various times before PHA addition. The time of incubation after PHA addition was kept constant at 42 hours. It was found that the fraction of cells which could be stimulated by PHA was drastically reduced, but the fraction of TG^r cells was not (Table 2).

The second approach was to kill cycling cells selectively by pre-incubating the cells in 50 μM cytosine-1-β-D-arabinofuranoside (ara-C) for 16 hours, then washing out the ara-C before adding PHA to stimulate blastogenesis and DNA synthesis. Although treatment with ara-C did produce a ten-fold reduction in the Vf (Table 3), pretreatment of RSLs with ara-C also resulted in a decrease in the fraction of PHA-responsive cells from 0.28-0.009. A larger fraction (0.031) of RSLs were stimulated by 1 μg/ml concanavalin-A after ara-C pretreatment.

Table 2
Effect of Preincubation with TG on Background Vf in RSLs

Preincubation period (days)	Labeling Index		Vf
	+PHA	−PHA	
0	0.100	0.044	3.2×10^{-3}
1	0.060	0.041	2.4×10^{-3}
4	0.005	0.005	4.3×10^{-3}

Table 3
Treatments Which Lower the Background Vf

Cells	Treatment	Vf initial	Vf treated
RSL	Ara-C	1.5×10^{-3}	0.14×10^{-3}
RSL	Freezing	1.5×10^{-4}	0.17×10^{-4}

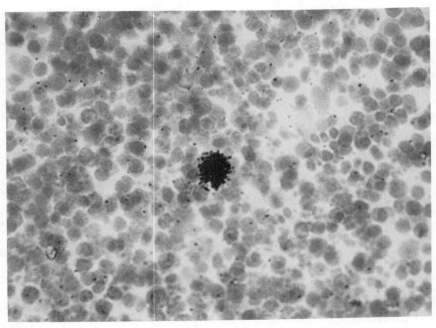

Figure 1
RSLs incubated with TG and plated at high density on a 9 mm × 9 mm coverslip. (•) cells covered with exposed silver grains after autoradiography (i.e., labeled cells) have incorporated radioactive thymidine and are presumptive HGPRT-deficient cells.

Albertini (pers. comm.) has observed that the background Vfs of PBLs frozen prior to assay appears lower than for the same freshly isolated lymphocytes. Therefore, the effect of freezing in 7.5% DMSO on the background of Vf of RSLs was examined. Holding cells for several days at $-80°C$ prior to thawing and assaying for TG^r was found to lower the Vf by a factor of ten (Table 3). While either freezing or pretreatment of spleen lymphocytes with ara-C had the desired effect on background Vf, there appeared to be at least a 10-fold variation in the RSL background Vfs between individual animals (Table 3), which is greater than that observed by Albertini (1980) in rat PBLs.

Early attempts to enumerate TG^r lymphocytes revealed a practical drawback to the procedure. The scoring, as it was then being done, was extremely time-consuming, as slides from one rat took 13 hours to score. The problem of tedious scoring was alleviated by placing all the +TG cells on a single 9 mm × 9 mm coverslip rather than three 18 mm × 18 mm coverslips, thereby reducing the area to be scanned under the microscope by 12-fold. The use of still smaller coverslips should speed up scoring even further. The ability of the assay to detect TG^r lymphocytes was improved by culturing 3-7 times as many cells +TG

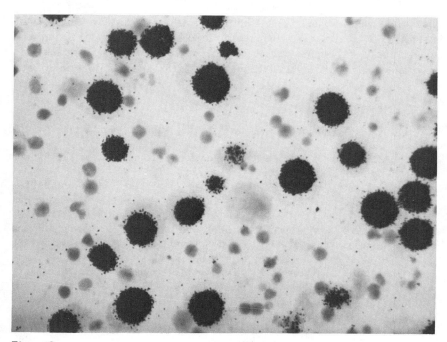

Figure 2
RSLs incubated without TG and plated at low density on an 18 mm × 18 mm coverslip. The ratio of labeled cells to total cells (LI) is a measure of the ability of the culture to respond to mitogen stimulation.

(up to 2.1×10^7 cells). Although the cells become quite crowded on the small coverslips (Fig. 1), labeled cells are still easily detected. Cells cultured −TG must still be plated on the 18 mm × 18 mm coverslips (Fig. 2), since the number of unlabeled, as well as labeled, cells must be visually quantitated for these cultures to determine the LI.

Although the techniques we have described should be useful in adapting the human lymphocyte assay for use in rats, important questions, including those of response to known mutagens, optimal expression time, cell, organ, sex, and species specificities remain to be answered. In addition, the complex relationships between lymphocytes in the spleen, bone marrow, peripheral blood, lymph nodes, thymus, and other organs with respect to pool sizes, specialized cell types, metabolic capabilities, circulation, and turnover rates make the quantitative interpretation of assay results a potentially complex task. While, in our hands, the unreliability of the yields of PBLs precluded their routine use, simultaneous determination of the Vfs in spleen and PBLs from individual animals is frequently possible, thus permitting the investigation of some of these parameters.

ACKNOWLEDGMENTS

The authors wish to thank R. J. Albertini for helpful discussions, as well as for inviting H. D. Garcia to his laboratory to observe and learn procedures for enumerating TG^r lymphocytes. A preliminary report of this work has been presented at the 13th Annual Meeting of the Environmental Mutagen Society.

REFERENCES

Albertini, R.J. 1979. Direct mutagenicity testing with peripheral blood lymphocytes. *Ban. Rep.* 2:359.

Albertini, R.J. 1980. Drug resistant lymphocytes in man as indicators of somatic cell mutation. *Teratog. Carcinog. Mutagen* 1:25.

Albertini, R.J. and R.DeMars. 1974. Mosaicism of peripheral blood lymphocyte populations in females heterozygous for the Lesch-Nyhan mutation. *Biochem. Genet.* 11:397.

Capizzi, R.L., B. Papirmeister, J.M. Mullins, and E. Chang. 1974. The detection of mutagens using the L5178Y/asn⁻ murine leukemia cells *in vitro* and in a host-mediated assay. *Cancer Res.* 34:3073.

Dean, B.J. and K.R. Senner. 1977. Detection of chemically induced somatic mutation in Chinese hamsters. *Mutat. Res.* 46:403.

Gabridge, M.G. and M.S. Legator. 1969. A host-mediated assay for the detection of mutagenic compounds. *Proc. Soc. Exp. Biol. Med.* 130:831.

Gocke, R., K. Eckhardt, M.-T. King, and D. Wild. 1982. Autoradiographic detection of point mutations induced *in vivo* in somatic mammalian cells. *Mutat. Res.* 85:442.

Kligerman, A.D., J.L. Wilmer, and G.L. Erexson. 1981. Characterization of a rat lymphocyte culture system for assessing sister chromatid exchange after *in vivo* exposure to genotoxic agents. *Environ. Mutag.* 3:531.

Legator, M.S. and H.V. Malling. 1971. The host-mediated assay, a practical procedure for evaluating potential mutagenic agents in mammals. *Chem. Mutagens* 1:569.

O'Neill, J.P., P.A. Brimer, R. Machanoff, G.P. Hirsch, and A.W. Hsie. 1977. A quantitative assay of mutation induction at the hypoxanthine-guanine phosphoribosyl transferase locus in Chinese hamster ovary cells (CHO/ HGPRT system): Development and definition of the system. *Mutat. Res.* 45:91.

Strauss, G.H. and R.J. Albertini. 1977. 6-Thioguanine-resistant lymphocytes in human peripheral blood. In *Progress in Genetic Toxicology. Developments in Toxicology and Environmental Sciences* (eds. D. Scott et al.), p. 327. Elsevier/North Holland, New York.

Strauss, G.H. and R.J. Albertini. 1979. Longitudinal determination of 6-thioguanine resistant peripheral blood frequencies in individuals receiving 8-methoxypsoralen and long-wave ultraviolet light treatments (PUVA). *Environ. Mutagen* 1:152.

Strauss, G.H., R.J. Albertini, P.A. Krusinski, and R.D. Baughman. 1979. 6-Thioguanine resistant peripheral blood lymphocytes in humans following psoralen-long-range UV light therapy. *J. Invest. Dermatol.* 73:211.

Thilly, W.G., J.G. Deluca, N.H. Hoppe, and B.W. Penman. 1978. Phenotypic lag and mutation to 6-thioguanine resistance in diploid human fibroblasts. *Mutat. Res.* 50:137.

van Zeeland, A.A. and J.W.I.M. Simons. 1976. Linear dose-response relationships after prolonged expression times in V79 Chinese hamster cells. *Mutat. Res.* 35:129.

COMMENTS

KLIGERMAN: What is the difference between the mutant frequency in human peripheral lymphocytes and the rodent peripheral lymphocytes?

GARCIA: From the limited number of experiments that I did before we switched to the specific pathogen-free animals, it appeared that they were similar in their spontaneous frequencies. I didn't look at induced frequencies. I didn't get to that point before we switched.

ALBERTINI: It is important to remember that what we measure with in vivo systems are mutant frequencies. But, we are interested, from a public health point of view, in mutation rates or mutation frequencies. Dr. Russell, in her '57 paper, which I have read several times, clearly states two difficulties with direct or in vivo tests. One is making sure of the mutant nature of the phenotype. This, I think, will no longer be a difficulty with cloning. But the other—to determine mutation rates from mutant frequencies, without making a lot of assumptions about pool sizes and cell kinetics—will be very difficult. Dr. Garcia tested not splenic lymphocytes; we test human peripheral blood lymphocytes. Cell division, cell traffic and other factors may be quite different between these two sources, and so might the relationship between mutants and mutations.

References

Russell, L.B. and M.H. Major. 1957. Radiation-induced presumed somatic mutations in the house mouse. *Genetics* 42:161.

Potential of Mutational Spectra for Diagnosing
the Cause of Genetic Change in Human Cell Populations

WILLIAM G. THILLY AND PHAIK-MOOI LEONG
Toxicology Group
Department of Nutrition and Food Science
Massachusetts Institute of Technology
Cambridge, Massachusetts 02139

THOMAS R. SKOPEK*
Department of Molecular Biophysics and Biochemistry
Yale University
New Haven, Connecticut 06520

The goal of genetic toxicology is to reduce the incidence of genetic diseases in society. Genetic defects in newborns occur at a frequency of approximately 2%, half of which can be attributed to gene locus mutation (Thilly and Liber 1980). It is a central premise of genetic toxicology practice that genetic change in humans is due to the action of environmental mutagens. However, this premise is untested. Whether human mutations arise spontaneously or are caused by environmental agents has not been determined.

One approach to assaying mutation in humans would be to obtain cell samples from individuals and to determine the frequency of recognizable mutations in each. We would expect higher frequencies of mutations in those persons exposed to mutagens and lower levels in those spared exposure if the central premise were correct. However, a serious problem may be foreseen for such an approach. Who has been spared exposure to environmental mutagens and would be available as negative control? If cell samples of all persons revealed the same frequency of mutants at a particular locus, one could not unambiguously differentiate between spontaneous or nonspontaneous etiology.

In this paper we explore another approach upon which we are working. We believe this approach does offer the ability to differentiate between mutation of spontaneous and nonspontaneous origin. The approach is based on three technical assumptions:

1. Spontaneous and induced mutations are nonrandomly distributed with regard to *kind* and *position* in the genome.
2. Analysis of mutation frequencies using a set of drug resistance genetic markers sensitive to a specific limited range of mutations with regard to

*Chemical Industries Institute of Toxicology, Research Triangle Park, North Carolina 27709

kinds and position in the genome would provide sufficient information to discriminate between spontaneous and induced mutations.

3. The markers used in cell culture systems could be applied to peripheral T-lymphocyte populations permitting us to analyze with sufficient precision and sensitivity to discriminate between spontaneous and induced mutations in human blood samples.

We review here the evidence which tests the truth of each assumption and outline our proposal for the diagnosis of the causes of human genetic change.

MUTATIONAL SPECTRA: SITE AND SEQUENCE SPECIFICITY OF MUTATIONS

The sequence specificity of spontaneous and induced mutations was first demonstrated in the rII locus of bacteriophage T4 by Benzer (1961). Benzer isolated a large number of rII mutants, both spontaneous and chemically induced, and located them to within a few base pairs in the rII gene by genetic crosses with deletion mutants in the region. Two key observations were made:

1. for a given agent several sites in the gene displayed higher frequencies of mutation than others, and
2. the distribution mutations as a function of base-pair position was different for spontaneous mutation and for induced mutations by a wide variety of chemicals.

For example, the site at map number 131 was more mutable for spontaneous mutation than for ethyl methane sulfonate (EMS)-induced mutation, while site EM5 was far more mutable for EMS mutation than for spontaneous (Table 1). By comparing the distribution of mutants at just these two sites, one could easily distinguish between a population of spontaneous mutants and one induced by EMS.

The phenomenon of mutagen site specificity was further studied in the lac I gene of *Escherichia coli* by Miller and coworkers (Coulondre and Miller 1977a,b; Schmiessner et al. 1977; Miller et al. 1977, 1978; Calos and Miller 1982). In this system, suppressible amber and ochre mutants in the lac I gene were selected and their approximate location in the gene determined by crosses

Table 1
Distribution of Mutations in Spontaneous and EMS-Induced rII Mutants of Bacteriophage T4 (%)

Map number:	131	117	N24	EM 5
Spontaneous	9.6 ± 0.01	18 ± 0.01	2.4 ± 0.01	0.3 ± 0.001
EMS	4.1 ± 0.02	8.8 ± 0.02	5.7 ± 0.02	5.6 ± 0.02

Data from Benzer (1961).

with deletion mutants. Since the lac I sequence was known (and, hence, every site which could possibly mutate to give rise to a nonsense codon), both the precise location of each mutation and the particular base pair change which had occurred could be deduced from its map location and its pattern of suppression.

Miller's data (Coulondre and Miller 1977a,b; Miller et al. 1977, 1978; Calos and Miller 1982) confirmed Benzer's (1961) conclusion that not all base pairs were subject to the same risk of mutation. Also, his data clearly established that the distribution of mutations at various sites was different for spontaneous and induced mutations (Fig. 1). Several sites in lac I display a two- to eightfold difference in frequency between spontaneous mutants and EMS-induced ones. For example, the ratio of the sum of mutations at am 51 and am 37 to sum of mutations at am 8 and am 5 is 6.7 ± 2.3 in spontaneous mutants compared to a ratio of 0.4 ± 0.05 in EMS-induced mutants, a difference of greater than 10-fold. Hence, by analyzing and comparing the frequency of mutation occurring at each site, it is possible to discriminate between spontaneous and EMS-induced mutation. In fact, the pattern of mutation induced by any mutagenic stimulus was significantly different from that seen for a spontaneously mutated population. Even the patterns of mutation induced by the alkylating agents, N-methyl-N'-nitro-N-nitrosoguanidine (MNNG) and EMS were differentiable by quantitative analysis of the data (P.-M. Leong et al., unpubl. results).

Recently, a system has been developed which analyzes the complete spectrum of observable forward mutation in the cI gene of bacteriophage lambda lysogen of E. coli using DNA sequencing techniques (Skopek and Hutchinson 1982). Unique mutation frequency spectra for various agents were observed in this gene and analyzed at the nucleotide level. Frameshift mutagens such as ICR-191, proflavin, and aminoacridine caused mutation predominantly at only two locations in the gene at runs of 4 and 5 guanines; two other runs of 4 guanines in the gene seemed refractory to frameshift mutagenesis (T. Skopek and F. Hutchinson, in prep.). In another example, bromouracil was observed to cause only AT → GC transitions; moreover, two-thirds of these transitions occurred at only four sites in the gene, all of which contained the sequence 5'ACGC3'. Conversely, spontaneous mutations included mostly transitions, transversions and IS1 and IS5 insertions. The IS1 and IS5 mutations comprised approximately one-third of the observed spontaneous mutations (Skopek and Hutchinson 1982).

CONSTRUCTION OF MUTAGENIC SPECTRA:
A PRACTICAL APPROACH

The work of Benzer, Miller, and Skopek (all references cited in section above) confirms the view that unique spectra exist due to the presence of mutagenic hotspots and that these spectra differ significantly enough to allow discrimination between populations of spontaneous- and chemically-induced mutants.

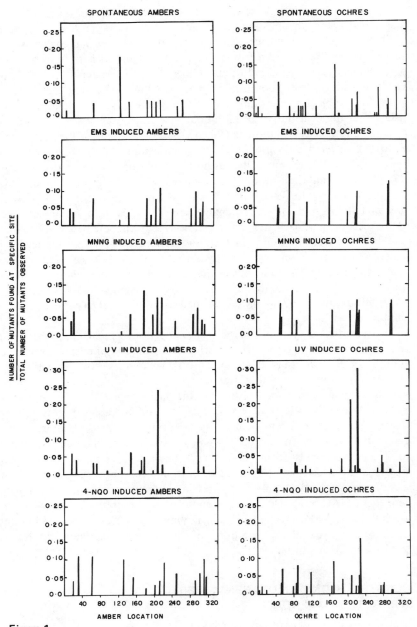

Figure 1
Distribution of spontaneous- and chemically-induced mutations in the lac I system of *E. coli*. Suppressible amber and ochre mutants in the lac I gene were selected and their approximate location in the gene determined by crosses with deletion mutants. The percentage of total amber and ochre mutations found at each site is illustrated. (Redrawn from Coulondre and Miller 1978.)

Theoretically, the approaches described for microbial systems could be applied to mammalian cells grown in culture or found in blood samples. If we could construct mutation spectra by sequencing mutants at the nucleotide level in a particular gene in mammalian cells, we would be fairly confident of distinguishing spontaneous mutants from induced ones. However, given present techniques, the amount of work necessary to carry out such analysis on a routine basis would be prohibitive.

Resolution of a mutation spectrum in a gene at the nucleotide level, however, would probably provide more information than is necessary to distinguish between spontaneous and induced spectra. Since mutation spectra are the result of relatively infrequent ($< 10\%$) mutation hotspots in the gene, one should be able to monitor mutations in blocks of nucleotides instead of by individual nucleotides, and still be able to see the characteristic hotspot nature of the spectrum. Thus, blocks not containing a hotspot would have low frequency of mutation, while those containing one or more hotspots would have a higher frequency of mutation.

To illustrate this approach, we have analyzed Miller's lac I data (Coulondre and Miller 1977a,b) for spontaneous and EMS mutants in the following fashion. The gene was divided into 32 blocks of 30 nucleotides each, and the fraction of total amber and ochre mutants in each block calculated (Fig. 2). Nine of these blocks show significant differences between spontaneous and EMS-induced mutants. For example, blocks B, H, L, S, W, and D each contributed greater than 2.5% (and as high as 10% in block D) of mutations found in EMS mutants, but less than 0.25% in spontaneous mutants. On the other hand, blocks C and N each contributed greater than 10% of mutations found in spontaneous mutants, yet less than 0.25% in EMS-induced ones. Quantitative analysis of the data further strengthens confidence in the feasibility of discrimination between spontaneous and EMS-induced mutants; the ratio of the sum of mutations in blocks C and N to the sum of mutations in blocks H, S, and D is found to be 2.8 ± 0.8 in spontaneous mutants and only 0.11 ± 0.03 in EMS-induced ones, a difference greater than 10-fold. Hence, the "approximate" spectra based on blocks of nucleotides still contain enough information to differentiate among different causes of mutation in a population.

In considering this approach, there is no reason why the blocks of nucleotides being monitored to construct the spectrum need be side-by-side in the same gene; if the blocks were located in different parts of the chromosome, the basic approach should still work to generate a unique spectrum. Thus, if we had a means to measure mutations within small blocks of nucleotides over the genome of a cell, we should be able to generate data sufficient to characterize the cause of mutation in a population of such cells.

Such means for measuring specific mutations within small blocks of nucleotides in human cells already exist, according to our hypothesis. Forward mutation assays which use selective agents that bind and inactivate *essential* cellular proteins should detect missense mutations at limited sets or blocks of

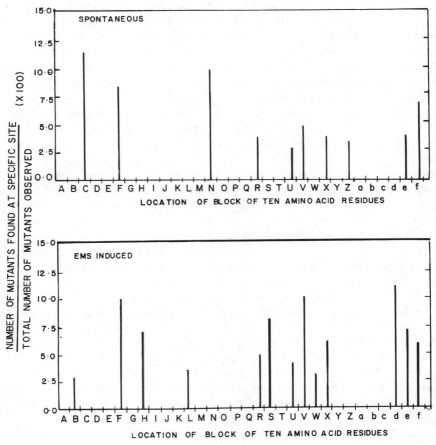

Figure 2
Distribution of spontaneous and EMS-induced mutations in the Lac I system of *E. coli* in blocks of 10 amino acid residues. The percentage of total amber and ochre mutations found at each block is illustrated. (Data from Coulondre and Miller 1978.)

nucleotides. In this type of assay the genetic target is putatively limited to a small number of codons which can mutate to prevent toxin binding but which have little effect on the enzyme's catalytic activity. Mutations affecting toxin binding but also resulting in loss of enzyme activity would not be detected because the enzyme's activity is vital for cell survival.

Therefore, the nucleotide target in this type of selection system and the number of permissible changes which can occur are small due to the specificity involved. Consequently, the response of these assays should reflect the specificity of *kind* and *sequence-dependence* of various mutagenic stimuli. By using a battery of these selective systems to monitor mutation in several discrete blocks of nucleotides, one should be able to generate meaningful mutation spectra by the method outlined for the lac I example.

Several examples of this specific type of forward mutation assay have been developed and characterized in mammalian cells, most frequently for Chinese hamster ovary (CHO) cells. The selective agents used include protein synthesis (ribosome) inhibitors such as cryptopleurine, emetine, trichodermin, and tylocrebine; an inhibitor of mRNA production, 5,6-dichloro(b)D-ribofuranosyl-benzimidazole; an oxidative phosphorylation inhibitor, venturicidin; and an inhibitor of Na^+/K^+ ATPase, ouabain. These selection systems can be applied directly to normal diploid cells due to the dominant nature of the mutant phenotype.

The type of assay being considered is distinctly different from selection systems involving the *loss* of a nonessential protein. An example of loss of activity is 6-thioguanine resistance (TG^r), which results from the *loss* of the HGPRT enzyme. Here the genetic target includes at least the entire *hgprt* structural gene. Since the target is so large (approximately 1000 base pairs) and diverse in mutable sequences, this system should respond similarly to virtually all mutagenic stimuli, regardless of their specificity.

Enough data exist for spontaneous and EMS-induced responses in CHO cells to these selective agents to demonstrate the feasibility of the proposed approach (Gupta and Siminovitch 1977, 1978, 1980; Caloche and Mulsant 1978; Whitfield et al. 1978; Lagarde and Siminovitch 1979). A summary of these studies is presented in Figure 3. Since actual mutation frequencies in induced populations were higher than spontaneous ones, results are presented as a ratio of the mutation frequency at a given specific forward assay to that observed at the *hgprt* locus under similar conditions to normalize the response in the various systems. As described previously, the response of the *hgprt* locus should not reflect the mutagenic specificity of spontaneous or EMS mutagenesis due to the large target size.

Three of the selection systems in CHO cells, cycloleucine resistance, trichodermin resistance, and venturicidin resistance, display a greater than 10-fold difference in the ratio (specific marker frequency)/(*hgprt⁻* frequency) when comparing spontaneous populations to EMS-treated ones (Table 2). A ratio of 3×10^{-2} is found for trichodermin resistance in spontaneous mutants and 0.06×10^{-2} for EMS-induced mutants; while venturicidin resistance gave 45×10^{-2} for spontaneous and 0.8×10^{-2} for EMS.

Currently, only a limited number of forward mutation systems based on drug resistance have been used in cell lines of human origin. However, responses to mutagens in human cells using ouabain resistance (Oua^r) as a specific marker have exhibited differences between spontaneous populations and chemically induced ones in human lymphoblasts (Fig. 4). Responses are again expressed as a ratio of Oua^r to TG^r for comparison purposes as discussed above for the CHO cell system.

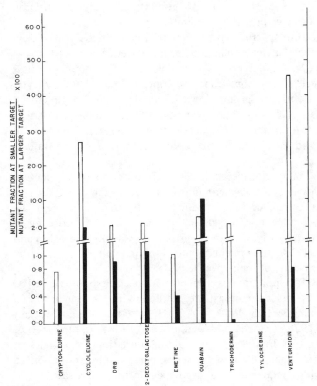

Figure 3
Mutational spectrum of CHO cells. Mutational frequencies in spontaneous and EMS-induced
(300 μg/ml × 20 hr) CHO cells from several laboratories are summarized and expressed as
the ratio of mutant frequency at the smaller target to mutant frequency at the larger target
(*hgprt* locus). (□) spontaneous mutational spectrum; (■) EMS-induced mutational spectrum;
(DRB) 5,6-dichloro(b)D-ribofuranosylbenzimidazole.

The ratios obtained from these differently treated populations are significantly different from one another (95% confidence). The frequency of spontaneous Oua^r and TG^r results in a ratio of 0.16×10^{-2}, compared to a ratio of
6×10^{-2} for EMS (a potent alkylating agent) and less than 0.08×10^{-2} for
ICR-191 (a potent frameshift mutagen) (Thilly 1979). Hence, the ratio of mutation frequencies observed in only two loci is sufficient to enable us to determine
if the population has been exposed to chemical mutagens. A series of such ratios
should increase sensitivity and resolution of the assay, hence permitting us to
illustrate the differential responses characteristic of the mutagenic stimuli.

Table 2
Differential Mutational Response at Selected Sites in CHO Cells

Selection System with Small Target (Drug Resistance)	$\dfrac{\text{M.F. at smaller target}}{\text{M.F. at larger target}}$ (X 10^2)	
	Spontaneous	EMS-induced
cryptopleurine	0.75[a]	0.3[a]
cycloleucine	26.5[b]	2.0[b]
DRB (5,6-dichloro(b)D-ribofuranosylbenzimidazole)	22.0[c]	0.86[c]
2-oxygalactose	3.25[d]	1.65[d]
emetine	1.0[c]	0.385[a]
ouabain	4.5	10.0[e]
trichodermin	3.0[f]	0.06[f]
tylocrebine	1.25[a]	0.33[c]
venturicidin	44.5[g]	0.82[g]

[a]Gupta and Siminovitch (1977)
[b]Caloche and Mulsant (1978)
[c]Gupta and Siminovitch (1980)
[d]Whitfield et al. (1980)
[e]Campbell and Warton (1979)
[f]Gupta et al. (1978)
[g]Lagarde and Siminovitch (1979)

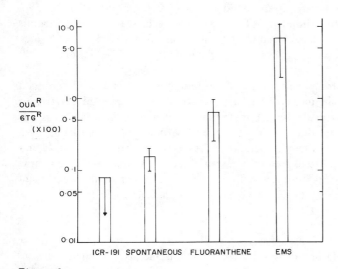

Figure 4
Differential response to different mutagenic stimuli in human lymphoblasts. Mutation frequencies were measured by 6TG[r] and OUA[r] and the responses illustrated as the ratio OUA[r]/6TG[r]. (Data from Thilly et al. 1980)

APPLICATION OF MUTATIONAL SPECTRA TECHNIQUE
TO HUMAN SUBJECTS

If the selection systems described above can be reliably applied to human cells in culture, it should be possible to construct useful spectra for cells isolated from human subjects. Human peripheral T-lymphocytes appear to be a promising source of cells for determination of mutations in humans. A blood sample, which can be obtained easily from an individual, yields approximately 1×10^6 T-lymphocytes/ml blood (Guyton 1971). Hence we could expect to be able to screen up to 1×10^8 T-lymphocytes for mutations by drug resistance at one site per sampling (100 ml) from any individual.

Using the ouabain selection system as the specific target and TG selection system as the larger target, a spontaneous mutant frequency of at least 5×10^{-8} and 3×10^{-5}, respectively, could be expected in vivo. These frequencies predict the presence of 5 (upper 95% confidence of 9) Oua^r mutants and 3000 ± 100 TG^r mutants in this sample. Since several markers are required to generate a useful spectrum, these cells could be propagated using the new T-lymphocyte cloning technique of Albertini (Albertini and Wayne 1982; Albertini, this volume) to obtain sufficient cells for determination to all the markers. Alternatively, one could screen for a smaller number of cells per marker at the expense of lower sensitivity.

In order to generate the mutational spectra in humans, it is crucial that we have the appropriate drug resistance markers involving essential genes. Selection systems using compounds which are toxic to human lymphoblasts, and many of which have already been found useful in CHO mutation studies, are being explored in our laboratory (Table 3). We are in search of compounds which would yield spontaneous mutant frequencies of at least 5×10^{-8} since the number of T-lymphocytes available for screening would be limited to 10^8 per marker.

The mutation frequencies using these compounds would be determined in spontaneous and chemically-induced populations and expressed as ratio of mutant frequency at smaller target to mutant frequency at *hgprt* locus as discussed above. We could illustrate the mutagenic response of a population by a series of these ratios, and hence characterize response with respect to kind and position of mutations. We expect that careful selection of markers which show differential responses to mutagens will enable us to discriminate between induced and spontaneous mutations in vivo with considerable sensitivity.

SUMMARY

Evidence in the literature demonstrates that spontaneous and chemically-induced mutations are specific and differ in terms of the *kind* of mutations induced and their *location* in DNA. This fact may be used to devise a means to discriminate between spontaneous and chemically induced mutations in cell samples taken from humans. A possible approach to this end is the use of a set

Table 3
List of Compounds Being Screened as Potential Selective Agents

Compound	Mode of Inhibition
Protein synthesis inhibitors	
Anguidine	Primarily inhibits initiation; also causes partial polysome stabilization at high drug concentration (Doyle and Bradner 1980)
Cycloheximide	Inhibits translocation at 60s ribosomal subunit (Pestka 1971)
Emetine	Inhibits ribosome movement at 40s ribosomal subunit (Pestka 1971)
Gougerotin	Inhibits transfer of amino acids from aminoacyl s-RNA to protein (Clark 1967)
Inhibitors of DNA/RNA metabolism	
Cycloleucine	Inhibits S-adenosyl methionine (SAM) biosynthesis and RNA maturation (Caloche and Bachellerie 1977)
Daunomycin	Primarily inhibits RNA synthesis; also interferes with DNA synthesis and is postulated to inhibit DNA-dependent DNA polymerase (DiMarco 1967)
DRB	Inhibits mRNA synthesis; postulated to inhibit initiation of chromosomal heterogenous RNA synthesis (Egyhezi 1974)
Inhibitors affecting microtubules	
Colchicine	Binds tightly and specifically to tubulin resulting in disruption of microtubules (Wilson et al. 1974)
Podophyllotoxin	Binds to tubulin and prevents microtubule assembly (Wilson et al. 1974)
Inactivators of essential enzymes	
Ouabain	Binds to Na^+/K^+ ATPase pump resulting in its inactivation (Reichenstein and Reich 1946)

of specific forward mutation assays each of which detects only missense mutations at a small number of base pairs within the human genome. The amount of mutation induced by a mutagen or arising spontaneously for the set of loci chosen constitutes a "mutational spectrum" sufficient to characterize the cause of mutation in a cell population.

We review evidence testing our hypothesis from phage T4, *E. coli*, CHO cells, and human lymphoblastoid cells. We further suggest that the technology exists to extend such analysis to the T-cell populations of human blood samples and thus determine if human genetic change (in T-cells) arises predominately from spontaneous or nonspontaneous origins.

REFERENCES

Albertini, R.J. and R.B. Wayne. 1982. Cloning *in vitro* of human 6-thioguanine resistance (6TG^R) peripheral blood lymphocytes (PBLs) arising *in vivo*. *Program and Abstracts, 13th Annual Meeting, Environmental Mutagen Society*, p. 134. Environmental Mutagen Society, Washington, DC.

Benzer, S. 1961. On the topography of the genetic fine structure. *Proc. Natl. Acad. Sci. U.S.A.* **47**:403.

Caloche, M. and J.P. Bachellerie. 1977. RNA methylation and control of eukaryotic RNA biosynthesis. (Effects of cycloleucine, a specific inhibitor of methylation, on ribosomal RNA maturation). *Eur. J. Biochem.* **74**:19.

Caloche, M. and P. Mulsant. 1978. Selection and preliminary characterization of cycloleucine-resistant CHO cells affected in methionine metabolism. *Somatic Cell Genet.* **4**:407.

Calos, M.P. and J.H. Miller. 1982. Genetic sequence analysis of frameshift mutations induced by ICR-191. *J. Mol. Biol.* **153**:39.

Campbell, C.E. and R.G. Warton. 1979. Evidence obtained by induced mutation frequency analysis for functional hemizygosity at emt locus in CHO cells. *Somatic Cell Genet.* **5**:51.

Clark, J.M., Jr. 1967. Gougerotin. *Antibiotics* (ed. D. Gottlieb and P.D. Shar), vol. 1, p. 278. Springer-Verlag, New York.

Coulondre, C. and J.H. Miller. 1977a. Genetic studies of the lac repressor: III. Additional correlation of mutational sites with specific amino acid residues. *J. Mol. Biol.* **117**:525.

_____. 1977b. Genetic studies of the lac repressor: IV. Mutagenic specificity in the lac I gene of *Escherichia coli. J. Mol. Biol.* **117**:577.

DiMarco, A. 1967. Daunomycin and related antibiotics. *Antibiotics* (eds. D. Gottlieb and P.D. Shar), vol. 1, p. 190. Springer-Verlag, New York.

Doyle, T.W. and W.T. Bradner. 1980. Trichothecanes. In *Anticancer agents based on natural product models* (eds., J.M. Cassady and J.D. Douros), p. 43. Academic Press, New York.

Egyhezi, E. 1974. A tentative initiation inhibitor of chromosomal heterogenous RNA synthesis. *J. Mol. Biol.* **84**:173.

Gupta, R.S. and L. Siminovitch. 1977. Mutants of CHO cells resistant to protein synthesis inhibitors, cryptopleurine and tylocrebine. Genetic and biochemical evidence for common site of action of emetine, cryptopleurine, tylocrebine and tubulosine. *Biochemistry* **16**:3209.

_____. 1978. Genetic and biochemical characterization of mutants of CHO cells resistant to the protein synthesis inhibitor Trichodermin. *Somatic Cell Genet.* **4**:355.

_____. 1980. Genetic markers for quantitative mutagenesis studies in CHO cells. Characteristics of some recently developed selection systems. *Mutat. Res.* **69**:113.

Guyton, A.C. 1971. *Textbook of Medical Physiology*, 4th Ed. II. Blood cells, immunity, and blood clotting. W.B. Saunders, Philadelphia.

Lagarde, H.E. and L. Siminovitch. 1979. Studies on CHO mutants showing multiple cross resistance to oxidative phosphorylation inhibitors. *Somatic Cell Genet.* **5**:847.

Miller, J.H., C. Coulondre, and P.J. Farabaugh. 1978. Correlation of nonsense sites in the lac I gene with specific codons in the nucleotide sequence. *Nature* **274**:770.

Miller, J.H., D. Ganem, P. Lu, and A. Schmitz. 1977. Genetic studies of the lac repressor. I. Correlation of mutational sites with specific amino acid residues: Construction of colinear gene-protein map. *J. Mol. Biol.* **109**:275.

Pestka, S. 1971. Inhibitors of ribosome functions. *Annu. Rev. Microbiol.* **25**:487.

Reichstein, T. and H. Reich. 1946. The chemistry of steroids. *Annu. Rev. Biochem.* **15**:155.

Schmiessner, U., D. Ganem, and J.H. Miller. 1977. Genetic studies of the lac repressor. II. Fine structure deletion map of the lac I gene, and its correlation with the physical map. *J. Mol. Biol.* **109**:303.

Skopek, T.R. and F. Hutchinson. 1982. DNA base sequence changes induced by bromouracil mutagenesis of lambda phage. *J. Mol. Biol.* **159**:19.

Thilly, W.G. 1979. Study of mutagenesis in diploid human lymphoblasts. *Ban. Rep.* **2**:341.

Thilly, W.G., J.G. DeLuca, E.E. Furth, H. Hoppe, D.A. Kaden, J.J. Krolewski, H.L. Liber, T.R. Skopek, S.A. Slapikoff, R.J. Tizard, and B.W. Penman. 1980. Gene locus mutation assays in diploid human lymphoblast lines. *Chem. Mutagens* **6**:331.

Thilly, W.G. and H.L. Liber. 1980. Genetic toxicology. In *Toxicology: The basic science of poisons,* 2nd ed. (eds. J. Doull et al.), p. 139. Macmillan, New York.

Whitfield, C.D., B. Buchsbaum, R. Bostedor, and E.H.Y. Chu. 1978. Inverse relationship between galactokinase activity and 2-deoxygalactose resistance in Chinese hamster ovary cells. *Somatic Cell Genet.* **4**:699.

Wilson, L., J.R. Bamburg, S.B. Mizel, L.M. Grishan, and K.M. Greswell. 1974. Interaction of drugs with microtubule proteins. *Fed. Proc.* **33(2)**:158.

COMMENTS

SHODELL: If you increase the concentration of EMS, is there a proportional increment in the mutation rate? Does the pattern hold over all the concentration ranges?

SKOPEK: We haven't looked at that, because, needless to say, the construction of one mutagenic spectrum is a great deal of work. But I see your point, and it is a very valid one: Do the spectra produced by low levels of agent reflect the same specificity as high levels? This is a question that must be tested.

EVANS: It is almost impossible to determine, because your background level becomes such a major part of what you are looking at that you would have to do an enormous experiment.

SKOPEK: If there is something very characteristic about your spectrum, it is very easy to distinguish between spontaneous and induced populations, even though there has been a very small increase over background. For example, if we test 300 mutants in *E. coli* and we find that 100 are insertions, then we conclude that we are looking at a normal spontaneous spectrum. However, if you induce a two-fold increase over background, you would expect to see the fraction of IS1 and IS5 halved.

EVANS: Well, that was essentially my question. But I will add a bit of information on it. If you look at chromosome structural changes induced by mutagens you will also get specificity for various kinds of alkylating agents; EMS is quite different from MMS, and so on, and you see a different spectrum. It is very dependent upon concentration. If you look at mitomycin C, at low concentrations, you get a lot of things involving chromosome 9 in man. As you raise the concentration, you get a much bigger variety of changes turning up. I was wondering, in fact, how fixed your spectra were with regard to concentration.

SKOPEK: I think we have to keep in mind that, in applying this system, we are especially interested in being able to distinguish spontaneous from induced. We are not at the point where we are hoping to distinguish between two different alkylating agents.

WYROBEK: On that point, can you relate the hotspots with the functional difference in the gene product? If you can't do it now, can you imagine being able to do it?

SKOPEK: Needless to say, the mutants that we are generating are of great interest to the people who study protein structure. We are just handing them over, because we are not experts in protein structure. We will have to rely on their expertise to make sense of the types of mutations we are seeing. However, the hotspots that we are observing, say, with bromo-uracil and frameshifts—especially frameshifts—are not due to the specificity regarded in knocking out a protein. Spontaneous is different from bromouracil, which is different from UV light. If the point that you mentioned is operating, one would expect all to be the same.

WYROBEK: Well, I was just wondering whether one could perhaps have diminished enzyme activity or altered enzyme activity. Do you imagine that the spontaneous would be different than the induced in that respect?

SKOPEK: I don't think so.

RUSSELL: Just a point of information. You can't do this beautiful intra-genic mapping, but just the spectrum among different loci does change with different mutagens in terms of germinal mutations in the specific locus test. You get a different distribution of mutations among the loci with different mutagens.

SKOPEK: That is the approach that Bill Thilly is about to discuss.

ALBERTINI: I may say that we have the beginnings of a mutational spectrum *in vivo*—at least as concerns different loci. I didn't give frequencies for the diphtheria toxin-resistant T-cells, but we could not generate a P_o class where we seeded 10^5 cells per well. The frequencies for TdR PBLs are going to be much higher than for thioguanine-resistant cells. Of course, several loci may be involved.

BRIDGES: What worries me about it is that you seem to have made the assumption that the spontaneous spectrum is going to be the same from one cell type to another, from one individual to another, quite apart from the effects of one environment, i.e., in vivo, to another environment in vitro. I think all those assumptions are questionable.

THILLY: I didn't get the point across. We have not made those assumptions. Using lymphocyte cloning assay, we will take multiple cells from the same person, grow them up and determine the spectrum of mutation from a set of cells from an individual. We will then take multiple samples from another individual, until we understand what the pattern of spontaneous mutation is within an individual and among individuals.

BRIDGES: And if there are big differences, what then?

THILLY: Then we have found that there are big differences.

BRIDGES: I bet that there are going to be big differences.

THILLY: I think we have already gone far enough to be reasonably sure that there are going to be big differences. But for the person with *spontaneous pattern A* and we will measure the pattern in their real T-cells, the cells that have not grown many generations in culture. Will those cells' pattern differ from the spontaneous pattern *of this individual*? We are not going to be comparing the real spectrum of person A to the spontaneous spectrum of some random group B.

BRIDGES: If you take biopsies from the same individual, (fibroblasts, just as an example) and you culture them, they are not all the same. Even within an individual, you will find differences, as the somatic cells have evolved their properties. So I wonder whether you will ever find a baseline. The first lot that you take out to get the spontaneous spectrum in vitro is not going to be controlling the next lot that you go in for.

THILLY: We will have to determine that. Sequential samples will permit us to find out if we are dead wrong or reasonably correct. Selecting for the T-lymphocytes which are T4-positive, for instance, would restrict the kinds of cells that we are going after.

ALBERTINI: The cloning assay itself really can turn out to be a fluctuation test, if you then simply let it go and then look for the P_0 class of TG^r cells in replated cells from several wells that were started at one cell per well. It is possible that there are different sorts of lymphocytes in vivo as regards mutability. For example, if there is a sensitive cell type and a resistant cell type, and some physiological or pathological condition changes the ratio of these two types, you may not have mutation *per se*, but rather selection of pools which give differences in mutant frequencies. To a limited extent, we are looking for this when we determine surface markers. Several other markers of heterogeneity are available.

STRAUSS: Presumably we will be able to biopsy more specifically by looking at specifically reactive cells—cells that all came from the same stem cell circulating around, since those can be selected for in these functional cells. Also B-cells are now culturable for long periods of time without trans-formation in the absence of virus. These should be capable of being used the same way and can probably be chosen on the basis of specific antibody production.

GERMAN: I don't know whether they have been mentioned at the conference, but there is a group of humans that certainly would be interesting to study in your system and some of other systems. These are people that are homozygous for some rare genes that produce syndromes that are cancer predisposing. All of these syndromes, we know, have chromosome instability, so they have an increased chromosome mutation rate. At least one of them, Bloom's syndrome, has been reported by two labs to have increased spontaneous mutation rate in vitro. Along with your medical students or whoever you are going to bleed, it would be nice to stick in a few of these Bloom's syndromes and Fanconi's AT syndromes and XPs. We have lymphoblastoid lines on all of them, also in heterozygotes.

THILLY: An interesting technical addition is that Victor Goldmacher in our lab found that the TK-6 B lymphoblast cells which form colonies well at 1 cell/μwell serve as feeder cells for some other B-lymphoblastoid lines which show poor low-density colony forming ability. Just send us a stamped envelope and you can have TK6. For non-commercial use.

BUTTERWORTH: The original premise that Tom Skopek introduced was the concern about genetic disease, and yet all of the things you are measuring are in lymphocytes, which are somatic cells. It seems to me that you are not measuring the right cell type.

THILLY: We are measuring the only cell type we think we can sample ethically in sufficient numbers for analysis. I am sure many of us have thought about Sandberg's haploid human cell lines from lymphomas. These things have a few more than 26 chromosomes. One keeps wondering if it will be possible to come up with a sperm fusion assay by first taking those pseudo haploids, giving whatever markers we want to put into them, fusing them with sperm, and actually working directly on human genetic material in that way. Do you want to look ahead?

BUTTERWORTH: Just what I was going to say. You have to begin to deal with the germ cells if you are worried about mutation since the clinical significance of somatic mutations can be very minimal.

THILLY: Let's not get carried away here, now. Your point is a hypothesis, too. It is possible that somatic mutations lead to a plethora of diseases. We are sure only that somatic mutations do not lead to germinal mutation.

MALLING: I would like to know how we will deal with sperm cells when we are dealing with women. We have to have the somatic cells to be able to measure the exposure in women.

THILLY: It is not going to work. This method will not work on oocytes. There are only on the order of 10^5 oocytes. This kind of mutational spectral analysis requires on the order of 10^8 cells for analysis.

ALBERTINI: One population of humans to look at for these kinds of studies are individuals who show a great increase in skin cancers after an important protective element has been eliminated. This may reveal individuals who have an increased susceptibility to mutations, as well as the role of somatic cell mutations in the genesis of their cancers. For example, kidney transplant recipients are effectively immunosuppressed. Most of these people do not have skin cancer, and a few have one, two, or three. Then, there are some patients who have bursts—multiple skin cancers. The same thing, I think, is happening with the PUVA-treated patients (for psoriasis),—that there are skin cancer "bursts" in some individuals. The "burst people" do not have any known breakage syndrome. Using such a "burst individual" allows us to ascertain a family. We can then look at cells of the patients or—better—the unaffected family members with these kinds of studies, and compare them with the population at large. This may be a way to begin studies on individual differences in mutability—and the heritability of these differences.

STRAUSS: That is true, and I think you can connect the systems we have been talking about by using animals. Certainly we have learned methods for doing that.

THILLY: I think it is opening up a new era in genetic toxicology with the ability to clone.

STRAUSS: Let's extrapolate backwards instead of trying to extrapolate forwards.

THILLY: Amen. We would differ in philosophy. I am going straight to people—if it works there, fine—and then back to animal models.

STRAUSS: That is what I said.

ALBERTINI: There has been a request to get back to the germ cells, so Harvey Mohrenweiser, University of Michigan, is going to discuss biochemical approaches to monitoring human populations for germinal mutation rates.

Models to Man:

Establishment of Reference Points for

Estimating Genetic Risk in Man

HARVEY W. MOHRENWEISER AND JAMES V. NEEL
Department of Human Genetics
University of Michigan Medical School
Ann Arbor, Michigan 48109

In recent years a significant number of test systems have been developed that have varying capabilities in identifying genotoxic agents. A major component of this conference has been devoted to describing the metabolic processing of potential mutagens in various somatic cell systems, the DNA adducts formed and removed in such systems, and the resulting mutagenic events, especially as these areas relate to the human situation. Although significant progress has been made, none of these somatic test systems will provide the necessary information for regulatory agencies to make cost and benefit decisions (unless a Delaney Clause approach is utilized) until such time as relevant reference data on germinal mutation rates and associated health costs in human populations are available. This should not be taken to imply that in vitro toxicity tests are not appropriate and(or) only in vivo tests are relevant but rather that it will be impossible to conduct meaningful extrapolations without studies that build the necessary bridges between the estimates from in vitro tests, in vivo somatic cell systems, animal germinal studies, and estimates of background and induced germinal mutation rates in human populations.

This presentation is limited to the current status of studies of germinal mutation rates. We adopt the position that the current uncertainties in extrapolation to people are such that we must make a major effort to obtain reference data from human populations in order to provide the intellectual bridge that will ultimately be the basis for extrapolation from other systems. We further argue that the most secure extrapolation for germinal mutation rates among species is provided by a biochemical approach, wherein comparable proteins are examined in the different species. Finally, we will urge that because of strain and species differences in mutation rates, it may be inappropriate to continue to extrapolate from the study of selected, highly inbred mouse strains (or *Drosophila melanogaster*) to heterogeneous human populations; the foundations for extrapolation must be developed with a goal of providing estimates of human health risk obtained through coordinated studies.

CURRENT STATUS OF ESTIMATES OF GERMINAL MUTATION RATES IN OTHER ORGANISMS

Background Mutation Rates

Estimates of background and induced germ-line mutation rates derived from studies of protein markers are available for both *Drosophila* and the mouse. The background mutation rate to electrophoretic mobility variants has been reported at 0.5×10^{-5} per locus/generation by Tobari and Kojima (1972), 0.18×10^{-5} by Mukai and Cockerham (1977), and 0.13×10^{-5} by Voelker et al. (1980). (The data of Voelker et al. are a continuation of the study of Mukai and Cockerham.) The latter studies also monitored for the frequency of mutation to null alleles and found that the background null mutation rate is at least three times the rate of mutations involving the interchange of amino acids with dissimilar charge and retention of catalytic function.

Less data are available on the background mutation rate(s) for biochemical markers in mice. Only one electrophoretic mobility mutant has been detected in studies in three different laboratories involving approximately 469,000 locus tests (Pretsch and Narayanan 1979; Pretsch and Charles 1980; Bishop and Feuers 1982; Casciano 1982; Johnson and Lewis 1982). No null mutations have been detected in these mouse studies but the number of locus tests would be only about 160,000 (Johnson and Lewis 1982). The frequency of spontaneous mutations in mice as determined with the visible specific locus method is $0.75\text{-}1.0 \times 10^{-5}$ (Searle 1974; Russell and Kelly 1982). This estimate is higher than the preliminary data on deficiency and mobility variants would suggest. This is somewhat surprising as one would not expect most of the mutations of the electrophoretic mobility type to be scored in the visible specific locus system.

Induced Mutation Rates

The chronic exposure of *Drosophila* to low-dose rate γ radiation (0.15 rad/min for 15 generations) increased the frequency of mutation by at least a factor of 5-10 over the background rate observed by Voelker et al. (1982) (nonconcurrent controls) and all of the mutations were to null variants rather than electrophoretic mobility variants (Racine et al. 1980). Similarly, Malling and Valcovic (1977) identified four null variants and no mobility variants in mice exposed to 1000 r of X-ray (2×500 r at 24-hour intervals). They calculated an induced mutation for biochemical markers rate which would appear to be at least an order of magnitude above the current background estimate obtained with similar techniques. Three of the five hemoglobin variants in offspring of acutely radiated mice identified by Russell et al. (1976) were associated with the absence of functioning α hemoglobin chains. Thus, not only did radiation increase the mutation rate but most of the mutations detected following exposure were associated with the loss of functional gene products; that is, a class of mutations were induced that would be expected to be physiologically significant.

Treatment of mice with ethylnitrosourea (ENU) increases the mutation rate by approximately two orders of magnitude (Russell et al. 1979; Johnson and Lewis 1981) with the frequency of null variants being 2-4 times the frequency of induced electrophoretic mobility variants (Johnson and Lewis 1982). Procarbazine (Pretsch and Charles 1980; Johnson et al. 1981), triethylenemelamine (Soares 1979), and ethylmethanesulfonate (Feuers et al. 1982) also induce both null and electrophoretic mobility variants in the biochemical specific locus tests and also induce mutations in the visible specific locus tests (Russell et al. 1979), although the potency of these latter compounds as germ-cell mutagens is much less than observed for either ENU or radiation. The real question is *how* mutagenic are these chemicals and what is the expected human health risk associated with given levels of exposure. It is apparent that chemicals can induce a class of variants that often result in significant health impairment in homozygous individuals, if the genetic defect is not associated with reproductive failure.

CURRENT STATUS OF BIOCHEMICAL APPROACHES TO DETECTING GERMINAL MUTATIONS IN HUMAN POPULATIONS

Methodology

Early estimates of the germinal mutation rate in humans were based either on alterations in population characteristics, such as sex ratio, stillbirth frequency, or infant mortality rates, or the rate of appearance of several easily recognized sentinel phenotypes (e.g., retinoblastoma, neurofibromatosis, hemophilia, etc.). The shortcomings of these approaches include, for the sentinel phenotypes, such problems as: ascertainment bias, nonrepresentativeness of the loci, and heterogeneity of syndromes. The shortcomings for the population characteristics approach include the issue of differentiating between genetic and nongenetic causes of changes in the indicator endpoints. These matters have been discussed in other reviews (Vogel 1970; Neel 1971).

With the development of various biochemical techniques, it has become possible to examine structural features of many proteins. Electrophoretic techniques utilizing a solid support media as developed by Smithies (1955) have been used by Harris and colleagues (Harris and Hopkinson 1972; Harris et al. 1974) at the Galton Laboratory in a large study of the genetic variability of a human population. Similar electrophoretic studies, for the identification of variants at 30-40 different loci expressed in erythrocytes or plasma with either starch or polyacrylamide as the support media, have been developed by Neel and colleagues (1979) for utilization in efforts to determine the background mutation rate in the United States population (Neel et al. 1980a) and the background and radiation induced mutation rate in Japan (Neel et al. 1980b). These techniques have also been used in studies which provide data for both direct and indirect estimates of the mutation rate among the unacculturated Amerindian populations (Neel and Rothman 1978; Neel et al. 1980a). A similar approach, using isoelectrifocusing techniques for the study of the hemoglobins has been developed by Vogel and Altland (1982) for studies of the background mutation rate among the West German population.

Standard one-dimensional electrophoretic techniques normally use enzyme specific staining techniques for identifying the protein of interest. Thus, for a genetic alteration to be identified by these electrophoretic techniques, the amino acid substitution must involve residues with nonidentical charges and the variant protein must retain catalytic function. But, many genetic events, be they base substitutions, insertions, or deletions will involve codons not only in the coding region of a gene but also in the flanking or intervening sequences and involving binding and splicing sites. Many of these alterations will result in either the presence of a nonfunctional protein or the absence of recognizable protein. Because of this, we have developed quantitative techniques to identify deficiency variants occurring at 12 erythrocyte enzyme loci (Mohrenweiser 1981, 1982a,b). It should be appreciated, as indicated previously, that occurrence of the various alleles responsible for this latter class of variants in the homozygous state is the basis for many of the inborn errors of metabolism.

Results of Surveying Human Populations

Four different industrialized populations have been studied with electrophoretic techniques: United Kingdom (Harris et al. 1974), United States (Neel et al. 1980a), Japan (Neel et al. 1980b), and West Germany (Altland 1982). To date, 725,803 direct equivalent locus tests have been completed. One mutation, involving the α chain of hemoglobin, has been observed in this series (Altland 1981). Although it would not be wise to calculate a mutation rate from this data, as it would have a numerator of one, if one should do it, they would obtain an estimate of 0.14×10^{-5}. Other data for direct estimates of the mutation rate for electrophoretic variants include 94,796 locus tests in the Amerindians (Neel et al. 1980a) and 1897 locus tests in an unexposed group of Marshall Islanders (Neel et al. 1982). In neither group was a putative mutation detected. One other substantial group, the offspring of individuals exposed to radiation from the atomic bombs in Japan, has also been studied. Currently, 419,666 equivalent locus tests have been conducted and two putative mutations have been observed (Neel et al. 1982). No mutations were detected in 1835 locus tests in children of radiation-exposed Marshall Islanders (Neel et al. 1982).

Preliminary data are available from three studies which have employed quantitative assays to detect mutations resulting in the loss of functional protein. Mohrenweiser (1981, 1982b) identified 22 individuals among 6142 determinations (675 newborns×9 enzymes) with levels of enzyme activity consistent with the individual being heterozygous (carrier) for an enzyme deficiency allele. Twelve (12) of the variants involved triosephosphate isomerase or glucose-6-phosphate dehydrogenase, both of which existed in polymorphic frequency in the Black sample, while 10 of the deficiencies were classified as rare enzyme deficiency variants. Satoh et al. (1982) at the Radiation Effects Research Foundation (RERF) detected 34 deficiency variants in 11,852 determinations (approximately 50% from children of exposed parents) at a similar series of erythrocyte enzyme loci. A study in Germany involving only five loci yielded seven variants with levels of enzyme activity consistent with our definition of an enzyme deficiency (Eber et al. 1979). Fifty-two (52) of the enzyme deficiency variants were inherited while one variant was associated with a three system paternal exclusion, thus no putative mutations were identified in these initial studies involving approximately 33,000 determinations.

Although it is not possible to compare the frequency of mutations to electrophoretically identifiable variants with mutations to enzyme deficiency variants in a human population at this time, it is possible to compare the relative frequency of these inherited variants, as they exist in human populations. In both the Ann Arbor and the RERF study, the *rare* deficiency variants are 1.5-3.0 times more frequent than electrophoretic variants (Mohrenweiser 1981; Mohrenweiser and Neel 1981; Satoh et al. 1982). This ratio of deficiency to mobility variants is observed both when the frequency of electrophoretic variants is calculated from all loci studied (20-23) and when the comparison is restricted to the seven loci studied with both techniques (Mohrenweiser and Neel 1981). In the absence of an assumption that the enzyme deficiency variants confer a selective advantage to the affected individual, this higher frequency of enzyme deficiency variants should reflect a higher rate of mutation for this class of variants.

The data on the frequency of electrophoretic variants have also been used to obtain an indirect estimate of the mutation rate in three unacculturated populations, the Amerindians, the Australian aborigines, and the natives of Papua New Guinea. The estimated mutation rate for electrophoretically detectable variants (unweighted average) is 0.9×10^{-5} (Neel et al. 1982). Although the many uncertainties in this estimate would imply a significant error term for this estimate, it has been argued that the mutation rate in these tropical, tribal, and nonindustrialized populations is higher than in the temperate-dwelling and industrialized populations (Neel and Rothman 1981).

New Methods

The study of mutation rates in higher eukaryotes is laborious and especially so when one seeks to establish differences of approximately 50% between groups exposed to potential mutagens and control groups (cf. Neel 1980). Currently the best possibility of rendering the search for mutations more efficient involves the development of computer routines for examining 2-dimensional (2-D) polyacrylamide gel preparations of various human cell types (or fractions thereof) for possible mutational events (Hanash et al. 1982; Rosenblum et al. 1982; Skolnick 1982; Skolnick et al. 1982). Such preparations when stained with the newer and sensitive silver stains exhibit as many as 1000 protein moieties, not all, of course, in sufficient quantity to render them proper objects for the study of mutation. Since screening the gels visually is extremely time-consuming, the principal challenges are:

1. the development of the most efficient computer programs possible for the search for rare variant proteins such as might result from mutation;
2. the development of the necessary micro-techniques for the recovery and partial characterization of such proteins; and
3. to determine with the necessary certitude, the affinities of any possibly mutant protein.

It should be appreciated that the 2-D gel electrophoresis approach could increase the amount of data obtainable from each individual by at least an order of magnitude as compared to the standard electrophoretic techniques, thereby resulting in a significant decrease in the number of offspring required for estimating a mutation rate.

PROBLEMS IN EXTRAPOLATION FROM MODEL SYSTEMS TO HEALTH RISK IN MAN ASSOCIATED WITH INDUCED GERMINAL MUTATIONS

The problems of extrapolating from in vitro test results to in vivo response have been extensively discussed in previous presentations in this conference and include differences in exposure, metabolism, pharmacokinetics, molecular biology, etc., among cells, tissues, and species. Although these various aspects have been extensively described with respect to many in vitro toxicological testing systems, it is not clear that we have significantly increased our ability to make decisions regarding genetic risk, as opposed to cataloging genotoxic agents. (See discussions in McElheny and Abrahamson 1979.)

Similar problems remain in extrapolating from the in vivo genetic test systems (*Drosophila* and mouse) to man. Mutator genes are known to exist in

Drosophila (Kidwell et al. 1973; Yamaguchi and Mukai 1974), thus the question of appropriate strains for estimating quantitative responses arises. Similar differences are observed among strains of mice exposed to several mutagenic agents (Generoso 1982) and, as expected, among different animal species (Lyon and Smith 1971). A similar situation is observed in man where an apparent difference in background mutation rate is observed between urban and industrialized and unacculturated and tropical populations (Neel and Rothman 1981). (Different methods were used in obtaining the two estimates, but the discrepancy is sufficiently large to argue against its difference being entirely due to the methodologies.) Other possible explanations for the apparent difference in mutation rate between human populations include differences in "background" exposure to mutagenic agents (Neel 1973) or differences in DNA repair capability, where large differences among individuals are already known (Arlett and Lehmann 1978).

Most of the current germinal mutation rate data has been obtained with treated males, the data being much more difficult to obtain with treated females. One should expect differences in pharmacokinetics and metabolism in the germinal tissues of males and females as well as differences associated with the development pattern of the germ cell in each sex. Thus, it seems probable that the mutation rate will be sex specific. An additional problem for extrapolation between species would be the ultrastructural differences in the blood-testis barrier between rodents and man (Waites et al. 1972; Setchell and Waites 1975) which would again emphasize the necessity of target cell dosimetry studies.

Current preliminary estimates of the background mutation rate in man, mouse and Drosophila are in the range of 0.1-0.5×10^{-5} for electromorphs. Given differences in life cycle, body temperature, and level of mutagen exposure in "wild" humans vs. "captive" mice and *Drosophila,* this similarity in rate seems somewhat paradoxical. Additional data are necessary to solidify these estimates, but possible differences in DNA repair capability (Neel 1978) metabolic pathways and pharmacokinetics are exciting possibilities for explaining the unexpected similarity in background rates.

The only data base on induced mutation rates that is available in a human population is that currently being acquired at the RERF in Hiroshima which is studying the offspring of the atomic bomb blast survivors. Preliminary evidence, based upon the frequency of untoward pregnancy outcome, early death, and sex chromosome aneuploidy suggests that the doubling dose for radiation-induced damage in man may be as much as four times higher than is estimated for the mouse (Schull et al. 1981) and provides additional cause for concern regarding our current ability to estimate human risk. Obviously, the data base being built in the RERF studies, on the mutation rate for protein variants, both mobility and deficiency, will be important in providing a further reference for mouse-man comparisons.

Future Directions

Sufficient data have been accumulated in several laboratories indicating the feasibility of utilizing the biochemical approach for monitoring human populations for the frequency of electrophoretic mobility and enzyme deficiency variants. Similar approaches with many of the same structural loci are in place in the mouse and *Drosophila*. Decisions on the feasibility of the 2-D gel techniques should be made within the near future which would provide for an additional interface between the human and mouse studies and increase the feasibility of studies in exposed human populations. Thus, if the goal of the genetic toxicologist is to provide a data base for risk estimation, it is time to address the issues of individual and species specificity (or differences) as described in the preceding section. Within this context, it is critical that our estimates of the background germinal mutation rate(s), with the expectation that the rate may be a function of methodology and endpoint, be refined with special emphasis directed toward understanding the basis for the differences in rates within and between species, assuming the differences which seem apparent at this time are confirmed. Further, we need to establish collaborative efforts that insure that we make maximum utilization of the hopefully small data base that can be derived from studies of offspring of populations that have received significant exposure to suspected mutagenic agents and from appropriate studies in model systems. This should include parallel studies with inclusion of efforts to relate exposure levels via chronic (or repeated) exposure, which will be the usual human experience, to somatic indicators of damage and subsequently to germinal mutation rates in both humans and mice.

ACKNOLWEDGMENTS

Supported in part by contract E(11-1)2828 from the Department of Energy.

REFERENCES

Altland, K. 1982. Monitoring for changing mutation rates using blood samples submitted to PKU screening. *Proc. VI Int. Congr. Hum. Genet.* (in press).
Arlett, C. F. and A. R. Lehmann. 1978. Human disorders showing increased sensitivity to the induction of genetic damage. *Annu. Rev. Genet.* 12:95.
Bishop, J. B. and F. J. Feuers. 1982. Development of a new biochemical mutation test in mice based upon measurement of enzyme activities. II. Test results with ethyl methanesulfonate (EMS). *Mutat Res.* 95:273.

Casciano, D. 1982. Detection of specific activity variants in mice. In *Utilization of mammalian specific locus studies in hazard evaluation and estimation of genetic risk* (ed. F. de Serres). Plenum Press, New York. (In press).

Eber, S. W., B. H. Belohradsky, and W. K. G. Krietsch. 1982. A case for triosephosphate isomerase testing in congential non-spherocytic hemolytic anemia. *J. Pediatr.* (in press).

Eber, S. W., M. Dunnwald, B. H. Belohradsky, F. Bidlingmaier, H. Schievelbun, H. M. Weinmann, and W. K. G. Krietsch. 1979. Hereditary deficiency of triosephosphate isomerase in four unrelated families. *Eur. J. Clin. Invest.* 9:195.

Generoso, W. 1982. Variability in genetic response resulting from genetic heterogeneity. In *Utilization of mammalian specific locus studies in hazard evaluation and estimation of genetic risk* (ed. F. de Serres). Plenum Press, New York. (In press).

Hanash, S. M., D. G. Tubergen, R. M. Heyn, J. V. Neel, L. Sandy, G. S. Stevens, B. B. Rosenblum, and R. Krzesicki. 1982. Two-dimensional electrophoresis of cell proteins in childhood leukemia. *Clin. Chem.* 28:1026.

Harris, H. and D. A. Hopkinson. 1972. Average heterozygosity per locus in man: An estimate based on the incidence of enzyme polymorphisms. *Ann. Hum. Genet.* 36:9.

Harris, H., D. A. Hopkinson and E. B. Robson. 1974. The incidence of rare alleles determining electrophoretic variants: Data on 43 enzyme loci in man. *Ann. Hum. Genet.* 37:237.

Johnson, F. M. and S. E. Lewis. 1981. Electrophoretically detected germinal mutations induced in the mouse by ethylnitrosourea. *Proc. Natl. Acad. Sci. U.S.A.* 78:3138.

Johnson, F. M. and S. E. Lewis. 1982. The detection of ENU-Induced mutations in mice by electrophoresis. In *Utilization of mammalian specific locus studies in hazard evaluation and estimation of genetic risk* (ed. F. de Serres). Plenum Press, New York. (In press).

Johnson, F. M., G. T. Roberts, R. K. Sharma, F. Chasalow, R. Zweidinger, A. Morgan, R. W. Hendren, and S. E. Lewis. 1981. The detection of mutants in mice by electrophoresis: Results of a model induction experiment with procarbazine. *Genetics* 97:113.

Kidwell, M. G., J. F. Kidwell, and M. Nei. 1973. A case of high rate of spontaneous mutation affecting viability in *Drosophila melanogaster*. *Genetics* 75:133.

Lyon, M. F. and B. D. Smith. 1971. Species comparisons concerning radiation-induced dominant lethals and chromosome aberrations. *Mutat. Res.* 11: 45.

Malling, H. V. and L. R. Valcovic. 1977. Biochemical specific locus mutation system in mice. *Arch. Toxicol.* 38:45.

McElheny, V. K. and S. Abrahamson. 1979. Assessing chemical mutagens: The risk to humans. *Banbury Rep.* 1.

Mohrenweiser, H. W. 1981. Frequency of enzyme deficiency variants in erythrocytes of newborn infants. *Proc. Natl. Acad. Sci. U.S.A.* 78:5046.

_____ . 1982a. Frequency of rare enzyme deficiency variants: Search for mutational events with human health implications. *Prog. Mutat. Res.* **3**:159.

_____ . 1982b. Biochemical approaches to monitoring human populations for germinal mutation rates. II. Enzyme deficiency variants as a component of the estimated genetic risk. In *Utilization of mammalian specific locus studies in hazard evaluation and estimation of genetic risk* (ed. F. de Serres). Plenum Press, New York. (In press).

Mohrenweiser, H. W. and J. V. Neel. 1981. Frequency of thermostability variants: Estimation of total "rare" variant frequency in human populations. *Proc. Natl. Acad. Sci. U.S.A.* **78**:5729.

Mukai, T. and C. C. Cockerham. 1977. Spontaneous mutation rates at enzyme loci in *Drosophila melanogaster*. *Proc. Natl. Acad. Sci. U.S.A.* **74**:2514.

Neel, J. V. 1971. The detection of increased mutation rates in human populations. *Perspect. Biol. Med.* 522.

_____ . 1973. "Private" genetic variants and the frequency of mutation among South American Indians. *Proc. Natl. Acad. Sci. U.S.A.* **70**:3311.

_____ . 1978. Mutation and disease in man. *Can. J. Genet. Cytol.* **20**:295.

_____ . 1980. Some considerations pertinent to monitoring human populations for changing mutation rates. *Proc. XIV Int. Congr. Genet.* **1**:225.

Neel, J. V. and E. D. Rothman. 1981. Is there a difference between human populations in which the rate which mutation produces electrophoretic variants? *Proc. Natl. Acad. Sci. U.S.A.* **78**:3108.

Neel, J. V., H. W. Mohrenweiser, and M. M. Meisler. 1980a. Rate of spontaneous mutation at human loci encoding protein structure. *Proc. Natl. Acad. Sci. U.S.A.* **77**:6037.

Neel, J. V., H. W. Mohrenweiser, C. Satoh, and H. B. Hamilton. 1979. A consideration of two biochemical approaches to monitoring human populations for a change in germ cell mutation rates. In *Genetic Damage in Man Caused by Environmental Agents* (ed. K. Borg), p. 29. Academic Press, New York.

Neel, J. V., C. Satoh, H. B. Hamilton, M. Otake, K. Goriki, T. Kagoeka, M. Fijita, S. Neriishi, and J. Asakawa. 1980b. A search for mutations affecting protein structure in children of atomic bomb survivors: Preliminary report. *Proc. Natl. Acad. Sci. U.S.A.* **77**:4221.

Neel, J. V., H. W. Mohrenweiser, S. Hanash, B. Rosenblum, S. Steinberg, K. H. Wurzinger, E. Rothman, C. Satoh, T. Krasteff, M. Skolnick, and R. Krzesicki. 1982. Biochemical approaches to monitoring human populations for germinal mutation rates: I. Electrophoresis. In *Utilization of mammalian specific locus studies in hazard evaluation and estimation of genetic risk* (ed. F. de Serres) Plenum Press, New York. (In press).

Pretsch, W. and K. R. Narayanan. 1979. Exfassung von Genmutationen bei mansen durch isoelectric fokussierung. *Hoppe-Seyler's Z. Physiol. Chem.* **360**:345.

Pretsch, W. and D. Charles. 1980. Genetical and biochemical characterization of a dominant mutation of mouse lactate dehydrogenase. In *Electrophoresis*

1979: Adv. methods, biochemical clinical appl. (ed. B. J. Radola), p. 817. DeGruyter, Berlin.

Racine, R. R., C. H. Langley, and R. A. Voelker. 1980. Enzyme mutants induced by low-dose-rate γ-irradiation in Drosophila: Frequency and characterization. *Environ. Mutagen.* 2:167.

Rosenblum, B. B., S. M. Hanash, N. Yew, and J. V. Neel. 1982. Two-dimensional analysis of red cell membranes. *Clin. Chem.* 28:925.

Russell, W. L. and E. M. Kelly. 1982. Specific locus mutation frequencies in mouse stem-cell spermatogonia at very low radiation dose rates. *Proc. Natl. Acad. Sci. U.S.A.* 79:539.

Russell, L. B., W. L. Russell, R. A. Popp, C. Vaughan, and K. B. Jacobson. 1976. Radiation-induced mutations at mouse hemoglobin loci. *Proc. Natl. Acad. Sci. U.S.A.* 73:2843.

Russell, W. L., E. M. Kelly, P. R. Hunsicker, J. W. Bangham, S. C. Maddux, and E. L. Phipps. 1979. Specific-locus test shows ethylnitrosourea to be the most potent mutagen in the mouse. *Proc. Natl. Acad. Sci. U.S.A.* 76:5818.

Satoh, C., A. A. Awa, J. V. Neel, W. J. Schull, H. Kato, H. B. Hamilton, M. Otake, and K. Goriki. 1982. Genetic effects of atomic bombs. *Proc. VI Int. Congr. Hum. Genet.* (in press).

Schull, W. J., M. Otake and J. V. Neel. 1981. Genetic effects of the atomic bombs: A reappraisal. *Science* 213:1220.

Searle, A. G. 1974. Mutation induction in mice. *Adv. Radiat. Biol.* 4:131.

Setchell, B. P. and G. M. H. Waites. 1975. The blood testis barrier. In *Handbook of physiology: Reproductive biology of the male* (ed., R. O. Greep and D. W. Hamilton), vol. 5. p. 143. American Physiology Society.

Skolnick, M. M. 1982. An approach to completely automatic comparisons of 2-dimensional electrophoresis gels. *Clin. Chem.* 28:979.

Skolnick, M. M., S. R. Sternberg, and J. V. Neel. 1982. Computer programs for adapting two-dimensional gels to the study of mutation. *Clin. Chem.* 28:969.

Smithies, O. 1955. Zone electrophoresis in starch gels: Group variations in the serum proteins of normal human adults. *Biochem. J.* 61:629.

Soares, E. R. 1979. TEM-induced gene mutations at enzyme loci in the mouse. *Environ. Mutagen.* 1:19.

Tobari, Y. N. and K. Kojima. 1972. A study of spontaneous mutation rates at ten loci detectable by starch gel electrophoresis in Drosophila melanogaster. *Genetics* 79:397.

Voelker, R. A., H. E. Schaffer, and T. Mukai. 1980. Spontaneous allozyme mutations in *Drosophila melanogaster*: Rate of occurrence and nature of the mutants. *Genetics* 94:961.

Vogel, F. 1970. Spontaneous mutation in man. In *Chemical Mutagenesis in Mammals and Man.* (eds., F. Vogel and G. Rohrborn), p. 16, Springer-Verlag, New York.

Vogel, F. and K. Altland. 1982. Utilization of material from PKU-screening programs for mutation screening. *Prog. Mutat. Res.* 3:143.

Waites, G. M. H., A. R. Jones, S. J. Main, and T. G. Cooper. 1973. The entry of antifertility and other compounds into the testis. *Adv. Bioscience* **10**:101.

Yamaguchi, O. and T. Mukai. 1974. Variation of spontaneous occurrence rates of chromosomal aberrations in the second chromosomes of *Drosophila melanogaster. Genetics* **78**:1209.

COMMENTS

BUTTERWORTH: When you do find a mutation, assuming that you will sometime in the future, what kinds of things will you do to follow up? Are you going to explore the smoking and employment history of the parents?

MOHRENWEISER: Well, the obvious thing we have to do first is make sure it is a real mutation and not nonpaternity, which is a significant job unto itself. Also remember that the generation time necessary to prove genetic transmission of this new phenotype is long. The obvious question which I know you are asking is: Is there anything in the parents' background which would suggest that they are at high risk? I am sure you are aware that this is a risky business because mutations are "naturally" occurring events. In point of fact, one doesn't know at this point in time whether any mutational event is induced or whether it is a spontaneous event. In reality, what you are really looking at is a statistical increase in the frequency of a set of events, between an appropriate control—and the key is an appropriate one—and a group of individuals (offspring) the parents of which have been exposed.

We are obviously going to look at Thilly's work very carefully, because these results will be a real help.

It is important to remember that, at this time, we are really looking for a statistically significant change in the frequency of events some of which are "naturally" occurring.

MALLING: How can you possibly increase your number of loci? One of the problems you have in this system is that each individual gives you so little information.

MOHRENWEISER: I did not discuss our work on the two-dimensional gel electrophoresis systems, which the group in Ann Arbor is also working on. It has the potential—and the key word is "potential" at this point—to increase the number of loci which can be examined in any individual, conservatively by an order of magnitude over what we can currently do with the standard one-dimensional electrophoretic techniques. It cuts down significantly, obviously, on the number of individuals you need to look at.

RUSSELL: I think the ratio of nulls to non-nulls—not necessarily electrophoretic, because you can't always do that—is going to change not only with the mutagen and with the locus, but also with the germ cell stage and germ cell type in which the mutation in induced. There is some evidence that when you analyze the specific locus mutations by complementation studies you can get spectra within a set of mutations at the same locus. For instance, mutations that are induced in spermatogonial stem cells have a much higher frequency of hypomorphs as compared to nulls than those that are induced in post-spermatogonial stages. Different loci differ also in the relative frequencies of hypomorphs to amorphs.

MOHRENWEISER: The triosephosphate isomerase story is an excellent example of that. There has been one electrophoretic variant seen in probably 25,000 humans that have been looked at. But when we use quantitative techniques we find four individuals who are carriers for a null allele among every 100 black newborns. We know from other studies that the structural constraints on TPI molecule are severe. I am sure there are other enzymes which will accept a goodly number of amino acid substitutions, which are probably going to give one a high frequency of electrophoretics and a low frequency of nulls. I am sure it is compound- and chemical-specific also.

ALBERTINI: I would like to make one comment. Much time has been spent discussing mammalian and human mutagenicity studies. Many of these studies used in vivo or direct assays. Perhaps we can also look at the relevance of these studies to human disease. It is probably worthwhile to say the obvious. The obvious is that disease is a threshold. We have to keep in mind the possibility that a large effect of the environment in causing disease may have to do with its effects on homeostatic systems that are actually protecting us from the consequence of initiating and other mutagenic events. For example, powerful immunosuppressives that interfere with this strong homeostatic mechanism can result in cancer.

One of the largest populations of mutagen exposed humans that we can study with our tests are individuals being treated with cancer chemotherapy. Unfortunately, cancer is not. Cancer chemotherapy is an exposure to known mutagens which is an ethical exposure, a mandatory exposure. But, second cancers occur in these treated groups at frequencies where epidemiologists can say something. Many cancer patients are cured of their original malignancy, or achieve a long-time remission, and these benefits far outweigh the risk of the second cancer caused by therapy. But, we have a large population of mutagen-exposed persons—some of whom will develop a disease probably resulting from this exposure. We should use our direct tests in studying these people, compare our results over time with clinical outcomes, and attempt to validate these tests as regard their performances as health risk predictors.

The point is this: we have tests now that are clearly imperfect. New ones are going to come along. But we can store much of the blood samples and other biological material obtained from treated cancer patients and create a repository. The epidemiological and clinical studies being performed on these patients are exquisite. Most patients are cared for and studied in large-scale chemotherapy trials in this country and in Europe. The medical groups conducting these clinical and epidemiological studies have the ability to share samples and information.

What it is going to take is for the genetic toxicological community, with an emphasis on disease, to see if we can validate our tests as predictors of human disease. We have mechanistic models by which we relate mutagenicity and human disease. They may not predict human disease. I think at this stage we still have to keep that open.

HEDDLE: I think that in a population study it is necessary to assume that a human population is homogeneous. In fact we know that it is extremely variable. I think that the rate of mutation will turn out to be variable from person to person, and possibly not only somatic, but also germinal. We must keep this in mind in trying to evaluate the data in the future. It may happen that, in fact, a significant fraction of mutations comes from a small fraction of the individuals, not necessarily because of their environmental exposure, but because of their genetic background.

SHODELL: If you are preselecting in the sort of test defined from your comments, are you preselecting for a population, since these are people who already have cancer or had cancer, with a different sort of mutagenic spectrum?

ALBERTINI: You might be. I am talking, though, about a large fraction of the people in Western countries—those who develop cancer. It is quite a large subpopulation. In such a study population we should be asking about the relative frequencies of detectable, induced somatic cell and possibly germinal genetic events. In some tests—and this is why I asked yesterday about sister chromatic exchanges—all exposed individuals may be evaluated. Do all exposed individuals show large elevations of SCEs, or mutations, or whatever, or are there differences? If there are differences, even if the baseline is high, do these have any disease predictability? Do all of us being treated for cancer have a 5% chance of developing acute leukemia, or do 5% of us have a 100% chance, which really bears on the question.

I think these are the kinds of things we can begin to approach with that population.

Some regimens may totally overcome individual differences. We may have to look at persons receiving lower doses.

All that I am really saying is that the human material is available. The infrastructure is there for sharing samples; the clinical and epidemiological studies are being done; and the people who are running these clinical studies are concerned also with these questions. What are predictors or indicators of second tumors? We have this entire body of mutagenicity tests. If we could somehow wed these two together—"creative epidemiology" and human mutagenicity testing—we may be able to validate our tests as public health monitors.

The Need to Develop a Short-Term In Vivo Assay for Specific Locus Mutation in the Chinese Hamster

A. W. HSIE, R. L. SCHENLEY, K. R. TINDALL, R. MACHANOFF,
P. A. BRIMER, S. W. PERDUE, J. R. SAN SEBASTIAN,
AND E.-L. TAN
Biology Division, Oak Ridge National Laboratory, and
the University of Tennessee-Oak Ridge Graduate School of
Biomedical Sciences
Oak Ridge, TN 37830

RICHARD J. ALBERTINI
Departments of Medicine, Surgery, and Medical Microbiology
University of Vermont
College of Medicine
Burlington, Vermont 05405

The current rapid progress being made in experimental biology and medicine is due significantly to recent advances in the genetics of mammalian somatic cells. Our mutagenesis studies using Chinese hamster ovary (CHO) cells are summarized here and we explain our use of CHO cells for studies of mutagenic effects at the cytogenetic level as well as mutagen-induced changes in DNA nucleotide sequences. From the viewpoint of mutagenesis (but not necessarily carcinogenesis), the existence of a large data base on gene mutation (Hsie et al. 1981), chromosome aberration (Preston et al. 1981), and sister chromatid exchange (Latt et al. 1981) in cultured Chinese hamster cells argues favorably for the need to develop an in vivo gene mutational assay. It is known that the Chinese hamster is a suitable animal for studying mutagen-induced cytogenetic effects in vivo (Preston et al. 1981). We, therefore, propose to develop a short-term in vivo specific locus mutation assay using the Chinese hamster as a model animal.

A method often used to study genetic risk is comparative mutagenesis. The proposed Multiphasic CHO System measures mutational events in the same (or similar) gene, the same chromosome in the same animal, and will eliminate much of the uncertainty generated when different mutational events in such

This paper is dedicated to the memory of our late colleague, Dr. Paul F. Mullaney.

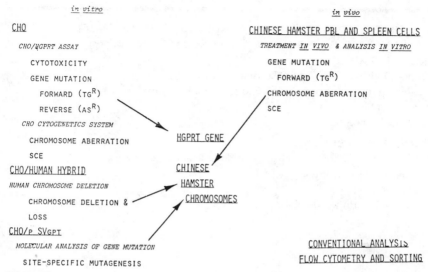

Figure 1
Program flow chart of Chinese hamster mutagenesis in vitro and in vivo. Details of individual assays are discussed in the text.

evolutionally divergent organisms as bacteria, insects, and mammals are compared. Our integrated approach to mammalian cell genetic toxicology in vitro and in vivo at the molecular-, gene- and chromosomal-level using both conventional analysis and the potential extension to the use of a cell sorter is outlined in Figure 1.

THE CHO/HGPRT ASSAY

The CHO/HGPRT assay has been discussed in detail elsewhere (Hsie et al. 1979). We used CHO-K_1-BH$_4$ cells (Hsie et al. 1975a), a subclone of the near-diploid CHO cell line, standard experimental procedure for cell culture, treatment with chemicals, and measurement of cytotoxicity and gene mutation (Hsie et al. 1975a; O'Neill et al. 1977; Hsie et al. 1978). Gene mutation was measured by quantifying the frequency of mutants resistant to a purine analogue, 6-thioguanine (TG). A metabolic activation system derived from Aroclor 1254-induced male Sprague-Dawley rat livers was used to determine mutagenicity of promutagens (O'Neill et al. 1977; Hsie 1980; Machanoff et al. 1981). We term cytotoxicity as the effect of chemical treatment on the cloning efficiency (CE) of cells relative to the untreated control (CE routinely 80% or higher). Mutagenicity is expressed as the number of TG-resistant (TG^r) mutants per 10^6 clonable cells. The spontaneous mutation frequency is usually 0-15 $\times$ 10^{-6} mutants/cell.

Questions concerning the genetic vs epigenetic origin of phenotypic variants have been raised earlier for studies of mutagenesis with mammalian cells. In the absence of direct evidence of gene mutation through analyses of the nucleotide sequence of the *hgprt* gene and the amino acid sequence of the HGPRT protein for TG^r mutants, we have exhaustively accumulated genetic, biochemical, and physiological evidence that indicates that the CHO/HGPRT system fulfills the criteria for a specific-locus mutation assay (Hsie et al. 1978).

We demonstrated that mutation induction by a direct-acting chemical mutagen, ethyl methanesulfonate (EMS) (Hsie et al. 1975a), or a physical agent, ultraviolet (UV) light (Hsie et al. 1975b), is quantifiable. Treatment of CHO cells with EMS (25-800 μg/ml) for 16 hours causes a concentration-dependent exponential killing after a distinct shoulder region where there is no appreciable decrease in cell survival. Mutation induction occurs over the entire concentration range tested (Hsie et al. 1975a). Studies using treatment times of 2-24 hours and varying EMS concentrations demonstrated the existence of a limited reciprocity of EMS mutagenesis; that is, when different combinations of EMS concentrations (0.05-3.2 mg/ml) are multiplied by varying treatment times (2-10 hours), the product (mg/ml)·hour yields a constant mutation frequency and cytotoxicity (O'Neill and Hsie 1977). Such a reciprocity effect was also demonstrated for S-9-coupled benzo[a]pyrene (B[a]P) mutagenesis and lethality (Machanoff et al. 1981).

We have extended our quantitative analysis of mutagenesis and survival to include the effect of interaction of a physical agent, near UV light, and a chemical, the photosensitizer 8-methoxypsoralen (8-MOP) (Schenley and Hsie 1981). Exposure of cells to either 8-MOP up to 20 μg/ml (93 μM) or near-UV light up to 40,000 J/m^2 had no effect on either survival or mutation frequency. Survival decreased and mutation frequency increased linearly when either the 8-MOP concentration or the near UV-light fluence was increased while the other factor was held as a constant. Mutation frequency appears to show reciprocity relative to the product of 8-MOP concentration times fluence of near-UV light $(\mu g/ml)\cdot(J/m^2)$ throughout a range apparently limited only by high cell lethality. The observed pooled data on mutation as a function of $(\mu g/ml)$ (J/m^2) fit a linear dose-response line. Cell survival, however, does not appear to exhibit such reciprocity.

These results suggest that exposure of CHO cells to near-UV light in the presence of 8-MOP induces both toxic and mutagenic effects. These effects are probably associated with the formation of DNA cross-links that contribute to lethal events (Ben-Hur and Elkind 1973) and mutagenic cross-links and(or) monoadducts (Grant et al. 1979). Although a direct relationship between DNA cross-links and cytotoxicity has been indicated (Ben-Hur and Elkind 1973), no direct measurement of the extent of mutation induction due to cross-links or monoadducts has been made with a bifunctional agent (Grant et al. 1979).

Having established the quantitative nature of the CHO/HGPRT system, we then extended our study to determine the effect of the alkylating moiety of 10 alkylating agents on the mutagenicity and cytotoxicity. Our results showed that, when comparisons are made at equimolar rather than equitoxic concentrations, the cytotoxic and mutagenic effects of the alkylating chemicals decrease with increasing size of the alkylating group. These results also reinforce our suggestion that the cytotoxic and mutagenic effects of EMS are dissociable (Hsie et al. 1975a; O'Neill and Hsie 1977). In addition, mutagenicity does not appear to correlate directly with S_N1 reactivity, because isopropylmethanesulfonate (iPMS) has the highest S_N1 reactivity (Lawley 1974) but is the least mutagenic among alkanesulfonates studied (Couch and Hsie 1978; Couch et al. 1978).

We also examined the correlation between the degree of DNA alkylation and the mutagenicity of selected chemicals. For this purpose, we chose methylnitrosourea (MNU) and ethylnitrosourea (ENU) because of the existence of well-documented mutagenicity in microbial systems (Lawley 1974; Neale 1976), and the CHO/HGPRT assay (Couch and Hsie, 1978), animal carcinogenicity (Pegg 1977), and chemical reactivity (Veleminsky et al. 1970).

In cultures treated with MNU and ENU, we found that both alkylation and mutation induction by MNU and ENU increase linearly with increasing nitrosamide concentrations over the concentration range tested. On an equimolar basis, MNU has 15 times the alkylating activity of ENU, but only 3 times the mutagenic activity. In terms of mutation induction per unit alkylation, ethylation of DNA by ENU appears to result in fivefold more mutagenic lesions than does methylation by MNU. This may reflect either a higher miscoding frequency, such as O^6-guanine alkylation, or an effect of ethylation per se (Thielman et al. 1979). Further studies will be directed toward determination of DNA adducts (alkylated nucleosides and phosphotriesters) and their relationship to mutation induction.

CHO CYTOGENETICS SYSTEM

We have recently incorporated the cytogenetics of CHO cells into the CHO/HGPRT assay. This Multiplex CHO Mutagenesis System (San Sebastian et al. 1981) is used to determine chromosome aberration and SCE in addition to cytotoxicity and gene mutation. Because most carcinogens are mutagens, the Multiplex CHO Mutagenesis System with a broad mutagen screening capacity will be more useful than the CHO/HGPRT system alone as a valid prescreen for chemical carcinogens. To validate this system, we have used four carcinogenic and noncarcinogenic pairs to study the interrelationship of the four biological endpoints. These compounds are the direct-acting carcinogens N-methyl-N'-nitro-N-nitrosoguanidine (MNNG) and ICR-170, their noncarcinogenic analogue N-methyl-N'-nitroguanidine and ICR 170-OH, the procarcinogens B[a]P and dimethylnitrosamine (NDMA) and their noncarcinogenic analogues pyrene and dimethylamine.

We observed that carcinogens MMNG, B[a]P, and NDMA, but not their noncarcinogenic counterparts, showed all four biological effects. Cytotoxicity, however, does not appear to correlate with any of the other end points. On a molar basis, pyrene shows similar toxicity to B[a]P. The other two noncarcinogenic analogues, dimethylamine and MNNG, exhibit minimal toxicity at concentrations significantly higher than cytotoxic concentrations of the carcinogens, NDMA and MNNG. Sister chromatid exchange (SCE) induction appeared to be the most sensitive indicator of "genetic activity." Therefore, MNNG appears to be the most active, followed by B[a]P and NDMA as ranked on a molar basis. This ranking is consistent with the expressed activity of these compounds in regard to both chromosomal aberration and gene mutation (San Sebastian et al. 1980a,b).

We also showed that ICR-170 is highly toxic and mutagenic (at 0.1-2.0 μM) to CHO cells and that ICR-170-OH exhibits a similar cytotoxic effect but does not induce gene mutations. ICR-170, but not ICR-170-OH, induces both SCE (at 0.05-0.5 μM) and chromosome aberrations. A striking increase in chromosome aberrations, primarily of the chromatid type, is induced by ICR-170 within the concentration range of 1.0-2.0 μM. At 2.0 μM, 100% of the cells treated with ICR 170 exhibit chromosome aberrations. Chromosome aberrations are not observed at concentrations <1.0 μM. In addition, ICR-170 induces mitotic delay at concentrations >0.15 μM; ICR-170-OH does not induce this effect even at highly toxic concentrations (San Sebastian et al. 1981).

These results indicate that the Multiplex CHO System is capable of discriminating divergent structural classes of carcinogenic and noncarcinogenic compounds. Both cytogenetic and gene mutation assays proved useful in quantifying the mutagenic effects of these carcinogens. Future studies will include continued verification of the utility of these end points in the detection of environmental carcinogens as well as a search for physiological events common to the induction of SCEs, chromosome aberrations, and gene mutations in mammalian cells. These studies should lead to a better understanding of molecular events mediated by carcinogens in general.

CHO/HUMAN HYBRID

The CHO/human hybrid cell system (A_L-J1 containing human chromosome 11) permits determination of a partial to total loss of a specific chromosome in addition to various single gene mutations (Waldren et al. 1979). Use of this system will help identify environmental agents associated with both cancer and genetic diseases.

CHO/pSVgpt

Most systems used to determine mutagen induced alterations in DNA have not analyzed DNA directly, but have extrapolated from altered protein primary sequence. Thus, little is known of the molecular aspects of mutations induced in mammalian cells.

The plasmid vector pSV2gpt (Mulligan and Berg 1980) carries the *Escherichia coli gpt* gene and expresses the purine salvage enzyme, xanthine/guanine phosphoribosyl transferase (XGPRT). This enzyme is the bacterial equivalent of the mammalian enzyme, HGPRT. Using specific inhibitors to block de novo purine biosynthesis in HGPRT-deficient CHO cells, selected subclones of CHO cells transfected with the pSV2gpt vector have been shown to carry and express the *E. coli gpt* gene (Tindall and Hsie 1982). *E. coli* XGPRT activity can be demonstrated electrophoretically in extracts from these cells. Genomic DNA derived from several stable gpt^+ CHO transfectants has been analyzed by Southern blot hybridization indicating the presence of one to several copies of pSV2gpt in the high-molecular-weight DNA. Because the CHO recipient (a presumptive deletion mutation of *hgprt*) is resistant to TG (10 μM), sensitive to azaserine (AS) (10 μM), nonrevertable to $hgprt^+$, lacking in detectable HGPRT activity and in immunoprecipitable HGPRT protein, it provides a suitable host for the study of both forward and reverse mutations at the *gpt* locus transfected into CHO cells.

Treatment of one gpt^+ CHO clone with EMS followed by selection in TG produced a dose dependent increase of phenotypically gpt^- TG^r colonies. Several spontaneous and EMS-induced TG^r clones have been isolated and shown to carry sequences which hydridize to the pSV2gpt vector. These results indicate that the gpt^- phenotype is not due to the segregation and loss of *gpt* sequences. Recent analyses of spontaneous and X-irradiation induced gpt^- mutants indicate that *gpt* sequences are deleted in some mutants (K. R. Tindall, unpubl. results).

FLOW CYTOMETRY AND SORTING

EMS mutation induction can be measured in CHO cells incubated in unattached cultures in low serum. Initial results have produced mutation frequencies significantly higher than those obtained using our CHO/HGPRT assay in which cells are grown in an "attached" monolayer. Using 0.5% serum during the expression period allows us to sample essentially 100% of the progeny of the treated population because the cells are in a growth arrested state (R. L.

Schenley and R. Machanoff, unpubl. results). This would be of particular value to assay for low mutagenic response.

Flow cytometry will be used to analyze mutagen-treated cells grown in low serum. The application of flow cytometry may necessitate the use of a purine or pyrimidine base analogue with good fluorescent characteristics that can be incorporated into the DNA of TG^r cells; however, our initial proposal is to use the thymidine analogue, 5-bromodeoxyuridine (BudR). This analogue would be incorporated into the DNA of growing cultures in the presence of TG. When stained with Hoest 33258 dye, cells substituted with BudR fluoresce with 1/4 the intensity of nonsubstituted cells. A difference of this magnitude can be quantified and sorted by flow cytometry. The adaptation to flow cytometry, as well as the modifications to low serum continuous cultures, will require extensive testing experiments.

CHINESE HAMSTER IN VIVO SPECIFIC LOCUS MUTATION ASSAY

Recently, the autoradiographic determination of purine analogue resistant (AG^r; TG^r) peripheral blood lymphocytes (PBLs) arising in man has been proposed as a direct mutation test (Strauss and Albertini 1979; Albertini 1980). Although it is feasible to measure variant frequency (Vf) of TG^r PBLs in cancer patients, the determination of direct mutagenic effects of putative environmental agents has to rely on a model mammalian system. The existence of a large data base on induced mutation at the *hgprt* locus, chromosome aberration, and SCE in CHO cells and other Chinese hamster cell lines makes it logical to develop a Chinese hamster TG^r system. The successful development and validation of this Chinese hamster direct mutational system would provide a good comparison between mutation in the hamster and its somatic cells, and a more realistic assessment of environmentally induced genetic risks.

In collaboration with R. J. Albertini of the University of Vermont, we have investigated parameters for Vf determination of Chinese hamster spleen cells resistant to TG. PBLs were found unsuitable because recovery from the animal and response to mitogens are poor. Spleen cells were chosen for subsequent study due to ease of obtaining sufficient responsive cells.

Using a proliferation assay measuring incorporation of [^{3}H]-TdR into macromolecules, optimal concentrations of serum, mitogen, and TG were determined. Concanavalin A (2.1 μg/ml) in RPMI medium containing 10% heat inactivated fetal calf serum produced a maximal proliferation response; where 10 ng/ml PHA-P in RPMI containing 20% heat inactivated fetal calf serum showed a maximal response with a labeling index of 0.46, as determined by autoradiography. The proliferation response was completely inhibited by 10^{-6} M TG for spleen cells treated with PHA-P. The Vf of spontaneous TG^r in spleen cells stimulated with PHA-P in the presence of 2×10^{-6}, 2×10^{-5}, 2×10^{-4}, and 2×10^{-3} M TG was 8.3×10^{-2}, 2.1×10^{-5}, 4.7×10^{-5}, and 1.8×10^{-6}, respectively. Although cryopreservation of human PBL's lowers the spontaneous Vf of TG^r, preliminary trials with hamster spleen cells have been unsuccessful.

Progress has been made in this preliminary work to establish a direct assay for mutation in vivo, but further studies to define proper assay conditions are required before validation of the assay can commence. These include optimal conditions for culturing hamster spleen cells and the optimal TG concentration to yield a stable, low spontaneous Vf of TG^r spleen cells. Later, experiments will involve induction of TG^r spleen cells, chromosome aberration, and SCE by the model mutagens EMS, MNNG, B[a]P, and NDMA. Screening of environmental agents will be performed following a proper validation of this system with appropriate carcinogenic/noncarcinogenic pairs.

CONCLUDING REMARKS

We believe that the successful development and use of these mammalian cell systems in vitro and in vivo will enable us to elucidate the relationship between chemical induced gene mutation and cytogenetic aberration in the mammal and its somatic cells as well as to evaluate the mutagenic, and predict the carcinogenetic, hazards of environmental agents with a high certainty.

ACKNOWLEDGMENTS

Research sponsored by the Office of Health and Environmental Research, U.S. Department of Energy, under contract W-7405-eng-26 with the Union Carbide Corporation.

REFERENCES

Albertini, R. J. 1980. Drug-resistant lymphocytes in man as indicators of somatic cell mutation. *Teratog. Carcinog. Mutagen* 1:25.

Ben-Hur, E. and M. M. Elkind. 1973. Psoralen plus near ultraviolet light inactivation of cultured Chinese hamster cells and its relation to DNA crosslinks. *Mutat. Res.* 18:315.

Couch, D. B. and A. W. Hsie. 1978. Mutagenicity and cytotoxicity of congeners of two classes of nitroso compounds in Chinese hamster ovary cells. *Mutat. Res.* 57:209.

Couch, D. B., N. L. Forbes, and A. W. Hsie. 1978. Comparative mutagenicity of alkylsulfate and alkanesulfonate derivatives in Chinese hamster ovary cells. *Mutat. Res.* 57:217.

Grant, E. L., R. C. von Borstel, and M. J. Ashwood-Smith. 1979. Mutagenicity of cross-links and monoadducts of furocoumarins (psoralen and angelicin) induced by 360 nm radiation in excision-repair-defective and radiation-insensitive strains of Saccharomyces cerevisia. *Environ. Mutagen.* 1:55.

Hsie, A. W. 1980. Quantitative mutagenesis and mutagen screening with Chinese hamster ovary cells. In *The predictive value of short-term screening tests in the evaluation of carcinogenicity.* (Eds. G. M. Williams et al.), p. 89. Elsevier/North-Holland, Amsterdam, Netherlands.

Hsie, A. W., J. P. O'Neill and V. K. McElheny (eds.). 1979. Mammalian cell mutagenesis: The maturation of test systems. *Ban. Rep.* 2.

Hsie, A. W., P. A. Brimer, T. J. Mitchell, and D. G. Gosslee. 1975a. The dose-response relationship for ethyl methanesulfonate-induced mutations at the hypoxanthine-guanine phosphoribosyl transferase locus in Chinese hamster ovary cells. *Somat. Cell Genet.* 1:247.

_____. 1975b. The dose-response for ultraviolet light-induced mutations at the hypoxanthine-guanine phosphoribosyl transferase locus in Chinese hamster ovary cells. *Somat. Cell Genet.* 1:383.

Hsie, A. W., D. A. Casciano, D. B. Couch, D. F. Krahn, J. P. O'Neill, and B. L. Whitfield. 1981. The use of Chinese hamster ovary cells to quantify specific locus mutation and to determine mutagenicity of chemicals: A report of the Gene-Tox Program. *Mutat. Res.* 86:193.

Hsie, A. W., J. P. O'Neill, D. B. Couch, J. R. San Sebastian, P. A. Brimer, R. Machanoff, J. C. Riddle, A. P. Li, J. C. Fuscoe, N. L. Forbes, and M. H. Hise. 1978. Quantitative analyses of radiation- and chemical-induced cellular lethality and mutagenesis in Chinese hamster ovary cells. *Radiat. Res.* 76:471.

Latt, S. A., J. Allen, S. E. Bloom, A. Carrano, E. Falke, D. Kram, E. Schneider, R. Schreck, R. Tice, B. Whitfield, and S. Wolff. 1981. Sister-chromatid exchanges: A report of the Gene-Tox Program. *Mutat. Res.* 87:17.

Lawley, P. D. 1974. Some chemical aspects of dose-response relationships in alkylation mutagenesis. *Mutat. Res.* 23:283.

Machanoff, R., J. P. O'Neill, and A. W. Hsie. 1981. Quantitative analyses of cytotoxicity and mutagenicity of benzo(a)pyrene in mammalian cells (CHO/HGPRT system). *Chem.-Biol. Interact.* 34:1.

Mulligan, R. C. and P. Berg. 1980. Expression of a bacterial gene in mammalian cells. *Science* 209:1422.

Neale, S. 1976. Mutagenicity of nitrosamides and nitrosamidines in microorganisms and plants. *Mutat. Res.* 32:229.

O'Neill, J. P., and A. W. Hsie. 1977. Chemical mutagenesis of mammalian cells can be quantified. *Nature* 269:815.

O'Neill, J. P., P. A. Brimer, R. Machanoff, G. P. Hirsch, and A. W. Hsie. 1977. A quantitative assay of mutation induction at the hypoxanthine-guanine phosphoribosyl transferase locus in Chinese hamster ovary cells: Development and definition of the system. *Mutat. Res.* 45:91.

Pegg, A. E. 1977. Formation and metabolism of alkylated nucleosides: Possible role in carcinogenesis by nitroso compounds and alkylating agents. *Adv. Cancer Res.* 2:195.

Preston, R. J., W. Au, M. A. Bender, J. G. Brewen, A. V. Carrano, V. A. Heddle, A. F. McFee, S. Wolff, and J. S. Wassom. 1981. Mammalian in vivo and in vitro cytogenetic assays. A report of the U.S. EPA's Gene-Tox Program. *Mutat. Res.* **87**:143.

San Sebastian, J. R., J. P. O'Neill, and A. W. Hsie. 1980a. Induction of chromosome aberrations, sister chromatid exchanges, and specific locus mutations in Chinese hamster ovary cells by 5-bromodeoxyuridine. *Cytogenet. Cell Genet.* **28**:47.

San Sebastian, J. R., J. P. O'Neill, A. Johnson, and A. W. Hsie. 1980b. Induction of specific locus mutations and sibling ("sister") chromatid exchanges by bromodeoxyuridine in Chinese hamster ovary cells. *Environ. Mutagen.* **2**:299.

San Sebastian, J. R., J. C. Fuscoe, K. R. Tindall, and A. W. Hsie. 1981. A study of ICR 170 and ICR 170-OH in the CHO Multiplex Genetic Toxicity System: Relationships between cytotoxic, mutagenic, and cytogenetic effects. *Environ. Mutagen.* **3**:323.

Schenley, R. L. and A. W. Hsie. 1981. Interaction of 8-methoxypsoralen and near-UV light causes mutation and cytotoxicity in mammalian cells. *Photochem. Photobiol.* **33**:179.

Strauss, G. H. and R. J. Albertini. 1979. Enumeration of 6-thioguanine-resistant peripheral blood lymphocytes in man as a potential test for somatic cell mutations arising in vivo. *Mutat. Res.* **61**:353.

Thielmann, H.-W., C. H. Schroder, J. P. O'Neill, P. A. Brimer, and A. W. Hsie. 1979. Relationship between DNA alkylation and specific-locus mutation induction by N-methyl- and N-ethyl-N-nitrosourea in cultured Chinese hamster ovary cells (CHO/HGPRT system). *Chem.-Biol. Interact.* **26**:233.

Tindall, K. R. and A. W. Hsie. 1982. A system for the molecular analysis of mutation in pSV2gpt transfected CHO cells. *Proc. 13th Annu. Meet. Environ. Mutagen. Society,* p. 91.

Veleminsky, J., S. Osterman-Golkar, and L. Ehrenberg. 1970. Reaction rates and biological action of MNUA and ENUA. *Mutat. Res.* **10**:169.

Waldren, C., C. Jones, and T. T. Puck. 1979. Measurement of mutagenesis in mammalian cells. *Proc. Natl. Acad. Sci. U.S.A.* **76**:1358.

COMMENTS

CARRANO: I admire your approach. It is an approach that many of us are taking, eventually trying to extrapolate to a biological end point, such as cancer induction. How do you propose to do this with the Chinese hamster?

HSIE: There is so much data based on gene mutation and cytogenetic data available about Chinese hamster cells in culture, it would be interesting from the mutagenic point of view to develop a Chinese hamster model. Rat and mouse models will be very useful from a carcinogenic point of view. I think these two animal model systems are complementary. Each serves its own purpose.

HEDDLE: Your cloning out the gene and saying it was 1000 base pairs long, and so forth, made me wonder whether there was any evidence that the frequency of mutation in the gene is any way influenced by the presence of introns. Does anyone know that?

HSIE: The *gpt* gene is bacterial in origin; there is no intron. In fact, we are still characterizing the physical state of this integrated foreign gene in the cell. We have isolated several transfectants which are responsive to mutagen treatment in a dose-response fashion. In what physical state, we don't know.

THILLY: John Heddle's question is difficult since we haven't cloned *hgprt* out of CHO or any of the other cells that people study for mutagenesis. We do have Tom Caskey's work, and work of others, who have looked for cross-reacting material in HGPRT-deficient cells (See Caskey and Kruh 1979; Kruh et al. 1979). Approximately 50% of the TG^r mutants that are isolated still have cross-reacting material. So one can say that it can't be worse than two times the size of the expectation from the structural gene. That is the kind of limit that we have amongst all of us, I think, in terms of thinking about the size of the hgprt target. I guess that may have been what motivated your question.

HSIE: There is recent work by Tom Caskey (pers. comm.) to show that the hgprt genes from mammalian cells are much bigger than gpt genes from bacteria. Functionally they are equivalent, but not identical.

BRIDGES: I wanted to just comment on this question regarding the relation of SCE to gene mutations. In yeast and bacteria one of the ways in which you learn about the relationship of different end points is by looking at repair-deficient mutants. We are studying a human cell mutant that we have from an agammaglobulinemia patient. It is a fibroblast strain, with a repair deficiency which we believe to be at the ligation step of repair. We have a large data base on this. I want to mention just a couple of observations.

First, if you expose it to UV, Leigh Henderson has shown that the SCE is hypersensitive. It is like an XP line. In this particular line, SCEs shoot up. If you look for HGPRT mutations, there aren't any. I mean, there really aren't any induced ones. The spontaneous ones are there, but no induced ones.

With ataxia-telangiectasia cells, we have suspected that induced mutations may occur. But because the cells are so darned sensitive, to gamma rays in that case, you can't really show that they are immutable. But in this case you can, because there are plenty of survivors, and the survivors do not mutate to HGPRT resistance with UV or with gamma rays.

Now, one interpretation of that is that it clearly shows that gene mutations and SCEs are quite wildly different; they are not the same thing. What it does suggest is that the repair deficiency is affecting both end points, but in a different way. One interpretation is that there is a small proportion of repair events which could go either way, which could, in the presence of this gene function that is mutated in our particular line, give rise to SCEs, whereas in the wild-type it might give rise to a gene mutation event.

So clearly they are linked in one repair pathway, but they are different. I think that by studying these mutants, we will begin to get a handle on what these events are.

CARRANO: In our lab Larry Thompson (pers. comm.) has isolated several Chinese hamster repair-deficient mutants, and we have compared relative mutation and SCE frequency. We find similar results. We don't get one where there is absolutely no mutation, but the relationship between SCE and mutation will uncouple depending upon the specific repair efficiency. There is one particular strain that is hypersensitive to BudR. It has an SCE frequency of over 100 per cell, but yet its mutation frequency is absolutely normal. The implication is that this SCE response is due to an effect of incorporated BudR. It is not a reflection of any induced mutagenesis. If you expose the cells to a·mutagen to induce mutation or SCE, you get the same relationship. There is also a series of repair-deficient mutants where you will get a hypersensitivity to SCE and a slight hypersensitivity to mutation, but the mutation does not follow the SCE.

But that doesn't mean to say that these end-points reflect the same type of lesion. It says that the repair processes that handle the lesions are different. But you can uncouple those curves I showed you correlating SCE induction and mutation by using cell strains with differing repair capacities.

HEDDLE: Bryn likened the cell line he referred to XP, but, in fact, most XP cells are not sensitive, as far as SCE production is concerned, to most chemical agents. The SV40 transformed line, which is sensitive, is unusual, and the characteristic it has is basically that it is a mar⁻ line. That is the reason that it is sensitive, rather than the fact that it is XP.

I wasn't contradicting what Tony said; I was adding to it.

THILLY: Wasn't it reasonably well established, even before we had mammalian mutants, when we are talking about chromosomal level changes as opposed to gene mutation changes, that we were dealing with at least one, if not many, independent steps on the way to the biological event? For instance, I remember Kao and Puck's work (1969) in which they simply showed that caffeine will bust chromosomes, but is a lousy gene mutagen, and things at the other extreme, like ICR-191, were really good at causing gene locus mutations, and, at reasonable levels, didn't cause any detectable chromatid aberrations.

Aren't we faced with a plethora of potential things that chemicals could do that end up on different pathways and with what must be a large number of gene products that are responsible for what happens after treating a cell?

The matrix of chemicals we end up studying is facing a matrix of mutants. The experiments that have always impressed me most were Louise Precocia's experiments (pers. comm.) where she just picked up 41 random variants in terms of MMS-UV sensitivity and mutability and found out that that first 41 isolated represent 38 complementation groups, by Poisson distribution, leading to a prediction of more than 100 independent complementation groups.

I kind of get a little nervous, because everywhere you go you hear about mer⁻/mer⁺. Sometimes at the start of a problem we imagine it is going to be simple and we are going to deal with a couple of mutants. However, I think we are going to be dealing with literally hundreds of complementation groups, if people are at least as simple as *Saccharomyces*—never mind all those wonderful chemicals to be studied.

BRIDGES: What you say is obviously true, and we have known it for years. But what this particular cell line shows is that one particular repair blockage can switch the way the cell deals with lesions from a mutagenic pathway to an SCE pathway. Now, that is not dealing with different chemicals.

THILLY: Just in terms of the fun of modeling—you only have one mutation, and you are asking it to do two different things—why not have your mutation, let's say, delay the cell a little in S-phase, so you could possibly have more SCE, but now your independent repair system will take care of your premutagenic lesions? I have trouble, when you start with a model where one mutation, presumably one gene product in this isolate, is going to go up and do two different molecular things. Do you see what I mean in terms of a model?

BRIDGES: If you get small effects, then obviously you do worry very much about things like cell cycle effects. These are dramatic, big effects. I don't think they can be explained in that way, but time will tell.

HEDDLE: One only has to imagine competing repair systems in a single mutation.

EVANS: Our discussion is obviously drawing to a close, but before concluding the session I want to make one comment that relates to much of our discussion on detecting the effects of mutagens on man himself.

The ease of obtaining peripheral blood samples, and the fact that we can analyze DNA, genes, and chromosomes in the cells in these samples, has not unnaturally led to considerable use of blood lymphocytes as sample cells for detecting in vivo exposure to mutagens, and as experimental cells for undertaking in vitro experiments. We have all realized over many years that the small lymphocytes in blood are a heterogeneous population of cells, even though the majority of them are in a nonproliferating G_0 phase with regard to the cell cycle. The recent development on the immunological front in distinguishing functionally different lymphocytes by virtue of their different cell surface properties—which can be readily ascertained through the use of specific monoclonal antibodies—has revealed a whole new range of differences within these cell populations. Moreover, we are already aware of differences in mutagen response, at least at the SCE level, as between B and T cells and it is probable that different subsets of these classes may also differ in their sensitivity. Since we are often confronted with quite small differences between lymphocyte 'damage' between exposed and nonexposed individuals, it is evident that we shall need to define our sample cells, and their relative sensitivities, with greater precision in the future.

References

Caskey, C. T. and G. D. Kruh. 1979. The HPRT locus. *Cell* **16**:10.

Kruh, G. D., R. L. Nussbaum, and C. T. Caskey. 1979. Tryptic peptic analysis of wild-type and mutant hypoxanthine guanine phosphoribosyl transferase from cultured Chinese hamster cells. *Fed. Proc.* **38**:736.

Kao, F. T. and T. T. Puck. 1969. Genetics of somatic mammalian cells. IX. Quantitation of mutagenesis by physical and chemical agents. *J. Cell Physiol.* **74**:245.

SESSION VII:

GERM CELL EFFECTS

DNA Repair in Spermatocytes and Spermatids of the Mouse

GARY A. SEGA
Biology Division
Oak Ridge National Laboratory
Oak Ridge, Tennessee 37830

The first evidence for unscheduled DNA synthesis (UDS) in mammalian somatic cells was provided by Rasmussen and Painter (1964) when they demonstrated that ultraviolet (UV) radiation induced the uptake of labeled thymidine (TdR) into the DNA of cultured HeLa and Chinese hamster cells grown in culture. Later, Kofman-Alfaro and Chandley (1971) also using in vitro procedures, were able to demonstrate UDS in spermatogenic cells of the mouse after exposure to X-rays or UV. Similar findings were observed in rat germ cells (Gledhill and Darzynkiewicz 1973) and in human germ cells (Chandley and Kofman-Alfaro 1971) after in vitro u.v. exposure.

The induction of UDS in mammalian germ cells can also be studied by in vivo procedures. In the male mouse it is possible to study UDS in vivo in meiotic and post-meiotic germ-cell stages (primary spermatocytes through mature spermatozoa) by making use of the sequence of events that occurs during spermatogenesis and spermiogenesis. In the developing germ cells of the mouse the last scheduled DNA synthesis takes place during a 14-hour period in preleptotene spermatocytes (Monesi 1962). After DNA synthesis these spermatocytes continue to develop through a series of germ cell stages for about a 28- to 30-day time period before leaving the testis as late spermatids (Oakberg 1956a,b). After leaving the testis, the late spermatids (now immature spermatozoa) enter the caput epididymis. They reach the caudal epididymis 2-3 days later and in 2-3 days after that they reach the vas as functional spermatozoa (See Fig. 1).

If the DNA from any meiotic or post-meiotic germ-cell stage is damaged by a physical or chemical agent and UDS is induced, it can be detected by an unscheduled incorporation of [^{3}H]-TdR into the affected germ cells. The unscheduled uptake of [^{3}H]-TdR can be measured either directly, by examining the affected germ-cell stages through autoradiography (Sotomayor et al. 1979), or indirectly, by waiting until the germ cells have matured into sperm in the caudal epididymis and vas, and then assaying the [^{3}H]-TdR activity contained in several million of these sperm by using liquid scintillation counting (LSC) (Sega 1974; Sega et al. 1976, 1978).

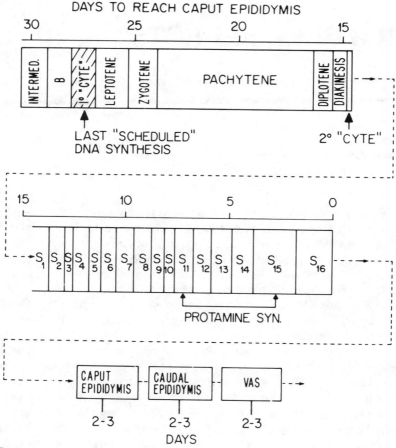

Figure 1

Timing of spermatogenesis and spermiogenesis in the mouse. The last scheduled DNA synthesis takes place in pre-leptotene primary spermatocytes. Mutagen treatment can induce UDS in germ-cell stages from leptotene through mid-spermatids. (Spermatid stages are labeled from S_1-S_{16}.) No UDS can be detected in later stages.

METHODS

Male mice are exposed to a chemical or physical agent in exactly the same way they would be in a genetic experiment. The general method of administration of chemicals is by intraperitoneal (i.p.) injection, but other routes such as inhalation, gavage and direct testicular injections have also been used (Sega and Sotomayor 1982; G. A. Sega, unpubl. results). $[^3H]$-TdR is injected directly into the testes either at the same time or at different times after mutagen treatment. For the testicular injections the mice are anesthetized with Metofane. A small incision is made in the scrotum to visualize the testes, and 36 μl of water containing ~ 18 to 36 μ Ci of [^{3}H]-TdR is injected into each testis. The incision heals in a few days without any special treatment.

To make direct observations of the germ-cell stages undergoing UDS, autoradiographic procedures are used (Sotomayor et al. 1979; Sega and Sotomayor 1982). These techniques also provide information on the uniformity of the UDS response within a particular germ-cell stage and can be used to study the distribution of [^{3}H]-TdR labeling from UDS throughout the chromosomes in diakinesis. However, autoradiographs are hard to quantitate and are not as sensitive as LSC for detecting low levels of induced UDS.

UDS occurring in any germ-cell stage can be studied indirectly by making use of the timing of spermatogenesis and spermiogenesis in the mouse and recovering sperm from the caudal epididymis or vas at the appropriate time after treatment. For example, sperm recovered from the caudal epididymis 16 days after treatment represent germ-cells that were mostly in early spermatid stages at the time of treatment. If a "mutagen exposure—UDS response" curve is desired for a particular germ-cell stage, then at the appropriate time after each level of exposure the mice are killed and sperm are recovered from the caudal epididymis or vas. If it is desired to study all of the meiotic and post-meiotic germ-cell stages undergoing a UDS response to a particular agent, sperm samples are recovered from treated animals every few days for 4-5 weeks after treatment.

For sperm recovery, the caudal epididymides or vasa are dissected from the animals, diced, and sonicated to destroy all cell types except for the sperm heads. The tails and midpieces are sheared off in this process. After several washes in a sucrose-SDS solution, a pure population of sperm heads is obtained free of sperm tails, midpieces, somatic cells, or cellular debris (Sega 1974; Sega et al. 1976; Sega and Sotomayor 1982). Typically, 7×10^6 sperm heads are recovered from the vas of one male and about twice that number are recovered from the caudal epididymis. Several million of these sperm are assayed by LSC to determine [^{3}H]-TdR activity, and the number of sperm present in each scintillation vial is determined by hemacytometer counts of a diluted sample of the same sperm stock. Finally, the [^{3}H]-TdR dpm/10^6 sperm heads is calculated. Although LSC is an indirect means of measuring the UDS that had occurred in various meiotic and post-meiotic germ-cell stages, it has the advantage of sensitivity to low levels of induced UDS since millions of sperm heads are sampled.

RESULTS AND DISCUSSION

Germ-Cell Stages that Undergo a UDS Response to Mutagenic Agents

In our studies that looked at all of the meiotic and post-meiotic germ-cell stages undergoing a UDS response to mutagen treatment we have used methyl methanesulfonate (MMS), ethyl methanesulfonate (EMS), cyclophosphamide (CPA), mitomen (DMO), and X-rays (Sega 1974; Sega et al. 1976; Sotomayor et al. 1978, 1979). Other workers have studied UDS induced by X-rays and UV light in an in vitro system with mouse germ cells (Kofman-Alfaro and Chandley 1971); UV in vitro with rat germ cells (Gledhill and Darzynkiewicz 1973); MMS and procarbazine (Schmid et al. 1978; Bürgin et al. 1979) in vivo with the rabbit. In all of these studies the basic observation is that mutagens can induce UDS in meiotic stages and post-meiotic stages up to about mid-spermatids. No UDS is detected in late spermatids and sperm cells.

The absence of UDS in the late spermatid stages to mature spermatozoa treated with MMS and EMS is not due to failure of these chemicals to alkylate DNA in these stages. Our chemical dosimetry studies have shown that DNA in these germ-cell stages is being alkylated (Sega et al. 1974; Sega et al. 1978; G. A. Sega, unpubl. results). As we have discussed previously (Sega 1974), the germ-cell stages failing to exhibit a UDS response are those in which protamine has either replaced, or is in the process of replacing the chromosomal histones. At the time of protamine synthesis an extensive condensation of the spermatid nucleus begins. It is possible that the DNA lesions present in these cells have become inaccessible to the enzymatic system which gives rise to UDS in the earlier germ-cell stages. Also, much of the cytoplasm is lost from the spermatids as they develop from mid- to late-spermatid stages, and the enzymatic system which produces the UDS response may be lost at this time.

UDS Response With Different Test Agents

A number of agents which have been tested for their ability to induce a UDS response in the germ cells of male mice are shown in Table 1. Generally, if a mutagen produces a positive UDS response in mouse germ cells, the UDS can be detected using lower exposures than those required to measure genetic end-points. Also, when exposures to a test agent are below toxic levels, there is usually a linear relationship between the exposure level and the UDS response of the germ cells (Sega et al. 1976, 1978, 1981). Figure 2 shows an example of the linear relationship obtained between i.p. exposures to MNU and the UDS response observed in early spermatid stages of the mouse. Such "chemical exposure-UDS response" curves can be an aid in determining how reasonable it is to linearly extrapolate genetic results obtained at high-exposure levels to what would be expected at low-exposure levels.

Not all mutagens tested have given a positive UDS response in the germ cells. For example, nitrosodimethylamine (NDMA) and nitrosodiethylamine (NDEA) both require metabolic activation in tissues such as liver, kidney, and lung to be biologically active (Weekes and Brusick 1975). The inability of these chemicals to induce a measurable UDS response in mouse germ cells may be explained by the low levels of mixed function oxidase activity found in the testes (Magee and Barnes 1967) and by the failure of active metabolites, derived from other tissues, to reach the testes.

B[a]P also requires metabolic conversion to be biologically active. We have found that after i.p. injection of [³H]-labeled-B[a]P dissolved in corn oil, the

Table 1

Agents Tested for Their Ability to Induce Unscheduled DNA Synthesis (UDS) in Mouse Germ Cells

Agent	Exposure	UDS response[a]
MMS[b]	5-100 mg/kg, i.p.	+
EMS[b]	10-300 mg/kg, i.p.	+
propyl methanesulfonate (PMS)[b]	50-700 mg/kg, i.p.	+
iPMS[b]	10-170 mg/kg, i.p.	+
CPA[c]	200 mg/kg, i.p.	+
DMO[c]	40 mg/kg, i.p.	+
ETO[d]	300-500 ppm X 8 h	+
MNU[e]	1-80 mg/kg, i.p.	+
ENU[f]	10-250 mg/kg, i.p.	+
hycanthone[f]	150 mg/kg (free base), i.p.	+
triethylenemelamine (TEM)[f]	0.4-2 mg/kg, i.p.	+
X-rays[g]	200-1200 R	+
EDB[h]	50-250 mg/kg, i.p.	−
NDMA[f]	4-8 mg/kg, i.p.	−
NDEA[f]	125 mg/kg, i.p.	−
B[a]P[f]	250-500 mg/kg, i.p.	−
caffeine[f]	200 mg/kg	−
MNNG[f]	25-50 mg/kg, i.p.	−
mitomycin C[f]	4 mg/kg, i.p.	−

[a]The criterion for a positive UDS response was that the treated germ cells had to show at least 2 X the incorporation of [^{3}H]-TdR as did the controls. For those agents giving a positive response, UDS was detected over the entire exposure range indicated.
[b]Sega et al. (1976)
[c]Sotomayor et al. (1978)
[d]Cumming and Michaud (1979)
[e]Sega et al. (1981)
[f]G. A. Sega, unpubl. results
[g]Sega et al. (1978)
[h]Sega and Sotomayor (1980)

B[a]P is present to the same extent in testis homogenates as it is in liver homogenates from the same treated animals. However, there is no clearly demonstrable binding of B[a]P (or any of its metabolites) to sperm DNA or testicular DNA, while there is extensive binding of [^{3}H]-labeled metabolites of B[a]P to liver DNA (See Fig. 3). Failure to detect binding of B[a]P or its metabolites to testicular DNA is in agreement with our negative UDS results with this compound.

Two other chemicals, that are currently of great environmental interest, Ethylene oxide (ETO) and ethylene dibromide (EDB), have also been studied for their ability to induce a UDS response in mouse germ cells (See Table 1). ETO is an important industrial chemical and is widely used as a sterilant and fumigant. EDB is used as a gasoline additive and is also used extensively as a

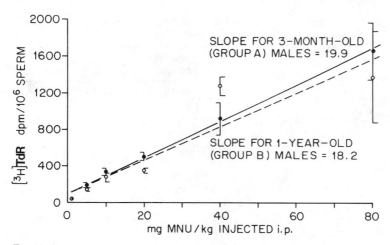

Figure 2
Initial level of UDS (occurring within ~ 4 hr after treatment) induced in early spermatids of (C3H × 101)F_1 mice as a function of i.p. exposure to MNU. Error bars are ± 1 S.E.M. The least-squares linear fit curves are shown by a solid line for the 3-mo. old males (•) and a dashed line for the 1-yr old males (○) (Reprinted, with permission, from Sega et al. 1981).

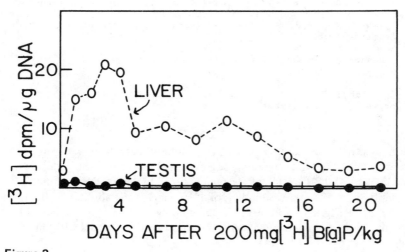

Figure 3
Binding of [³H] B[a]P and its metabolites to liver DNA (○- - -○) and testicular DNA (•——•) in (C3H × 101)F_1 hybrid mice as a function of time after i.p. exposure to 200 mg/kg of the chemical. The specific activity was 4.4×10^{-11} disintegrations per min./molecule of B[a]P.

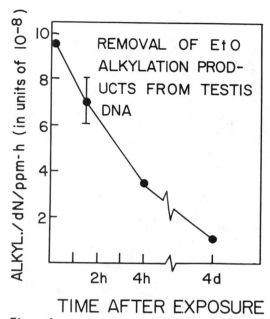

Figure 4
Alkylation level of testicular DNA of the mouse at various times after inhalation exposure to 3 ppm-hr of ETO.

pesticidal fumigant. Recently, it has been used to help control the Mediterranean fruit fly in California (Walsh 1982).

After inhalation exposure to ETO, UDS is induced in mouse germ cells (Cumming and Michaud 1979). Chemical dosimetry experiments using [³H]-labeled ETO have also been carried out (G. A. Sega and R. B. Cumming, unpubl. results). The dosimetry experiments show a gradual removal of alkylation products from testicular DNA with time after inhalation exposure to 3 ppm-h of [³H] ETO (See Fig. 4). Only about 10% of the adducts remain in the testicular DNA by 4 days after exposure. Thus, with ETO, there is agreement between the positive UDS response of the germ cells and removal of adducts from testicular DNA.

As Table 1 and Fig. 5 indicate, EDB does not induce a measurable UDS response in mouse germ cells. Our chemical dosimetry experiments carried out using [³H] EDB have shown that after exposing mice to EDB, DNA adducts are formed in liver, testis, and sperm (See Fig. 6). While adducts are gradually removed from liver DNA, no measurable removal of adducts from testicular DNA is observed over at least a 12-day period following exposure. In the case of EDB there is also agreement between the negative UDS response of the germ cells and the persistence of DNA adducts in the testis.

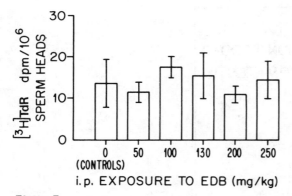

Figure 5
Absence of a UDS response in early spermatids of (C3H × 101)F₁ mice after various i.p. exposures to EDB.

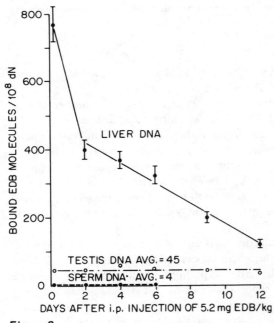

Figure 6
Alkylation of DNA from liver, testis and sperm of (C3H × 101)F₁ mice as a function of time after exposure to ³H-labeled EDB.

Relation Between UDS and Repair of Genetic Damage

Because not all germ-cell stages in the mouse undergo a UDS response after mutagen exposure, it has been possible to use this fact to study what effect UDS may have in altering the ultimate expression of genetic damage. The results to date have been ambiguous. Chemicals such as EMS and MMS produce dominant lethals and translocations only in mid- to late-spermatids and spermatozoa, where no UDS occurs (Ehling et al. 1968; Generoso and Russell 1969). In germ cell stages where UDS is induced by these chemicals no dominant lethals or translocations are induced. However, mutagens such as isopropyl methanesulfonate (iPMS) and X-rays produce dominant lethals in germ-cell stages where UDS has been induced (Ehling 1971; Schröder and Hug 1971; Ehling et al. 1972). In fact, with X-rays, the dominant lethal frequency is about twice as high in postgonial germ-cell stages where UDS occurs as it is in the more advanced stages where no UDS can be detected.

With X-rays, there is also no reduction in specific locus mutation frequencies in postgonial germ cell stages where UDS occurs compared to those postgonial stages in which no UDS occurs (Sega et al. 1978). Clearly, much more work will be required to understand the relationship between UDS and the repair of genetic damage in mammalian germ cells.

SUMMARY

When male mice are exposed to chemical agents that reach the germ cells several outcomes are possible in terms of the germ-cell UDS response and removal of DNA adducts. It is possible that:

1. The chemical binds to the DNA and induces a UDS response with concomittant removal of DNA adducts (e.g., ETO).
2. The chemical binds to the DNA but no UDS response is induced (e.g., EDB).
3. The chemical does not bind to DNA and no UDS is induced (e.g., B[a]P).

Many mutagens have already been shown to induce a UDS response in postgonial germ-cell stages of the male mouse up through about mid-spermatids, but the relationship between this UDS and the repair of genetic damage within the germ cells is still unknown. While some mutagens appear to have an effect only in germ-cell stages where no UDS occurs, others are able to induce genetic damage in stages where UDS has been induced.

ACKNOWLEDGMENT

Research sponsored by the Office of Health and Environmental Research, U.S. Department of Energy, under contract W-7405-eng-26 with the Union Carbide Corporation.

REFERENCES

Bürgin, H., B. Schmid, and G. Zbinden. 1979. Assessment of DNA damage in germ cells of male rabbits treated with isoniazid and procarbazine. *Toxicology* **12**:251.

Chandley, A.C. and S. Kofman-Alfaro. 1971. "Unscheduled" DNA synthesis in human germ cells following UV irradiation. *Exp. Cell Res.* **69**:45.

Cumming, R.B. and T.A. Michaud. 1979. Mutagenic effects of inhaled ethylene oxide in mice. *Environ. Mutagen.* **1**:166.

Ehling, U.H. 1971. Comparison of radiation and chemically-induced dominant lethal mutations in male mice. *Mutat. Res.* **11**:35.

Ehling, U.H., R.B. Cumming, and H.V. Malling. 1968. Induction of dominant lethal mutations by alkylating agents in male mice. *Mutat. Res.* **5**:417.

Ehling, U.H., D.G. Doherty, and H.V. Malling. 1972. Differential spermatogenic response of mice to the induction of dominant lethal mutations by n-propyl methanesulfonate and isopropyl methanesulfonate. *Mutat. Res.* **15**:175.

Generoso, W.M. and W.L. Russell. 1969. Strain and sex variations in the sensitivity of mice to dominant lethal induction with ethyl methanesulfonate. *Mutat. Res.* **8**:589.

Gledhill, B.L. and Z. Darzynkiewicz. 1973. Unscheduled synthesis of DNA during mammalian spermatogenesis in response to UV irradiation. *J. Exp. Zool.* **183**:375.

Kofman-Alfaro, S. and A.C. Chandley. 1971. Radiation-initiated DNA synthesis in spermatogenic cells of the mouse. *Exp. Cell Res.* **69**:33.

Magee, P.N. and J.M. Barnes. 1967. Carcinogenic nitroso compounds. *Adv. Cancer Res.* **10**:163.

Monesi, V. 1962. Autoradiographic study of DNA synthesis and the cell cycle in spermatogonia and spermatocytes of mouse testis using tritiated thymidine. *J. Cell Biol.* **14**:1.

Oakberg, E.F. 1956a. A description of spermiogenesis in the mouse and its use in analysis of the cycle of the seminiferous epithelium and germ cell renewal. *Am. J. Anat.* **99**:391.

―――. 1956b. Duration of spermatogenesis in the mouse and timing of stages of the cycle of the seminiferous epithelium. *Am. J. Anat.* **99**:507.

Rasmussen, R.E. and R.B. Painter. 1964. Evidence for repair of ultraviolet damaged deoxyribonucleic acid in cultured mammalian cells. *Nature* **203**:1360.

Schmid, B., I.P. Lee, and G. Zbinden. 1978. DNA repair processes in ejaculated sperm of rabbits treated with methyl methanesulfonate. *Arch. Toxicol.* **40**:37.

Schröder, J.H. and O. Hug. 1971. Dominante Letalmutationen in der Nachkommenschaft bestrahlter männlicher Mäuse. I. Untersuchung der Dosiswirkungsbeziehung und des Unterschiedes zwischen Ganz- und Teilkorperbestrahlung bei meiotischen und postmeiotischen Keimzellenstadien. *Mutat. Res.* **11**:215.

Sega, G.A. 1974. Unscheduled DNA synthesis in the germ cells of male mice exposed in vivo to the chemical mutagen ethyl methanesulfonate. *Proc. Natl. Acad. Sci. (USA)* **71**:4955.

Sega, G.A. and R.E. Sotomayor. 1980. Chemical dosimetry and unscheduled DNA synthesis studies of ethylene dibromide in the germ cells of male mice. *Environ. Mutagen* **2**:274.

Sega, G.A. and R.E. Sotomayor. 1982. Unscheduled DNA synthesis in mammalian germ cells.–Its potential use in mutagenicity testing. *Chem. Mutagens* **7**:421.

Sega, G.A., R.B. Cumming, and M.F. Walton. 1974. Dosimetry studies on the ethylation of mouse sperm DNA after in vivo exposure to [³H]ethyl methanesulfonate. *Mutat. Res.* **24**:317.

Sega, G.A., J.G. Owens, and R.B. Cumming. 1976. Studies on DNA repair in early spermatid stages of male mice after in vivo treatment with methyl-, ethyl-, propyl-, and isopropyl methanesulfonate. *Mutat. Res.* **36**:193.

Sega, G.A., R.E. Sotomayor, and J.G. Owens. 1978. A study of unscheduled DNA synthesis induced by X-rays in the germ cells of male mice. *Mutat. Res.* **49**:239.

Sega, G.A., K.W. Wolfe, and J.G. Owens. 1981. A comparison of the molecular action of an S_N1-type methylating agent, methyl nitrosourea and an S_N2-type methylating agent, methyl methanesulfonate, in the germ cells of male mice. *Chem. Biol. Interact.* **33**:253.

Sotomayor, R.E., G.A. Sega, and R.B. Cumming. 1978. Unscheduled DNA synthesis in spermatogenic cells of mice treated in vivo with the indirect alkylating agents cyclophosphamide and mitomen. *Mutat. Res.* **50**:229.

Sotomayor, R.E., G.A. Sega, and R.B. Cumming. 1979. An autoradiographic study of unscheduled DNA synthesis in the germ cells of male mice treated with X-rays and methyl methanesulfonate. *Mutat. Res.* **62**:293.

Walsh, J. 1982. Spotlight on pest reflects on pesticide. *Science* **215**:1592.

Weekes, U. and D. Brusick. 1975. In vitro metabolic activation of chemical mutagens. II. The relationships among mutagen formation, metabolism and carcinogenicity for dimethylnitrosamine and diethylnitrosamine in the livers, kidneys and lungs of BALB/cJ, C57BL/6J and RF/J mice. *Mutat. Res.* **31**:175.

Chromosome Aberrations in Decondensed Sperm DNA

R. JULIAN PRESTON
Biology Division
Oak Ridge National Laboratory
Oak Ridge, Tennessee 37830

If it is determined that an individual, group, or population has possibly been exposed to a clastogenic or mutagenic agent, there are three questions that need to be considered:

1. can it be established whether or not there was an exposure?
2. can the mean exposure dose be established?
3. is it possible to estimate any increased deleterious health effects from such an exposure?

In the case of a radiation exposure, it is possible to provide a rather reliable estimate of the exposure from the analysis of chromosome aberrations in in vitro cultured peripheral lymphocytes. This method of estimation is applicable to an individual, group, or population. Of course, the reliability of this estimate will be dependent upon several factors such as the homogeneity of the exposure and the time of sampling after exposure. Furthermore, in the case of exposed populations, but *not* for exposed individuals, it is possible, from data available in humans and laboratory animals, to make some reasonable estimate of any increased genetic or carcinogenic risk from such an exposure (e.g., Report of BEIR III Committee). This type of estimation is made feasible because of the reliable dosimetric methods available, and the fact that dose distributions are determined by physical factors, rather than biological ones.

If the suspected exposure is to a chemical agent, the same three questions can be asked, but the answers are much more equivocal. Chromosome aberrations can be analyzed in in vitro cultured peripheral lymphocytes from possibly exposed individuals, groups, or populations. The observed frequency is compared to the frequency in a partially "matched" control, and a significant increase in the potentially exposed group is indicative of an exposure to the suspect agent. This is a reasonable approach, but it has to be realized that the relative insensitivity of the peripheral lymphocyte to the induction of chromosome aberrations means that the lack of a significant increase in aberration frequency in an individual compared to a "matched" control does not indicate the absence of an exposure. Thus the assay is only applicable for determining an exposure to groups or populations and not to individuals, and then only for indicating exposure and not for excluding an exposure. The frequency of aberrations analyzed in cultured lymphocytes is not directly related to exposure as it is following radiation. This argument is based on the fact that, in contrast to radiation-induced aberrations, the aberrations induced by the great majority of chemicals are induced not in the peripheral lymphocytes in vivo, but rather in the DNA synthesis phase that occurs following mitogenic stimulation and in vitro culture. As with radiation, the aberration frequency will also be dependent upon homogeneity of exposure and time of sampling, as well as several other factors that need not be discussed here. In fact, the measurement of sister chromatid exchanges (SCEs) is generally a more sensitive indicator of chemical exposure than aberration analysis. However, no genetic significance has yet been associated with the presence of SCE, and so their analysis is only applicable to indicating or estimating an exposure. There is a relative paucity of data on the genetic and carcinogenic risk of chemical exposures to laboratory animals, and an almost complete lack of human data, such that an estimation of the deleterious effects in humans of chemical exposures is virtually impossible at this time. Such an estimation is further complicated by the fact that dosimetry of chemical agents is not well-defined and variations in exposure are largely determined by biological and not physical factors, and consequently will always be less well defined.

In order to be able to estimate the genetic consequences of a chemical exposure to human populations, it is inevitable that studies must be performed on germ cells. This means that the exposure to germ cells needs to be estimated, and genetic consequences of such a germ cell exposure need to be determined. Much of this information will, through necessity, be obtained in laboratory animals, but some means of extrapolating to man must be determined. This paper addresses one part of such an approach, that is, can human sperm be analyzed for chromosome alterations, and can the frequency of chromosome alterations be related directly or indirectly to exposure to a clastogenic agent. There is also a discussion of some of the parameters that could influence the aberration frequency or aberration types, based upon results obtained with laboratory animals.

METHODS OF STUDYING DECONDENSED HUMAN SPERM DNA FOR CHROMOSOME ABERRATION ANALYSIS

Direct Sperm DNA-decondensation

Several methods are described for decondensing sperm DNA (Calvin and Bedford 1971; Kvist 1980; and Incharoensakdi and Panyim 1981) using treatments with dithiothreitol or sodium dodecyl sulfate. Subsequent analysis has been at the electron-microscope level. The types of preparation obtained are quite unsuitable at this time for any analysis of chromosome alterations, or numerical changes. It is also to be noted that such analysis of spermatozoa indicate effects in premeiotic germ cells, since spermatozoa do not pass through a DNA synthesis phase until after fertilization, and DNA repair is absent in mature sperm, and so aberrations would not be produced in treated sperm. This is in contrast to the methods described under the methods mentioned below that could also detect aberrations produced from damage induced in sperm DNA. However, this direct decondensation approach is currently not usable for indicating an exposure either by chromatin changes or chromosomal alterations. However, it is a viable research area, and could provide a relatively straightforward way of analyzing a large number of human germ cells.

Fusion of Human Sperm with Somatic Cells

Over the past 10 years there have been many reports describing attempts to reactivate sperm nuclei by fusing spermatozoa with cultured somatic cells (Brackett et al. 1971; Gledhill et al. 1972; Bendich et al. 1974; Higgins et al. 1975; Elsevier and Ruddle 1976). It is probably reasonable to say that there has to this point been very little success. A variety of methods of fusing sperm and fibroblasts have been employed, including spontaneous fusion, lysolecthin, glyceryl monooleate, and inactivated Sendai virus. Van Meel et al. (1981) interpret the lack of reactivation of the sperm nuclei as resulting from the fact that the majority of "fusion" occurs by phagocytosis, with the subsequent disintegration of the sperm chromatin. In their experiments, using Sendai virus for fusion, they observed that about 80% of sperm were incorporated by phagocytosis, and subsequently degenerated. However, some 20% of the sperm integrations appeared to be by fusion of the sperm plasmalemma and the fibroblast plasma membrane. The sperm heads incorporated by membrane fusion showed "chromatin decondensation in the absence of nuclear envelope, which fits the morphologic criteria of nuclear reactivation." However, there has been no evidence of a mitotic division of the sperm nucleus, and so chromosomal analysis is not yet feasible. There is clearly some progress towards such an end, and with a better understanding of both the membrane fusion process and the subsequent development of the sperm nucleus this could be a method of assessing chromosome alterations in human germ cells. The ease of obtaining large numbers of the host cells (fibroblasts) make this a rather attractive approach.

In Vitro Fertilization Using Human Sperm

At this time, the in vitro fertilization of isolated oocytes by human sperm appears to be the most successful of the three approaches described for analyzing chromosome aberrations in the decondensed or reactivated sperm nucleus. It is important to add that there is a high frequency of fertilization when hamster oocytes are used, but none when mouse or rat oocytes are employed, indicating the specificity of the plasma membrane in the latter two species.

The basic technique was first described by Yanagimachi et al. (1976), and with some slight modifications has been utilized for a variety of studies by several laboratories (Rudak et al. 1978; Barros et al. 1979; Binor 1980; Hall 1981). Details of the methods used can be found in these references. Only a few generalizations will be given here.

In order to obtain a large number of eggs, female golden hamsters (Mesocricetus auratus) are superovulated with pregnant mare's serum gonadotrophin (PMG) and human chorionic gonadotrophin (HCG) administered 2 or 3 days after the PMG. The ovulated eggs are collected from the oviduct about 17 hours after the HCG and isolated from the cumulus cells. An important step, in order to obtain sperm penetration, is to remove the zona pellucida of the isolated eggs, usually using 0.1% trypsin. Sperm samples are prepared by a series of filtrations and washings, and incubated for 4-7 hours to allow capacitation to take place. Final sperm concentrations should be greater than 6×10^5 motile sperm/ml (Binor et al. 1980). Fertilization takes place by mixing sperm and zona-free eggs in specific medium (Biggers et al. 1971) under mineral oil. Incubation at $37°C$ in a 95% air/5% CO_2 atmosphere for 3 hours results in 75% of eggs being penetrated by one or more sperm (Rudak et al. 1978).

In order to allow the fertilized eggs to proceed to the first cleavage division, Rudak et al. (1978) transferred them to fresh medium under mineral oil, and continued the incubation for 12-13 hours. They were then transferred to medium containing colcemid and metaphases accumulated for 6-7 hours. Fixation and slide preparation were as described by Tarkowski (1966). The method of Rudak et al. (1978) appears to be the only one where first cleavage division metaphases were obtained in an analyzable condition. Even then, despite the elegance of the technique, it is clear that the number of analyzable human haploid genomes is rather low—a maximum of about 10 per 150 fertilized eggs that reached the fixation stage. Rudak et al. (1978) also reported that 20-40 additional human complements could not be analyzed because of inadequate spreading of the chromosomes. It is quite possible that modifications and improvements in the technique could greatly increase the proportion of analyzable cells. The low frequency does not appear to be the result of a low yield of fertilized eggs. It hsould be added that the fertilized eggs that reached the fixation stage had between one and three human complements. However, in some of these cases, each sperm complement was discrete, and could be analyzed with adequate spreading.

The results reported by Rudak et al. (1978) show that human complements containing an X or Y chromosome occurred with about equal frequency, and that the frequency of aneuploidy was about 5%, although only 60 cells were analyzed. It should also be noted that the frequency of aneuploidy could be biased by the loss of chromosomes during preparation, for which control is rather difficult.

There is clearly more information to be obtained before this in vitro fertilization technique is usable as a relatively reproducible assay for estimating chromosomal alterations induced in human germ cells, or produced subsequent to fertilization from DNA damage induced in post-meiotic germ cells. Even with improvements in technique, however, it will always be a time-consuming and fairly intricate assay that cannot be considered as a routine test, but rather to be employed in specific circumstances. Furthermore, it is a technique that is only applicable to the study of effects in males.

STUDIES WITH LABORATORY ANIMALS

There are a large number of studies in laboratory animals that provide information on factors that could influence the interpretation of results obtained in in vitro fertilization or fusion assays. Only what appear to be the most important ones will be considered in this report.

Fertilization or fusion by chromosomally normal vs. abnormal sperm

If the intention is to develop the in vitro fertilization or fusion techniques for use as assays for detecting exposure to clastogenic agents in humans, it is clearly essential to determine that chromosomally abnormal sperm have an equal probability of fusing with or fertilizing the host cells as chromosomally normal sperm.

There is some evidence that suggests that this might not be the case. It has been reported by Ford et al. (1964) and Brewen et al. (1974) that the frequency of reciprocal translocations induced by radiation in mouse spermato-gonial stem cells and recovered in the F_1 is only 50% of that expected from the frequency observed directly in spermatocytes derived from treated stem cells. The reasons for this are not known, but it was postulated by Ford et al. (1969) that the translocation-bearing sperm were less likely to be involved in fertiliza-tion than chromosomally normal sperm. This needs to be considered further, particularly with regard to chemically treated animals for which there is little or no information. It is perhaps interesting to note that with the radiation studies the frequency of transmission of translocations was about 50% of the expected value irrespective of the dose, and hence the induced translocation frequency, perhaps indicating the operation of a mechanism of selection other than simply the lower probability of fertilization by translocation-bearing sperm.

This question of whether chromosomally abnormal sperm have a reduced probability of fertilization will inevitably be a difficult one to answer. It will be difficult, it not impossible, to determine the frequency of aberrations present in sperm following exposure of pre-meiotic germ cells. The frequency in sperm will obviously be lower than the frequency induced in the pre-meiotic germ cells, because there is a loss of aberrant cells as a consequence of the cell divi-sions that occur between aberration induction in pre-meiotic cells and recovery in sperm. The extent of this reduction is not known. The frequency of aberra-tions can be analyzed in spermatocytes following treatment of pre-meiotic germ cells, but these spermatocytes have to undergo a further two divisions when selection could occur. If the aberration frequency in sperm is unknown, then it is not possible to determine whether the frequency observed at first cleavage division is close to the frequency expected based upon no selection against aberrant cells for fertilization. Furthermore, for cells treated as post-meiotic cells the aberrations will not be induced until after fertilization, such that any selective pressure can only operate against cells containing unrepaired DNA damage of a variety of different types. Whether such selection occurs is not experimentally determinable at this time.

It might be that one approach is to use sperm samples with known pro-portions of cells from chromosomally normal and abnormal individuals, and determine whether or not the expected frequencies of normal vs. abnormal complements, based on no selection against chromosomally abnormal sperm, are recovered at the first cleavage metaphase. Further work is clearly necessary using laboratory animal models as well as human in vitro fertilization techniques.

Effect of Maternal Genotype on Aberration Frequency from Treated Sperm

The DNA damage induced by exposure of spermatozoa to a chemical or physi-cal agent remains unrepaired until after fertilization. This lack of repair is possibly due to the condensed nature of the chromatin in sperm, involving the production and association of protamine to the DNA, making it less accessible to repair, and also perhaps as a result of much reduced repair enzyme availability, because of the very small amount of cytoplasm associated with sperm. As a consequence of this lack of repair in sperm, the DNA damage induced in sperm will be repaired after fertilization, and will be dependent upon the repair capabilities of the egg. Chromosome aberrations, resulting from DNA damage induced in sperm, will also be produced after fertilization. Since the majority of aberrations induced by most chemical agents are produced during or after DNA replication, the frequency of aberrations will be dependent not only upon the repair capabilities of the egg (i.e., how much repair of the damage, that can be converted into aberrations, takes place before the S-phase), but also upon the length of the G_1 stage prior to the first cleavage cycle S-phase (i.e., the shorter this G_1 the less time for repair and the more damage remaining unrepaired at the time of replication). The influence of these potential variables (rate of repair and length of G_1) on aberration frequency will, of course, be dependent upon the particular damaging agent (Preston 1982).

The importance of this in relationship to the present discussion is: Do different strains or species have different repair capabilities in the fertilized egg, and are the durations of the cell cycle stages different in different strains or species? These characteristics, that are probably dependent upon maternal genotype, need to be defined for any technique involving analysis at the first cleavage division. The information available for laboratory animals is almost exclusively for the mouse.

It was reported by Generoso et al. (1979) that the frequency of dominant lethal mutations from treatment of males with isopropyl methanesulfonate (IMS) was different when these males, all of one stock, were mated to females of different stocks. The authors interpreted this as being due to the fact that different female stocks had different capacities for repairing IMS damage induced in sperm DNA; the high dominant lethal frequencies being associated with a repair deficiency and the low dominant lethal frequency with a repair proficiency. Since repair of DNA damage induced in sperm takes place after fertilization, the efficiency will be determined in part by the maternal genotype. This appears to be a plausible explanation, but the fact that much smaller differences in dominant lethality between stocks were observed following treatment with triethylenemelamine (TEM) ethyl methanesulfonate (EMS) and benzo[a] pyrene B[a]P and no difference with X-rays indicate that an alternative explanation might be sought. Preston (1982) presented an alternative explanation of the results. It was suggested that if the different strains of mice had different lengths of first cleavage division G_1 then the time for repair of DNA damage before replication would be different. Thus when G_1 is short more damage would be present at the time of replication, and the probability of inducing chromatid-type aberrations (and hence dominant lethality) would be high. For a longer G_1, the probability of inducing aberrations during or after replication would be lower, as more repair would take place before replication. It is interesting to note that for X-rays there was no difference in dominant lethality for different female strains. X-ray induced DNA damage is repaired rapidly, and unless G_1 is very short, there would be little effect of length of G_1 on the aberration or dominant lethal frequency. There is some experimental evidence in support of this explanation. Shire and Whitten (1980b) showed that there were large differences in the duration of the first cleavage depending upon the strain of female used. Much smaller differences could be attributed to paternal genotype (Shire and Whitten 1980a). It was not determined if the differences in cleavage time resulted specifically from differences in G_1 length.

The possible effects of maternal genotype on the frequency of chromosome aberrations analyzed at first cleavage metaphase from damage induced in sperm are important to determine. It is also necessary to determine whether differences in DNA repair capabilities or cell cycle times are responsible for differences in sensitivity. This question is appropriate when considering the use of first cleavage analysis for determining exposures to germ cells or predicting the genetic consequences of such an exposure.

Time of Sampling after Exposure, and the Aberration Types Observed

A great deal of the information pertinent to this section can be found in Brewen and Preston (1978) and Adler and Brewen (1982). Some general comments will be given here.

As discussed above, the majority of chemically induced aberrations are produced during or shortly after DNA replication and are of the chromatid type. Thus chromatid-type aberrations can be induced in pre-meiotic germ cells. However, the frequency that will be recoverable in the sperm will be considerably less than the induced frequency. This results from the fact that most aberration types are cell lethal, because of the loss of acentric fragments at division—in this case spermatogonial cell divisions. In those cases where lethality does not occur, it should be noted that the induced chromatid-type aberrations will be converted into chromosome-type aberrations as a result of cell division and a subsequent DNA replication. In the case of chromatid-type reciprocal translocations, 25% of the possible segregants will allow for recovery of the translocation in a daughter cell. The subsequent DNA replication converts this into a chromosome-type reciprocal translocation that will be transmitted to both daughter cells by segregation at a mitotic division. Since there is a fairly high probability of transmitting an induced chromatid-type translocation it might be expected that this is the most likely aberration type to be recovered in the sperm following exposure of pre-meiotic germ cells. However, when spermatogonia are treated with agents that are known to induce aberrations, no, or very few, translocations are observed in spermatocytes (Leonard et al. 1971). The reasons for this have been discussed (Brewen and Preston 1973). The important point is that if sperm samples are taken from individuals who were acutely exposed to an agent some weeks before sampling, when the sperm will have been derived from treated spermatogonial cells, the frequency of aberrations observed at first cleavage will be very much lower than the induced frequency. It might well be impossible to detect an effect of even quite high exposures, and even if this were possible the observed aberration frequency will be very indirectly related to the exposure. Further experiments with laboratory animals would certainly be needed to establish a relationship between induced and recovered aberration frequency. It can be argued, in fact, that there is quite probably a higher probability of recovering aberrations, particularly translocations, following chronic or low-dose exposures, but this requires further testing (Preston 1982).

If samples are taken shortly after acute exposures, where, in a sperm sample, spermatozoa will be the treated cell type, the aberrations observed at first cleavage will be of the chromatid type, produced during the first cleavage S-phase. Since there is little or no repair in sperm, the aberration frequency will be influenced by the repair that takes place during G_1. The observed aberration frequency is the induced frequency because the analysis is at the first division after induction. Furthermore, the assay will be rather sensitive, and a relationship between aberration frequency and exposure is calculatable, with information obtainable in controlled experiments with laboratory animals.

It was recently reported by Tanaka et al. (1981) that if cells were treated at the late spermatid stage with methyl methanesulfonate (MMS) the aberrations observed at first cleavage were of the chromosome-type. (MMS produces chromatid-type aberrations at first cleavage when spermatozoa are treated [Brewen et al. 1975]). The conclusion is that the aberrations are formed prior to the S-phase of first cleavage, perhaps as a result of non-enzymatic conversion of the induced damage into some new damage during sperm storage that can result in chromosome-type aberrations in G_1 (Tanaka et al. 1982). This needs to be studied for other agents, because the consequences of chromosome-type aberrations in terms of a genetic hazard are different from those for chromatid-type observations, particularly for reciprocal translocations.

If the suspected exposure has been received chronically, as is likely to be the case for human exposures, all the problems discussed are relevant, since cells in all stages will be exposed. However, the cells receiving the largest accumulated exposure will be differentiating spermatogonia or spermatogonial stem cells. In this case, as already discussed, chromatid-type aberrations will be induced, and selection against aberrant cells will occur. The frequency recovered in a sperm sample will be lower than the induced frequency, and the relationship to exposure is difficult to determine at this time. Further work using animal models could clarify this.

CONCLUSIONS

There are clearly other factors that could influence the aberration frequency observed at first cleavage following in vivo exposure of germ cells, but the ones discussed above seem to be the most important to consider initially. The techniques of chromosome aberration analysis following sperm DNA condensation by in vitro fertilization or fusion would seem to be viable research areas for providing information on human germ cell exposures. However, the potential sensitivity of the assay needs to be better understood, and factors that can influence this sensitivity require a great deal of further study using animal models.

ACKNOWLEDGMENTS

Research sponsored by the Office of Health and Environmental Research, U.S. Department of Energy, under contract W-7405-eng-26 with the Union Carbide Corporation.

REFERENCES

Adler, I.-D. and J. G. Brewen. 1982. Effects of chemicals on chromosome aberration production in male and female germ cells. *Chem. Mutagens* 7:1.

Barros, C., J. Gonzales, E. Herrera, and E. Buskos-Obregon. 1979. Human sperm penetration into zona-free hamster oocytes as a test to evaluate the sperm fertilizing ability. *Andrologia* 11:197.

Bendich, A., E. Borenfreund, and S. S. Sternberg. 1973. Penetration of somatic mammalian cells by sperm. *Science* 183:857.

Biggers, J. D., W. K. Whitten, and D. G. Whittingham. 1971. The culture of mouse embryos in vitro. In *Methods in mammalian embryology* (ed. J. D. Daniel) p. 86. W. H. Freeman and Co., San Francisco.

Binor, Z., J. E. Sokoloski, and D. P. Wolf. 1980. Penetration of the zona-free hamster egg by human sperm. *Fertil. Steril.* 33:321.

Brackett, B. G., W. Baranska, W. Sawicki, and H. Koprowski. 1971. Uptake of heterologous genome by mammalian spermatozoa and its transfer to ova through fertilization. *Proc. Nat. Acad. Sci. USA* 68:353.

Brewen, J. G. and R. J. Preston. 1973. Chromosome aberrations as a measure of mutagens: Comparisons in vitro and in vivo and in somatic and germ cells. *Environ. Health Perspect.* 6:157.

_____ . 1978. Analysis of chromosome aberrations in mammalian germ cells. *Chem. Mutagens* 5:127.

Brewen, J. G., R. J. Preston, and W. M. Generoso. 1974. *X-ray-induced translocations: Comparison between cytogenetically observed and genetically recovered frequencies.* Biology Division Annual Progress Report, ORNL-4993:74.

Brewen, J. G., H. S. Payne, K. P. Jones, and R. J. Preston. 1975. Studies on chemically induced dominant lethality. I. The cytogenetic basis of MMS-induced dominant lethality in premeiotic male germ cells. *Mutat. Res.* 33:239.

Calvin, H. I. and J. M. Bedford. 1971. Formation of disulfide bonds in the nucleus and accessory structures of mammalian spermatozoa during maturation in the epididymis. *J. Reprod. Fertil.* (Suppl.) 13:65.

Elsevier, S. M. and F. H. Ruddle. 1976. Haploid genome reactivation and recovery by cell hybridization: Induction of DNA synthesis in spermatid nuclei. *Chromsoma* 56:227.

Ford, C. E., A. G. Searle, E. P. Evans, and B. J. West. 1969. Differential transmission of translocations induced in spermatogonia of mice by irradiation. *Cytogenetics* 8:447.

Generoso, W. M., K. T. Cain, M. Krisha, and S. W. Huff. 1979. Genetic lesions induced by chemicals in spermatozoa and spermatids of mice are repaired in the egg. *Proc. Natl. Acad. Sci. USA* 76:435.

Gledhill, B. L., W. Sawicki, C. M. Croce, and H. Koprowski. 1972. DNA synthesis in rabbit spermatozoa after treatment with lysolecithin and fusion with somatic cells. *Exp. Cell Res.* 73:33.

Hall, J. L. 1981. Relationship between semen quality and human sperm penetration of zona-free hamsters ova. *Fertil. Steril.* 35:457.

Higgins, P. J., E. Borenfreund, and A. Bendich. 1975. Appearance of fetal antigens in somatic cells after interaction with heterologous sperm. *Nature* 257:488.

Incharoensakdi, A. and S. Panyim. 1981. In vitro decondensation of human sperm chromatin. *Andrologia* 13:64.

Kvist, U. 1980. Sperm nuclear chromatin decondensation ability. *Acta Physiol. Scand.* (Suppl.) 486:1.

Leonard, A., G. Deknudt, and G. Linden. 1971. Failure to detect meiotic chromosomal rearrangements in male mice given chemical mutagens. *Mutat. Res.* 13:89.

Preston, R. J. 1982. The induction of chromosome aberrations in germ cells: A discussion of factors that can influence sensitivity to chemical mutagens. In *Proceedings of 3rd International Conference on Environmental Mutagens*, (eds., T. Sugimura) p. 463. University of Tokyo Press, Japan and Alan R. Liss, New York.

Report of the BEIR III Committee, 1980. The Effects on Populations of Exposure to Low Levels of Ionizing Radiation. National Research Council, National Academy of Sciences, Washington, D.C.

Rudak, E., P. A. Jacobs, and R. Yanagimachi. 1978. Direct analysis of the chromosome constitution of human spermatozoa. *Nature* **274**:911.

Shire, J. G. M. and W. K. Whitten. 1980a. Genetic variation in the timing of first cleavage in mice: Effect of paternal genotype. *Biol. Reprod.* **23**: 363.

_____. 1980b. Genetic variation in the timing of first cleavage in mice: Effect of maternal genotype. *Biol. Reprod.* **23**:369.

Tanaka, N., M. Katoh, and S. Iwahara. 1981. Formation of chromosome-type aberrations at the first cleavage after MMS treatment in late spermatids of mice. *Cytogenet. Cell Genet.* **31**:145.

Tarkowski, A. K. 1966. An air-drying method for chromosome preparations from mouse eggs. *Cytogenetics* **5**:394.

VanMeel, F. C. M., G. C. Beverstock, P. L. Pearson, and W. T. Daems. 1981. Electron microscopic studies on the incorporation of human spermatozoa into mouse fibroblasts following Sendai virus fusion. *J. Ultrastruct. Res.* **75**:142.

Yanagimachi, R., H. Yanagimachi, and B. J. Rogers. 1976. The use of zona-free ova as a test system for the assessment of the fertilizing capacity of human spermatozoa. *Biol. Reprod.* **15**:471.

Human Sperm Morphology Testing:

Description of a Reliable Method and Its Statistical Power

ANDREW J. WYROBEK, LAURIE A. GORDON,
GEORGE WATCHMAKER, AND DAN H. MOORE II
Lawrence Livermore National Laboratory
Biomedical Sciences Division, L-452
University of California
Livermore, California 94550

Sperm tests for counts, motility, and morphology have a long history in the diagnosis of fertility, especially in domesticated animals and man. Recently it has been shown that these tests may also be used as indicators, and, under certain circumstances, as dosimeters of sperm damage induced in males by exposure to physical or chemical agents (for reviews see Wyrobek et al. 1982a, b).

Both impaired fertility and the induction of heritable genetic mutations are major concerns in males with induced spermatogenic damage. Though the human and animal literature both suggest that major reductions in sperm quantity and quality are clearly associated with reduced fertility, the effects of *subtle* sperm changes on fertility remain uncertain and the topic of ongoing research. There is also increasing evidence, mainly from the mouse, of a relationship between induced sperm changes and heritable genetic damage (Wyrobek et al. 1982b). Thus, sperm tests are receiving increased attention in the assessment of chemical mutagenicity and infertility.

There are numerous examples in the human literature in which sperm tests were employed to assess spermatogenic damage in men exposed to chemical toxins; the effects of at least 90 different exposures, representing experimental or therapeutic drugs, occupational or environmental agents, and recreational drugs have been studied (for a review see Wyrobek et al. 1982a). The four most commonly used human sperm tests are sperm counts, motility, morphology (seminal cytology), and double Y-bodies (a new test thought to measure Y-chromosomal nondisjunction during meiosis in the human male). As described below, these tests differ markedly in frequency of use and in our understanding of their genetic and reproductive implications.

An evaluation of the literature on the use of human sperm tests to assess spermatogenic damage in men exposed to chemical or physical agents emphasizes the difficulties in applying these tests to groups of exposed men in a nonclinical setting. Questions arise as to the reliability of the scoring methods, the statistical sensitivities of each test, and the sample size requirements to detect subtle changes. In this paper we briefly review the background of the human sperm tests, describe a reliable method for visually scoring human sperm morphology, and compare the relative statistical sensitivities and sample size requirements for the human sperm count and morphology tests.

HUMAN SPERM TESTS AND APPLICATIONS TO MONITORING

Motility and Counts

Although sperm motility may be one of the best performance evaluations of spermatogenic function in relation to fertility, its measurement requires a fresh semen sample and is very sensitive to time and temperature after sample collection. Thus sperm motility is a difficult and unreliable measure of semen quality, unsuitable for large-scale cross-sectional study, especially when samples are collected at home. Motility will not be discussed further in this paper. Counts, morphology, and double Y-bodies, which can be scored from frozen cells or air-dried smears, do not have these limitations.

Sperm count is reported as the number of sperm per milliliter of ejaculate (or as the total number of sperm ejaculated), and is usually determined by hemocytometer. The measurement is technically easy to make and automated methods are available. Sperm count has been the single most commonly employed test to assess the effects of physical and chemical agents on human spermatogenesis (for a review of radiation effects, see Ash 1980; for a review of chemical effects, see Wyrobek et al. 1982a). However, interpretation of results may be confounded by a number of factors, such as variable continence time prior to sample collection, frequency of sexual contact, and collection of incomplete ejaculate (Schwartz et al. 1979). The measure is variable even in normal unexposed men, and there has been considerable controversy as to what constitutes a normal sperm count in relation to fertility. Though the underlying variability in this measure leads to statistical problems in detecting small induced changes, there are many clear examples of agent-induced reductions in sperm counts (Wyrobek et al. 1982a).

Sperm Morphology

Sperm morphology (also referred to as seminal cytology) is the visual assessment of the shapes of sperm in an ejaculate. Although sperm-head shape is usually emphasized, some assessments also incorporate midpiece and tail abnormalities as well as immature forms. The visual assessment of sperm morphology is very subjective, and is critically dependent on the classification scheme used. In contrast to the highly angular shapes of the sperm heads of laboratory rodents, in which it is relatively easy to detect visually subtle deviations in shape, the typical human sperm head is more ovoid, making deviations in shape harder to perceive. Large laboratory-to-laboratory differences exist in morphology criteria, and in general there is little agreement on what constitutes a "normal" sperm (Freund 1966; Fredricsson 1979). However, studies of MacLeod (1974), David et al. (1975), Eliasson (1971), and others have shown that quantitative approaches to the visual assessment of morphology can be used with considerable success. Human sperm morphology has been used to study the effects of approximately 40 different chemical exposures (for a review see Wyrobek et al. 1982a). In addition, there has been some recent success in measuring sperm morphology in rodents with automated image analysis (Moore et al. 1982; Young et al. 1982).

Though the mechanisms await clarification, there is increasing evidence that induced changes in sperm shape may be related to induced genetic damage and impaired fertility. In the mouse, which has been studied in detail, it has been shown that an agent's ability to induce sperm-shape abnormalities is related to its ability to act as a germ cell mutagen, as judged by dominant lethal, heritable translocation, or specific locus tests (Wyrobek et al. 1982b). In the human literature there are supporting examples of possible relationships between semen quality and adverse reproductive outcome, including correlations between high proportions of shape abnormalities and habitual abortions (Jöel 1966; Czeizel et al. 1967; Jöel and Chayen 1971) or spontaneous abortions (Furuhjelm et al. 1962). More recently, however, another study did not find any relationship between spontaneous abortions and semen quality of the father (Homonnai et al. 1980).

Certain categories of sperm-shape abnormalities may be related to reduced fertility in man (e.g., Nistal et al. 1978). However, it is not yet clear whether all the categories that are classed abnormal or non-oval in shape are also abnormal in their fertilizing capacity. In the mouse we know that the shape of the normal sperm, as well as the percent and types of abnormalities seen are remarkably genotype-specific (Wyrobek et al. 1976; Wyrobek 1979; Krzanowska 1981). Likewise, individual men seem to produce remarkably consistent proportions and patterns of sperm shapes even when sampled over a period of more than a year (MacLeod 1965; A. Wyrobek et al., in prep.).

As we show in this paper, human sperm morphology can be visually scored reliably, and seems to have unique statistical characteristics as a test of induced chemical effects on sperm production.

Double Y-Bodies

The double Y-body test is based on scoring the frequency of fluorescent spots in human sperm stained with quinacrine dye. Studies in somatic cells suggest that these spots represent Y chromosomes. The double Y-body test scores the frequency of sperm with two spots. These sperm are thought to contain two Y chromosomes due to meiotic nondisjunction, though there remain major uncertainties in this interpretation. Unlike the other sperm tests (counts, motility, and morphology), the double Y-body test has no direct counterpart in common laboratory or domestic animals. The Y-chromosomal fluorescence seems to be unique to man and the higher apes (Seuànez 1981). It is interesting to note that some field voles, *Microtus*, have sex chromosomes that are uniquely heterochromatic and can be readily identified in haploid spermatids. This system may provide a promising animal model of chemically-induced nondisjunction in male germ cells (Tates 1979).

The double Y-body test is very new and only a few populations of exposed men have been analyzed with it (Kapp 1979; Wyrobek et al. 1981). An evaluation of the criteria used in scoring Y-bodies, and the statistical characteristics of this test are described elsewhere (A. Wyrobek and G. Watchmaker, in prep.).

A NEW METHOD FOR SCORING HUMAN SPERM MORPHOLOGY

Description of the Method

The human sperm morphology test is performed by using air-dried smears, which are fixed with 95% ethanol and then stained with a modified Papanicolaou method. Figure 1 gives several examples of each of the 10 categories of sperm shape we use in our assessment. The literature in the mouse and other mammals suggests that the shape of sperm heads is genetically controlled and is generally insensitive to abstinence period and technical differences in preparation of smears (for review see Wyrobek et al. 1982b). Therefore, our method focuses primarily on the shape of the sperm head. Figure 2 describes the decision process we employ in assigning each sperm to a shape category.

For scoring purposes, we consider only those cells with tails attached. Any unusually high prevalence of loose sperm heads, immature germ cells, and nongerminal cells are described in the comments section of the worksheet. As shown in Figure 2, for each sperm, the first decision is whether it is "double" (see Fig. 1, No. 5). This includes cells with more than one head or tail and is the only example of a category that is based in part on the tail characteristic. If the cell is not "double," the next decision is whether it is of the "ghost" category. These cells generally have very large, diffuse sperm heads which stain less densely than those of the other categories. The next three decisions, which are for the "normal," "small," and "large" categories (see Fig. 1, Nos. 1, 2, and 3) deal with sperm that show the characteristically oval shape but differ in length of the major axis of the sperm head. An eyepiece reticle is used to

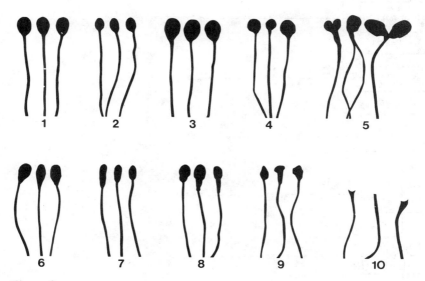

Figure 1
Variations in shape of human sperm. The head shapes of the sperm in category 1 are oval and are considered by us as normal. Those in categories 2 to 10 are scored as morphologically abnormal. The sperm in category 2 are small; 3, large; 4, round, 5, doubles, 6, narrow-at-base (of head); 7, narrow; 8, pear-shaped; 9, irregular; and 10, ghost. Tail lengths are uniformly shortened (see text for further details).

distinguish cells with lengths of < 3 μm (small), 3-5 μm (normal), and > 5 μm (large). The next category in the decision process deals with cells that are "round" regardless of size. Continuing with the flow diagram, the next three decisions deal with sperm that have some part of the head that is proportionately narrower than for the ovals. We have categorized these as pear-shaped (an abrupt indentation and narrowing at the distal end of the head), narrow (uniformly narrow along the entire width of the head), and narrow-at-base (smoothly tapered at the distal end of the head). The last category (irregular) includes sperm with head shapes not covered by the previous descriptions.

Though this scoring logic is very reproducible, it has several unique characteristics. It should be noted that although sperm can have more than one defect, this scheme uses a priority system to assign each sperm to only one shape category. For example, sperm in the "double" category are variable in their head shapes. Furthermore, shape is generally given priority over size and thus shape category can contain differing size distributions, depending on the donor.

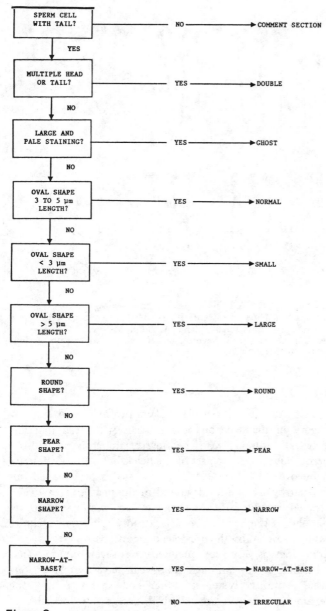

Figure 2

Schematic of the decision process used in assigning human sperm to shape categories. Smears of undiluted ejaculates are made, air-dried, fixed, stained, and observed under a light microscope (see text for details). Entries are made on worksheets that contain columns for slide code, date, scorer, each of the 10 morphology categories, and comments. The division process shown is applied individually to typically 500 sperm per ejaculate. For each ejaculate abnormal prevalences of nonsperm cells are noted in the comments section.

Table 1
Summary Statistics for Repeat Readings of Reference
Slides for Human Sperm Morphology

Slide number	Number of readings[a]	Mean[b]	Observed variance	Ratio[c]	Chi-square[d]	p-Value[e]	Residual variance
1	30	41.99	11.97	2.46	71.25	<0.001	7.10
2	30	31.67	6.16	1.42	41.24	0.066	1.83
3	29	26.01	6.28	1.63	45.70	0.019	2.43
4	29	66.51	5.50	1.23	34.58	0.182	1.05
5	28	50.62	7.66	1.53	41.37	0.038	2.66
6	25	62.72	15.88	3.40	81.48	<0.001	11.20
7	19	47.62	8.77	1.76	31.67	0.024	3.78
8	19	38.22	5.93	1.26	22.59	0.207	1.20
9	12	92.55	1.39	1.01	11.09	0.426	0.01
10	11	55.18	39.93	8.07	80.73	<0.001	34.99
11	11	56.97	7.57	1.54	15.44	0.117	2.67
12	9	25.89	8.05	2.10	16.78	0.032	4.21
13	9	35.18	15.24	3.34	26.74	0.001	10.68
14	8	57.56	11.71	2.40	16.78	0.019	6.83

[a]All readings were done blind on coded slides by the same scorer (L.G.) over a 4-yr period.
[b]Average percent of sperm in classes 2-10 (Fig. 1) based on scoring 500 sperm per reading.
[c]Ratio = observed variance/expected variance, where expected variance = mean $(100 - mean)/500$.
[d]$\chi^2 = (N - 1) \cdot$ Ratio, where N = number of readings.
[e]p-Values based on Cochran's test for homogeneity of a binomial proportion (Snedecor and Cochran 1967). Statistical significance should be judged by comparing against a Bonferroni-adjusted level of significance (e.g., 0.05/14 = 0.004; Miller 1966).

Reference Slides

A set of reference slides is used to reduce the subjectivity inherent in the visual method for scoring morphology. The set was selected to represent examples of the various types of sperm-head shapes seen in a typical human study. These reference slides are coded and used for practice, and for warm-up before scoring a series of study slides, and are interspersed randomly at regular intervals in each study.

Table 1 shows the results for the 14 slides in the reference set used by one of us (L.G.) over nearly a 4-year period. Each slide has a characteristic mean proportion of sperm that do not fall into the normal shape category (i.e., percent morphologically abnormal by our criteria). In Table 1 we compare the observed variance in the repeat readings of each slide with the variance we would expect from repeated random sampling of 500 sperm. For 10 of the 14 slides, the ratio of the observed to expected variance was not statistically different from unity (see Table 1, column 7 and footnote e), suggesting that counting statistics could account for all the variation observed. Four slides (Nos. 1, 6, 10, and 13) showed statistically elevated ratios. An inspection of the raw data suggests that these smears contain sperm that are more difficult to categorize. For example, slide No. 10, which has the largest discrepancy, contains a major proportion of sperm with oval head shapes, but with a size very close to the cut-off length (judged by an eyepiece reticle) for distinguishing small from normal. The other slides with elevated ratios contained major proportions of sperm that are very close to the visual cut-off for classifying sperm as oval vs. narrow (Nos. 6, 13), or oval vs. round (No. 1).

A summary appraisal of the results on repeated scoring of the reference set suggests that scorer subjectivity and counting statistics contributed approximately equally to the overall observed variance. The fact that such a small scorer subjectivity component was observed even when slides were scored over nearly a 4-year period and under so many different experimental conditions demonstrates that human sperm morphology can be scored reliably and reproducibly.

STATISTICAL DESCRIPTION OF SPERM COUNTS AND MORPHOLOGY, AND SAMPLE-SIZE CALCULATIONS

Summary descriptive statistics for the distribution of sperm counts and percent morphologically abnormal sperm (by our criteria) for three groups of normal, healthy men from different geographic locations are shown in Tables 2 and 3. A comparison of the summary statistics for each group clearly shows that sperm counts are considerably more variable than sperm morphology. There is great variability in sperm count both within each group (measured by the minimum and maximum values and summarized by the coefficient of variation statistic) and among the groups (measured by the spread of the means). It is interesting to note, however, that the Kolmogorov–Smirnov test (Conover 1971) could not distinguish among the three groups. This suggests that the results could be pooled into a single sample. We kept them separate to show the effects of differing means and variances on sample-size calculations.

The last column in each table shows the result of testing each distribution of measurements for normality using Filliben's normal order statistic correlation test (Filliben 1975). The results show that sperm counts for groups 1 and 2 and percent abnormal sperm for group 1 fail to pass the test for normality.

Table 2
Descriptive Statistics for Sperm Counts in Groups of Normal, Healthy Men[a]

Group	Number of subjects	Mean (10^6)	Median	Minimum value	Maximum value	Standard deviation	Coefficient of variation	P for normality[b]
1	34	128.7	87.5	10	664	137.4	1.07	<0.005
2	26	66.2	42.0	0	183	56.0	0.85	<0.01
3	20	90.4	77.2	8	200	59.9	0.66	0.50

[a]Number of sperm per ml of ejaculate (10^6), one sample per subject.
[b]P for normality based on Filliben's tables (1975). Values of $p > 0.05$ represent acceptable fits to normal distributions.

Table 3
Descriptive Statistics for Sperm Morphology in 3 Groups of Normal, Healthy Men[a]

Group	Number of subjects	Mean[a]	Median	Minimum value	Maximum value	Standard deviation	Coefficient of variation	P for normality[b]
1	34	41.9	40.5	25.2	72.6	12.4	0.30	<0.01
2	25	47.7	46.6	21.2	67.0	12.7	0.27	>0.25
3	20	47.5	47.8	26.6	82.7	13.9	0.29	>0.10

[a]Based on scoring 500 sperm per sample, one sample per subject; percent morphologically abnormal by criteria described in text, Figs. 1 and 2.
[b]P for normality based on Filliben's tables (1975). Values of $p > 0.05$ represent acceptable fits to normal distributions.

Table 4

Sample Sizes for 90% Probability of Detecting a Decrease
in Sperm Counts at 5% Level of Significance[a]

Group	Percent decrease in counts					
	10	20	30	40	50	60
1	797	190	80	43	26	17
2	1664	394	166	88	53	34
3	747	178	75	40	25	16

[a]Sample size is number required in each of two groups: controls and exposed. Numbers are based on t-test applied to normalized data (square root for sperm counts).

Table 5

Sample Sizes for 90% Probability of Detecting an Increase
in Percent Abnormal Sperm at 5% Level of Significance[a]

Group	Percent increase in abnormally shaped sperm[b]					
	10	20	30	40	50	60
1	128	30	13	7	5	3
2	116	27	12	7	4	3
3	123	29	12	7	4	3

[a]Sample size is number required in each of two groups: controls and exposed. Numbers are based on t-test applied to normalized data (log for percent abnormal).

[b]For example, a 10% increase in percent morphologically abnormal sperm in Group 1 (Table III) would represent a change from 41.9 to 46.1% in the mean.

The family of power transformations, $y = x^p$, was applied to each data set to determine whether or not the data could be normalized using the same transformation for each sperm test. (It is necessary to normalize the data in order to determine sample sizes required for detecting changes in the mean values.) We found that the square root transformation ($p = \frac{1}{2}$) worked best for the sperm counts from all three groups (although for group 1, the log transformation provided a better correlation with the normal distribution, as we reported in Wyrobek et al. 1981). The log transformation (corresponding to $p = 0$, which is by convention redefined to be the log transformation), worked best for percent abnormal sperm.

After applying these transformations sample sizes required to detect various percent decreases in sperm count or increases in percent abnormal sperm can be calculated based on using the two-sample t test (Guenther 1981) (Tables 4 and 5). With the assumption that human sperm counts and morphology have similar biological sensitivity to the agent(s) being tested, these tables show that there would be 5- to 15-fold savings in the number of subjects required to show similar percent relative changes in sperm morphology compared to sperm counts.

DISCUSSION

Human sperm tests for counts, motility, and morphology have been used to assess the effects of at least 90 different exposures on human sperm production (Wyrobek et al. 1982a). These published studies emphasize the need for major improvements in various aspects of these tests before application to monitoring exposed men can become more routine. In this paper we have summarized our initial attempts to improve the visual scoring of human sperm morphology and to understand some of the statistical characteristics of this test.

We have shown that human sperm morphology can be scored reproducibly when scoring logic is well defined and when coded reference slides are used for standardization and random comparisons. Although the data on the reference slides were collected over a period of several years, there was little increase in variation caused by scorer subjectivity. Overall, we estimate that the component of the variation due to scorer subjectivity was not statistically larger than the component of the variation due to sampling statistics. The residual variation is mainly due to the readings from four reference slides which were difficult to score. On these slides there seemed to be a small proportion of sperm with shape and size characteristics such that classification into one of two categories was ambiguous, e.g., normal vs. narrow, normal vs. small, and normal vs. round. Automated image-processing methods will be required to further improve the scoring reliability in such cases.

A comparison of the results for sperm counts and sperm morphology for three different groups of normal, healthy controls shows that in each group the coefficient of variation for sperm morphology is considerably smaller than that for counts, although there is some variation among the cohorts. In cases where the biological sensitivity of these two tests might be similar, we would predict that the sperm morphology test requires 1/5-1/15 as many mean as the test for sperm counts. This would suggest that in situations of small exposures, where little or no sperm damage is expected, the sperm morphology test may be more sensitive than the test for sperm counts, especially when sample sizes are limited.

Caution is required when applying such sample-size calculations to a human study. Though the human data are limited, we suspect that the assumption that human sperm counts and morphology will have similar biological sensitivities will not be true for all chemical and physical agents. As examples,

1. The studies of workers exposed to dibromochloropropane show that counts were reduced with little change in morphology;
2. There are numerous examples of agents that seem to have similar effects on both tests;
3. There are several agents (e.g., lead, carbaryl, and tobacco) that seem to affect human sperm morphology more than counts (for references see Wyrobek et al. 1982a).

In the mouse where sperm morphology dose-effect data are available for a broad spectrum of agents (Wyrobek et al. 1982b), a comparison of count depression with increases in percent morphologically abnormal sperm in epididymal sperm showed that the measure of morphology was generally more sensitive to dose than counts (A. Wyrobek, unpubl. results). Sample-size calculations for the human sperm morphology test will need to be adjusted for biological sensitivity as data for more agents become available. Further research is also needed to understand the quantitative impact of agent-induced changes in human sperm morphology on fertility and reproductive outcome.

The data presented in this paper show that a reliable, visual human sperm morphology test can be developed and should be incorporated along with other sperm tests (e.g., counts) into studies of the effects of physical and chemical agents on sperm production.

SUMMARY

We have developed a reliable method for scoring human sperm morphology that has unique advantages for assessing the effects of physical and chemical agents on sperm production. Though it has been relatively easy to develop sperm morphology tests for laboratory rodents, the heterogeneity of human sperm shapes has made it difficult to develop a dependable test in man. With our method, human sperm are visually classified into 10 shape categories based mainly on the contour of the sperm head. The method relies on the use of a set of coded reference slides that is used both for standardization and random comparisons. With this approach we demonstrate that human sperm can be reliably and reproducibly assigned to individual shape categories.

Studies of both sperm counts and sperm morphology in three different groups of normal, healthy men show that the human sperm morphology test has considerably greater statistical power than sperm counts; it may require much smaller sample sizes than counts to detect induced changes. Furthermore, sperm morphology is considerably less sensitive to abstinence and preparative factors than are sperm counts and motility. These results suggest that a reliable visual sperm morphology test can be developed, and should be incorporated along with other sperm tests into studies of the effects of physical and chemical agents on sperm production in man.

ACKNOWLEDGMENTS

We thank L. Dobson, D. Pinkel, and B. Brandriff for incisive suggestions in the preparation of this manuscript. We also thank J. Cherniak and A. Riggs for their detailed help in editing, formating, and typing this manuscript.

Work performed under the auspices of the U.S. Department of Energy by the Lawrence Livermore National Laboratory under contract number W-7405-ENG-48. This document was prepared as an account of work sponsored by an agency of the United States Government. Neither the U.S. Government nor the University of California nor any of their employees, makes any warranty, expressed or implied, or assumes any legal liability or responsibility for the accuracy, completeness, or usefulness of any information, apparatus, product, or process disclosed, or represents that its use would not infringe privately owned rights. Reference herein to any specific commercial products, process, or service by trade name, trademark, manufacturer, or otherwise, does not necessarily constitute or imply its endorsement, recommendation, or favoring by the U.S. Government or the University of California. The views and opinions of authors expressed herein do not necessarily state or reflect those of the U.S. Government thereof, and shall not be used for advertising or product endorsement purposes.

REFERENCES

Ash, P. 1980. The influence of radiation on fertility in man. *Br. J. Radiol.* **53**: 271.

Conover, W. J. 1971. *Practical nonparametric statistics.* p. 309, John Wiley and Sons, Inc., New York.

Czeizel, E., M. Hancsok, and M. Viczian. 1967. Examination of the semen of husbands of habitually aborting women. *Orvosi. Hetilap.* **108**:1591.

David, G., J. P. Bisson, F. Czyglik, P. Jouannet, and C. Gernigon. 1975. Anomalies morphologiques du spermatozoïde humain. 1) Propositions pour un système de classification. *J. Gynecol. Obstet. Biol. Reprod.* **4** *(Suppl. 1)*:17.

Eliasson, R. 1971. Standards for investigation of human semen. *Andrologia* **3**:49.

Filliben, J. J. 1975. The probability plot correlation coefficient test for normality. *Technometrics* 17:111.

Fredricsson, B. 1979. Morphologic evaluation of spermatoza in different laboratories. *Andrologia* 11:57.

Freund, M. 1966. Standards for the rating of human sperm morphology. A cooperative study. *Int. J. Fertil.* 11:97.

Furuhjelm, M., B. Jonson, and C. G. Lagergren. 1962. The quality of human semen in spontaneous abortion. *Int. J. Fertil.* 7:17.

Guenther, W. C. 1981. Sample size formulas for normal theory t-tests. *American Statistician* 35:243.

Homonnai, Z. T., G. F. Paz, J. N. Weiss, and M. P. David. 1980. Relation between semen quality and fate of pregnancy: Retrospective study on 534 pregnancies. *Int. J. Androl.* 3:574.

Jöel, C. A. 1966. New etiologic aspects of habitual abortion and infertility, with special reference to the male factor. *Fertil. Steril.* 17:374.

Jöel, C. A. and R. Chayen. 1971. Pathological semen as a factor in abortion and infertility. In *Fertility disturbance in men and women*, p. 496. Karger, Basel, Switzerland.

Kapp, R. W. 1979. Detection of aneuploidy in human sperm. *Environ. Health Perspect.* 31:27.

Krzanowska, H. 1981. Sperm head abnormalities in relation to the age and strain of mice. *J. Reprod. Fertil.* 62:385.

MacLeod, J. 1965. Human seminal cytology following the administration of certain antispermatogenic compounds. In *A symposium on agents affecting fertility* (eds. C. R. Austin and J. S. Perry), p. 93. Little Brown and Co., Boston.

_____. 1974. Effects of environmental factors and of antispermatogenic compounds on the human testis as reflected in seminal cytology. In *Male fertility and sterility*. Proceedings Serono Symposium (eds., R. E. Mancini and L. Martini), vol. 5, p. 123. Academic Press, New York.

Miller, R. G., Jr. 1966. *Simultaneous statistical inference*, p. 67. McGraw-Hill Book Co., New York.

Moore, D. H. II, D. E. Bennett, D. Kranzler, and A. J. Wyrobek. 1982. Quantitative methods of measuring the sensitivity of the mouse sperm morphology assay. *Anal. Quant. Cytol.* (in press).

Nistal, M., A. Harruzo, and F. Sanchez-Corral. 1978. Toratozoospermia absoluta de presentacion familiar. Espermatozoides microcefalos irregulares sin acrosoma. *Andrologia* 10:234.

Schwartz, D., A. Laplache, P. Jouannet, and G. David. 1979. Within-subject variability of human semen in regard to sperm count, volume, total number of spermatozoa, and length of abstinence. *J. Reprod. Fertil.* 57:391.

Seuànez, H. N. 1980. Chromosomes and spermatozoa of the African great apes. *J. Reprod. Fertil. (Suppl.)* 28:91.

Snedecor, G. W. and W. G. Cochran. 1967. *Statistical methods*, p. 240. The Iowa State University Press, Ames, Iowa.

Tates, A. D. 1979. *Microtus* oeconomus (Rodentia), a useful mammal for studying the induction of sex-chromosome nondisjunction and diploid gametes in male germ cells. *Environ. Health Perspect.* **31**:151.

Wyrobek, A. J. 1979. Changes in mammalian sperm morphology after x-ray and chemical exposures. *Genetics (Suppl.)* **91**:5105.

Wyrobek, A. J., G. Watchmaker, L. Gordon, K. Wong, D. Moore II, and D. Whorton. 1981. Sperm shape abnormalities in carbaryl-exposed employees. *Environ. Health Perspect.* **40**:255.

Wyrobek, A. J., L. A. Gordon, J. G. Burkhart, M. C. Francis, R. W. Kapp, G. Letz, H. V. Malling, J. C. Topham, and M. D. Whorton. 1982a. An evaluation of human sperm as indicators of chemically induced alterations of spermatogenic function: A report of the U.S. Environmental Protection Agency GENE-TOX Program. *Mutat. Res.* (in press).

_____. 1982b. An evaluation of the mouse sperm morphology test and other sperm tests in nonhuman animals: A report of the U.S. Environmental Protection Agency GENE-TOX Program. *Mutat. Res.* (in press).

Wyrobek, A. J., M. L. Meistrich, R. Furrer, and W. R. Bruce. 1976. Physical characteristics of mouse sperm nuclei. *Biophys. J.* **16**:811.

Young, I. T., B. L. Gledhill, S. Lake, and A. J. Wyrobek. 1982. Quantitative analysis of radiation-induced changes in sperm morphology. *Anal. Quant. Cytol.* (in press).

Sperm Morphology in Cigarette Smokers

H. JOHN EVANS
Medical Research Council
Clinical and Population Cytogenetics Unit
Western General Hospital
Edinburgh EH4 2XU, Scotland

Following the original demonstration by Keir et al. (1974) that metabolically activated cigarette smoke condensates were mutagenic to bacteria, there have been a large number of reports that have confirmed and extended these findings (Evans 1981). In our own studies on the effects of cigarette smoke condensate on the chromosomes of human blood lymphocytes exposed in vitro, we were able to show that these condensates were extremely potent inducers of sister chromatid exchanges (SCEs), in that the condensate from a fraction of one puff of a cigarette added to 10^6 lymphocytes in 10 ml of culture medium gave a threefold increase in SCE frequency (Hopkin and Evans 1979; Vijayalaxmi and Evans 1982). This response is equivalent to that observed if the cells are exposed to ethylmethanesulfonate (EMS) at a concentration of 1×10^3 M. These chromosome damaging effects of cigarette smoke condensates have now been confirmed in a number of laboratories (de Raat 1979; Madle et al. 1981). However, unlike the induction of mutations in bacteria, the agents in cigarette smoke condensates that damage human chromosomes in vitro are direct-acting and do not require metabolic activation to produce their effects (Hopkin and Perry 1980).

The evidence that pyrolysis of the components in cigarettes yields potent mutagens whose effects can be readily demonstrated in vitro is therefore overwhelming, but what evidence do we have concerning possible genetic effects produced in vivo within the cigarette smoker?

The most direct evidence of an in vivo genetic effect in man would be the demonstration of an increase in heritable abnormalities among the products of conception and in the newborn of cigarette smoking parents. Although Mau and Netter (1974) claimed a small, but significant, increase in perinatal mortality and congenital abnormalities in the newborn to couples where the male partner smoked more than 10 cigarettes per day, this finding has not been confirmed. A second approach would be to look for increased chromosome damage in somatic cells of cigarette smokers relative to nonsmokers. There are now many reports describing small, but significant, increases in the frequencies of SCEs in blood lymphocytes of cigarette smokers and although there are contradictory findings there is no question that cigarette smokers may show up to a 50% increase in SCE frequencies relative to nonsmokers (Vijayalaxmi and Evans 1982). Whether these increases in SCE reflect an in vivo DNA damaging effect of cigarette smoke is open to debate and I do not discuss this question here. Putting aside the SCE data there is now, however, good evidence of increased chromosome damage in blood lymphocytes of cigarette smokers. In a double blind cytogenetic analysis in our laboratory of peripheral blood lymphocyte (PBL) chromosomes from a group of 96 individuals, including 55 cigarette smokers, there is a very clear increase in chromatid and chromosome-type aberrations in cells from the smokers relative to their matched controls (Table 1). The majority of smokers in this population smoked less than 20 cigarettes per day, but the incidence of chromosome aberrations in smokers is seen to be two to four times greater than in nonsmokers. A significant excess of exchange aberrations in lymphocytes from cigarette smokers has also been recently reported in another large survey by Obe et al. (1982) so that there is no question that cigarette smokers have an increased incidence of genetic damage in somatic cells relative to nonsmokers.

We had earlier obtained evidence for an increased aberration frequency in cells from smokers, as indeed had Obe and Herha (1978) in a study of very heavy smokers, and this was a major factor that prompted us to ask whether the damaging effects of cigarette smoke would also be evident in the form of an increased incidence of abnormalities of sperm morphology in smokers relative to nonsmokers. Two other important reasons for studying sperm morphology in smokers were: first, the work of Wyrobek and Bruce (1975, 1978) on the mouse, showing clear correlations between degree of exposure to chemical mutagens and incidence of morphologically abnormal sperm; second, an early report by Viczian (1969) of an increased incidence of morphologically abnormal sperm in cigarette smokers relative to nonsmokers.

Viczian (1969) studied a population of 120 cigarette smokers who had smoked varying amounts for more than one year prior to study, and 50 nonsmokers. Seminal fluid samples were analyzed for sperm count, motility, and for morphologically abnormal forms and the frequency of morphologically abnormal spermatozoa was found to be significantly higher in the smokers (28%) than in the nonsmokers (19%). Moreover, the incidence of abnormal sperm appeared to correlate with the reported numbers of cigarettes smoked per day (Fig. 1). I should emphasize that the variances of the data are quite large, such that the proportion of morphologically abnormal sperm in the seven smokers who smoked >31 cigarettes per day is not significantly different from the proportions found in the other groups including controls. The data are indeed somewhat unusual in two respects: first, the exceedingly good

Table 1

Chromosomal Aberrations in the PBLs[a] of Cigarette Smokers
and Nonsmokers

	Nonsmokers	Smokers	Corrected X^2
Total subjects	41	55	
Females	22	27	
Males	19	28	
Mean age (range)	45.8 (20-75)	48.2 (25-75)	
Females	48.7 (20-75)	49.4 (25-75)	
Males	43.0 (25-65)	47.0 (25-60)	
Total metaphases analyzed	4,100	5,500	
Cells with damaged chromosomes (%)	91 (2.22)	176 (3.20)	8.0***
Chromatid gaps	82 (2.00)	140 (2.55)	2.9*
Chromatid breaks	22 (0.54)	19 (0.35)	1.6
Chromatid interchanges	1 (0.02)	12 (0.22)	5.2**
Chromosome gaps	10 (0.24)	31 (0.56)	4.9**
Chromosome breaks (fragments)	14 (0.34)	37 (0.67)	4.3**
Chromosome interchanges (dicentrics)	6 (0.14)	23 (0.42)	4.9**
Abnormal monocentrics (including symmetrical interchanges)	7 (0.17)	11 (0.20)	0.01

[a]100 metaphases scored from each individual.
*, **, ***Significance at the 5%, 1%, and 0.1% levels.

correlation between degree of abnormality and apparent numbers of cigarettes smoked; second, the unusually low incidence of abnormal shaped sperm in controls.

In our own laboratory we have studied sperm counts, motility, and morphology in ejaculates from a series of males attending our subfertility clinic, many of whom of course have a perfectly normal sperm profile. From this population we attempted to compare cigarette smokers with nonsmokers and tried to eliminate any other factors that were likely to cause sperm abnormalities (Evans et al. 1981). We therefore excluded all patients with a history of alcoholism, exposure to radiation, toxic chemicals, drugs, hormones, or who had undescended testicles at any age, varicocele, or any operation associated with possible infertility. We ended up with 43 cigarette smokers and 43 matched

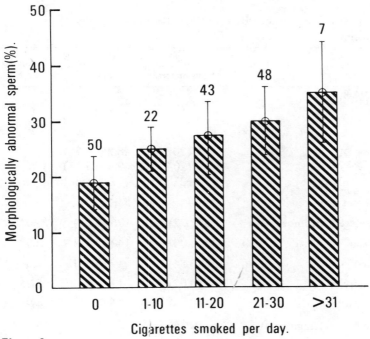

Figure 1
Proportions of sperm that are morphologically abnormal in samples from 50 nonsmokers and 120 cigarette smokers (Viczian 1969).

nonsmokers. Matching was undertaken on the basis of sperm count and, in analyzing our results, individuals were allocated to groups according to sperm count as shown in Table 2. The reason for this matching is that sperm count is sometimes correlated with morphological abnormality, particularly when there are systemic disturbances due for example to viral infection (MacLeod 1971).

The preparations analyzed were smears of undiluted semen fixed in alcohol:ether and stained with the Papanicolaou technique. All the slides were kept and later coded and analyzed by one observer. On each slide 100 spermatazoa were classified as being morphologically normal or abnormal. The abnormal sperm were categorized by one or more of the following features: large, small, tapered, dense, vacuolated, amorphous or multiple heads; abnormal midpiece; abnormal tails and immature forms. Following completion of the scoring, the slides were decoded and the results are summarized in Table 2.

The data in Table 2 show that within each of the four groups matched on the basis of sperm count, the proportion of morphologically normal sperm is

Table 2

Proportions of Morphologically Normal Sperm (%)
in Cigarette Smokers and Nonsmokers

Sperm count	Controls		Smokers	
	Number of patients	Normal sperm (% ± SE)	Number of patients	Normal sperm (% ± SE)
<20 m/ml	3	48.7 ± 3.2	3	43.7 ± 5.47
20-40 m/ml	5	61.2 ± 2.73	3	45.3 ± 1.51
40-60 m/ml	12	54.3 ± 2.06	14	55.1 ± 2.73
>60 m/ml (no child)	13	59.6 ± 2.06	13	53.4 ± 1.55
>60 m/ml (previous child)	10	59.9 ± 1.57	10	54.4 ± 3.45
Total	43	57.7 ± 1.18	43	52.9 ± 1.47
Total of >60 m/ml	23	59.7 ± 1.38	23	53.8 ± 1.78

lower in smokers than in nonsmokers. If we consider only those men with sperm counts $>60 \times 10^6$ per ml, then the proportion of normal sperm in the 23 smokers is $53.8 \pm 1.78\%$ as compared with $59.7 \pm 1.38\%$ for the 23 nonsmokers, a difference that is significant at the 1% level. Taking the overall data the 43 smokers had $52.9 \pm 1.47\%$ normal forms whereas the 43 nonsmokers had $57.7 \pm 1.18\%$ sperm with normal morphology; again significant at the 1% level. Smokers therefore have approximately 10% more abnormal sperm than nonsmokers.

An alternative way of comparing smokers and nonsmokers is to contrast the numbers of individuals falling into seven categories defined by the proportions of normal sperm in the range 36-75%. The resulting histograms are shown in Figure 2 and the different pattern between smokers and nonsmokers is quite distinctive.

Based on information on smoking habits probided by the patients when they attended the clinic, we then examined the data to see if there was any correlation between the alleged number of cigarettes smoked per day and the frequency of sperm abnormalities. The results are summarized in Table 3 and, in contrast to the earlier report of Viczian (1969), they reveal no clear association between degree of abnormality and numbers of cigarettes smoked. It may be relevant here to note that Viczian also reported that sperm motility of cigarette smokers was lower than that of nonsmokers and we confirm this finding (Table 4). Our own data reveal a similar reduction in motility (54.9%-46.7%), but the

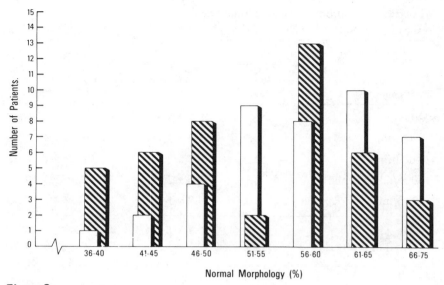

Figure 2

Numbers of cigarette smoking and nonsmoking patients falling into categories defined by the proportions of sperm of normal morphology (■) smokers; (□) controls.

variances are large, the differences are not significant, and I shall comment no further on this.

The publication of some of our findings (Evans et al. 1981) stimulated Godfrey (1981) to analyze her data on sperm morphology in smokers which were accumulated over a period of 8 years and not, as in our case, in a single blind scoring exercise undertaken over a period of weeks. In this analysis of data accumulated from 75 smokers and 74 nonsmokers, there was also no association between numbers of cigarettes smoked per day and percentage of abnormal shaped sperm, but neither was there any significant overall difference between smokers and nonsmokers, although the smokers had a slightly lower percentage of normal shaped sperm (61.6 ± 2.4%) relative to the nonsmokers (65.4 ± 1.8%) and the degree of difference between the groups was similar to that recorded in our own study.

The main conclusion from our work is that in a controlled, blind study we find a small, but highly significant, excess of abnormally shaped sperm from cigarette smokers as compared with nonsmokers. Our own data give no indication of a correlation between the incidence of abnormal forms and numbers

Table 3

Percentage of Normal Sperm in Relation to Cigarettes Smoked Per Day

Cigarettes smoked/day	Number of patients	Normal sperm (%)
0	43	57.7 ± 1.18
$\leqslant 15$	6	50.7 ± 5.9
20	24	51.8 ± 9.4
25	3	62.3 ± 13.9
$\geqslant 30$	10	54.2 ± 10.3

of cigarettes smoked per day, but show that a much larger study would be necessary to demonstrate whether or not such a correlation exists. The more important question to ask, however, is what is the meaning of this increased incidence of abnormal sperm in cigarette smokers?

We undertook our study on the premise that an increased incidence of abnormal shaped sperm might, at least in some cases, be indicative of exposure to a mutagenic agent and, in the light of firm evidence of an increased incidence of genetic damage in somatic cells of cigarette smokers, the implication might therefore be that our sperm studies are indicative of a genetic effect of cigarette smoke products on male germ cells. However, we must be very wary indeed of drawing such a conclusion, for there is no direct evidence of any sort in support of this possibility and, moreover, we are very well aware that in addition to genetic factors there is a whole host of conditions which are not in themselves mutagenic and which can result in increasing the incidence of abnormally shaped sperm. These may include, e.g., vitamin deficiency (Blair et al. 1968; Leathem 1970), mumps orchitis (Bartak 1973), and hyperthermia (Van Demark and Free 1970), which may of course be associated with a wide variety of febrile diseases. Moreover, in the mouse it has been shown that dietary restrictions can result in large increases in the incidence of abnormal sperm and Komatsu et al. (1982) emphasize that the mouse sperm abnormality test is not a relevant mutagenicity assay for chemicals that may severely inhibit the animals' food intake. Our cigarette-smoking population were selected healthy individuals who were certainly not undernourished, or suffering from febrile disease, but the obvious high sensitivity to disturbance of the processes governing sperm morphology behoves us to err on the side of caution in interpreting our findings.

Table 4
Sperm Motility (%) in Cigarette Smokers and Nonsmokers

Sperm count	Controls			Smokers			't'
	Number of patients	Motility (% ± S.E.)		Number of patients	Motility (% ± S.E.)		
<20 m/ml	3	35.3 ⎫		3	41.0 ⎫		
20-40 m/ml	5	49.0 ⎬ 46.0 ± 5.7		3	40.0 ⎬ 40.0 ± 5.0		0.79
40-60 m/ml	12	47.3 ⎭		14	39.8 ⎭		
>60 m/ml	23	62.6 ± 4.7		23	52.5 ± 4.8		1.51
Total	43	54.9 ± 3.9		43	46.7 ± 3.6		1.55

REFERENCES

Bartak, V. 1973. Sperm count, morphology and motility after unilateral mumps orchitis. *J. Reprod. Fertil.* **32**:491.

Blair, J. H., H. E. Stearns, and G. M. Simpson. 1968. Vitamin B$_{12}$ and fertility. *Lancet* i:49.

de Raat, W. K. 1979. Comparison of the induction by cigarette smoke condensates of sister-chromatid exchanges in Chinese hamster cells and of mutations in *Salmonella typhimurium. Mutat. Res.* **66**:253.

Evans, H. J. 1981. Cigarette smoke induced DNA damage in man. In *Progress in mutation research* (ed. A. Kappas), vol. 2, p. 111. Elsevier/North-Holland, Amsterdam, Netherlands.

Evans, H. J., J. Fletcher, M. Torrance, and T. B. Hargreave. 1981. Sperm abnormalities and cigarette smoking. *Lancet* i:627.

Godfrey, B. 1981. Sperm morphology in smokers. *Lancet* i:948.

Hopkin, J. M. and H. J. Evans. 1979. Cigarette smoke condensates damage DNA in human lymphocytes. *Nature* **279**:241.

Hopkin, J. M. and P. E. Perry. 1980. Benzo(*a*)pyrene does not contribute to the SCEs induced by cigarette smoke condensate. *Mutat. Res.* **77**:377.

Keir, L. D., E. Yamasaki, and B. N. Ames. 1974. Detection of mutagenic activity in cigarette smoke condensates. *Proc. Natl. Acad. Sci. U.S.A.* **71**:59.

Komatsu, H., T. Kakizoe, T. Niijima, T. Kawachi, and T. Sugimura. 1982. Increased sperm abnormalities due to dietary restriction. *Mutat. Res.* **93**:439.

Leathem, J. H. 1970. Nutrition. In *The testis: Influencing factors* (eds. A. D. Johnson et al.), vol. III, p. 169. Academic Press, New York.

MacLeod, J. 1971. Human male infertility. *Obstet. Gynecol. Survey* **26**:335.

Madle, S., A. Korte, and G. Obe. 1981. Cytogenetic effects of cigarette smoke condensates in vitro and in vivo. *Hum. Genet.* **59**:349.

Mau, G. and P. Netter. 1974. Auswirkungen des väterlichen Zigarettenkonsums auf die perinatale Sterblichkeit und Missbildungshäufigkeit. *Dtsch. Med. Wochenschr.* **99**:1113.

Obe, G. and J. Herha. 1978. Chromosomal aberrations in heavy smokers. *Hum. Genet.* **41**:259.

Obe, G., H.-J. Vogt, S. Madle, A. Fahning, and W. D. Heller. 1982. Double-blind study on the effect of cigarette smoking on the chromosomes of human peripheral blood lymphocytes in vivo. *Mutat. Res.* **92**:309.

Van Demark, N. L. and M. J. Free. 1970. Temperature effects. In *The testis: Influencing Factors* (eds. A. D. Johnson et al.), vol. III, p. 223. Academic Press, New York.

Viczian, M. 1969. Ergebnisse von Spermauntersuchungen bei Zigarettenrauchern. *Z. Haut-Geschlechtskr.* **44**:183.

Vijayalaxmi and H. J. Evans. 1982. In vivo and in vitro effects of cigarette smoke on chromosomal damage and sister-chromatid exchange in human peripheral blood lymphocytes. *Mutat. Res.* **92**:321.

Wyrobek, A. J. and W. R. Bruce. 1975. Chemical induction of sperm abnormalities in mice. *Proc. Natl. Acad. Sci. U.S.A.* **72**:4425.

_____. 1978. The induction of sperm-shape abnormalities in mice and humans. *Chem. Mutagens* **5**:257.

COMMENTS

MOHRENWEISER: Did you divide your controls into those which cohabit with a smoking female and those which don't?

EVANS: No, we didn't. You are implying that there might be an effect of passive smoking, and I think there could well be. I don't think we would ever see this in terms of sperm morphology but we might possibly see it in terms of aberrations. I should perhaps emphasize the fact that the smoking members of the population of 100 people in which we analyzed aberration frequencies were light cigarette smokers smoking an average of 10 cigarettes per day. Heavy smokers show higher aberration frequencies, so that it seems quite possible that passive smoking might also result in some chromosome damaging effects.

WYROBEK: I have a general question about the kind of data you raised at the end. We know from the human literature as well that spermatogenesis breaks down if you raise the testicular temperature. The same effect occurs if the individual is suffering from disease such as viral disease or has a fever. It is possible that the abnormalities seen with alcoholics might be a dietary issue in some way. To jump from saying that because spermatogenesis broke down under these conditions there are no mutations underlying these events I think may be somewhat dangerous. I have not seen any studies where they have done any kind of a heritable mutation test for spermatogenesis under these kinds of conditions.

EVANS: No, my only point here is that I personally have no evidence at all. I have no knowledge that there is any direct link between abnormal sperm morphology and mutation incidence in those abnormal sperm. I think that is true.

WYROBEK: That is true.

EVANS: I think the indirect evidence that you produce is very pressing and very strong. But we haven't got *direct* evidence. What does concern me— and you made the point, which I should have made at the end—is that any febrile disease, a virus infection, any slight rise in temperature, then the proportion of sperm with normal morphology drops; you get more abnormal sperm. If you had mumps orchitis, or all matter of other infectious disease, you can show very clearly an increased abnormal sperm frequency.

The processes responsible for sperm shape are obviously very sensitive phenomena which may be influenced in a number of ways. Sperm morphology is therefore a very sensitive parameter which may be disturbed by a number of things. This means that we have to carefully control our experiments and our analysis, paying regard to temperature, disease, feeding, etc., if we want to make use of this as a measure of any genotoxic agent.

COMBES: Does frequency of ejaculation have an effect on the sperm number and morphology?

EVANS: Yes, sperm number is diminished and there is some correlation between numbers and morphology, but I don't know if the proportion of abnormal forms is significantly increased with increasing frequency of ejaculation. What we do know, and as Andrew Wyrobek indicated, is that the sperm "profile" of any given individual is relatively constant over periods of at least a few months.

Indicators of Genotoxic Exposure:

Status and Prospects

BRYN A. BRIDGES
MRC Cell Mutation Unit
University of Sussex
Falmer, Brighton BN1 9QG
England

In the final session of this meeting I would like to return to the theme of the first session, the linking of laboratory studies and epidemiology. At a meeting in Rome in 1979 I wrote "Despite the extremely large number of screening tests that have been performed, remarkably little rational thought has been devoted to the use that should be made of the results of such tests to man of a potential mutagenic or carcinogenic agent is to be estimated in anything like a quantitative manner". I went on to suggest that "we should adopt an approach involving the calibration of human response for a few model mutagens and carcinogens utilizing markers for effective dose and effective biological response both at the cellular and whole animal levels. This will involve the linking of epidemiology and laboratory investigations into both the calibration of human response against tissue exposure and the calibration of man against rodents for the same tissue exposures" (Bridges 1980). Figure 1 shows a schematic representation of the sort of parallelogram approach that I envisaged, first for calibration and later for risk estimation. It has grown out of the parallelogram that Fritz Sobels first proposed for estimating genetic risk from ionizing radiation.

My particular interest in this conference has been to review the current approaches to the determination of biologically significant dose and markers of genetic damage, both of which are necessary for this approach to risk estimation. In 1979 I saw particular merit in two methods for estimating biologically significant dose. One, the measurement of phosphotriesters in DNA, appears to have become completely overlooked for reasons which are not clear. The other, the alkylation of hemoglobin, continues to be developed in a profitable way. Both methods are in principle well suited to the estimation of chronic exposures since the alkylations are accumulated with time and are not removed by repair processes. The hemoglobin method has been developed to an extremely high degree of sensitivity. The ability to detect alkylation of histidine in hemoglobin by an agent whose ability to alkylate DNA is below the level of detection (although theoretically inevitable) raises questions both practical and philosophical about the use of the term genotoxic.

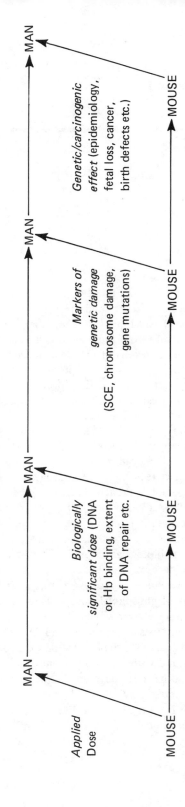

Figure 1
Suggested linkage of laboratory and epidemiology studies necessary for the calibration of human response to DNA damaging agents.

Since 1979 the emphasis has swung back towards the measurement of DNA adducts, mainly because of the development of methods of analysis that do not depend on the injection of the human subject with high levels of radioactivity to label the DNA. The application of high pressure liquid chromotography (HPLC) and fluorometry, monoclonal antibodies, and the labeling of DNA after extraction have been among the most exciting advances reported here. The monoclonal antibody technique promises the ability to study the distribution of DNA adducts within the various cells of a tissue and to detect the theoretical minimum of one DNA adduct in a cell. The method does, however, share with the hemoglobin alkylation method, the need to know well in advance exactly what adduct is required to be detected. The so called post-labeling technique is less constrained by this limitation. All DNA adduct estimations, however, suffer from the often unknown effect of DNA repair which imposes severe limitations on their use with chronically exposed individuals.

Using DNA repair elicited as a measure of DNA damage continues to be well studied. Like DNA binding, it is not particularly suitable for chronic exposures and seems to be finding its best use in testing animal models for tissue specificity. In certain instances the products of repair may be detected in the urine and provide a measure of exposure. Urine may also be used (along with other body fluids) for the determination of mutagenic activity excreted thereon. There seems to be no reason why the activity detected in urine should necessarily be due to the presence of the same mutagenic substances taken into the body. It would seem to be prudent to include chemical analyses whenever urine is used to measure mutagenic exposure.

The second level of markers are those needed to indicate some sort of genetic response in the whole animal or in man. Cytogenetic analysis of peripheral blood lymphocytes (PBLs) has been used operationally for several years. Perhaps the most interesting recent development has been the use of micronucleus counting to estimate chromosomal damage in a variety of other tissues including colonic epithelium in rodents and human cord blood. Applied to animals this should enable hypotheses to be tested concerning the importance of chromosomal damage in species and tissue specificity. The ability to monitor for gene mutations in PBLs is on the verge of becoming a practical proposition. The original protocol of Strauss and Albertini suffered from two major disadvantages. One, immediately apparent, was the tedium involved in counting small proportions of labelled nuclei. The other, seen with hindsight, was the presence of a major artefact in that cycling cells may be detected as thioguanine-resistant (TG^r) phenocopies. Freezing the lymphocytes to a large extent overcomes this problem but even more attractive is adding a feeder layer and T-cell growth factor so that clonogenic survival in the presence of thioguanine may be used as the end-point. This has the additional advantage that the mutant clones are available for cytogenic, biochemical, or molecular analysis. The use of metaphases to score TG^r cells has also proved to be a feasible approach.

In trying to apply these dose and response markers to epidemiological studies of genetic disease I was, perhaps, rather more optimistic in 1979 than I am now. The difficulties in the use of spontaneous fetal loss rates to estimate dominant lethal rates were clearly pointed out by Dr. Hook. There is, moreover, the fact that fetal loss in man (in contrast to rodents) is weighted towards numerical chromosomal aberrations since many structural aberrations are lost too early in embryonic life to be detected. I remain, however, impressed by the fact that the "latent period" for fetal loss is minimal and sample populations need not be enormous so that the introduction of, say, a mutagenic (and thus potentially carcinogenic) drug could be detected much earlier in a human population than if one waited for cancers to appear. In addition, I continue to be frustrated by the absence of studies on the association of cigarette smoking with heritable disease (Bridges et al. 1979). If one cannot detect heritable effects in the offspring of cigarette smoking males one is perhaps unlikely to be able to detect them in any other group with the possible exception of patients successfully treated with cytostatic drugs.

The linking of laboratory dose and genetic response data with cancer incidence seemed in 1979 to be dependent upon overcoming problems associated with tissue specificity. While it would be too soon to argue that these have been solved, we can, perhaps, now see several ways in which the solutions may be approached. Perhaps the major remaining problem lies in the area of promotion. Some mutagenic carcinogens are almost purely initiators, others (sometimes termed "complete" carcinogens) also have promoting properties. Any correlation of mutagenic events in vivo with cancer incidence can be completely devastated by lack of knowledge about the promotional status of the animal colony or human population being studied. The hypotheses we are testing and the inter species calibrations we are making relate specifically to the mutagenic and initiating action of agents; differences in the efficiency of promotion could easily mask real correlations at the mutational and initiation level. Nevertheless, we should remember that when we come to make risk estimates, promotional interaction is likely to be important, and could at some sites be the dominant factor in determining cancer risk.

References

Bridges, B.A., J. Clemmesen, and T. Sugimura. 1979. Cigarette smoking—does it carry a genetic risk? *Mut. Res.* **65**:71.

Bridges, B.A. 1980. An approach to the assessment of the risk to man from DNA damaging agents. *Arch. Toxicol.* (Suppl) **3**:271.

NAME INDEX

SUBJECT INDEX